Billing Dictionary

Published By:
Althos Publishing

404 Wake Chapel Road
Fuquay-Varina, NC 27526 USA
Telephone: 1-800-227-9681
1-919-557-2260
Fax: 1-919-557-2261
email: Success@Althos.com
web: www.Althos.com

ALTHOS
SIMPLIFYING KNOWLEDGE

Althos

Copyright © 2008 By Althos Publishing
First Printing

Printed and Bound by Lightning Source Printing, La Vergne, TN USA

International Standard Book Number: 1-932813-38-1

Editors and Advisors

Avi Ofrane is the president and CEO, and a master instructor of The Billing College. Mr. Ofrane founded The Billing College in 1996 to address the converging market trends associated with telecommunications Billing and Customer Care. Mr. Ofrane began his career in 1977 as an analyst with the IBM Corporation, designing and implementing manufacturing systems. Throughout his extensive career, Mr. Ofrane has been involved in all aspects of the industry, including strategic planning, RFP processing, vendor evaluation and selection, business process engineering, business/systems analyses, project management, implementation, operations, quality assurance, and executive management. Since 1982, Mr. Ofrane has concentrated exclusively on the telecommunications industry, in which he is now a recognized expert and master instructor in Billing and Customer Care. Mr. Ofrane lectures extensively in the US and in Europe on Billing and Customer Care issues, strategies, methodologies, and practices and he is a frequent speaker at major industry conferences. He has authored several leading books on billing systems. Mr. Ofrane holds a BS, Computer Science, from Pennsylvania State University.

Mr. Harte is the president of Althos, an expert information provider which researches, trains, and publishes on technology and business industries. He has over 29 years of technology analysis, development, implementation, and business management experience. Mr. Harte has worked for leading companies including Ericsson/General Electric, Audiovox/Toshiba and Westinghouse and has consulted for hundreds of other companies. Mr. Harte continually researches, analyzes, and tests new communication technologies, applications, and services. He has authored over 60 books on telecommunications technologies and business systems covering topics such as mobile telephone systems, data communications, voice over data networks, broadband, prepaid services, billing systems, sales, and Internet marketing. Mr. Harte holds many degrees and certificates including an Executive MBA from Wake Forest University (1995) and a BSET from the University of the State of New York, (1990).

Contributing Editors and Advisors

Ben Levitan is a wireless, cellular and Internet telephony expert. Levitan's career has spanned more than 25 years in the field of cellular communication and includes 27 patents in cellular technology innovations, including four in the area of GPS applications. Ben Levitan is one of the most sought after telecommunications consultants and also acts as legal expert witnesses in State and Local courts providing expertise on multiple telephone and data technological. Mr. Ben Levitan is an engineer and an expert on new and developing wireless standards for Satellite, Cellular and land mobile radio (LMR) Systems including ANSI-41, GSM, and 3rd Generation systems. Mr. Levitan is an active participant for the US and international development of ANSI-41, GSM, 802.11, All-IP (3G Standards), Interstandard Roaming and FCC and US Government mandated features and requirements. Mr. Levitan has represented COMSAT, Intelsat and Aeronautical Radio, Inc. in the development of satellite standards and was a member of the United States Delegations to the United Nation's International Telecommunications Union conferences for worldwide standardization of telecommunications for eight years. Mr. Levitan has consulted and is an expert witness for several major wireless companies and is a frequent speaker at wireless industry forums and conferences.

Ms. Dunham joined the Billing College in 1999 as an instructor, developer and consultant. With over 33 years of experience in project leadership, resource management, negotiations, analysis (business and systems), her work experience ranges from data center operations to root cause analysis of work processes and information technology systems. Ms. Dunham  holds an MBA from the University of Alabama at Birmingham (UAB) and a BBA from the University of North Florida with emphasis in Management studies and Information Systems and has developed courses on topics such as CI&B, Interconnection, Utilities Metering, Settlement and Billing, and Revenue Assurance.

Eric Hill has been the Director of Industry Relations for VeriSign since 2004, where he is responsible for marketing and business development within leading information and communication technology trade associations and standards bodies. From 2000 to 2004 Eric was Director, Interoperability Programs for the GSM Association, developing strategies for GSMA interaction with standards and certification bodies and various industry groups. He  oversaw billing and interoperability requirements for the GSM Global Roaming Forum, established to nurture interstandard roaming between GSM and TDMA, CDMA and iDEN carriers. From 1996 to 2000, Eric was the Director of Roaming Services for Iridium, a pioneering provider of global mobile satellite services and interstandard roaming. Eric oversaw the commercial launch of billing, data clearing, and financial settlements with over 180 roaming partners.From 1989 to 2006 Eric worked with the then-named Cellular Telecommunications Industry Association (CTIA) as Director for Industry Affairs, overseeing strategic development of roaming, fraud management, and billing programs. He oversaw the launching of CTIA's Fraud Task Force in 1991 and its daily investigative and training operations with law enforcement and research labs. Eric also worked for CTIA's CIBERNET subsidiary, developing and managing the industry's CIBER roaming billing standard and its net financial settlement program.

Ms. Dahlen is responsible for Highdeal's strategy and operations in the Americas region. Previously, Ms. Dahlen had global account management responsibilities for EHPT, a joint venture between Ericsson and Hewlett-Packard, where she served customers throughout Europe, South America and North America. She helped build EHPT's South American operations into one of the most successful software businesses in the region. Prior to EHPT, Ms. Dahlen led strategic marketing and competitive analysis initiatives at Telia, the telecommunications leader in Scandinavia and northern Europe. A frequent speaker at industry events, Camilla holds a B.Sc. in Business Administration from Boston College and a bilingual MBA in Spanish/English from Instituto de Empresa in Madrid, Spain.

Denise Caron is a modern visionary with over 12 years of international experience in Information Technology with focus in the communications sector. She is an expert in OSS and BSS technologies for telephony (wired, fixed wireless and wireless), Internet, cable and satellite product lines. She specializes in technology and business transformation and is currently leading a large program for the cable industry in North America.

MV Balasubrahmanyam (Balu) has over 24 years of industry experience in R&D, software services and product engineering with expertise in Telecom OSS/BSS solutions for the communication service providers world wide. His work areas include strategic planning, solution architecture, project management, product engineering and implementation, quality assurance and executive management. His interesting work areas are Telecom Billing, Telecom Fraud Control, Revenue Assurance, NGN, NGOSS and Lawful Interception. Balu is heading the OSS/BSS Global Development Centre for Xalted at Bangalore, India. Balu is a member of GRAPA (Global Revenue Assurance Professionals Association) and a regular speaker in international forums on Telecom billing and Revenue Assurance. Balu holds a Bachelors Degree in Electronics & Telecommunications Engineering and currently doing Doctoral Research in Telecom Management strategies.

Nitin Bhandari has over a decade of experience in Telecommunications industry and has led the IT strategy and consulting activities at various Wireless and Wireline Service Providers across the globe. Nitin has been instrumental in proposing technology, process and organisational changes to successfully transform communication service providers into service-centric business-aware management organisations. Nitin is a known speaker at Telecom events globally and specializes in service management, lean transformation, benchmarking, service provider maturity modelling, and customer experience management paradigms. He was an active member of TMForum's service model framework working group that published "GB924 – Service Model Framework" guidebook in 2004. Nitin holds a Bachelor of Technology (with Honours) degree in Computer Science and Engineering from the Institute of Technology, Banaras Hindu University in India. He is currently a Principal Consultant at Wipro Technologies and leads their telecom consulting activities for Europe.

Aart Bakker has 22 years of billing experience in the European Telecommunication Industry. Due to his experience as project- and interim manager on various billing projects national and international for KPN Headquarters and AT&T/Unisource Services, Aart has an excellent knowledge of the Telecom Etom/NGOSS organization, required systems, software and hardware selection and various processes from a functional and technical perspective. He founded Moba Consulting in 1999, which has supported companies in the telecom industry to solve billing issues, implement new billing systems, and redesign billing and collections processes. Supporting MVNO's with business support processes is a new service provisioning model which Moba Consulting has developed for the Telecom market. Moba Consulting takes care of the daily operations enabling customers to concentrate on Marketing and Sales.

Katrin Tillenburg is a billing and revenue assurance expert who has worked in the IT area for more than 27 years. Over the past 13 years, she has focused on the telecommunication industry, including fixed networks, mobile systems and cable operators. Tillenburg founded ABITEL Consulting GmbH in 1999 with Frank Siebert, focusing on the CC&B area including all the interfaces from network to the accounting. Tillenburg's expertise comprises a very detailed view on this topic from point of view of the processes to supporting IT-systems. Her experience includes product definition, system integration, migration projects and QS projects. Katrin Tillenburg has managed international projects dealing with the whole revenue chain from network to invoice generation as well as implying new product development processes, dealer commissioning and multiple types of reconciliations and audits. Her revenue assurance work also included definition and implementation of controls (e.g. SOX), metrics (also according to TMF proposals), key performance indicators (KPIs), and the setup of management reporting in an international environment.

Gnanapriya C is a Principal Architect, heading the architecture group with the Communications, Media and Entertainment Business Unit at Infosys Technologies Limited. She has more than 14 years of software and telecom expertise. She has been involved in the Wireless and B/OSS related projects / programs implemented for various OEMs and service providers. She consults transformation programs, architecture, reviews, E2E solution. Gnanapriya holds an Electronics & Communications Engineering Degree from Thiagarajar College of Engineering, Madurai Kamaraj University and Masters in Communication Engineering from PSG College of Technology, Bharathiar University and MBA from Madurai Kamaraj University. She has been in the teaching for 4 years. She is a Open Group certified TOGAF architect. She has published international papers and co-authored a book on Digital Signal processing published by McGraw Hill.

Jeremy Cowan is the founder, editorial director & publisher of Prestige Media Ltd. He is the editor of VanillaPlus magazine (www.vanillaplus.com), and managing editor of Stream magazine (www.streammag.com), launched in 1999 and 2004 respectively. Prior to setting up PML in 1998, he was managing editor of the communications division at Nexus Communications with editorial & business development responsibility for magazines such as Mobile Europe, Mobile & Cellular, Communications News, Cabling World, Network Europe, and WLL Directory.

Dae Decker is a billing services expert with 17 years experience in the telecommunication industry ranging from operator services to implementing professional services. She has extensive knowledge in outsourcing services with duties including ensuring SDLC processes are followed, monthly revenue forecasting, maintaining customer relationships, monitoring project activity status, and training. Dae Decker has negotiated contracts for back office support services that increased departmental revenue by 4 million dollars. She managed team projects that developed billing system protocols, GUI application interfaces, outsourcing programs, user acceptance testing, and internal user acceptance testing procedures. Dae Decker is certified in LD Billing and Customer Care, Wireless Billing, IP Application Billing, and Interconnect Billing from the Billing College at Rutgers University.

Table of Contents

Numbers

.Com Company-Dot Com Company

.MPG-MPEG

.ZIP-A filename extension that identifies that a file is compressed by the WinZip or PKZip utilities.

?-Question Mark

μ, Mu-Greek letter lower case Mu. (1-metric prefix) One millionth or 0.000001. (2-length unit) One micrometer, 0.000001 meter (formerly called one micron). (3-digital code designator) The algorithm for compression or mapping of measured voltage amplitude to binary code value used in conjunction with the DS-1 (T-1) digital PCM encoding of telephone channel waveforms is designated Mu-law or μ-law, in contrast to a similar but distinct algorithm called A-law and used in conjunction with the 2.048 Mbit/s PCM system.

Note that the English letter u is often substituted for μ when the Greek character cannot be produced due to limited typographic capability. Understandable, but often ambiguous!

0800 Freephone-A service that allows callers to dial a telephone number without being charged for the call. The toll free call is billed to the receiver of the call. In the Americas and other parts of the world, Freephone numbers are sometimes called "Toll Free" and they typically begin with 800, 888, 877, or 866.

1 Dimension Barcode (1D Barcode)-A 1 dimension barcode (1D barcode) is an image composed of graphic bar shapes that represent information where the barcodes have a single length so the data is stored in 1 dimension.

100/5-A contract stating rights to payment of additional 100 percent royalties to the instated author at which time the first five episode broadcasts are repeated. The predetermined amount will be paid in installments of 20% according to time of re-run.

100/50/50-Formula used to calculate bonuses to a service provider in correlation to a film subsequent its theatrical debut. (100 percent before first television broadcast, 50 percent after, and an additional 50 percent if the movie is exhibited in theaters overseas before or after first television broadcast.)

1024 (K)-Widely used but unofficially, the number 1024 (equal to 2 raised to the 10th power) is represented by the capital letter K. This number normally occurs only when describing file size, memory size or other numbers that are integral powers of 2 because of internal use of the binary number system. Take care to distinguish capital K from small k, which represents 1000.

112-An emergency call number that is used in the European Union.

119-An emergency telephone number that is used in Japan's telecommunication system. It is the equivalent of the 911 emergency number used in the United States.

1992 CATV Subscriber Protection and Competition Act-The 1992 CATV subscriber protection and competition act defined new regulatory requirements for the cable television industry including the regulation of rates and access requirements that allowed alternative provides (third-party) to obtain fair access to content.

1996 Telecommunications Act-Legislation designed to spur competition in the telecommunications industry. Resulting deregulation affects local, long distance, and wireless carriers. Direct competition between these suppliers and other 'non-traditional' competitors (cable, utilities, etc.) should result in better prices and more services for the consumer. The act was signed into law by President Clinton Feb.8, 1996.

1D Barcode-1 Dimension Barcode

1FB-One Flat Business Line

1G-First Generation

2 Dimension Barcode (2D Barcode)-A 2 dimension barcode (2D barcode) is an image composed of graphic bar shapes that represent information where the barcodes have different lengths or shapes so the data is store in 2 dimensions.

20/60/10/10 Formula-A set budget formula based on paying fees at fixed percentages of 20,60,10 and 10 percent of the negotiated price over a period of time. This usually means 20 percent for pre-production, 60 percent during period of production, 10 percent is received for the delivery of the directors cut, and the last 10 percent is received for the delivery of the final print.

24/7-Service that is available 24 hours a day, 7 days a week.

24x7 Support-24x7 support is the providing of people or systems that can provide information that enables people or companies to setup or maintain the operation of a product or service.

2500 Telephone Set-The term "2500 telephone set" is used in North America to describe a standard analog telephone having a 12-button touch-tone (DTMF) dial. The dial is used both for dialing and for end-to-end signaling via the voice channel. This name was originally a particular model number used by the first manufacturer of this type of telephone set, but has become a generic term today. The design and utility patents covering this set, first manufactured in 1964, have elapsed and many different manufacturers now make virtually indistinguishable replicas of the original design. This device usually provides plain old telephone service (POTS). The 2500 telephone set has replaced the earlier rotary dial 500 set and the briefly manufactured 1500 type set (which had a 10-button touch-tone dial) in many parts of the world.

This figure shows a block diagram of a standard POTS telephone (also known as a 2500 series phone). This telephone continuously monitors the voltage on the telephone line to determine if an incoming ring signal (high voltage tone) is present. When the ring signal is received, the telephone alerts the user through an audio tone (on the ringer). After

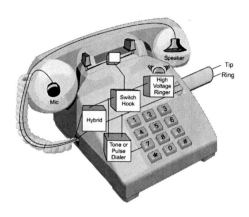

POTS Telephone Block Diagram

the customer has picked up the phone, the hook switch is connected. This reduces the line connection resistance (through the hybrid) and this results in a drop in line voltage (typically from 48 VDC to a few volts). This change in voltage is sensed by the telephone switching system and the call is connected. When the customer hangs up the phone, the hook switch is opened increasing the resistance to the line connection. This results in an increase in the line voltage. The increased line voltage is then sensed by the telephone switching system and the call is disconnected.

2B+D-The combination of two 64 kbps basic rate DS0 (B) and one 16 kbps (D) signaling channel that forms the ISDN basic rate interface (BRI).

2B4Q Line Coding-Two Binary to Four Quaternary line coding method used for ISDN subscriber lines (the U interface) in North America. In this method of line coding, each two binary bits are mapped into one pulse symbol having one of four distinct voltage levels on the transmission wires. A sequential method of encoding is used to ensure that the same voltage level will not be transmitted in consecutive symbols, so that the waveform has constant alternation and will be accurately transmitted via coupling transformers and coupling capacitors.

2D Barcode-2 Dimension Barcode
3DES-Triple Data Encryption Standard
3G-Third Generation
3G Wireless-The third generation of technology of the mobile wireless industry. Third generation (3G) systems use wideband digital radio technology as compared to 2nd generation narrowband digital radio. For third generation cordless telephones, 3G wireless describes products that use multiple digital radio channels and new registration processes allowed some 3rd generation cordless phones to roam into other public places.
3GPP-3rd Generation Partnership Project
3GPP2-3rd Generation Partnership Project 2
3GSM-3rd Generation GSM
3rd Generation GSM (3GSM)-The third generation GSM system uses wideband code division multiple access (WCDMA) technology to provide voice, data and broadband multimedia services. The radio portion of the WCDMA system uses wideband radio channels (5 MHz wide) which are different and not compatible with 200 kHz GSM channels. The WCDMA infrastructure is compatible with GSM mobile radio communication system.
3rd Generation Partnership Project (3GPP)-The 3GPP oversees the creation of industry standards for the 3rd generation of mobile wireless communication systems (W-CDMA). The key members of the 3GPP include standards agencies from Japan, Europe, Korea, China and the United States. More information about 3GPP can be found at www.3GPP.org.
3rd Generation Partnership Project 2 (3GPP2)-The 3GPP2 is a collaborative group that is working on the creation of 3rd generation industry global standards that provide for high-speed multimedia wireless services. The key members of the 3GPP2 include standards agencies from North America and Asia. The 3GPP2 is similar to the 3GPP project which is developing 3rd generation WCDMA specifications. More information about 3GPP2 and its standards can be found at www.3GPP2.org.

3rd Party Services-3rd party services are the providing of processes such as information transfer, data manipulation or application processing by other companies or people.
3WC-Three Way Calling
411-The standard service code for local directory assistance in the United States.
4G-Fourth Generation
4-Wire Circuit-A communication circuit that uses separate pairs for each direction of transmission. Normally associated with synchronous, dedicated communications where simultaneous two-way, full duplex transmission is required
511 - North American Traffic Report Number-511 was designated by the US FCC in the year 2000 as the universal telephone number for vehicular traffic reports.
56K Line- A telephone circuit that has a data transmission rate of 56 Kbps. It is sometimes called dataphone digital service (DDS).
611-The universal service code for telephone repair service in the United States.
64 Clear Channel Capability-A channel that allows the end user to transmit 64 kbps without any other constraints such as a maximum ones density or number of-consecutive-zeros restrictions.
64K Line- A digital communication line that allows the end user to transmit 64 kbps.
700 Service Access Code-One of several non-geographic numbering plan area (NPA) codes.
711 - TRS Access Number-711 was designated by the US FCC in the year 2000 as the universal access number for state Telecommunications Relay Service (TRS) centers for the deaf. Most TRS centers also continue to use distinct 800 and local numbers as well as 711.
800 Data Base Service-An intelligent network service that facilitates a more efficient provision of the 800 exchange access service through "number portability." The service enables subscribers to change carriers without changing their 800 number. Subscribers can

also select any unassigned 800 number and use it with any participating carrier's 800 Service.

800 Number Administration and Service Center-A communication center that allows toll free service providers (800, 888, 877, 866) to access an administrative computer system that provides toll free data base services.

800 Service-A communication service where the receiving party pays for incoming calls. 800 services are often called "Toll Free" In the Americas and "Freephone" in Europe and the dialing digits are 0800. In the United States, additional toll free numbers include 888, 877, and 866.

800 Service Access Code-A non-geographic numbering plan area (NPA) code that indicates that a called party will be charged for a call rather than the calling party being charged.

802.16-802.16 is a family of standards from the IEEE Working Group on Broadband Wireless Access Standards. The common name for this family of standards is WiMAX or the official name, WirelessMAN. 802.16 is split into various subgroups that contain specifications for both fixed and mobile WiMAX.

811-A standard feature code used in the Americas that permits' the user to reach the business office of a local telephone service provider.

900 Service Access Code-An area code designated for special services, such as information services, tech support lines, polling or other pay-per-use services.

911-The standardized service number assigned for emergency calls in the United States and Canada, and the islands that are included in the North American Numbering Plan.

976 service-A calling party pays (CPP) service that uses the local exchange code (the first 3 digits of the 7 digit local exchange telephone number) to route the calls and identify it is a CPP service.

99.999%-Five Nines

999-The emergency telephone number that is used by customers to contact public safety officials in countries such as Hong Kong (the countries that are and were part of the British Commonwealth).

9-Track Tape-A 1/2" magnetic tape storage that stores data from computer systems. Nine track tapes sore data on eight tracks and parity information on the ninth track.

A La Carte-A la carte is an option or process that is independent (not bundled) with other items.

A/V-Audio Visual

A+ Certification-A+ Certification is a certification program that is designed to ensure the competence in basic troubleshooting and repair for computer technicians. The certification is controlled by the Computer Technology Industry Association (CompTIA). A+ certification requires the passing two tests: a core exam and a specialty exam. The core exam tests the general knowledge of PCs technology including installation, configuration, software and hardware upgrading, diagnostics, maintenance, repair, interaction with customers, and safety procedures. The specialty exam tests specific operating system knowledge.

AAA-Authentication, Authorization, Accounting

AAA Server-An AAA server is a computing system that can receive, process and send results of authentication, authorization or accounting requests. The AAA server holds or has direct access to the information necessary to validate the credentials of a user or device.

AABS-Automated Attendant Billing System

AAC-Advanced Audio Codec

AAR-Automatic Alternate Routing

Abandoned Attempt-Abandoned attempt is a call attempt that is aborted by the caller before completion.

Abandoned Call-Abandoned call is a call setup request (origination request) that is aborted by an originator before the call is answered.

Abandoned Call Cost-Abandoned call cost is the amount of revenue that is lost as a result of a caller abandoning a call attempt. Abandoned call cost can be calculated by the number of calls attempted to how many calls are connected.

Abandoned in place (AIP)-Abandoned in place (AIP) are cable facilities that are left in place (e.g. in underground conduit or lashed to aerial strand after the old facilities have been deactivated). The new cable facilities are "live" after splicing is complete and may run parallel in the same conduit or in the case of aerial plant may be lashed to the abandoned in place cable.

Abbreviated Address Dialing-Abbreviated address dialing is a call-processing feature that allows a user to initiate dialing to a person using a lesser number of digits than the telephone number is composed of. Abbreviated address dialing may automatically insert (pre-append) digits (such as country and area codes) to the digits dialed by the caller.

ABC-ABC is the term that was used for North American Numbering Plan (NANP) to designate the first three digits of a 7-digit telephone number within a specific Numbering Plan Area (NPA). Each ABC code within an NPA was assigned to an end-office switching system. A single switching system could have had more than one ABC assigned to it. The ABC code is now called the NNX/NXX code, office code, or prefix.

ABC-Activity Based Costing

ABC digits-ABD digits is the term that was used for North American Numbering Plan (NANP) to designate digits in a telephone number. In the past, the first three digits of a 7-digit telephone number within a specific Numbering Plan Area (NPA) were designated by the letters ABC, or alternatively called the central office (CO) code. Each ABC code within an NPA was assigned to an end-office switching system. A single switching system could have had more than one ABC assigned to it. The ABC code is now also called the NNX or NXX code, office code, or prefix. More recently, the 10 digits in a North American telephone number have been represented in some documents by the first 10 letters of the alphabet thus: ABC-DEF-GHIJ, in which the letters ABC represent the area code and DEF represent the central office code.

ABEC-Alternate Billing Entity Code

Abilene-Abilene is a high speed communication backbone network used in the Internet 2 system.

ABR-Available Bit Rate

Absolute Address-Absolute address is a data memory address that is used in a computing device to uniquely identify a specific storage location. An absolute address usually requires a larger address word than a relative data address reference.

Absolute Addressing-Absolute addressing is the use of the specific identification codes or address numbers in a LAN or computer system that is permanently assigned to a storage register, location, or device. The antonym for this term is "relative addressing," a method in which the stated relative address must be added to a so-called base number value to obtain the absolute address.

Absolute Category Rating (ACR)-Absolute category rating is a subjective rating method for audio quality. See mean opinion score (MOS).

Absolute Link-An absolute link is a resource identifier that points to a location or resource that can be found without making reference to the current location or address of the absolute link.

Absolute Time-Absolute time is a time unit that is referenced to a permanent or long-term metric of time (such as GMT).

Abstract Rules-Abstract rules are the ability of a system or person to apply rules to sale of products or services where the rules do not exactly match the products or services offered.

Abstract Syntax Notation (ASN)-Abstract syntax notation is an industry standard that describes data formats that are transmitted in communication systems. ASN is a data format that is used for X.509 security certificates.

Abuse Of Privilege-Abuse of privilege is an action of a user that was not allowed based upon a law or organizational policy.

AC-Access Customer

ACADIA Validation-ACADIA validation is a certification process that helps to ensure broadband network equipment and software reliably deliver the services and applications for system operators.

ACC-Access Control Center

Accelerated Depreciation-Accelerated depreciation is a depreciation method or period of time, including the treatment given cost of removal and gross salvage, that is used when calculating depreciation deductions (asset usage) on income tax returns which is different from the depreciation method or period of time prescribed by the Commission for use in calculating depreciation expense recorded in a company's books of accounts.

Acceleration Clause-An acceleration clause is a term in an agreement (such as a marketing or sales representative agreement) that causes a specific action or new terms to be applied when a specific criterion is reached.

Accelerator Clause-An accelerator clause is a term in a sales agreement that increases (accelerates) the amount (percentage) of sales commission that is earned. Accelerator clauses are used to help motivate sales people to sell more in a shorter period of time.

Acceptable Use Policy (AUP)-An acceptable use policy is the rules and requirements for the use of equipment and services by a person, company or agency. AUPs typically restrict the use of equipment and services to actions that are related to the performance of job duties.

Acceptance-Acceptance is the process of authorizing the use or acquisition of products or services.

Acceptance Credit-Acceptance credit is the providing of value or guarantees of payment for the acceptance of products or services.

Acceptance Test-Acceptance test is a test that evaluates the successful operation and/or

A

performance of an electronic assembly or communication system. Acceptance tests usually have specific operation requirements and test measurements. Acceptance tests are often used as a final product approval and may authorize a product for production or purchase.

Acceptance Trial-An acceptance trial is a set of tests conducted by users of a product or service to determine if the services or systems are operating or performing to specification levels that are acceptable to the buyer of the system or services.

Access-Access is the ability for a user or device to connect to a system or service. An example of access is the requesting and competition of a mobile telephone to get the attention and service access to a cellular radio system.

Access Charge-Access charge is telecommunications service charges that are approved by the Federal Communications Commission (FCC) that compensate a local exchange carrier (LEC) for connection of local customers to a long distance telephone service company (IXC).

Access Code-Access codes are numbers that are dialed by a user to select the access provider for services assigned that access code. The access code usually changes the interexchange (long distance) service provider on a per-call basis. When no access code is entered, the default carrier (or no service provider) is selected. At present, North American access codes have the format 1010xx...x and they are dialed as a prefix to the destination number.

Access Control-Access control is the actions taken to allow or deny use of the services and features of a communication system to individual users.

Access Control Center (ACC)-An access control center is a facility or system that can setup, control and terminate access rights for devices and users that are connected to a communication system.

Access Control Information (ACI)-Access control information (ACI) is a standard directory access control process that is defined by the ITU-T X.500 directory service model where the ACI is used to identify and access a file or directory.

Access Control List (ACL)-Access control list (ACL) is a table or list of users, processes, or objects that are authorized to access files or objects. ACL specify access permissions and other parameters such as read, write, execute, and delete.

Access Coordination Fee (ACF)-Access coordination fee is a value that is assessed for the management of access services.

Access Coupler-Access coupler is a device that allows signals to enter or be extracted (access) from a transmission medium (such as fiber lines).

Access Customer (AC)-Access customer is a user that purchases end user access services (e.g. leased line) from a communication carrier.

Access Customer Number Abbreviation (ACNA)-Access customer number abbreviation is a number that is assigned to a Customer for use in the provisioning and billing of services through an access network.

Access Customer Terminal Location (ACTL)-Access customer terminal location is a code (the CLLI code) that identifies the location of the switch that provides network access to a customer.

Access Devices-Access devices are any types of equipment that can be connected at the end of a communications system or circuit. Access device types include telephone sets, television sets and computers that have network access.

Access Fee-Access fee is a fee that is paid for the use of another network to originate, route, or terminate calls. Access fees are commonly paid by a long distance service provider to a local access provider for allowing calls to enter or terminate through the local network.

Access Gateway (AGW)-An access gateway is a device or assembly that transforms data that is received from a device or user into a format that can be used by a network. An access gateway can adjust the modulation, protocols and timing between two dissimilar communication devices or networks.

Access Independent-Access independent is the ability of a device or service that uses communication links to perform its functions or processes independent of the type of communication link it is using.

Access Level-An access level is a set of permissions that are assigned or associated with one or more files, programs or resources.

Access Line-Access line is the physical link (typically a copper wire or fiber) between a customer and a communications system (typically a central office) that allows a customer to access local and toll switched networks. Access lines may include a subscriber loop, a drop line, inside wiring, and a jack.

Access Link (A-Link)-Access link (A-Link) is a communications line (link) that provides access from a service switching point (SSP) and signaling control point (SCP) to signaling transfer points (STPs) in a SS7 network.

Access List-Access lists are a group of items that are stored in switching devices (such as routers) to control access to or from users or devices that want to access network resources or applications.

Access Log File-An access log file is a list of access attempts that have occurred for when potential users request access to software programs or services. The access log file is continually updated (added to) as new access events occur. Access log files can be used to analyze marketing criteria or to identify problems or fraudulent activities that have or may occur for applications or services.

Access Methods-Access methods are the processes used by a communication device to gain access (obtain services) from a system. Some systems allow communication devices to randomly compete for access (contention based) while other systems assign periods of time or setup events (such as token passing). They precisely control (non-contention based) the access times and methods.

Access Minutes-Access minutes is the length of time that telecommunications facilities are used for long distance service (interstate or international service). This is for both originating and terminating calls. Access minute timing stops when one of the parties disconnects.

Access Multiplexing-Access multiplexing is a process used by a communications system to coordinate and allow more than one user to access the communication channels within the system. There are four basic access multiplexing technologies used in wireless systems: frequency division multiple access (FDMA), time division multiple access (TDMA), code division multiple access (CDMA) and space division multiple access (SDMA). Other forms of access multiplexing (such as voice activity multiplexing) use the fundamentals of these access-multiplexing technologies to operate.

This figure shows the common types of access channel-multiplexing technologies used in wireless systems. This diagram shows that FDMA systems have multiple communication channels and each user on the system occupies an entire channel. TDMA systems dynamically assign users to one or more time slots on

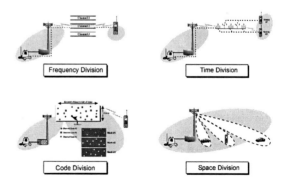

Access Multiplexing Operation

A

each radio channel. CDMA systems assign users a unique spreading code to minimize the interference received and caused with other users. SDMA systems focus radio energy to the geographic area where specific users are operating.

Access Network (AN)-Access network is a portion of a communication network (such as the public switched telephone network) that allows individual subscribers or devices to connect to the core network.

Access Node (AN)-Access nodes are access points or concentration points that allow one or more users to connect to a communication network. Access nodes are usually network devices that allow customers to connect to their communication devices to data networks via wires or wireless connections that use standard transmission protocols such as Ethernet or 802.11 wireless networks.

Access Number-Access number is (1-Wireless) a phone number that can be used to directly dial a cellular or PCS customer in the local wireless system when they are roaming. To contact the roaming customer, the access number is dialed first and when a tone is received, the complete mobile phone number (area code and number) of the roaming customer is entered. The local system will then page the roaming customer. Access numbers were very important in the first few years of cellular service because the systems were not automatically connected. Because most cellular and PCS systems can automatically deliver calls to roaming customers, access numbers are rarely used. (2-calling card) A telephone number of a gateway that allows customer's to enter information that permits the connection of a call or access to information services.

Access Permission-Access permission is the process of assigning permission for a user to gain access to a communication facility.

Access Point (AP)-Access point (AP) is typically, a point that is readily accessible to customers for access to a wireless or wired system. May be called a radio port or access node.

Access Point Name (APN)-Access point naming (APN) is the assigning of identifying names (universal resource locators-URLs) to access points that allow names to identify node IP addresses within GSM and GPRS networks. An APN is composed of two parts; a network ID and an operator specific ID. The network ID identifies the service requested by a user in a network and the operator ID identifies the specific routing information.

Access Point Selection-Access point selection is the process used to select which communication node (such as a wireless access point) a device will connect to.

Access Priority-Access priority is a priority level (user priority level) that is assigned to users or devices within a communication network that is used to coordinate access privileges based on network activity or other factors.

Access Privileges-Access privilege is the authorization to connect, process, use or transfer products or services.

Access Probe-An access probe is a process of sending a message or alert signal to an access point or a system to discover if access is possible or to request access to the system.

Access Process Parameter (APP)-Access process parameters (APP) are the parameters used or assigned to a port (logical channel point) in a packet switching, network that coordinate the operation of the port.

Access Protection-Access protection is the processes that are used to coordinate and restrict access to content or services to unauthorized users.

Access Protocol-Access protocol is the set of procedures and rules that are used by communication devices to coordinate access to a shared communication media.

Access Provider-Access provider is the company that provides and controls access to communication devices or to other networks.

Access Request Information System (ARIS)-Access request information system (ARIS) is the software system that is used by a customer service center at a carrier that processes access service orders (service activation).

Access Rights-Access rights are the properties or access right assignment associated with files, directories, or communication services that define the ability of users to change, interact or access services.

Access Router Card (ARC)-An access router card (ARC) is a packet data switching device that allows users to connect different geographically located computer systems and local area data networks. The ARC can adapt between different types of traffic and different protocols integrating a data network into a single enterprise-wide network.

Access Server-An access server is a computer (or computers) that coordinate access for end users who connect to a communication system. The function of access servers can vary from simple access control to advanced call processing services.

Access Service-Access service is a service that connects customers that are located within a local access and transport area (LATA) to interexchange carriers (long distance telephone service providers).

Access Service Area-Access service area is a geographic area established for the provision and administration of communications service. An access service area encompasses one or more exchanges, where an exchange is a unit of the communications network consisting of the distribution facilities within the area served by one or more end offices, together with the associated facilities used in furnishing communications service within the area.

Access Service Group (ASG)-Access service group (ASG) is the group of switching systems associated with the subtending of a tandem switch.

Access Service Request (ASR)-Access service request (ASR) is a process or form by a local carrier to request services from a local network operator. A form for ASR was developed by the alliance for telecommunications industry solutions (ATIS) in 1984 to help define access requests for the long distance service providers.

Access Services-Access services are providing of communication capabilities to users at the entry (access point) to a network or system.

Access Software-Access software are software programs that are used to transmit and receive information to and from communication systems.

Access Surcharge-An access surcharge is a fee that is added to a communication service charge for the providing of access to systems or services (such as providing long distance service to local telephone access lines).

Access System-An access system is a portion of a communication system that coordinates requests for services and enables the transfer of information when authorized.

Access Tandem (AT)-Access Tandem (AT) is a high level switching system that interconnects low level (local exchange) switching systems. An access tandem can also provide access for nonconforming end offices such as for equal access to other long distance service providers.

Access Tariff-Access tariff is a tariff imposed by federal or state regulators on carriers (typically local exchange carriers) that offer access to telephone exchange services to customers or other companies.

Access Technology-Access technology is the use of systems and features to enable users or devices to connect to a system or service and transfer data or information.

Access Terminal-Access terminals are data input and output devices that are used to communicate with an access point or remotely located computers. Access terminals frequently consist of a keyboard, display monitor, and communication circuitry that can connect the access terminal with the remotely located computer.

Access Time-Access time is the amount of time required to retrieve information from a device or system.

Access Unit (AU)-(1-MPEG-1) An access unit is a coded representation of a presentation unit. (2-LAN) An access unit is a wiring concentrator in a local area network.

Accessibility-Accessibility is the ability of network elements, such as servers, trunks, or ports that can be accessed by users within a group.

Accidental Click Fraud-Accidental click fraud is the selection of links ("clicks") by users for a purpose other than its intended purpose without the intent of defrauding the system or for monetary gains.

Account-(1-billing) A billing account is a unique identifier that designates a customer or company that is used to associate billing charges for products and services. A single customer may have multiple billing accounts. (2-network) On LANs or multi-user operating systems, an account is set up for each user. Accounts are usually kept for administrative or security reasons. For communications and online services, accounts identify a subscriber for billing purposes.

Account Administrator-An account administrator is a person that is responsible or authorized to track, change and manage accounts.

Account Average-Account average is the average of assets for all telecommunications company assets associated with a particular account that include maintenance, repair experience, and salvage value expectancies.

Account Balance Management-Account balance management is a process that allows a user or account owner to monitor, view and pay balances on one or more of their accounts.

Account History-Account history is the set of records or event information that identifies or provides information on customer communication, transaction events or other information that is associated with a account.

Account Identifier (AccountID)-An account identifier is a unique code that can locate and guide transactions to where account transactions are recorded, grouped and managed.

Account Inquiry Centers (AICs)-Account inquiry centers (AICs) are call centers that receive and process calls from customers regarding the status of their accounts.

Account Lockout-In network operating systems (Noose's), an account lockout is the result of a count of the number of invalid logon attempts allowed before a user is locked out.

Account Management-Account management is the authorizing, recording, and assignment of costs to users and groups based upon their authorization and use of network resources.

Account Policy-Account policy is the set of rules or processes used by networks and multi-user operating systems that define how users are allowed to access the system based on their predefined account access privilege settings.

Account Profile-An account profile is a set of characteristics that are associated with an account. These characteristics may include rate plans, usage patterns and service limits along with many other characteristics.

Account Rendering-Account rendering is the displaying of the unpaid charges and unsettled credits associated with an account.

Account Top Up-Account top up is the process of adding funds to an existing prepaid account.

Account Types-Account types are the defining of customer accounts into groups that are based on characteristics such as location, distribution function and the products they carry. Examples of account types include wholesale, corporate, distributor, retailers, and house accounts. While it is desirable that account types are unique, in practice, some account types may overlap.

AccountID-Account Identifier

Accounting-Accounting is (1-general) the process of recording, assigning, and tracking the usage of resources on a network to specific accounts or users. (2-FCAPS) Accounting is one of the five functions defined in the FCAPS model for network management. The accounting level might also be called the allocation level, and is devoted to distributing resources optimally and fairly among network subscribers. This makes the most effective use of the systems available, minimizing the cost of operation. This level is also responsible for ensuring that users are billed appropriately.

Accounting Code-An accounting code is a label or identification number associated with accounting charge or revenue group category. The accounting code may be structured in a way that allows segmentation of the code to identify specific departments or functions that are associated with the cost or revenue.

Accounting Procedures-Accounting procedures are the rules and processes that are used to record, process and report on the creation and use of services or resources.

Accounting System-Accounting systems is a set of accounts, rules, processes, software programs, equipment, and other mechanisms that are used to necessary to operate and evaluate a business system for an operations (business decisions), financial (investor reporting), and regulatory (taxes) perspective.

Accounts Payable (AP)-Accounts payable is a summary rendering of bills that have not been paid. An accounts payable report typically identifies the open (unpaid) balances and how long the amounts have been due to be paid (account aging).

Accounting System Connectivity-Accounting system connectivity is the ability of an accounting system to send and/or receive data from other applications or services.

Accreditation-Accreditation is recognition from a perceived or defined authority that a product or service meets specific criteria that is beneficial (e.g. meets UL approval) or required by regulatory laws.

Accredited Standards Development Organization-Accredited standards development organization is a group that is composed of members that have been accredited by an institution that is responsible for standards accreditation within an industry.

Accrual Basis-Accrual basis is an accounting process that records transactions when they are confirmed or recognized. An example of accrual basis is the recording of sales revenue when the order is received from the customer, even though payment did not occur with the order.

Accruals-Accruals are charges or fees that have been assessed or estimated up to a specific time or event.

Accumulation-Accumulation is a process that counts the number of times individuals are exposed to a communication message over a period of time.

Accuracy-Accuracy is (1-General) a comparison of an actual signal or measured value as compared to the theoretical or pre-established limits. (2-Network Management) Accuracy is the measurement of interface traffic that does not result in error as used in Network Management. Accuracy can be expressed in terms of a percentage that compares the success rate to the total packet rate over a period of time.

ACD-Automatic Call Delivery

ACD-Automatic Call Distribution

ACF-Access Coordination Fee

ACH-Automated Clearinghouse

ACI-Access Control Information

ACIS-Automatic Customer Identification System

ACK-Acknowledgment

Ack-Acknowledgment Message

Acknowledgment (ACK)-Acknowledgement (ACK) is a process or control code that is used

A

to confirm a message has been received or a process has been started or completed. An example of an acknowledgement process is the sending of an ASCII 06 character during data transmission between computers to indicate that data has been received without any errors.

Acknowledgment Message (Ack)-Acknowledgement message is a message that is responded by a communications device to confirm a message or portion of information has been successfully received. If a communications device is supposed to send ack messages back to the originator and an acknowledgement message is not received, the system will typically resend the message. See also negative acknowledgement message (Nack).

ACL-Access Control List

ACNA-Access Customer Number Abbreviation

Acoustic Coupler-Acoustic coupler is a device that is placed against a telephone handset to allow the transmission of audible modem signals over telephone lines.

Acquiring Bank-An acquiring bank is a financial institution that processes credit card transactions.

Acquisition-(1-General) Acquisition is the process of tuning and decoding a signal from a system. (2-Satellite) The process of a satellite receiver device locking on a satellite's global positioning system (GPS) signal. The acquisition includes tuning to the signal frequency and applying automatic gain control (AGC), synchronization (time alignment) and decoding (processing) of the data signal. (3-customer) Acquisition is the process of acquiring new customers from a population of prospects who were not customers in the past. (4-billing) Acquisition is the process of obtaining goods, services or the rights to use good or services.

Acquisition Agreements-Acquisition agreements are documents or recordings that define the terms for the transfer (acquisition) of products or services between a person or company who owns assets and the person or company that the assets will be transferred to.

Acquisition Cost-Acquisition cost is the combined costs that are associated with marketing and adding new customers to a system or service.

Acquisition Planning-Acquisition planning is the process of identifying, evaluating and projecting when equipment or systems will be purchased in the future.

Acquisition Point-(1-general) An acquisition point is a location where information is captured and/or modified. (2-digital rights management) An acquisition point is a location where content is encrypted and usage rights and restrictions are added.

Acquisition Point Certificate-An acquisition point certificate is an authenticated electronic key that is signed to ensure the acquisition point is compliant with required specifications.

Acquisition Rate-Acquisition rates are the percentages of customers or potential customers who newly subscribe (acquire) to new products or services.

ACR-Absolute Category Rating

ACR-Address Correction Request

Acronym-(1-general) An acronym is an artificial word that is formed from the initial letters or groups of letters of words in a set phase or series of words. (2-annoying people) Acronyms are excessively used by annoying people.

Action Words-Action words are expressions or statements that inspire an action or emotion in a reader.

Actionable Measurement-An actionable measurement is the detection of a person's actions (such as mouse over an object on a web page) that can be measured or recorded.

Activate-Activate is the process or action (such as an operator's service screen entry) that starts or reactivates a service.

Activated Channels-Activated channels are those channels engineered at the head end of a cable system for the provision of services generally available to residential subscribers of the cable system, regardless of whether such services actually are provided, including any channel designated for public, educational or governmental use.

Activation-Activation is the process of inputting specific information into a telephone network database to authorize a service account. For a telephone network to provide services, the system must be informed of the services and account codes. Activation of services usually requires that certain customer financial criteria must also be met. After this information is input, the service or telecommunications device will become activated and the customer can request and receive services.

Activation Code-An activation code is a value that is provided to a user or a device that enables the device to begin or to upgrade its operation from a trial mode to a licensed mode of operation.

Activation Commission-Activation commission is a commission paid to a retailer or other entity for activating a new customer, or adding additional lines of service to an existing customer. The activation commission is typically paid after a contract with the customer is completed.

Activation Fee-Activation fee is a one-time fee that is charged for the initial setup of communication service. Activation fees are also called "setup fees."

Active Campaign-Active campaign is the marketing or sales campaign that is currently in process at a company or call center.

Active Directory-Active directory is a system of directory services that allows a client to make an IP telephone call via their PC and the active directory. The PC asks the Active directory for the phone number of a person within the organization. The Directory finds the number and passes it to the PC who dials the number. The Active Directory is fluid tracking clients around the organization.

Active Mediation-Active mediation is a process of receiving, reformatting and processing information between network elements. Active mediation devices are commonly used for billing and customer care systems as these devices can take non-standard proprietary information (such as proprietary digital call detail records) from switches and other network equipment and reformat them into messages billing systems can understand.

Active Members-Active members are users or devices in a system that indicate or are recognized by the system as being active.

Active Monitoring-Active monitoring is the process of modifying information or processes within a system and listening to or viewing a communication service for the purpose of determining its quality or whether it is free from trouble or interference.

Active Physical Link-An active physical link is a defined connection between devices on a physical channel (such as radio or optical) that can maintain direct control between the devices.

Active Server Pages (ASP)-An active server page (ASP) is a script interpreter (real time program operation) that allows the development and use of software program modules to operate on a web server (such as a Microsoft Internet Information Server).

Active Subscriber-An active subscriber is an end user of a service that is authorized to use the service.

Active Window-An active window is the graphic display area in which the user is currently interacting with.

ActiveX-ActiveX is a development of Microsoft's COM that adds network capabilities by creating a set of component-based Internet- and intranet-oriented applications.

ActiveX Control-ActiveX control is a software control module that requires an ActiveX container (such as a spreadsheet or Web browser) to provide a specific function. ActiveX control allow database access or file accesses that can communicate with another ActiveX containers, ActiveX controls, and to interface with the underlying Windows operating system. ActiveX can directly access files using a

A

security that contain digital certificates that authenticate the source and validity of the control.

Activity Based Costing (ABC)-Activity based costing is an accounting system process that associates costs with the activities they are associated with. ABC accounting can help managers to better understand their financial performance and key business drivers.

Activity Factor-The activity factor is the ratio or percentage of time that a device is performing an operation or service. An example of activity factor is the percentage of time a communication device is actively transmitting data on a communication channel compared to the total time the device spends attached (logically connected) to the communication channel.

Activity Report-Activity report is a report of call records that contain the call identifier, date, usage amount (e.g. transmission time, kbytes of data transferred), destination address (e.g. telephone number or IP address), and other pertinent information.

ACTL-Access Customer Terminal Location

ACTS-Advanced Communications Technologies And Services

Actual Price Index (API)-An actual price index is a listing of rates in a group of products or services (e.g. a basket) that are used to calculate prices.

Ad Actions-Advertisement Actions

Ad Aggregator-An ad aggregator obtains the rights from multiple content providers to resell and distribute advertising messages through other communication channels. An ad aggregator typically receives and reformats media content, stores or forwards the media content, controls and/or encodes the media for security purposes, accounts for the delivery of media and distributes the media to the systems that provide the media to specific types of customers.

Ad Application-Advertising Application

Ad Bartering-Ad bartering is an exchange of items or services that are used as value instead of the use of money to pay for advertising programs.

Ad Behavior-Ad behavior is the actions that media messages or programs perform during certain conditions or events. Examples of ad behavior include displaying a graphic when a software application has started or initiating a call when the ad viewer has selected click-to-call.

Ad Bidding-Ad bidding is the process of assigning bid amounts that are associated with a specific advertising message along with the criteria that will be used to enable the bidding process. When the criteria is matched (such as matching a search word in an online search), the bid amounts are reviewed, the highest bids are selected and the advertising message(s) are displayed in the order of bid amount.

This figure shows how IPTV ad bidding may work for selling cars. This diagram shows that bidding for advertising messages may occur for particular age groups, income ranges, program types, geographic regions. This example shows that IPTV advertisers may bid for ads that may appear on a variety of programs throughout various geographic regions. The advertiser sets the maximum bid they are willing to offer and a maximum number of impressions that may be selected to ensure advertising budgets can be maintained. This example shows that the advertiser may also be able to select if the same ad should be sent to the same person more than one time.

Ad Name	Age Group	Income Range	Program Type	Regions	Ad Repeats	Bid per Impression	Max Impressions
Utility Vehicle	25-39	70k+	Sports	Nationwide	Yes	0.10	10,000
Status Auto	40-54	Any	Entertainment	Nationwide	No	0.18	20,000
Luxury Car	55-69	70k+	Travel	Florida	Yes	0.07	10,000

IPTV Ad Billing

Ad Blocking-Ad blocking is the process of blocking ads from being received or viewed. An example of ad blocking is the restriction of pop-up windows in an Internet browser.

Ad Booking-Ad booking is the process of reserving a media time slot or media program resources so that an advertising message can be inserted.

Ad Campaign-Advertising Campaign

Ad Capabilities-Advertisement Capabilities

Ad Click Rate-Ad click rate is the ratio of how many times an ad is clicked (selected) as compared to the number of time it is viewed. Ad click rate is also called click-through rate (CTR).

Ad Click Report-An ad click report is a table, graph or images that represent specific aspects of advertising campaigns where the selection (clicking) of an ad can be monitored and recorded. Ad click reports usually consist of the number of impressions and the number of clicks that occur during a period of time.

Ad Clicks-Ad clicks are the recorded number of clicks that have resulted from users selecting an ad banner or link.

Ad Copy-Ad copy is the text portion of an ad.

Ad Credits-Advertising Credits

Ad Expansions-Ad expansions are the number of conversions of advertising messages (selected or automatically expanded) from a smaller and/or shorter version of an ad to a larger and/or longer version of an ad.

Ad Expiration Date-An ad expiration date is the day that an authorization, validity or ad insertion process will end.

Ad Frequency-Ad frequency is the number of times an ad is inserted or displayed over a period of time.

Ad Hoc (Ad-Hoc) Network-Ad hoc (Ad-Hoc) network is a wireless network (typically temporary) comprising only stations without access points that receive and retransmit between stations.

Ad Hoc Query-Ad hoc query is a filtering action or query that is created dynamically or temporarily.

Ad Impression-An ad impression is a single presentation of an advertising message or image to a media viewer.

Ad Insert-AD Insertion

AD Insertion (Ad Insert)-Ad insertion is the process of inserting an advertising message into a media stream such as a television program. For broadcasting systems, Ad inserts are typically inserted on a national or geographic basis that is determined by the distribution network. For IP television systems, Ad inserts can be directed to specific users based on the viewer's profile.

Ad Insertion Module-Ad insertion modules a process used in cable television or broadcast radio networks that allow the insertion of advertising during pre-determined time segments. This process allows different advertising messages to be inserted in different geographic regions.

Ad Link-Ad links are portions of an web based advertising message that can be selected or used to obtain addition information or perform a process such as sending an email that can request additional information.

Ad Listing-Ad listing is the entire contents of an ad that is visible. This includes the title, copy, and graphics.

Ad Management-Ad management is the process of selecting, monitoring and adjusting advertising messages (ads) for use in marketing programs such as adword campaigns, banner advertising and print media or other communication channels.

Ad Model-Advertising Model

Ad Profit-Ad profit is the difference between sales revenue generated by a specific ad and the cost of developing the ad and the media (advertising) cost.

Ad Response Capabilities-Ad response capabilities are the processes that a device or service can use to respond to advertising messages.

Ad Rotation-Ad rotation is the changing of ads that are displayed using a list of ads. Ad rotation may be performed for all visitors or for repeat visitors.

Ad Selection-Ad selection is the process of selecting ("clicking") a link on a link, button or graphic image in an advertising message.

Ad Space-Ad space is the amount of display space that is dedicated for an ad on a media display (such as an Internet web page).

Ad Splicer-Advertising Splicer

Ad Sponsor-Ad Sponsor is the company or person who pays for the media (advertising) cost of an ad.

Ad Supported Business Model-An ad supported business model is a set of rules or sets of relationships or formulas that can be used to define how advertising will generate revenues or profits.

Ad Supported Content-Ad support content is the providing of media where the creation, distribution and/or consumption of the content is paid for by an entity in return for including advertising messages with the content.

Ad Syndication-An ad syndication is the promotion of an advertising message by group of companies in the media industry that agree to work together to promote ads that are accepted by the syndication. Companies in the ad syndication agree to promote ads and share information that is necessary to accomplish advertising objectives.

Ad Telescoping-Telescoping advertisements are extended advertising messages (selected or automatically expanded) from a smaller and/or shorter version of an ad to a larger and/or longer version of an ad. Ad telescoping allows the viewer to immediately obtain more information about a product or service by selecting an interactive option on the advertising message.

Ad Trackers-Ad trackers are programs or services that are used to track the selection, interaction or response to advertising messages.

Ad Tracking-Ad tracking is the process of recording or measuring the results of the sending or placement of advertising messages. For web ad messages, ad tracking may include the number of views (impression) and items clicked on an advertising page.

Ad Unit-Advertising Unit

Ad View-An ad view is a single presentation of an advertising message or image to a media viewer.

Ad Views-Ad views are the number of times an advertising message has been viewed. For web pages, impressions are the number of times a web page has been requested and displayed.

Adapter-(1-computer board) An adapter is a printed circuit board that plugs into a computer's expansion bus to provide added capabilities. Common adapters include video adapters, joy-stick controllers, and I/O adapters, as well as other devices, such as internal modems, CD-ROMs, and network interface cards. One adapter can often support several different devices. (2-cable) A cable or connector adapter converts and/or adjusts the physical and/or electrical properties from one connection to another connection.

Adaptive Spare-Adaptive spare is a spare or standby piece of equipment in a communication network that can replace a failed piece of equipment and adaptively change its configuration or performance to match the characteristics of the failed piece of equipment. Use of adaptive spares reduces the number of backup equipment assemblies that are required in a communications network.

ADC-Analog to Digital Converter

ADC-Automated Data Collections

Addendum-An addendum is a supplemental item (such as a case study) that is associated with a document.

Additional Information Field (AIF)-An additional information field is an area within a data record or a form that allows for the inclusion of additional (usually uncategorized) information.

Additions (Adds)-Additions (Adds) are the number of customers (subscribers) or devices that are added to a system over a time period.

Add-On-Add-on is a telephone call center sales process designed to encourage customers to purchase additional products or services secondary to the primary sale. Occasionally, it is incorrectly referred to as up-selling, which means the quantity or quality of the primary product is offered at a premium and higher cost.

Add-On Conference-Add-on conference is a conference call feature (usually in a PBX system) that allows additional participants to be added to the conference call. To add-on a conference participant, a participant (or moderator) places the existing call or on hold, obtains system dial tone, and connects to the add-on participant. After the new participant has agreed to participate, the originating participant reactivates (re-connects) to the conference call in progress.

Address-Address is a grouping of numbers that uniquely identifies a station in a local area network, a location in computer memory, a house on a street, etc. For a local area network station or a computer memory location, electronic logic can be arranged to ignore messages not bearing the appropriate address and accept messages that do bear the appropriate address.

Address Accuracy-Address accuracy is a measure or value that indicates how the information in an address (such as a prospective sales lead) matches an address in another list or database (such as a company database).

Address Block-An address block is an area (a block of data) that is used to print or display address information.

Address Correction Request (ACR)-Address correction request is the process of requesting that information about an address be verified by another system or company.

Address Field-Address field is a section of a message, generally at the beginning, in which the addresses of the message source and destination are found.

Address Management-Address management is the process of assigning and controlling unique identifying numbers or names for devices in communications system.

Address Resolution-Address resolution is a process that is used to specifically identify differences between computer addressing schemes. The address resolution process may be implemented by mapping channels on layer 3 (network layer) to specific addresses on layer 2 (data link layer).

Address Selection-Address selection is the process that is used to identify and assign an address to a device or service.

Address Signaling-Address signaling is a process of sending the signaling message that includes the dialed telephone number to a exchange carrier. For the public switched telephone network (PSTN), this is performed by dial pulsing (rotary phone) or touch-tone (TM) signals.

Address Space-Address space measures the amount of memory that can be installed on a computer. A computer that utilizes 24 bits to designate a memory location is said to have a 24-bit address space; it can address twice as much memory as a computer having a 23-bit address space.

Address Standardization-Address standardization is the process of converting or adjusting address information into a format that is more uniform. Address standardization may be used to help identify duplicate records or to assist in the delivery and routing of items.

Address Verification Service (AVS)-Address verification service is a process that can be used to verify that the billing address information matches the credit cardholders mailing address. AVS validates the address by using the street address number and the zip code.

Addressable-Addressable is the ability to uniquely identify a device or service enabling it to send or receive information based on its unique identification address.

A

Addressable Advertising-Addressable advertising is the communication of a message or media content to a specific device or customer based on their address. The address of the customer may be obtained by searching viewer profiles to determine if the advertising message is appropriate for the recipient. The use of addressable advertising allows for rapid and direct measurement of the effectiveness of advertising campaigns.

This diagram shows how addressable advertising may operate. In this example, a program source is sent with an ad insert period. When the video distribution point detects the beginning of the ad insert period, it may select and insert ads for specific customers. This allows different ads to be inserted for different viewers, even in the same geographic area.

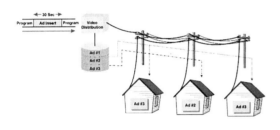

Addressable Advertising

Addressable IPTV Advertising-Addressable IPTV advertising is the process of enabling separate streams to be sent to different TV sets during the same advertising time slot. Because of this, ads can be tailored according to individual demographics; location; interests; viewing habits; time of day; language and a raft of other factors.

Addressing Authority-Addressing authority is the authority responsible for the unique assignment of Network addresses within a network address domain.

Adds-Additions

Adds, Moves and Changes (AMC)-Adds, moves and changes is the process of adding, changing the configuration of equipment and removes (disconnection) of equipment and services for new, existing, and terminated customers.

Adhesion Contract-An adhesion contract is an agreement that is entered into by a licensee that is typically non-negotiable and it binds (adheres) the user to the terms of the agreement to gain access to the underlying product and/or service no matter how restrictive or unbalanced the agreement is.

Adhoc Group-An adhoc group is a temporary group of members or companies that is typically formed to solve a specific problem or function.

Ad-Hoc Network-Ad Hoc

Adjudicative Proceeding-Adjudicative proceedings involve future rates or practices, initiated upon the Commission's own motion or upon the filing of an application, a petition for special relief or waiver, or a complaint or similar pleading that involves the determination of rights and responsibilities of specific parties.

Adjustments-Adjustments are changes to item charges, invoices or account balances. Adjustments are typically made to correct errors or to provide satisfaction to customers for billing errors or services that did not meet expectations. Some of the common adjustments include charge waivers and refunds.

Adminisphere-An adminisphere is an organizational layer that begins above the unskilled worker layer and ends somewhere between the managerial and executive layers. Adminisphere layer is responsible for organizing, processing and delaying the distribution of data to people who need it.

Administered Roaming-Administered roaming is the process of managing the access rights for customers that move from one service provider's system service area to another service provider's service area and obtain service.

Administration Panel-An administration panel is a form that can be used to control features, services and other functions (such as billing) that apply to an account or service.

Administrative Alerts-Administrative alerts are informational messages sent to specific accounts, groups, or computers to announce security events, impending shutdown due to loss of server power, performance problems, and printer errors. When a server generates an administrative alert, the appropriate message may be sent to a predefined list of users and computers.

Administrative Domain-An administrative domain is a region, area or portion of a system or network, which can be controlled or modified by an owner and/or operator. Administrative domain commonly refers the control of security, accounting and services.

Administrative Services-Administrative services are the supporting, clerical and other management duties that are necessary to perform business tasks or processes.

Administrator Account-Administrator account is an account on a communications system or software program (such as Microsoft Windows) that is provided with a level of authority and permission that allows the administrator to assign and remove permission to users or groups within the system.

Admission-Admission is the process of a device requesting and receiving authorization to obtain service from a communication system or network.

Admission Control-Admission control is the process of reviewing the service authorization level associated with users and determining how much network resources will be allocated if the network resources are available. Admission control is used to adjust, limit or assign the use of limited network resources to specific types or individual users. Access control may allow for the assignment of higher access level priority for specific types of users such as public safety users.

Adoption Process-Adoption process is the sequence of actions or experiences that a person or company may perform before they accept and use new processes.

ADRMP-Auto Dialing Recorded Message Player

ADSL-Asymmetric Digital Subscriber Line

ADSL Forum-ADSL is a forum that was started in 1994. This forum assists manufacturers and service providers with the marketing and development of ADSL products and services. The ADSL forum has been renamed the DSL forum.

ADSL Modem-ADSL modem is an electronics assembly device that modulates and demodulates (MoDem) asymmetric digital subscriber line (ADSL) signals. ADSL signals are usually transmitted on a twisted pair of copper wires. An ADSL modem may be in the form of an internal computer card (e.g. PCI card) or an external device (Ethernet adapter). Most ADSL modems have the ability to change their data transfer rates based on the settings that are programmed by the DSL service provider and as a result of the quality of the communication line (e.g. amount of distortion).

ADSL Router-ADSL router is a packet routing device that is commonly used to interfaces a computer or other residential data communication devices with an ADSL telephone line. The use of a router (as opposed to a hub) allows multiple computers within a home to be assigned different Internet protocol (IP) addresses.

ADSL-Lite-ADSL-Lite is a limited version of the standard ADSL transmission system. This limited version of ADSL allows for a simpler filter installation that can often be performed by the end user. The limitation of ADSL-Lite is a reduced data transmission rate of 1 Mbps instead of a maximum rate of 8 Mbps.

This figure shows that an ADSL-lite system is similar to the ADSL network with the primary difference in how the end user equipment is connected to the telephone network. The ADSL-lite system does not require a splitter for the home or business. Instead, the end user can install microfilters between the telephone line and standard telephones. These microfilters block the high speed data signal from interfering with standard telephone equipment. The ADSL-Lite end user modem contains a filter to block out the analog signals.

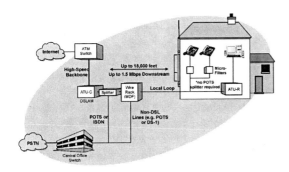

ADSL Lite System

Advance Payment-Advance payment (prepayment) is the transfer of value for products or services that will be provided or authorized for use in the future.

Advance Premium-An advance premium is an incentive that is offered to a consumer who agrees to make a purchase at a future time or to continue or upgrade an existing service.

Advanced 800 Services-Advanced 800 services is a group of toll free (800, 888 number) services that includes call routing that can be based on parameters such as the originating phone number or area code, time of day, or volume of calls. This allows toll free services to be redirected to handle overflow from call centers or redirecting to call centers that are available specific times within a day.

Advanced Audio Codec (AAC)-Advanced audio codec (AAC) is a lossy audio codec standardized by the ISO/IEC Moving Picture Experts Group (MPEG) committee in 1997 as an improved but non-backward-compatible alternative to MP3. Like MP3, AAC is intended for high-quality audio (like music) and expert listeners have found some AAC-encoded audio to be indistinguishable from the original audio at bit rates around 128 kbps, compared with 192 kbps for MP3.

Advanced Client-An advanced client is a computer, hardware device or software program that is configured to request services from a network which can process some of the information that it requests or receives.

Advanced Communications Technologies And Services (ACTS)-Advanced communications technologies and services is a European research program that was started in 1995 to assist with advancements in mobile communication technologies and services.

Advanced Encryption Standard (AES)-Advanced encryption standard (AES) is a data encryption standard promoted by the United States government and based on the Rijndael encryption algorithm. The AES standard is supposed to replace the Data Encryption Standard (DES).

Advanced Mobile Communication Services (AMCS)-Advanced mobile communication services are the providing of voice, data or video services that go beyond the basic mobile voice and data communication services. These include short messaging, telemetry, and other information services.

Advanced Mobile Phone Service (AMPS)-Advanced mobile phone service (AMPS) is an analog cellular communications system that uses frequency-division multiple Access (FDMA) for control and frequency division duplex (FDD) for two transmission. The AMPS radio channel types include 30 kHz FSK control channels and 30 kHz voice channels and it operates in the 825 MHz to 890 MHz frequency range.

Advanced Planning and Scheduling (APS)-Advanced planning and scheduling is a process or system that is used to estimate future needs and to schedule acquisitions or projects to satisfy the needs. APS systems may use information from ERP systems.

Advanced Radio Data Information Service (ARDIS)-Advanced radio data information service (ARDIS) is a two-way wireless data network that was announced in January 1990. The ARDIS system allows simulcast radio transmission to increase the penetration and reliability of wireless data signals into buildings. The data rate used by an ARDIS device is approximately 9.6 kbps and the frequency range commonly used for ARDIS systems are in the 800-900 MHz range.

Advanced Replacement-Advanced Replacement is a process that allows the end user or repair facility to obtain a replacement component (transmitter, receiver, software, etc.) before returning the defective equipment or software. This process usually requires the requesting person or company to obtain a reference number for the advanced replacement component that is requested. The person or company often uses the box from the replacement equipment to return the defective equipment or software.

Advanced Shipping Notice (ASN)-Advanced shipping notice is a message that is sent to a customer by a vendor that provides information about a shipment that is about to occur for their purchase of goods or services. ASNs may be provided for large shipments to customers where some preparation must be performed to receive the goods or services.

Advanced Television (ATV)-Advanced television is television technology that provides audio and video signals that have quality that is better than the quality of existing television broadcast systems (e.g. NTSC or PAL).

Advanced Television Services-Advanced Television Services are television services that are provided using alternative technologies (such as digital television). Advanced televi-sion services are defined the report: "Advanced Television Systems and Their Impact Upon the Existing Television Broadcast Service", MM Docket 87-268.

Advanced Video Coding (AVC)-Advanced video coding is a video codec that can be used in the MPEG-4 standard. The AVC coder provides standard definition (SD) quality at approximately 2 Mbps.

Advances-Advances are the providing of funds or assets for products or services that will be provided or authorized for use in the future.

Advertainments-Advertainments are programs or media messages that that provide entertainment to viewers while communicating an advertising message.

Advertiming-Advertiming is the comparison of the results of advertising programs over time periods (such as from year to year).

Advertisement Actions (Ad Actions)-Advertisement actions are the processes that can be performed by an advertising item. These items may include animation, linking to web pages, click to SMS, click to outside application, click to call, or other media processing actions.

Advertisement Capabilities (Ad Capabilities)-An advertisement capability is the ability of a media item to perform an action such as to perform animation or to include links to other services (for electronic ads).

Advertisement Effectiveness-Advertisement effectiveness is a measure of a desired response from the presentation of an advertising message.

Advertiser-An advertiser is a company or person that pays for providing of services to other people in return for the inclusion or presentation of marketing messages.

Advertising-(1-general) The communication of a message or media content to one or more potential customers. (2-Data Networks) The process by which services on a network inform other devices on the network of their availability. Novell NetWare uses the Service Advertising Protocol (SAP) for this purpose. Such a process is used in packet networks to inform switches and router of new routes and route changes.

Advertising Agency-An advertising agency is a company that represents or manages the services of other companies for the purposes of providing media communication services (advertising).

Advertising Allowance-An advertising allowance is an amount of value that is offered by a distributor or provider of products that is designated to be used in for advertising a product or brand.

Advertising Application (Ad Application)-An advertising application is a software program that can create advertising messages or media that is played to a display or device when certain conditions exist.

Advertising Availability (Avail)-An 'avail' is the time slot within which an advertisement is placed. Avail time periods usually are available in standard lengths of 10, 20, 30 or 40 seconds each. Through the use of addressable advertising, which may provide access to hundreds of thousands of ads with different time lengths, it is possible for many different advertisements going to different audiences to share a single avail.

Advertising Budget-An advertising budget is planned amounts of money that is designated to pay for advertising.

Advertising Campaign (Ad Campaign)-Advertising campaigns are marketing activities designed to send specific advertising messages to customers about products, services, and options offered by a company.

Advertising Credits (Ad Credits)-Advertising credits are values that are provided in return for watching or responding to advertising messages.

Advertising Elasticity-Advertising elasticity is the connection between advertising cost or budget to the resulting change of sales or effectiveness of the advertising message.

Advertising Income-Advertising income is revenue that is generated from inserting and distributing media communication messages to one or more potential listeners or viewers.

Advertising Material-Advertising material is any media in a program that is developed to assist in the promotion or sales of products or services.

Advertising Medium-An advertising medium is a communication channel or format that is used to communicate messages or promote a product or service. Advertising mediums include print, radio, television or the Internet.

Advertising Model (Ad Model)-An advertising model is a representation of the processes of how promotional communication messages are transferred and received by media viewers and listeners.

Advertising Network-Advertising network is a group of advertising systems (such as multiple web sites) that allow product or service sellers to reach a broad audience. Advertising networks typically provide discounts to companies or people for advertising to multiple networks.

Advertising on Demand (AoD)-Advertising on demand is a service that provides end users to interactively request and receive advertising messages. These advertising messages can be identified from previously stored messages (advertising bookmarks) or they can be as a result of an advertising message search.

Advertising Page-An advertising page is a web page that contains an advertising message.

Advertising Page Exposure-Advertising page exposures are the number of times a media message is available for viewing by a person regardless if the person perceived the ad or not.

Advertising Promotion (Promo)-Advertising promotions are messages that offer added incentives to potential customers for products or services.

Advertising Regulations-Advertising regulations are government rules that broadcasters must follow to be allowed to transmit media. Advertising regulations can include the types of advertising messages, the maximum number of advertising messages in a time period, the types of programs they can be inserted into, and a mix of content restrictions. Advertising regulations can dramatically vary from country to country.

Advertising Reports-Advertising reports are tables, graphs or images that represent specific aspects of advertising campaigns or the information or data that is created from advertising campaigns.

Advertising Research-Advertising research is the process of gathering and analyzing information that is used to determine the potential or achieved results for advertising campaigns.

Advertising Splicer (Ad Splicer)-An advertising splicer is a device that selects from two or more media program inputs to produce one media output. Ad splicers receive cueing messages (get ready) and splice commands (switch now) to identify when and which media programs will be spliced.

Advertising Stream-An advertising stream is a source of money or value received that is associated with the sale or providing of advertising services.

Advertising Supported-Advertising supported is the providing of media to viewers or users where the broadcasting or providing of the media is paid for by advertisers.

Advertising Unit (Ad Unit)-An advertising unit is a display area size that is used a reference value for media insertions. A standard advertising unit has a column width of 2 1/16 inches with a depth of one inch that is surrounded by 1/8-inch gutter area.

Advertising, Premiums and Incentives (API)-Advertising, premiums and incentives are the providing of promotional products or other items of value to encourage or reward for actions or performance.

Advertorial-An advertorial is a media message that created and paid for by an advertiser that looks like an editorial news story.

Advice Note-An advice note is a message that is sent to a customer by a vendor that provides information about their purchase of goods or services.

Advice Of Charge (AOC)-Advice of charge is the ability of a telecommunications system to advise the user of the actual costs of telephone calls either prior or after calls are made or services are used. For some systems, (such as a mobile phone system) the AOC feature is delivered by short message service.

Advisory Committee-An advisory committee is a group of people who review and provide information that helps a company and/or person to make decisions.

Advisory Tones-Advisory tones are audio tones that are provided from the telephone system to inform the customer of the call status or change in call status. Advisory tones include busy, dialtone, ringing, fast-busy, call-waiting, and other tones.

Adword-An adword is a key word that is used in an ad bidding marketing campaign.

Adword Marketing-Adword marketing is a process which uses key words that potential customers enter into search engines to find product or service information. Adword marketing is usually paid for by a fixed fee or bidding process. To Adword market, a list of keywords is selected and associated with a URL and a short message to accompany the listing. When the search term(s) matches the keyword, the URL and the descriptive text are displayed. These listings are called sponsored listings.

Adword Marketing Cost-Adword marketing cost is the combined costs of producing and presenting advertising messages on search engines. Adword marketing costs can include ad creation, keyword list selection, ad management and click fees.

Aerial Plant-Aerial plant is the physical property and facilities of a telephone, cable or company that uses a transmission system that is mounted above the ground. Aerial plant property includes the telephone poles, transmission cables, amplifiers, and channel multiplexing equipment.

Aeronautical Fixed Service-Aeronautical fixed service is a wireless communication service that is used for air navigation safety and to coordinate efficient operation of air transportation. These communication services are between fixed transceivers (stations).

Aeronautical Mobile Satellite Service-Aeronautical mobile satellite service is a satellite communication service that allows aircraft to directly communicate with or through orbiting satellites. For this service, mobile Earth stations are loaded on board aircraft. Other mobile satellite stations such as emergency position-indicating radio-beacon stations may be part of this service.

Aeronautical Mobile Service-Aeronautical mobile service is a mobile communication service between aeronautical transceivers (stations) and aircraft stations or directly between aircraft stations.

Aeronautical Radionavigation Satellite Service-Aeronautical radionavigation satellite service is a radio communication service that provides position information for aircraft navigation through the use of satellite radio transmission.

Aeronautical Radionavigation Service-Aeronautical radionavigation service is a radio communication service that provides position information for aircraft navigation.

AES-Advanced Encryption Standard

Affiliate-(1-general) A company or person that owns, controls or is owned or controlled by another company or person. Regulations may specify the definition of "affiliate" based on an equity percentage of ownership. Regulations may limit the maximum amount of ownership (e.g. 10%) by an affiliate that is involved in related industries (such as television and radio system ownership) or the amount of ownership by a foreign person or company. (2-marketing) A company or person who markets the products or services of another company or person in return for a fee or commission.

Affiliate Broker-An affiliate broker is a person or company who works with companies that participates in affiliate programs on a commission basis.

Affiliate Click Fraud-Affiliate click fraud is the selection of links ("clicks") by companies that have a co-marketing (affiliate) relationship with the link sponsor who has an interest in selecting the link for a non-intended purpose. These intentions may include inflating the value of the affiliate relationship (higher click rates) or getting paid for an increase of the affiliate leads that are provided.

Affiliate Directory-An affiliate directory is a listing of affiliate partners or programs.

Affiliate Fraud-Affiliate fraud is the selection of affiliate links by users who do not have an interest in using the link for its intended purpose. Affiliate fraud may result from users or affiliate partners who receive compensation for affiliate link clicks on their web sites or web sites they manage.

Affiliate Link-An affiliate link is an Internet hyperlink that redirects potential customers from one web site to another website. Affiliate link operation requires the sending of tracking codes that are sent when the user selects the link ("clicks the link") that identifies which affiliate partner the user has come from. These tracking codes may be part of the hyperlink (visible to the user) or they may be sent along with the link when the user clicks on the hyperlink (hidden from the user). These identification codes may be used to update the processing of orders such as discounts or to direct the affiliate visitor to preferred products.

This figure shows the basic operation of affiliate links. This example shows that affiliate links are provided to affiliate marketing partners and these links are inserted on pages that have a common interest with the affiliate's product. When visitors to the affiliate web sites click on the affiliate link, this link provides the web address of the destination page along with an affiliate ID code. When the destination page is accessed, the affiliate ID code is stored during the visitor's session. If the visitor purchases a product, the affiliate ID is stored along with the order information. This order information and affiliate commission table is used to calculate the commissions that are paid to the affiliate partner.

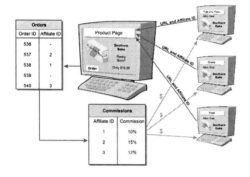

Internet Marketing Affiliate Link Operation

Affiliate Management-Affiliate management is the process that identifies, defines and tracks products and services that are sold by affiliated companies. Affiliate management may include the providing of URL links that can be placed on affiliate web sites, which contain embedded link tracking codes.

Affiliate Management System (AMS)-An affiliate management system is the combination of hardware and software that identifies, categorizes and manages products or services that are sold by affiliated companies.

Affiliate Marketing-Affiliate marketing is the process of sharing marketing and sales programs that compensates marketing affiliates (partners) for their role in communicating and selling to customers.

Affiliate Merchant-Affiliate merchant is the company or person in an affiliate marketing program that is performs the marketing and/or selling of products or services.

Affiliate Network-An affiliate network is a group of companies that work with each other to promote and sell products or services from companies in the affiliate group.

Affiliate Page-An affiliate page is a web page that is dedicated to promoting a product or service from an affiliated partner.

Affiliate Payment-Affiliate payment is a value that is provided to an affiliate partner for their promotion of a product or services.

Affiliate Payout-Affiliate payout is the transferring of funds for earned affiliate commissions.

Affiliate Program-Affiliate program is a marketing program that compensates a company that assists in the marketing of another company's products or services.

Affiliate Recruitment-Affiliate recruitment is a program that identifies affiliate candidate companies and proposes affiliate-marketing programs.

Affiliate Software-Affiliate software is an application program that is included and potentially linked with software programs produced or provided by another company.

Affiliated Companies-Affiliated companies are the companies that directly or indirectly own or control a company or resource. Regulations may specify the definition of an "affiliated company" based on an equity percentage of ownership. Regulations may limit the types of companies that may participate as an affiliated company based on an amount of ownership (e.g. 10%) and the actions these companies may perform or may be required to perform because they are considered an affiliate company.

Affiliation Through Common Facilities-Affiliation through common facilities is an affiliation that results when a company or person shares resources (such as a communication system or office space) with other companies or people. This affiliation is usually visi-

A

ble when more than one person or company who uses the shared resource can change the shared resources that have potential control of others involved in the use of the resource.

Affiliation Through Common Management-Affiliation through common management is the affiliation between entities (corporation or individuals) that arises where agents of the company serve as a controlling element of the management or board of directors of another entity.

Affiliation Through Contractual Relationships-Affiliation through contractual relationships is the affiliation between entities (corporation or individuals) that arises where the control of agents of an entity is dependent upon contractual terms of another entity.

Affiliation Through Stock Ownership-Affiliation through stock ownership is the affiliation between entities (corporation or individuals) that arises where equity owners (shareholders) own or have the power to control (e.g. vote proxies) of more than 50 percent of the voting stock.

Affinity Group-An affinity group is a set of people that share a degree of preference, trust or common interest.

Affinity Marketing-Affinity marketing is focusing of marketing projects and media materials to people or companies that have established (an affinity) buying patterns.

Affinity Programs-Affinity programs are the marketing processes that rewards customers for activities such as repeated purchases or multiple visits to a web site.

Afloat-Afloat indicates that products that are being shipped are currently in transit aboard a ship.

Aftermarket-Aftermarket is the market for related hardware, software, and peripheral devices created by the sale of a large number of computers of a specific type.

Agency-(1-general regulatory) A commission, board of commissioners, committee, or other group of commissioners who are authorized to act on behalf of the commission or regulatory department. (2-communications) The Federal Communications Commission (FCC) agency of the U.S. Government as defined by section 105 of title 5 U.S.C., the U.S. Postal Service, the U.S. Postal Rate Commission, a military department as defined by section 102 of title 5 U.S.C., an agency or court of the judicial branch, or and an agency of the legislative branch, including the U.S. Senate and the House of Representatives. (3-marketing) A marketing agency is a company that represents or manages the services of other companies for the purposes of providing media communication services.

Agency Commission-An agency commission is a fee that is assessed for the services provided by an agency.

Agenda Engine-An agenda engine is a software program that can receive, process and respond to an end user's (client's) request to schedule events.

Agent-(1-general) A person or a device that performs tasks for the benefit of someone or some other device. (2-software) A program that performs a task in the background and informs the user when the task reaches a certain milestone or is complete.

A program that searches through archives looking for information specified by the user. A good example is a spider that searches Usenet articles. Sometimes called an intelligent agent.

In SNMP (Simple Network Management Protocol), a program that monitors network traffic.

In client-server applications, an agent is a program that mediates between the client and the server.

Agent Channel-Agent Channel is the distribution channel that makes use of a third party sales force (individual and corporate) to deliver products and services directly to customers.

Agent Time-Agent time is the activity categories a call agent spends time performing and the amount of time in each category that is spent.

Aggregate Customer Information-Aggregate customer information is collective data that relates to a group or category of services or customers, from which individual customer identities and characteristics have been removed.

Aggregate Request-An aggregate request is a message that defines the amount of a resource (such as transmission bandwidth) that is requested to provide for a combined group of applications or services.

Aggregated Link-Aggregated link is a process of combining two or more physical links to provide a single high-speed interface to higher layer protocol layers. This process is also called inverse multiplexing.

Aggregating Usage-Aggregating usage is the combining of usage charges for a device, account or services. Usage aggregation is performed to create a bill and to determine appropriate usage discounts.

Aggregation-Aggregation is the process of combining multiple services or communication circuits into a higher-capacity service or system.

Aggregation Device-An aggregation device is used to combine (concentrate) two or more communication channels onto a higher-speed communication channel.

Aggregation Sites-Aggregation sites are web pages or media servers that receive, sort and distribute content from multiple sources to multiple recipients.

Aggregator-(1-service provider) A company or service provider that performs the operations required to make multiple physical links function as a combined (aggregated) link. Aggregators typically purchase network services in discounted bulk quantities and pass along the savings or support services to smaller users of the services. (2-billing) A billing aggregator is a company that gathers billing

records from one or more companies and posts them to another billing system. An example of a billing aggregator is the gathering of billing of information from many companies that provide information or value added services (e.g. news or messaging services) and transferring these to another basic services carrier for direct billing to a customer.

Aging-Aging is the process of removing entries from a database when those entities have become old or become inactive for a period of time (aging time).

Aging Report-An aging report is a presentation of information where the data is grouped into time periods.

AGRAS-Air-Ground Radiotelephone Automated Service

Agreement-An agreement is a set of terms and obligations that have been agreed to by two or more entities. Agreements can be active or passive. Active agreements are created when the parties actively in the creation, acceptance and passive agreements are entered into without the direct participation of the entities (parties) involved in the agreement.

AGW-Access Gateway

AHRA-Audio Home Recording Act

AI-Artificial Intelligence

AIAG-Automotive Industry Action Group

AICs-Account Inquiry Centers

AIF-Additional Information Field

AIP-Abandoned in place

AIP-Application Infrastructure Service Provider

Air Play-Air play is the amount of time that media is transmitted or broadcasted to listeners or viewers.

Air-Ground Radiotelephone Automated Service (AGRAS)-Air-ground radiotelephone automated service is a telephone service that is used for aircraft.

Air-Ground Radiotelephone Service-Air-ground radiotelephone service is a wireless communications service for subscribers in aircraft.

Air-Ground Service-Air-ground service is a wireless service between aircraft and a telephone or data network. There are two basic types of air-ground services; general aviation and passenger or commercial service.

AirTime-(1-radio) Air time is the length a communication service (typically radio transmission time) is used. (2-broadcasting) Time that a person or media source is being transmitted on a broadcast network.

Airtime Asset-An airtime asset is a program or data that will be broadcasted over a period of time (airtime).

AK-Authorization Key

AK Grace Time-Authentication Key Grace Time

Alarm-Alarm is an audible or visible indication of a trouble condition. Alarms are classified as minor, major, or critical, depending on the degree of service degradation or disruption.

Alarm Monitoring Service-Alarm monitoring system is a service or company that detects and responds to equipment failures, alarm triggers, or incidents.

A-law Encoding-A-law encoding is a digital signal companding process that is used for encoding/decoding signals in pulse-code-modulated (PCM) systems. This companding process increases the dynamic range of a binary signal by assigning different weighted values to each bit of information then is defined by the binary system. The A-law encoding system is an international standard. A different companding version is used in the Americas as u-Law.

Album-An album is a set or group of objects (such as photos or music programs) that are combined into a single storage device or area.

Alert Tones-Alert tones are the types of alert tones that are available to indicate a particular status of a telecommunications event. An example of an alert tone is a sound that alerts the user that a new short message has been received.

ALG-Application Layer Gateway

Algorithm-Algorithm is a set of well defined steps or rules that allows for the solution of a problem or processing of information. Algorithms are commonly the name for a portion of a software program or a function.

ALI-Automatic Location Identification

ALI-Automatic Location Identifier

Alias Domain-An alias domain is a URL address that points to another web site.

Aliasing-(1-network addressing) Providing temporary or alternative identification codes or names to identify a channel or service. Aliasing allows devices or services to be addressed using a shortened code or allows the user of a name that hides the underlying addressing information. (2-sampled data waveform processing) Signals appearing in the wrong part of the frequency spectrum due to insufficiently frequent sampling of the original waveform.

Alignment-Alignment is the process of adjusting a circuit or the status of system components that interact with each other and can be changed so the performance of the system can be enhanced. An example of alignment is the tuning of multiple tuned circuits in a transmitter or receiver assembly so the signal can be amplified in each stage. When all the amplifiers are tuned to the correct frequencies, the system is said to be aligned.

A-Link-Access Link

All Trunks Busy (ATB)-All trunks busy (ATB) is a measurement in a communication network of the amount of time or number of times that all trunks in a group are busy (unavailable for service because they are in-use with other channels).

Allocated Circuit-Allocated circuit is a circuit designed and reserved for the use of a particular customer.

Allocated Quantity-Allocated quantity is a number of products or resources that have been assigned to projects or tasks.

Allocation-Allocation is the assignment or reservation of resources (such as storage space or switching capacity) to allow for the performance of processes or tasks.

Allocation Group-Allocation group, in automated facility planning, is any message trunk group, specially defined special-service circuit group, or specially defined carrier system group that creates a demand for facilities and equipment.

Allocation of Resources-Allocation of resources is the assignment or authorization to use time and materials to perform specific assignments or projects.

Allocation Start Time-Allocation start time is the defined or commanded time that a device may begin to communicate or perform a process or service.

Allowances-Allowances are amounts of value that are given to offset a price or previously agreed upon term.

Almanac-An almanac is an annual publication or collection of a data file that contains time related information on the history of events that may be used to predict future events.

Alpha Paging-Alpha paging are text messages that are sent via operator or computer to an Alpha Pager.

This figure shows how an alpha paging system receives voice, text, or data messages from callers and forwards these messages in text form to an alphanumeric pager. In this example, a sender can access the system by voice or by sending email messages via the Internet. When accessing the system by voice, a caller dials a paging access number and is either connected to an interactive voice response (IVR) unit or to an operator. When connected to an IVR, the user may be given options for specific messages (canned messages) or their voice may be converted to text messages. When connected to an operator, the operator converts (keys in) their messages to text form. When messages are sent via the Internet, their format is changed to a form suitable for the alpha paging system. In any of these cases, the messages are placed in a message queue

that holds the message until the system is available (not other messages waiting) before it. When the message reaches the top of the queue (available time to send), it will be encoded (formatted) to a form suitable for transmission on a radio channel. In this example, the message is sent as part of group 4. Sending the messages in groups allows the pager to sleep during transmission of pages from other groups that are not intended to reach the alphanumeric pager. The text message includes the pager address along with the text message in digital form. During the reception of the message, it is stored into the message paging memory area so the pager can display the message after it is received.

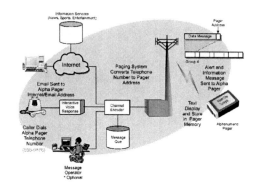

Alpha Paging Operation

Alpha Testing-Alpha testing is the first stage in testing a new hardware or software product, usually performed by the in-house developers or programmers. Alpha testing is the initial internal and possibly limited field testing process used to confirm the operation and performance of new hardware or software products. The key purpose of Alpha testing is to identify basic problems during typical operating conditions. The typical number of Alpha test participants is 10 to 50.

Alphanumeric-Alphanumeric is a generic term for alphabetic letters, numerical digits, and special characters that can be processed and displayed by a machine. Alphanumeric displays provide a character set that includes letters, numbers, and punctuation marks.

Alphanumeric Keyboard-An alphanumeric keyboard is a physical device that allows a user to enter numbers and text characters.

Alphanumeric Memory-Alphanumeric memory is a user storage feature that allows a user to store numbers and names in a memory area (such as storing telephone numbers with names in a mobile telephone).

ALS-Alternate Line Service

Alternate Billing Entity Code (ABEC)-Alternate billing entity code is an identifier that is used (e.g. by long distance service providers) to bill third parties' for services they provided.

Alternate Billing Services-Alternate billing services are calling cards, collect calls, and third-number-billed calls whose originating party does not pay for the call.

Alternate Line Service (ALS)-Alternate line service (ALS) is a multiple telephone number service where a customer can have two or more telephone numbers with a single telephone device. ALS is used to allow a business phone number and personal phone number to share the same telephone. Many telephones allow for the selection of different ring tones for different ALS numbers. ALS is also very important for telephone devices that may receive voice, data or fax information. For example, one ALS phone number could be used for a dedicated fax number the other for voice. If a call is incoming on a fax number, the telephone device knows to answer as a fax call.

Alternate Operator Services (AOS)-Alternate operator services (AOS) are operator services that are provided by another company (third-party operator). Alternative operator services are often used by interexchange carriers or small companies that do not have their own operators. Examples of common businesses that commonly use alternate operator services include small retail businesses, hotels, hospitals, and independent pay-phone companies.

Alternate Operator Services Provider (AOSP)-Alternate operator services provider (AOSP) is a company that provides operator (verbal call assisted) services other than the company providing that is providing the physical communication (transmission) service.

Alternate Route-Alternate route is an alternative path or connection of circuits between two points that is used as second or next-choice in the event a primary route is disconnected.

Alternate Routing-Alternate routing is a network switching feature that enables alternate routing of trunk or path assignments. Alternate routing may be enabled if a failure occurs in the primary route (path) or for least-cost routing service.

Alternate Voice Data Circuits-Alternate voice data circuits are circuits that are electrically conditioned to handle both voice and data signals.

Alternative Access Provider-Alternative access provider is a telecommunications service that provides an access connection between the end customer and a telecommunications network. This provider is a different company than the established LEC or PTT company.

Alternative Name (AN)-Alternative name (AN) is an identifier that is used instead of another identifier.

Alternative Route-Alternative route is a secondary communications path to a specific destination. An alternative route is used when the primary path is not available.

ALU-Arithmetic Logic Unit

Always Connected Devices-Always connected devices are communication assemblies or systems that can have continuous access to one or more networks when required.

Always-On-Always-on is a connection to a communications network (such as the Internet) that appears always on to the customer. Although always on connections appear as a dedicated connection to the end user (no need to initiate a dial up sequence), the connection may be temporary and automatically re-established each time the user accesses the network.

AMA-Automatic Message Accounting

Amateur Service-Amateur service is a radio-communication service for the purpose of self-training, intercommunication and technical investigations carried out by amateurs. Amateur operators are people that use radio communication for their own personal goals without financial or commercial service provider interest.

Ambiguity Resolution-Ambiguity resolution is the processes that are used to reduce the uncertainty or inaccuracies that occur in the acquisition and/or processing of information.

AMC-Adds, Moves and Changes

AMCS-Advanced Mobile Communication Services

Amendments-Amendments are instructions or values that indicate alterations or modifications to a document, specification or plan.

American National Standards Institute (ANSI)-American National Standards Institute (ANSI) is the US organization that sets the rules and procedures for, and also authorizes specific standards setting organizations. ATIS and EIA/TIA are two ANSI authorized standards setting organizations in the US in the subject area of telecommunications.

American Registry for Internet Numbers (ARIN)-American registry for Internet numbers is a not-for-profit organization that is responsible for the management of Internet protocol (IP) addresses in North America, South America, the Caribbean, and sub-Saharan Africa.

American Society of Composers, Authors and Publishers (ASCAP)-The American society of composers, authors and publishers is an organization that assists and represents people and companies that are involved in the creation and distribution of content.

American Standard Code for Information Interchange (ASCII)-American standard code for information interchange (ASCII) is a widely accepted standard for data communications that uses a 7-bit digital character code to represent text and numeric characters. When companies use ASCII as a standard, they are able to transfer text messages between computers and display devices regardless of the device manufacturer.

This table shows the symbols that can be represented by the 7 bit ASCII code. The columns indicate the upper 3 bits of the ASCII code (b5 - 67) and the rows indicate the lower 4 bits of the ASCII code (b1-b4). For example, to represent the letter A, the ASCII code would be 100 (upper) + 0001 (lower) resulting in ASCII code 1000001. Computers that receive the code 1000001 would display the capital letter A.

	000 (0)	001 (1)	010 (2)	011 (3)	100 (4)	101 (5)	110 (6)	111 (7)
0000 (0)	NUL	DLE	SP	0	@	P	`	p
0001 (1)	SOH	DC1	!	1	A	Q	a	q
0010 (2)	STX	DC2	"	2	B	R	b	r
0011 (3)	ETX	DC3	#	3	C	S	c	s
0100 (4)	EOT	DC4	$	4	D	T	d	t
0101 (5)	ENQ	NAK	%	5	E	U	e	u
0110 (6)	ACK	SYN	&	6	F	V	f	v
0111 (7)	BEL	ETB	'	7	G	W	g	w
1000 (8)	BS	CAN	(	8	H	X	h	x
1001 (9)	HT	EM	)	9	I	Y	i	y
1010 (A)	LF	SUB	*	:	J	Z	j	z
1011 (B)	VT	ESC	+	;	K	[	k	{
1100 (C)	FF	FS	,	<	L	\	l	
1101 (D)	CR	GS	-	=	M	]	m	}
1110 (E)	SOH	RS	.	>	N		n	~
1111 (F)	SI	US	/	?	O		o	DEL

ACK - Acknowledge
BEL - Bell
BS - Backspace
CAN - Cancel
CR - Carriage Return
DC - Direct Control
DEL - Delete Idle
DLH - Data Link Excape
EM - End of Medium
ENQ - Enquiry
EOT - End of Transmission
ESC - Escape
ETB - End of Transmission Block
ETX - End of Text
FF - Form Feed
FS - Form Separator
GS - Group Separator
HT - Horizontal Tab
LF - Line Feed
NAK - Negative Acknowledge
NUL - Null
RS - Record Separator
SI - Shift In
SO - Shift Out
SOH - Start of Heading
STX - Start of Text
SUB - Substitute
SYN - Synchronous Idle
US - Unit Separator
VT - Vertical Tab

American Standard Code for Information Interchange (ASCII)

American Standard Committee (ASC)-The American standard committee is a group within the American national standards institute that defines industry standards for electronic data interchange.

American Standards Association (ASA)- American Standards Association (ASA) is the predecessor organization to ANSI.

Amortization-Amortization is the assignment of cost of an asset or acquisition for distribution over the time period that the asset or acquisition will be used.

Amortized Cost-Amortized cost is valuation of an item or asset that is adjusted for amortized payments and discounts.

AMPS-Advanced Mobile Phone Service

AMR-Automatic Meter Reading

AMS-Affiliate Management System

AMS-Audience Measurement System

AN-Access Network

AN-Access Node

AN-Alternative Name

Analog-Analog is an information form that is represented by a continuous and smoothly varying amplitude or frequency changes over a certain range such as voice or music. Analog lines allow the representation of information to closely resemble the original information signal.

This figure shows a sample analog signal created by sound pressure waves. In this example, as the sound pressure from a person's voice is detected by a microphone, it is converted to its equivalent electrical signal. This

diagram shows analog audio signals continuously varies in amplitude (height, loudness, or energy) as time progresses.

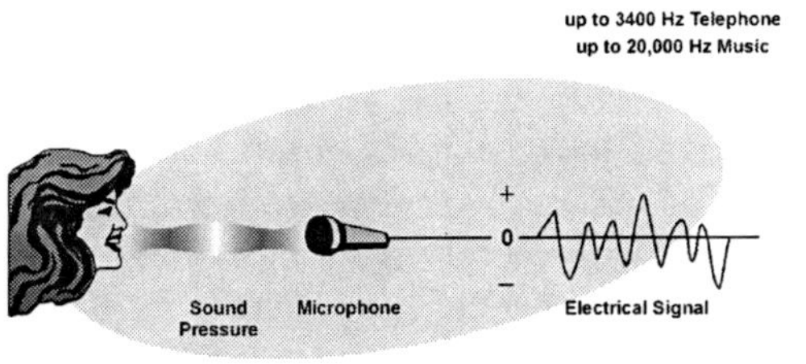

Analog Audio-Analog audio is the representation of a series of multiple sounds through the use of a rapidly changing signal (analog). This analog signal indicates the luminance and color information within the audio signal.

Analog Facsimile-Analog facsimile is a facsimile that transmits images on an analog communication line through the conversion of images or shades of images to analog signals (tones).

Analog Media-Analog media is the format of information that is used to express information (media) which is represented in a form that can have levels or signal composition of any level (analog).

Analog Signal-An analog signal is a direct representation of a physical process. For instance, an analog electromagnetic signal representing your voice on a telephone line is represented by continuous variations in voltage. Loud sounds are represented by large voltages, soft ones by small. High voices are represented by high frequency variations in the voltage, low ones by low frequencies. Analog signals provide the most nuanced and precise record of a physical process because of this exact representation. However, they are more easily distorted by noise and other factors than are digital signals.

This figure shows how an analog signal exactly matches each portion of the information source. This diagram shows a person who is creating sound pressure waves that are converted into an electrical signal via a microphone. This example shows that each portion of the sound pressure wave is represented by its own instantaneous electrical signal level.

Analog Audio Signal

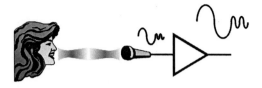

Analog Signal

Analog Switch-An analog switch is a connection device that is used in a communication system that can setup direct connections between input lines to output lines. The connections are continuous so there is no changing of the signals as they pass through the switch.

Analog Tape-Analog tape is a magnetic tape storage format that changes magnetic information on the tape to represent analog signals.

Analog Telephone Adapter (ATA)-Analog Telephone Adapter (ATA) is a device that converts analog telephone signals into another format (such as digital Internet protocol). These adapter boxes may provide a single function such as providing Internet telephone service or they may convert digital signals into several different forms such as audio, data, and video. When adapter boxes convert into multiple information forms, they may be called multimedia terminal adapters (MTAs) or integrated access devices (IADs).

Analog telephone adapters (ATA) must convert both the audio signals (voice) and control signals (such as touch tone or hold requests) into forms that can be sent and received via the Internet.

Analog Television (ATV)-Analog television is a process or system that transmits video images through the use of analog transmission. The analog transmission is divided into frequency bands or modulation types to transfer video and audio information.

Analog Television Adapter (ATVA)-Analog Television Adapter (ATVA) is a device that converts digital multimedia signals (such as MPEG) into analog television signals (such as NTSC or PAL). These adapter boxes may provide a single function such as providing Internet television service or they may convert digital signals into several different forms such as audio, data, and video. When adapter boxes convert into multiple information forms, they may be called multimedia terminal adapters (MTAs) or integrated access devices (IADs).

Analog television adapters (ATA) must convert video, audio, and control signals (such as requests for changing channels) into forms that can be sent and received via data networks such as the Internet.

Analog Terminal Adapter (ATA)-Analog terminal adapter (ATA) is a communications adapter that allows analog telephone devices (e.g. a computer modem) to interconnect to digital telephone systems.

Analog to Digital Converter (ADC)-Analog to digital conversion is a process that changes a continuously varying signal (analog) into a digital values. A typical conversion process includes an initial filtering process to remove extremely high and low frequencies that could confuse the digital converter. A periodic sampling section that at fixed intervals locks in the instantaneous analog signal voltage, and a converter that changes the sampled voltage into its equivalent digital number or pulses.

This diagram shows how an analog signal is converted to a digital signal. This diagram shows that an acoustic (sound) signal is converted to an audio electrical signal (continuously varying signal) by a microphone. This signal is sent through an audio band-pass filter that only allows frequency ranges within the desired audio band (removes unwanted noise and other non-audio frequency components). The audio signal is then sampled every 125 microseconds (8,000 times per second) and

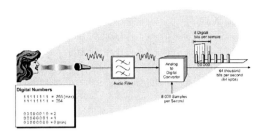

Signal Digitization Operation

converted into 8 digital bits. The digital bits represent the amplitude of the input analog signal.

Analog Transmission-Analog transmission is a system that is capable of transferring continuously varying signals (analog signals) between points. The system may directly transfer the analog signal or the analog signal may modify another carrier (such as a radio carrier).

Analog Video-Analog video is the representation of a series of multiple images (video) through the use of a rapidly changing signal (analog). This analog signal indicates the luminance and color information within the video signal.

Sending a video picture involves the creation and transfer of a sequence of individual still pictures called frames. Each frame is divided into horizontal and vertical lines. To create a single frame picture on a television set, the frame is drawn line by line. The process of drawing these lines on the screen is called scanning. The frames are drawn to the screen in two separate scans. The first scan draws half of the picture and the second scan draws between the lines of the first scan. This scanning method is called interlacing. Each line is divided into pixels that are the smallest possible parts of the picture. The number of pixels that can be displayed determines the resolu-

tion (quality) of the video signal. The video signal television picture into three parts: the picture brightness (luminance), the color (chrominance), and the audio.

Analysis Tools-Analysis tools are programs, systems or processes that can be used to evaluate the status or condition of information or processes.

Ancillary Services-Ancillary services are additional features or services provided, including call features, detail billing, voice mail, etc.

Angel Investor-An angel investor is a person or company who provides funding or resources to a person or company which is in an early stage of business (start-up). Typically an angel invests when the company is little more than an idea, a simple business plan and several management people, but rarely a full management team.

Angel Investors-Angel investors are people who provide money to businesses in their early growth stages. Angel investors may also provide additional resources such as counseling loan guarantees or providing access to facilities (such as office space).

ANI-Automatic Number Identification

ANI Identification-ANI identification is a software processing function at an end office (local) switch that forwards the billing number to the termination point. ANI is needed for the caller identification service offered by many service providers.

Animation-Animation is a process that changes parameters or features of an image or object over time. Animation can be a change in position of an image within a video frame to synthetically created images that change as a result of programming commands.

Announcement Server-An announcement server provides network announcements to callers (e.g. "please wait while your call is being connected").

Announcement Service-Announcement service is a service that provides a caller with a predefined message when an incoming call is received.

Announcement Trunk Group-Announcement trunk group is a trunk group used to inform customers or operators about call status or to access announcement services.

Annual Turnover-Annual turnover is the ratio of sales to inventory levels.

Annuity Loan-An annuity loan is the lending of money or assets for a fixed period of time (term) where the loan is repaid through a multiple of regularly spaced payments.

Anonymizer-An anonymizer is a company or service that is located between a user or consumer and a supplier of a product or service who isolates the identity of the user (such as an IP address) from the vendor.

Anonymous FTP-Anonymous FTP is a security process that allows general users to have limited access to file servers without the need for registering for account identification and passwords. The user enters anonymous as an unregistered user and the password is usually your e-mail address.

Anonymous Web Browsing-Anonymous web browsing is the process of navigating through web sites using an IP address or other information that is independent from the identity of the person who is performing the browsing.

ANSI-American National Standards Institute

Answer Mode-Answer mode is a call processing function that determines how a telephone device (e.g. a fax or a modem) will respond to incoming calls. The answer mode may be set to no answer, number of rings, or immediate answer.

Answer Supervision-Answer supervision is the sending of an off-hook supervisory signal back to an exchange carrier's point of termination to show that the called party has answered.

Answering Machine-An answering machine is a device that can automatically answer telephone calls, play a prerecorded greeting message, store audio information, and allow retrieval and deletion of messages.

Anti-Circumvention-Anti-circumvention is the inability or restrictions on assisting in the process of defeating or getting around privacy or security methods used in products or services.

AntiFraud-Antifraud is a process or system that can be used to identify, block or reduce fraudulent activities.

Anti-Rollback Clock-An anti-rollback clock is a time reference that can detect if the time reference has been modified.

Antitrust Laws-Antitrust law has the meaning given it in subsection (a) of the first section of the Clayton Act, except that such term includes section 5 of the Federal Trade Commission Act to the extent that such section 5 applies to unfair methods of competition.

Anti-Virus-Antivirus Program

Antivirus Program (Anti-Virus)-Antivirus program is a program that can detect the presence of and possibly eliminate a computer virus. Anti-virus program usually locate and identify viruses by looking for previously identified patterns or suspicious activity in the system.

AOC-Advice Of Charge

AoD-Advertising on Demand

AOS-Alternate Operator Services

AOSP-Alternate Operator Services Provider

AP-Access Point

AP-Accounts Payable

Apache HTTP Server-The Apache Web Server is the outgrowth of the original National Super Computing Applications (NSCA) http server. The NSCA released the source code under a freeware licensing scheme that allowed it to be extended by researchers and other computer programmers as long as the modified software was given back into the public domain for free use.

In its infancy the software was far from perfect and required a lot of software updates to fix bugs and to add needed functionality. These software updates came as small new sections of code referred to as patches. Just as Unix was a play on the Multics operating system the unix community named the NCSA web server Apache since it required so many patches to fix it.

Today the Apache Web Server is the most widely used http server of the world wide web and is supported on all of the major operating systems. The best part; it's still free.ost, excellent performance, good scalability, and great flexibility. Don't expect easy graphical configuration programs and hypertext help; you'll get the command line and the man pages instead, so it certainly helps to have staff with Unix experience.

Apache Server is available as part of the Red Hat Software Linux distribution, which also provides developers with full support for CGI, Perl, Tcl, a C or C++ compiler, an Apache server API, and a SQL database.

API-Actual Price Index

API-Advertising, Premiums and Incentives

API-Application Program Interface

APN-Access Point Name

APP-Access Process Parameter

Applet-An applet is a small software program that uses the Java programming language to request, transfer and process information in a computer.

Applicability Statement 2 (AS2)-Applicability statement 2 is an industry specification that defines how to securely and reliably send electronic data interchange (EDI) messages through public data communication systems (such as the Internet).

Applicable Rate-Applicable rates are a group of billing codes or product identifiers that are associated with a specific customer or service.

Applicant-An applicant is the entity that submits a form or application to participate in an event (such as a communication license auction) that may include all holders of partnership and other ownership interests and a percent of stock (equity) interest or outstanding stock in the entity (e.g. more than 5%) and officers and directors of that entity.

Application-An application is a software program that is designed to perform operations using commands or information from other sources (such as a user at a keyboard). Popular applications that involve human interface include electronic mail programs, word processing programs, and spreadsheets. Some applications (such as embedded program applications) do not involve regular human interaction such as automotive ignition control systems.

Application Architecture-Application architecture is the types of applications, the structure of functional elements used in the applications and how they relate to each other.

Application Based Call Routing-Application based call routing is a process that controls call routing of an incoming call based on the type of selected application such as sales, customer service, or order tracking.

Application Based ITSP-Application based ITSP is an Internet telephone service provider that provides customers with Internet telephone service by using a software application or adapter device.

Application Boundary-An application boundary is the limits or defined resources that an application may use or communicate with.

Application Connectivity-Application connectivity is the ability of applications to connect and communicate with other applications. Application connectivity includes the ability of the application to initiate connections, respond to connection request, terminate (end) connections, and manage communication in the event communication transfer is distorted or altered.

Application Development-Application development is the set of tasks and processes that are used to create software programs or applications.

Application Development Guide-An application development guide is a set of processes or procedures that can be used to assist a person or company to customize or develop new applications or services on an existing system.

Application Engine-An application engine is a software program that can receive, process and respond to an end user's (client's) request for information or information processing.

Application Environment-Application environment is the processes and configurations of other programs, systems and services that may effect and be affected by running an application.

Application Extension-An application extension is a set of features or processes that can be added to an existing application program. Application extensions are commonly used to extend the capabilities of an existing application without changing the underlying application or systems.

Application Firewall-An application firewall is a data filtering device that uses and analyzes application protocols to filter (block) unwanted data that is sent between a computer server or data communication device and a public network (e.g. the Internet). Application firewalls continuously look for data packet transmission patterns associated with a specific application (such as IP Telephony) that indicate authorized or unauthorized use to the server.

This figure shows how voice over data network (VoIP) telephone service can work through a firewall. This example describes how IP telephone registration sets up the firewall to allow call control messages (such as an incoming call) to reach the IP telephone device. When an IP telephone first senses it has been connected to a data network, it attempts to register with it's call server. In this example, the call server

VoIP Firewall Operation

is at a distant location outside the firewall. Because the IP telephone is part of the local area network (LAN), it is a trusted device and the firewall allows it to request a communication session with the data network (probably the Internet). This creates a communication session and the firewall remembers the details of the communication session. These communication session details include the Internet address of the call server and the IP telephone device address in the LAN. When packets are received in the future from the call server with the Internet address assigned to the IP telephone, these packets will be forwarded to the Internet telephone device.

Application For Service-An application for service is the submission of information along with authorization to setup and provide services. The application for service information usually includes contact and billing information, desired services and features along with supporting technical and other descriptive information, which enables the service provider to establish and provide the services.

Application Framework-An application framework is a hypothetical or functional structure that is used to plan or describe a system or software program that is designed to perform functions or processes that satisfy user needs.

Application Infrastructure Service Provider (AIP)-Application infrastructure service provider (AIP) are companies that provide communication application services such as email, web hosting, and voice communication.

Application Interface-An application interface is a specification that defines the messages, processes and/or hardware equipment that allow applications (such as software programs) to communicate and interact with each other.

Application Layer-The application layer coordinates the information interface between the communication device and the end user. The application layer receives data from the underlying protocols and processes this information into a form required or requested by the user or endpoint device. The application layer usually requests or responds to requests for a communication session. The location of the application layer is at the top of protocol stacks. The application layer is layer 7 in the open system interconnection (OSI) protocol layer model.

Application Layer Gateway (ALG)-An application gateway is a communications device or assembly that transforms data that is received from one network into a format that can be used by a different network for specific applications.

Application Lifecycle-Application lifecycle is the progression of usage of an application from its initial introduction to when people or companies stop using the application.

Application Management-Application management is the management of the installation, operation, and resource allocation for software and communication applications.

Application Manager-An application manager is a program or system that coordinates the initiation, operation and termination of application programs. The application manager is responsible for ensuring the resources (such as memory) are available for applications when they are initiated (launched) and

that these resources are released when the application has terminated (ended).

Application Notes-Application notes are descriptions or instructions that are provide to assist in the application or design of a device, product, or service into a system. Manufacturers commonly provide application notes to help their customers to use their products or services in their systems.

Application Portability-Application portability is the ability to transfer or use an application on other devices or systems.

Application Profile-Application profiles are particular implementation of protocols, feature operations, and/or processes that ensure applications operate in a specific manor.

Application Program-An application program is a software file that contains commands that performs application processes.

Application Program Interface (API)-Application program interface (API) are defined and documented entry points into a software application where other programs may interact with the application in order to provide customized extensions or perform special processing functions. Typically an API is a public function call that then itself calls on the services of the application. In this way the API hides the underlying details and complexities of the application software making it easier for programmers to add custom functionality.

Application Protocols-Application protocols are commands and procedures used by software programs to perform operations using information or messages that are received from or sent to other sources (such as a user at a keyboard). Application protocols are independent of the underlying technologies and communication protocols. The use of well-defined application protocols (agreed commands and processes) allows the software applications to interoperate with other programs that use the application protocol independent from the underlying technologies that link them together (such as wires or wireless connections.)

This figure shows that wireless personal area networks (WPANs) can use standard industry protocols to connect to standard communication applications. In this example, a laptop computer is communicating with a wireless mouse, a personal digital assistant (PDA), and an access node using a single WPAN PCMCIA card. When communicating with the mouse, the laptop uses the standard RS-232 protocol. When transferring (exchanging) items between an address book stored in the laptop and an address book stored in the PDA, it uses the standard Object Exchange (OBEX) protocol. To connect to the Internet, the laptop connects through an access node to a router using standard point-to-point protocol (PPP). This diagram shows that the PPP connection is only part of the communication link that reaches an email server that is connected to the Internet. The computer uses standard simple mail transfer protocol (SMTP) to send and retrieve email messages.

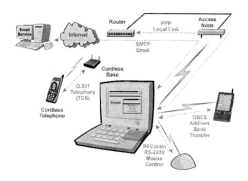

Application Protocols

Application Server (AS)-An application server (AS) is a computer and associated software that is connected to a communication network and provides information services (applications) for clients (users). Application servers are usually optimized to provide spe-

cific applications such as database information access or sales contact management.

Application servers are a key component of nextgen networks, and an enabler of IP-based enhanced services. These enhanced services will generate much-needed new revenue streams for service providers. Examples include all forms of conferencing, voice mail and unified messaging. Being softswitch-based, application servers have the flexibility to easily offer services that go beyond the feature set of legacy switched telephony. In terms of network configuration, the application server works in tandem with the media server, providing it with business logic and instructions for delivering enhanced services.

Application Service Element (ASE)-Application service element (ASE) is a software program or portion of a communication protocol that is part of an application layer of a protocol stack. Several ASEs may be combined to form a complete application protocol.

Application Service Provider (ASP)-Application service provider (ASP) is a company that provides an end user with an information service. An ASP owns or leases computer hardware and software system that allows one or more users to access information services on or through that computer systems.

This figure shows an example of an application service provider (ASP). This diagram shows that an end user is operating a multimedia capable computer that gains access to ASP via the telephone network and Internet service provider (ISP). In this example, the ASP provides weather and airline flight schedule information to the customer. The application service provider owns or operates software on a host computer that is connected to the Internet. The end user is operating web-browsing software over a dial-up data communications channel (a telecommunications service) that enables the ASP to receive and process requests for information services (the application service).

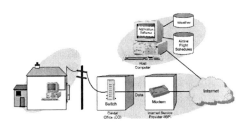

Application Service Provider (ASP) System

Application Sharing-Application sharing is the use of the processes and/or resources of an application by two or more users.

Application Software-Application software is a computer program or group of programs that provide a set of processes that can be used by devices, systems or people.

Applications-Applications are the software programs that perform processing functions for specific types of user needs.

Applications Partner-An applications partner is a company or person that writes software applications or modules that are used in applications or software programs.

Applications Processor-Applications processor is a computer system or information service provider that is dedicated to processing applications such as voice mail or billing systems.

Appreciation-Appreciation is the assignment of a value of an asset or acquisition that has increased in value over the time period that the asset or acquisition is owned or used.

Appropriation-An appropriation is a value or resource that is allocated or reserved for a future use.

Approval Authority-An approval authority is an entity (such as an agency or company) that is authorized (e.g. by a government or company) to approve the selection of a vendor or the issuance of a project.

Approved Customer List-An approved customer is a person or company that has been authorized to buy products or services from a company. Sales representatives may only be allowed to sell to companies on an approved customer list. The approved customer list may be attached to an appendix of the sales representative agreement or it may be a separate document that is continually updated.

Approved Vendor-An approved vendor is a person or company that has been authorized to provide products or services on behalf of another company.

Approved Vendor List-An approved vendor is a person or company that has been authorized to provide products or services to a company. Some companies may only be allowed to purchase products from companies on an approved vendor list.

APS-Advanced Planning and Scheduling

AQCB-Automated Quote Contract Billing

Arbitration-(1-general) Arbitration is a process or a set of rules that is used to manage conflicts where a designated third party can mediate disputes between people or companies. (2-computers) Arbitration is the process of competing for computer resources such as memory or peripheral devices, made by multiple processes or users.

Arbitration Rules-Arbitration rules are the set of requirements and processes that are used to help manage conflict resolution between people or companies who have chosen to use the arbitration process.

ARC-Access Router Card

Architecture-Architecture is the functional design of a network, computer or telecommunications system elements and the relationships between them. The architecture usually includes hardware and software components.

Archival-Archival is a process or media storage device that has a long-term minimum lifespans over which the information will not become corrupted.

Archive Management-Archive management is the processing of identifying, transferring and storing information or data into or onto a media system for long-term storage.

Archiving-Archiving is the processing of transferring information or data into or onto a media system from short-term memory to long-term data storage.

Archiving Files-Archiving files is a process where the information contained in an active computer file is made ready for storing in a non-active file, perhaps in off-line or near-line storage. Typically when files are archived, they are compressed to reduce their size. To restore the file to its original size requires a process know as unarchiving.

ARDIS-Advanced Radio Data Information Service

Area Code-An area code is a 3-digit number that generally identifies a geographic area of a switch that provides service to a telephone device. In North America, the Numbering Plan Area (NPA) is the area code. In countries other than North America, the area code may have any number of digits depending on the regulation of telecommunication in that country.

Area Code Restriction-Area code restrictions is the process of blocking dialed telephone numbers based on the area code dialed. Area code restrictions may be used to control (inhibit) the ability of telephone users (such as employees of a company) to dial numbers within specific area codes.

Area Exchange-Area exchanges are geopolitical areas that are defined as region authorized to provide local telephone services. Small metropolitan areas or a collection of towns often share a single area exchange.

Arguments-Arguments are a set of parameters (values, items or characteristics) that can be used or transferred to a service, function or software program under certain conditions such as when a software program is run or when a function is activated ("function call").

ARIN-American Registry for Internet Numbers

ARIS-Access Request Information System

Arithmetic Logic Unit (ALU)-Arithmetic logic unit (ALU) is a part of a central processing unit (CPU) that performs arithmetic and logical operations on data.

Arithmetic Unit-Arithmetic unit is the processing portion of a computing system that performs a majority of the arithmetic operations processed by the computer.

Arm's Length-Arm's length is a process or transaction that has processes that isolate or separate the relationships between the entities involved. An example of an arm's length transaction is the sale of assets from one division of a company to another where the company must review bids for the asset prior to making the decision to sell the asset from one division to the other.

ARN-Authentication Random Number

ARPC-Average Revenue per Customer

ARPM-Average Rate Per Minute

ARPU-Average Revenue Per User

ARQ-Automatic Retransmission Request

Array-(1-data) An array is a grouping of objects in multiple dimensions. (2-storage) A grouping of storage systems.

Arrears-Arrears is the assessment of charges for services that have been used (e.g. usage).

Arrival Rate-Arrival rate, as related to call centers, is the pattern at which incoming calls are received at the call center. Arrival rates are classified as steady, random or peaked. This term is also used in queuing theory to describe the rate at which entities enter into an ordered list to be processed.

Artificial Intelligence (AI)-Artificial intelligence (AI) is a deductive reasoning process that can be applied on an automated basis by computer processing. Artificial intelligence simulates the reasoning capabilities much like the human mind through user input. British mathematician Alan Turning introduced the Artificial intelligence term in the 1950's.

A

Artwork-Artwork is the illustrative materials that are used in media.

ARU-Audio Response Unit

AS-Application Server

AS-Authentication Server

AS-Autonomous System

AS2-Applicability Statement 2

ASA-American Standards Association

ASC-American Standard Committee

ASCAP-American Society of Composers, Authors and Publishers

ASCII-American Standard Code for Information Interchange

ASCII File-ASCII file is a text file that only contains text characters from the ASCII character set. An ASCII character set only includes letters, numbers, and punctuation symbols. The extended ASCII code contains non-standard graphics and characters that can be used as text-formatting codes.

ASCII Format-ASCII format is the representation of commands, data and/or information by a sequence of ASCII text codes (often in readable sequences). Because it may take several text codes to represent a single command or media block, ASCII formats can have much larger data transmission rate than binary equivalents.

ASE-Application Service Element

ASG-Access Service Group

ASN-Abstract Syntax Notation

ASN-Advanced Shipping Notice

ASN-Authentication Sequence Number

ASP-Active Server Pages

ASP-Application Service Provider

ASR-Access Service Request

ASR-Authorized Sales Representative

Assembly Language-Assembly language is a low-level programming language that each statement corresponds to a single machine language instruction that a processor can execute. Assembly language is very closely related to the actual instructions and architecture of the microprocessor being targeted and therefore assembly language instructions typically map one-for-one to machine level instructions. Assembly languages are specific to a given microprocessor.

Asset Library-An asset library is a collection of programs, digital media or dta files that can be stored, transferred or accessed.

Asset Lifecycle-An asset lifecycle is the duration from the initial acquisition of an asset to when its usage stops.

Asset Management-Asset management is the process of acquiring, maintaining, distributing and the elimination of assets. Assets may be in the form of hardware (e.g. equipment), software (e.g. applications) or information content (e.g. media programs).

Asset Storage-Asset storage is the maintaining of valuable and identifiable data or media (e.g. television program assets) in media storage devices and systems. Asset storage systems may use a combination of analog and digital storage media and these may be directly or indirectly accessible to the asset management system.

Asset Tracking-Asset tracking is the process of determining the location and or status of an item or asset.

Assigned Pairs-Assigned pairs are communication lines (wire pairs) that are assigned for customer service. These lines may be working or idle.

Assignment Of Authorization-Assignment of authorization is a transfer of authorization to provide communication services from one party to another.

Assignment Rights-Assignment rights are the authorization to allow some or all of the claims, rights, interest or property in an agreement to be transferred to another person or company.

Association List-Association lists are group of people who have registered to be part of an industry association. An industry association is an organization that represents the interests of an industry or group of people that share common interests. Members of an industry association share common interests.

Assumed Receipt-An assumed receipt is an indication or condition that indicates that a product or service has been received without direct confirmation from the purchaser of the products or services.

Assurance-Assurance is the process of making sure that customers receive the levels of service that they have purchased or agreed to purchase. Assurance may include commitments to a high-level of overall customer satisfaction of quality of service.

Assurance Level-Assurance level is the level of probability that a service or product will meet a specific criteria or range of limits. Assurance level is often expressed as a percent. For example, there is 99% assurance (probability) that a dialtone will be available in a subscriber loop.

Assured Delivery-Assured delivery is a protocol said to provide assured delivery if each packet is guaranteed to be delivered. The sender accomplishes this via receiver acknowledgement and retransmission when packets are not acknowledged. Examples of protocols that provide assured delivery are TCP/IP and IEEE 802.2 LLC connection oriented services.

Asymmetric-Asymmetric channels are two-way communication channels that allow for transmission rates that can vary by direction. For example, the downlink broadcast channel may be a high-speed channel (e.g. 1.9 Mbps) and an uplink (reverse direction) channel may only be 15 kbps.

Asymmetric Digital Subscriber Line (ADSL)-Asymmetric digital subscriber line (ADSL) is a communication system that transfers both analog and digital information on a copper wire pair. The analog information can be a standard POTS or ISDN signal. The maximum downstream digital transmission rate (data rate to the end user) can vary from 1.5 Mbps to 9 Mbps downstream and the maximum upstream digital transmission rate (from the customer to the network) varies from 16 kbps to approximately 800 kbps. The data transmission rate varies depending on distance, line distortion and settings from the ADSL service provider.

This figure shows that a typical ADSL system can allow a single copper access line (twisted pair) to be connected to different networks. These include the public switched telephone network (PSTN) and the data communications network (usually the Internet or media server). The ability of ADSL systems to combine and separate low frequency signal (POTS or IDSN) is made possible through the use of a splitter. The splitter is composed of two frequency filters; one for low pass and one for high pass. The DSL modems are ADSL transceiver units at the central office (ATU-C) and the ADSL transceiver unit at the remote home or business (ATU-R). The digital subscriber line access module (DSLAM) is connected to the access line via the main distribution frame (MDF). The MDF is the termination point of copper access lines that connect end users to the central office.

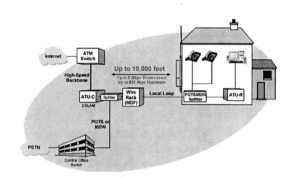

Asymmetric Digital Subscriber Line (ADSL) System

Asymmetrical-(1-General) Asymmetrical transmission is two-way communication that has different data transmission rates in send (forward) and receive (reverse) directions. (2-Bluetooth) Asymmetrical is a type of

Asynchronous Connectionless (ACL) link that operates at two different speeds in the upstream and downstream directions. An example of an asymmetrical connection is the Bluetooth ACL link. The Bluetooth specification specifies a maximum data rate of up to 723.2 Kbps in the downstream direction, while permitting 57.6 Kbps in the upstream direction. See also symmetrical.

Asymmetrical Private Virtual Circuit (Asymmetrical PVC)-Asymmetrical private virtual circuit is a virtual circuit that permits uneven (asymmetrical) data transmission rates for each direction of transmission.

Asymmetrical PVC-Asymmetrical Private Virtual Circuit

Asynchronous-Asynchronous channels are dynamically adjusted channels that do not have a fixed synchronization with some other reference signal. The communications on an asynchronous channel is not sequential and may appear random or unbalanced in nature. This diagram shows the process of data transmission using asynchronous (unscheduled) transmission. In this example, each message or block of data that is transmitted in an asynchronous data communication system must include indicators (delimiters) that identify the start and stop of a block of data. The blocks of data usually include some bits of information that are dedicated for flow control (e.g. routing and/or error protection).

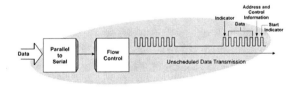

Asynchronous Data Transmission

Asynchronous Link-An asynchronous link is a communication connection that exists between devices that can send data at unscheduled time periods.

AT-Access Tandem

AT&T Consent Decree-AT&T consent degree is the order for AT&T to divest parts (the break-up) of its business and the restriction not to allow AT&T to provide local telephone service.

ATA-Analog Telephone Adapter

ATA-Analog Terminal Adapter

ATB-All Trunks Busy

ATM Address-ATM address is a 20 byte address used in asynchronous transfer mode (ATM) systems that identify the country, area, and end-system identifiers. ATM address formats are defined in the user network interface (UNI) specification.

ATM Forum-ATM forum is a forum that was started to assist manufacturers and service providers with marketing and development of ATM products and services. The ATM forum was started in 1991.

Attachment-An attachment to an electronic mail is non-text data included in an e-mail using Multipurpose Internet Mail Extensions (MIME). An e-mail message may contain any number of attachments. Each attachment has a "MIME type" property that suggests to the user's e-mail application the data type of the attachment, for example HTML text or a JPEG image. Depending on the software, in some cases the user's e-mail application will display attachments within the e-mail application (like JPEG images). In other cases, an attempt will be made to open an external application, sometimes requiring assistance or permission from the user, or the user may save the attachment as a file.

Attack-Attack, in network and computer security parlance, is an attempt to disable or gain unauthorized access to a computer system or network by exploiting weaknesses within the operating system or implemented security measures.

Attempt-Attempt is any process of requesting or demanding service from a communications system.

Attendant-An attendant is a person who answers, screens, or directs calls in a communication system.

Attendant Access Loop-An attendant access loop is a communication line that is connected to an attendants telephone station.

Attendant Call Waiting Indication-Attendant call waiting indication is the indication light or message on the attendant console that indicates that one or several calls are in queue to be answered. The indication may change (e.g. flash or ring) when additional thresholds (e.g. maximum number of waiting calls) are reached.

Attributes-Attributes are the characteristics of an object or information element (e.g. file) that can be used to identify, qualify, or assist in the control of that object or element.

Attrition-Attrition is the loss of customers or users of a service over a period of time.

Attrition Rate-Attrition rates are the percentages of customers or potential customers who discontinue their subscription for an application or service.

ATV-Advanced Television

ATV-Analog Television

ATVA-Analog Television Adapter

AU-Access Unit

Auction-An auction is a procedure for choosing buyers, license recipients or those involved in bidding for the rights to assets. For communication systems, auctions may be used to award licenses for the right to operate communication systems or the right to use frequency bands..

Auction Pricing System-An auction pricing system is a process of allowing the price to change until the product is sold.

Audience Measurement System (AMS)-An audience measurement system is a combination of equipment, protocols and transmission lines that are used to monitor, record and potentially analyze the selections and habits of television viewers.

Audio Blog (Audioblog)-An audio blog (audioblog) is a shared media resource on a web page that allows participants to contribute audio media that is related to topics or questions.

Audio Bridge-Audio bridge, in telecommunications, is a device that mixes multiple audio inputs then provides the composite audio back to each communication device, less that devices audio input. An example of an audio bridge is a conference call.

Audio Broadcast Services-Audio broadcast services are the transmission of program material (typically audio) that is typically paid for by advertising. Most commercial stations receive the bulk of their ad revenues from local advertising, as opposed to television, which gets most of its revenue from network advertising.

Audio Channel-An audio channel is a stream of audio information transmitted as part of a distinct communication, or conversation, or for a particular purpose or end use.

Audio Chat Room-Audio chat rooms are real-time communication services that allow several participants (typically 10 to 20) to interact act much like an audio conference session. Audio conference chat rooms may be public (allow anyone to participate) or private (restricted to those with invitations or access codes.)

Audio Conference-Audio conferencing (also called teleconferencing) is a process of conducting a meeting between two or more people through the use of telecommunications circuits and equipment.

Audio Digitization-Audio digitization is the conversion of analog audio signal into digital form. To convert analog audio signals to digital form, the analog signal is digitized by using an analog-to-digital (pronounced A to D) converter. The A/D converter periodically senses (samples) the level of the analog signal and creates a binary number or series of digital pulses that represent the level of the signal. The typical sampling rate for conversion of analog audio ranges from 8,000 samples per second (for telephone quality) to 44,000 samples per second (for music quality).

Audio Format-Audio format is the method that is used to contain or assign media within a file structure or media stream (data flow). Audio formats are usually associated with specific standards like MPEG audio format or software vendors like Quicktime MOV format or Windows Media WMA format.

Audio formats can be a raw media file that is a collection of data (bits) that represents a flow of sound information or it can be a container format that is a collection of data or media segments in one data file. A file container may hold the raw data files (e.g. digital audio and digital video) along with descriptive information (meta tags). Some of the common audio formats include AU, MIDI, CDA, Musicam, MP3, AAC and RIFF.

Audio Home Recording Act (AHRA)-The audio home recording act is regulation passed by U.S. Congress in 1992 that affirmed a consumer's right to make copies of materials for reasonable use (e.g. backup copies) and did not authorize the reproduction of copies. The system that was used to inhibit the copying of copies is called serial copy management system (SCMA). As a result of this regulation, royalties were assessed on sales of media players that could record content on blank media.

Audio Menu-An audio menu is a structured indexing system that identifies items in the index or selection options in audio form.

Audio Programming Services-Audio programming services are the providing of information content (programs) by a communication system operator (such as a radio broadcast station).

Audio Quality-Audio quality is the ability of an audio device or transfer system to recreate the key characteristics of an original audio signal or sound.

Audio Response Unit (ARU)-Audio response unit is a device or system that can translate data files (usually stored on a computer) into audio voice messages.

Audio Streaming-Audio streaming is a real-time system for delivering audio, typically over the Internet. Upon request, a server system will deliver a stream of audio (usually compressed) to a client. The client will receive the data stream and (after a short buffering delay) decode the audio and play it to a user. Internet audio streaming systems are used for delivering audio from 2 kbps (for telephone-quality speech) up to hundreds of kbps (for audiophile-quality music).

This figure shows how a media server can adjust its data transmission rate to compensate for different audio streaming data rates. This example shows how a media server is streaming packets to an end user (for an Internet audio player). Some of the packets are lost at the receiving end of the connection because of the access device. The receiving device (a multimedia PC) sends back control packets to the media server indicating that the communication session is experiencing a higher than desirable packet or frame loss rate. The media server can use this information to change its media compression and data transmission rates to compensate for the slow user access link.

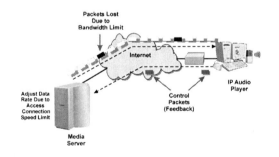

Audio Streaming

Audio Video (AV)-Audio video is equipment or signals that are used to capture, process or present sound and/or visual information.

Audio Visual (A/V)-Audio visual is the combination of audio and visual components.

Audio Watermarking-Audio watermarking is a process of adding or changing information in an analog or digital audio media tape, streaming media or other form of audio media to uniquely identify the media and/or its authorized uses. Audio watermarking may be performed by adding audio tones above the normal frequency or by modifying the frequencies and volume level of the audio in such a way that the listener does not notice the watermarking information.

This figure shows how digital watermarks can be added to digital audio to provide identification information. The digital watermark is added as a high frequency audio component that is typically not perceivable to the listener of view of the media.

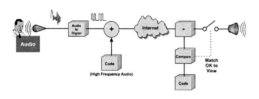

Audio Watermarking

Audioblog-Audio Blog

Audiovisual Objects-Audiovisual objects are parts of media images (media elements). Media images or moving pictures may be analyzed and divided into audiovisual objects to allow for improved media compression or audiovisual objects may be combined to form new images or media programs (synthetic video).

Audit-An audit is the review of information and/or processes to ensure the information and/or processes have been implemented correctly and successfully.

Audit File-An audit file is a set of records that are created or used to validate (audit) transactions or other information in a billing system.

Audit Trail-An audit trail is the availability of information elements (such as financial transactions and usage events) that can be linked to determine the origin and history of the usage of a product, service or information (content).

Auditing-Auditing is the process of reviewing records or information to ensure that events and transactions have been correctly acquired and/or processed. For communication systems, auditing may include transactions that involve the use of network resources.

Auditory Pattern Recognition-Auditory pattern recognition is the process of identifying specific audio patterns. These patterns can be used to identify people based on their audio patterns.

Augmentation Systems-Augmentation systems are a combination of equipment and services that can provide information or capabilities to other systems to enable them to improve their accuracy or performance.

Augmented Product-An augmented product is a modified version of a product or service. Consumers or companies may produce augmented products as a way to provide new or improved features and benefits.

AUP-Acceptable Use Policy

Aural-Aural is the ability of the ear to detect or sense sounds.

Authenticate-Authenticate is a process that is used to verify the identity of a user, device, or other entity.

Authentication-Authentication is a process of exchanging information between a communications device (typically a user device such as a mobile phone or computing device) and a communications network that allows the carrier or network operator to confirm the true identity of the user (or device). This validation of the authenticity of the user or device allows a service provider to deny service to users that cannot be identified. Thus, authentication inhibits fraudulent use of a communication device that does not contain the proper identification information.

This figure shows the operation of a basic authentication process used in a radio communication system. As part of a typical authentication process, a random number that changes periodically (RAND) is sent from the base station. This number is regularly received and temporarily stored by the mobile radio. The random number is then processed with the shared secret data that has been previously stored in the mobile radio along with other information in the subscriber to create an authentication response (AUTHR). The authentication response is sent back to the system to validate the mobile radio. The system processes the same information to create its own authentication response. If both the authentication responses match, service may be provided. This process avoids sending any secret information over the radio communication channel.

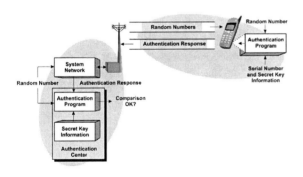

Basic Authentication Process

Authentication Code-An authentication code is a unique number or label that is used to identify a person, device or service.

Authentication Credentials-Authentication credentials are the information elements that are used to identify and validate the identity of a person, company or device. Authentication credentials may include identification codes, service access codes and secret keys.

Authentication Data-Authentication data is information that is used to perform or assist in the validity of the identity of a user or a device. Authentication may include authentication algorithm identifiers, nonce values or other authentication parameters.

Authentication Key Grace Time (AK Grace Time)-Authentication key grace time is an amount of time that an authorization key (AK) is valid during the transition or assignment of an authentication key. During an authentication key grace time, typically either of two AKs will work to decode or authenticate information.

Authentication Model-An authentication model is a representation of a functional part and/or processes that can be used to simulate or assist in the understanding of the identification and verification of users or devices.

Authentication Procedure-Authentication procedure is the sequence of steps carried out by two end points of a communication system to exchange the information necessary to insure that some aspect of the communication session are valid. This may include user validation, data validation or service validation.

Authentication Random Number (ARN)-Authentication random number is a random value that is sent prior or during an authentication procedure that ensures calculated results change over repeated authentication processes.

Authentication Sequence Number (ASN)-An authentication sequence number is a counted quantity that represents the number of authentications that have occurred. The use of an ASN can inhibit the ability of hackers to clone identification or authenticate information as the ASN count will change with fraudulent attempts.

Authentication Server (AS)-An authentication server is a computing system that can receive, process and send results of authentication (identity validation) requests. The authentication server holds or has direct access to the information necessary to validate the credentials of a user or device.

Authentication, Authorization, Accounting (AAA)-Authentication, Authorization, and Accounting (AAA) are the processes used for validating the claimed identity of an end user or a device, such as a host, server, switch, or router in a communication network. Authorization is the act of granting access rights to a user, groups of users, system, or a process. Accounting is the method to establish who or what performed a certain action, such as tracking user connection and logging system users.

Authenticator-An authenticator is a device or program that can interact with a user or device to determine the validity of their identity.

Authoring-Authoring is a process of creating or preparing text, graphics or video.

Authoring System-An authoring system is a combination of software and hardware that is used to help authors or developers design media without the need to use complex computer programming instructions.

Authorization-Authorization is the enabling of services to a device or customer that requests services. Authorization is often part of the billing and customer care (BCC) system and is maintained in a customer database service profile.

Services are initially enabled in a network as a result of provisioning. Provisioning is a process within a company that allows for establishment of new accounts, activation, termination of features, and coordinating and dispatching the resources necessary to fill those service orders. Provisioning is usually part of customer care systems.

Authorization Code-An authorization code is a group of numbers and/or characters that are provided or used by a person or device to gain access to a service or system. For some systems, the use of an authorization code initiates the creation of shared secret information (keys) that may be used after the initial access (authorization) attempt which removes the need to enter an authorization code again.

Authorization Key (AK)-An authorization key is a unique code that is generated by a system that is passed to a receiving device for the purpose of deriving other keys used in communication system.

Authorized Agent-An authorized agent is a person or company who is authorized to perform specific types of transactions for another person or company.

Authorized Billing Agent-An authorized billing agent is a person or company that is used by a telecommunications service provider to perform billing and collection services.

Authorized Deviation-An authorized deviation is a change in a process or a modification to the terms of agreement that have been authorized by the vendor or customer for the product or service.

Authorized Domain-An authorized domain is a group of users who are provided access to and possibly allowed to distribute media or information to other users within their authorized domain. Examples of an authorized domain are users or devices within a household.

Authorized Products-Authorized products are a list of products or services that a person or company is authorized to sell.

Authorized Sales Representative (ASR)-An authorized sales representative is a person or company who has received permission to perform sales functions for one or more companies.

Authorized Service-Authorized service is the conformance to usage provisions of the regulatory requirements for providing services.

Authorized User-An authorized user is a person or company that is authorized to obtain or use products or services.

Auto Available-Automatic Availability

Auto Bidding-Automatic Bidding

Auto Call-Automatic Calling

Auto Configuration-Auto configuration is the process of allowing a device or system to detect, initialize, update or program features and parameters without the direct assistance of a user or technician.

Auto Dialing Recorded Message Player (ADRMP)-An auto dialing recorded message player is a device that automatically dials a list of telephone numbers and plays a message when the call is answered.

Auto Discovery-Automatic discovery is a process where a network manager automatically searches through a range of network addresses and discovers specific types or all types of devices present in that range of addresses. The auto discovery process may be manually initiated by the network administrator or it may be initiated after a new device automatically registers after it is connected to the network.

Auto Greeting-Automatic Greeting

Auto Line Feed-Auto line feed is a process of inserting a line feed (LF) character during specific events such as when an "Enter" key has been pressed.

Auto Service Profile Identifier (Auto SPID)-Auto service profile identifier is the process of automatically discovering the SPID assigned to ISDN services.

Auto SPID-Auto Service Profile Identifier

Autodialer-An autodialer is a machine or assembly that is programmed to dial a list of telephone numbers that dialed automatically. In some cases, the autodialer may be used to dial numbers and automatically detect voice

activity (e.g. someone answers the call) and then connect the call to an available customer service representative.

Automated Attendant Billing System (AABS)-Automated attendant billing system is a set of hardware and software that allows collect and third-number billed toll calls to be placed on an automated basis. AABS uses an automated attendant system to prompt the caller to enter the information necessary to capture information necessary to bill the call or to provide services.

Automated Attendant System-Automated attendant system is a processor control system that performs telephone console attendant functions such as answering a call, transferring callers to specific user stations, directing callers to voice mail, or performing other related call-routing functions without the assistance of a live attendant. The caller's activation's of these features occurs through pressing keys that activate DTMF signaling.

This diagram shows how computer telephony systems can be used to create virtual (simulated) call attendants. In this diagram, a call is received to the main telephone number of the company to the computer telephony board. The automated telephony call processing soft-

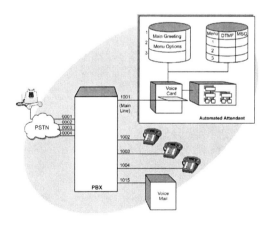

Automated Attendant Operation

ware detects a ring signal, answers the phone (creates an off-hook signal) and plays a pre-recorded message informing the caller of options they may choose to direct the call to a specific extension. In this example, the automated call attendant software decodes DTMF tones or limited list of voice commands to determine the routing of the call. The automated call attendant software then determines if the destination choice is within the option list and if the extension is available. If the extension is available, the automated attendant will send a command to the computer telephony board switching the call to the selected extension. If the extension is not valid or not available, the automated attendant will provide a new voice prompt with updated information and additional options.

Automated Bill Reconciliation-Automated bill reconciliation is the process of comparing billing records or usage data received to invoices or bills that were created to confirm the completeness or accuracy of the data.

Automated Clearinghouse (ACH)-An automated clearinghouse is a company or association that transfers billing records and/or performs financial clearing functions between companies that allow them to settle financial transactions between them.

Automated Data Collections (ADC)-Automated data collections are the use of technologies or devices that assist or automate the function of collecting data or information. ADC includes bar code readers, optical character recognition (OCR), voice recognition and smart cards.

Automated Machine Guidance-Automated machine guidance is the process of automatically adjusting the motion of a machine using position location information (such as from a GPS or land based positioning system).

Automated Purchasing System-An automated purchasing system is a set of accounts, rules, processes, software programs, equipment, and other mechanisms that can automatically identify suppliers, materials and services that need to be ordered, issuing purchase orders, tracking receipts of materials and services and issuing payments for these materials and services.

Automated Quote Contract Billing (AQCB)-Automated quote contract billing is process that allows for the automatic pricing of products and services.

Automated Voice Response Systems (AVRS)-Automated voice response systems is a system that will automatically answer an incoming telephone call and provide voice instructions or information to the caller. The caller's response to these instructions may be keypad tones or even spoken words and will be used by the system to route the call to the appropriate extension or to other sources of additional information.

Automatic Alternate Routing (AAR)-Automatic alternate routing is a network feature that allows the system to automatically reroute traffic when specific conditions occur. AAR typically allows the rerouting of calls without interrupting the flow of information.

Automatic Availability (Auto Available)-Automatic availability is an automatic call processing distribution feature that marks agents as available after they finish calls so additional calls can be automatically connected to them.

Automatic Backup-Automatic backup is the process of copying data or information for redundant storage where the process is initiated without the need for user actions. Automatic backup is typically performed at repeated time intervals.

Automatic Bidding (Auto Bidding)-An automatic bidding system is a process that adjusts bid amounts based on other information or criteria. Auto bidding systems can help marketing managers to set rules or criteria that dynamically adjust and control of adword marketing programs based on conditions such as the changing of ad position or changes in competitor bid amounts.

Automatic Call Delivery (ACD)-Automatic call delivery is the process of establishing a communication session using information already provided. ACD systems do not require assistance from the user or other people to setup and connect a call.

Automatic Call Distribution (ACD)-ACD is a system that automatically distributes incoming telephone to specific telephone sets or station calls based on the characteristics of the call. These characteristics can include an incoming phone number or options selected by a caller using an interactive voice response (IVR) system. ACD is the process of management and control of incoming calls so that the calls are distributed evenly to attendant positions. Calls are served in the approximate order of their arrival and are routed to service positions as positions become available for handling calls.

This figure shows a sample automatic call distribution (ACD) system that uses an interactive voice response (IVR) system to determine call routing. When an incoming call is initially received, the ACD system coordinates with the IVR system to determine the customer's selection. The ACD system then looks into the databases to retrieve the customers' account or other relevant information and transfer the call through the PBX to a qualified customer service representative (CSR). This diagram also shows that the ACD system may also transfer customer or related product information to the CSR.

Automatic Callback-Automatic callback is a CLASS service feature that allows a caller to complete a call to a busy station by dialing an activation code (usually a single digit) and hanging up. The system automatically rings both parties when the lines are available.

This figure shows the basic operation of automatic callback. To activate automatic callback service, after a call has dialed a number that is busy, the customer dials an automatic callback feature code and hangs up the telephone. The local switch (caller's phone carrier) informs the remote (distant) switch of automatic callback request. This reserves (blocks) the called number from receiving additional calls until the automatic callback service is completed. When the called number becomes available, the remote switch sends a message to the local

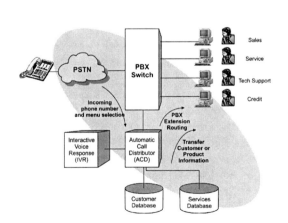

Automatic Call Distribution (ACD) Operation

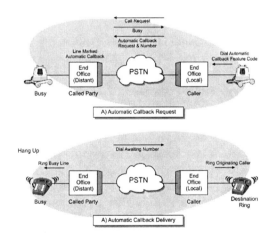

Automatic Callback

switch and this rings the original caller's number (possibly with distinctive ring feature.)

Automatic Calling (Auto Call)-Automatic calling is a telephone system call-processing feature that allows a computer, telephone station or transmission control unit to automatically dial a telephone number.

Automatic Cover Letter-An automatic cover letter is the creation of a letter that accompanies other documents which contains information that is unique to the recipient. This unique information may include the recipient's address and name or it may contain other customized information.

Automatic Customer Identification System (ACIS)-Automatic customer identification system is set of hardware and software that allows a program (such as a customer care system) to identify a customer from the information that has been already gathered or provided (such as by the incoming telephone number).

Automatic Forwarding-Automatic forwarding is a feature that automatically retransmits data or messages to another destination.

Automatic Greeting (Auto Greeting)-An automatic greeting is a pre-stored audio message that is played when a call is received or when a need the greeting message is determined.

Automatic Location Identification (ALI)-Automatic location identification is a database function that identifies and provides the location of a mobile telephone that has called an emergency number (e.g. 911) to a public safety access point (PSAP). In addition to providing the location of the mobile phone, the ALI must also identify the location of the most appropriate PSAP to route the emergency call.

Automatic Location Identifier (ALI)-Automatic location identifier (ALI) is a number (such as a telephone number or MAC address) that identifies a location of a device or assembly.

Automatic Message Accounting (AMA)-Automatic message accounting (AMA) is an automatic system for recording data describing the origination time of day, dialed number and time duration of a call for purposes of billing. The earliest systems used punched paper tape, later replaced by magnetic computer tape and then later magnetic computer disk. AMA is a term mostly used in the public network, and similar terms, some used in private, PBX, or inter-carrier systems are Call Detail Recording (CDR), Station Detail Message Recording (SMDR), and Automatic (calling) Number Identification (ACNI or ANI).

Automatic Meter Reading (AMR)-Automatic meter reading (AMR) is the process of reading meters (generally electric or water utility meters) via a communication systems such as wireless or wired technology.

Automatic Network Restoral-Automatic network restoral is process that allows a network to automatically reconfigure equipment after a failure condition.

Automatic Number Identification (ANI)-Automatic number identification (ANI) is the providing of the originating telephone number, including an extension number in a Centrex system or PBX system. The ANI is an administrative number provided by the telephone system and may not be the actual originating number.

Automatic Rate Card-An automatic rate card is a fee schedule that is automatically assessed or applied for services that are ordered, used or modified.

Automatic Replication-Automatic replication is the process of automatically duplicating and updating data. Automatic replication may be used in multiple computers in a network to help ensure that data is continuously available in the event that a device or network connection fails or operates differently than expected.

Automatic Retransmission Request (ARQ)-Automatic retransmission request is an acknowledgment process whereby the sending device can retransmit blocks of data that were received incorrectly at the receiving device.

Automatic Roaming-Automatic roaming is the ability of a customer (such as a mobile telephone user) to make and receive calls automatically outside of the customer's home service area.

Automatic Vehicle Location (AVL)-Automatic vehicle location is a process that can determine the location of vehicle as it moves within a given geographic area. The position of the vehicle can be determined by system sensing (such as the Teletrac system) or by the vehicle reporting its location using position locating systems (such as the Global Positioning System).

Automation-Automation is the process of using a system that has established rules or procedures that allows for the completion of tasks or processes while reducing the amount of processes performed by people.

Automotive Industry Action Group (AIAG)-The automotive industry action group is an organization that defines how systems operate for companies that are involved in producing or supporting automotive related goods or services. The AIAG defines the EDI formats for the automotive industry.

Autonomous Registration-Autonomous registration is a process where a mobile radio independently transmits information to a wireless system that informs it that it is available and operating in the system. This allows the system to send paging alerts and command messages to the mobile radio. The mobile radio may be stimulated to register with the system when it detections it has entered into a new radio coverage area or it detects a registration request message.

Autonomous System (AS)-Autonomous system (AS) is a system that operates autonomously from other systems. An example of an AS is an Internet Service Provider (ISP). Within the ISPs network, routers exchange information with each other as trusted devices and they are under the control of a single administration system. Autonomous systems can be setup to run protocols that only operate within its network. These protocols include Interior Gateway Protocol (IGP) and Open Shortest Path First (OSPF).The protocols used between different AS networks are usually different because the same level of control and trust does not exist. Examples of protocols used between AS networks Exterior Gateway Protocol (EGP), Border Gateway Protocol (BGP), and InterDomain Routing Protocol (IDRP).

Autoresponder-An autoresponder is a system or software program that can automatically respond to an event (such as the receipt of an email message) by returning a message. Autoresponders may be used to allow users to automatically request brochures, price lists, or catalogs.

Auxiliary Service Trunk Groups-Auxiliary service trunk groups or multi-circuit lines that can provide selected services that can terminate at announcement systems or desks. Examples include directory assistance, intercept, public announcement, and repair service.

AV-Audio Video

Avail-Advertising Availability

Availability-Availability is a measurement that indicates the connection status or a commitment to provide a minimum amount of connection status of a network during a period of time. Availability may be measured by a connection time or by a minimum performance measurement (e.g. at a minimum data transfer rate). Availability is often tied to reliability.

Availability Restriction-Availability restriction is the inability to assign or provision a service due to missing capabilities or incomplete resources that are required for the service.

Availability Time-Availability time is the duration between when a device or service becomes available for use and when it is no longer available.

Available Bit Rate (ABR)-Available bit rate (ABR) is a communications service category that provides the user with a data transmission rate that varies dependent on the availability of the network resources. ABR service may provide the user with feedback as to the changed data transfer rate and may have established minimum and maximum levels of data transmission rates.

Available Time-Available time is the duration that a device or service can be used.

Avatar-An avatar is a supervisor or administrator account.

AVC-Advanced Video Coding

Average-An average is the sum of items in a list divided by the number items in the list.

Average Delay In Queue-Average delay in queue is a measure of the customer responsiveness of a call center.

Average Holding Time-Average holding time the sum of the connection duration times divided by the number of connections.

Average Rate Per Minute (ARPM)-Average rate per minute is the value that indicates the average cost or fee that is charged per minute of service use or availability.

Average Revenue per Customer (ARPC)-Average revenue per customer is the revenue generated for each customer unit, such as a product purchaser or company service account. The average revenue per customer (ARPC) is an indicator of companies business's operating performance. Severely declining ARPC typically is a negative sign that may indicate a company is adding too many low-revenue generating customers to its customer base.

Average Revenue Per User (ARPU)-Average revenue per user (ARPU) is an indicator of a service operators business's operating performance. ARPU measures the average monthly revenue generated for each customer unit, such as a cellular phone or cable television customer that a carrier has in operation. Severely declining ARPU typically is a negative sign that may indicate a carrier is adding too many low-revenue generating customers to its subscriber base.

Average Variable Cost-Average variable costs are expenses that occur and increase as products or services are produced or provided over a period of time.

AVL-Automatic Vehicle Location

AVRS-Automated Voice Response Systems

AVS-Address Verification Service

AVS Code-An AVS code is an additional number located on a credit card that can be used to verify that a purchaser has possession of the card.

Award Criteria-Award criteria are a set of evaluation criteria elements (such as price and time commitments) along with the rating values or processes that are used to assign a value to each criteria element for the issuance of a contract or project.

Award Date-An award date is the day that a contract or project is authorized or a legal commitment occurs.

Awareness-Awareness is the ability of a person to remember a brand, message or other information and its associated characteristics.

B

b-Bidirectional

b-Bit

B-Byte

B 911-Basic 911

B2B-Business to Business

B2B Advertising-Business to Business Advertising

B2C-Business to Consumer

B2D-Business to Distributor

BAA-Blanket Authorization Agreement

Back Date (Backdate)-Back dating is the process of using a date within an order or some form of agreement that pre-dates the actual date of the document.

Back Door-In an otherwise secure system, an intentional way for a trusted party to circumvent the security with secret knowledge such as the nature of a designed-in security flaw. Since back doors might be found and exploited by malicious parties, they weaken the security of what may appear to be a well-protected system.

Back End (Backend)-A back end is a system or database server function that processes data via a network connection.

Back Hoe Fade-A reduction in the ability of a communication system to route calls due to the cutting of a buried communication cable (e.g. fiber optic cable). The reduction in capacity comes from the automatic re-routing of communication circuits through other systems that have a lower capacity than the original communication circuit.

Back Log (Backlog)-A backlog is a listing of products or services that have been ordered but not fulfilled yet.

Back Office Operations-Back office operations are processes and systems that are used to assist with the operation and management of communication systems. Back office operations may include billing and customer care systems, accounting, maintenance services and asset management.

Back Order (Backorder)-Back ordering is the process of initiating an order before a product or service is available.

Backbone-A network backbone is the core infrastructure of a network that connects several major network components together. A backbone system is usually a high-speed communications network such as ATM or FDDI.

Backbone Network-A communications network that connects the primary switches or nodes within the network. The backbone network is usually composed of high-speed switches and communication lines.

Backbone Service Provider (BSP)-A service provider that provides interconnection services (usually high-speed data transmission) to other carriers.

Backcharging-A process of charging for usage of a service when the service request is initiated rather than when the service is connected. An example of backcharging is the process of starting the billing time on a mobile telephone when the user initiates the call, not when the call is actually connected.

Backdate-Back Date

Backdoor Attack-A backdoor attack is an attempt to disable or gain unauthorized access to a computer system or network by exploiting a designed-in (backdoor) security flaw. Since back doors might be found and exploited by malicious parties, they may be difficult to detect.

Backend-Back End

Backend Functionality-Backend functionality is the capabilities of systems and people to perform supporting processes or functions. An example of backend functionality is the ability to accumulate and post award points for customer incentive programs such as frequent flyer programs.

Back-End Processor-This is another computer system or dedicated microprocessor that is optimized to perform a specialized task in order to offload work from the main processing resources.

Backend System-Backend systems perform supporting functions for business operations. Examples of backend systems include billing, customer service and inventory management.

Background Check-A background check is the process of obtaining information about a person or company to determine their ability to perform functions (such as a job assignment) or to provide specific types of services.

Backhaul-Backhaul is the process of transferring packets or communication signals over relatively long distances to a separate location for processing.

Backing Out-The process of returning the state of a system to its prior known state after a transaction fails to complete. This process guarantees the integrity of the information contained within the system. Used primarily for database and distributed systems.

Backlog-Back Log

Backorder-Back Order

Backronymed-Backronymed is the applying of an acronym to other terms or names. An example of a backronym is the use of the term podcasting to the iPod device produced by Apple computer. Podcasting was performed before the iPod was created.

Backroom Departments-Backroom departments are functional parts of a business that are necessary to its operation that usually operates independently from contact with customers or users.

Backup-(1-data) A copy of data information, usually on a storage device that is stable over relatively long periods of time. (2-process) The process of transferring files or information to a storage device or media.

Backup Domain Controller-A backup domain controller is a server (computer) that maintains a copy of company's or person's domain information. This includes lists of authorized users, account information, and lists of other primary and backup servers it may communicate with.

Backup Link-A backup link is a communication link that is used for communication in the event a primary or specified link becomes inactive or disabled.

Backup Plan-A backup plan is a set of actions or processes that may be used to copy data information, usually on a storage device that is stable over relatively long periods of time. A backup plan may include a combination of local storage, remote storage and online options.

Backup Power Supply-A redundant power supply that takes over if the primary power supply fails. This may or may not be automatic although when supplied as part of the overall system it usually is.

Backup Program-This is an application program that is used to make archival copies of the contents of a computer file system. Most backup programs provide methods for both full and incremental backups. Incremental backups only archive files that have been modified since the last full backup. Most commercial backup programs provide a means of also restoring the archived information if the need arises.

Backup Rights-Backup rights are the authorizations to copy and store information for the potential need to restore the installed version of the information.

Backup Server-A program that administers the copying of users' files so that at least two up-to-date copies exist.

Backward and Forward Compatibility-Backward and forward compatibility is the ability to use existing and future modes of operation of a system or service.

Backward Compatibility-Ability of new hardware or software to operate effectively with older versions of the same equipment or programs.

Backwards Compatible-Backward compatibility is the ability to use future modes of operation of a system or service on existing (legacy) systems.

Badged-An English term that represents that a product manufactured by one firm that is sold by another. The Original Equipment Manufacturer (OEM) produces the product with identification and/or brand of the selling firm.

BAF-Bellcore Automatic Message Accounting Format

BAIC-Barring of All Incoming Calls

Bailment-Bailment is the providing of products to a customer with the understanding that the products will eventually be returned to the supplier.

Bait Advertising-Bait advertising is promotion of goods or services that can motivate customers to visit a store or initiate an inquiry where the products promoted are not easily obtained or available and the customer is redirected to an alternative product or service (usually more profitable product for the seller).

Balance Due-Balance due is an amount of an invoice or account that is still requires payment.

Balance Forward-Balance forward is transferring of previous unpaid amounts or units to another invoice or account.

Balance Management-Balance management is a process that allows a user or account owner to monitor, view and pay balances on one or more of their accounts.

Balance Sheet-A balance sheet is a presentment of financial information that defines a company or individuals assets, liabilities and net worth on a specific date.

BAN-Billing Account Number

Banded Rate-A banded rate is a range of service fees or usage costs. Companies may be regulated to provide services and charge fees within the banded rates.

Bandwidth Allocation-Bandwidth allocation is the frequency width of a radio channel in Hertz (high and low frequency limits) that can be modulated (changed) to transfer information (voice or data signals). The amount type of information being sent determines the amount of bandwidth used and the method of modulation used to impose the information on the radio signal.

Bandwidth Awareness-Bandwidth awareness is the process and/or protocols that are used to allow for the recognition of the bandwidth that is available to devices that are connected to a network. For example, a media server needs to discover the connection speed of a multimedia computer that has requested to view a digital video stream. If the server has bandwidth awareness capability, it can determine an optimum compression and data transmission rate for the requested digital video stream.

Bandwidth Cost-Bandwidth cost is the combination of transmission or proportioned system equipment usage cost(s) between entry and exit points of a path on a communications network along with associated equipment usage and data processing costs divided by the amount of bandwidth used or reserved. For example, a leased line of 45 Mbps that costs $5,000 per month, this equates to a bandwidth cost of $111 per Mbps per month ($5,000/45).

Bandwidth Management-Bandwidth management is the process used to plan and organize the data transmission characteristics to devices or systems.

Bandwidth On Demand (BoD)-A system that allows different data transmission rates based on requests from the customer, their application (e.g. voice or video), and the data transmission capability of the system.

Bandwidth Reservation-Bandwidth reservation is a process that is used to reserve bandwidth capacity through devices or communication line for specific communication sessions or services.

Bandwidth Scalability-Bandwidth scalability is the ability of a system or program to provide a service to a number of users at different bandwidth rates.

Bangtail-A bangtail is an envelope that has a second flap that can be used for other purposes such as an order form or a change of address form.

Bank-(1-financial) A bank is a financial institution that receives, processes and lends money. (2-equipment) A bank is a row or rack of similar components, cards or devices.

Bankruptcy-Bankruptcy is inability of a person or company to pay their creditors from revenue and assets that they control. People or companies may file for bankruptcy protection to the government to help or force creditors to settle their financial disputes.

Bankruptcy Discharge-A bankruptcy discharge is a legal order that defines the amount of money or remedy that creditors have after a person or company has filed for bankruptcy protection.

Bankruptcy Petition-A bankruptcy petition is legal request from a person or company to be protected under bankruptcy laws.

Banner Ad-A banner ad is a graphic or image that is located on a web site, usually on the top of the page. The banner ad contains a hyperlink that links to another web page or web site and is typically used for advertising products or services. The purposes of banner ads include redirecting potential customers to a new web site and to create or to validate an awareness about a product or service.

Banner Ad Cost-Banner ad cost is the combined costs of producing and presenting banner ads on web sites. Banner ad costs can include banner graphics creation, impression fees, click fees and banner network exchange fees.

Banner Ad Network-A banner ad network is a company or group that acts on behalf of multiple advertisers to find and place banners on web sites that participate in the banner ad network program.

Banner Exchange-A banner exchange is the agreement by two (or more) companies to display banner ads on web sites for some form of value. The banner exchange may be a simple agreement for each of the participants to place banner ads that point to each others company's web site or it may involve additional incentives such as cash, email marketing or providing of a service.

Banner Exchange Program-A banner exchange program is an agreement to insert or display banner ads on a web site in return for the insertion of banner ads on another web site.

Banner Network-A banner network is a group of companies that are willing to insert banner ads from other companies. Banner network hosts receive money for banner ad insertion on their web page and the banner network operator coordinates the selling and billing for banner advertisers.

Banner Rotation-Banner rotation is the displaying of sequence of banner ads so multiple banner ad images can be displayed in the same banner ad area. The rotation order may be sequenced or random.

Banner Type-Banner ad type is the image file format of the banner and the type of display (static or animated).

BAOC-Barring of All Outgoing Calls

Bar Code (Barcode)-A standardized sequence of typically black vertical bars separated by white spaces that may be read with an optical decoder. The decoder reads the sequence of bars and interprets them as alphanumeric characters. The pattern of adjacent thick and thin bars is located on a contrasting background. A binary bit group in a bar code represents each decimal digit of an identification number and parity check digits are typically appended. An optical scanner device for input to a computer system for purposes of inventory control or the like can scan a bar code.

Barcode-Bar Code

Bare Bones-Bare bones are a product or service that contains none of the available options.

BARG-Billing and Accounting for Roaming Group

Barge In-Barge in is a call processing feature that allows another caller to be added to a communication line that is already in use. The barge in feature is sometimes called executive override.

Barge Out-A call processing feature that allows a user to leaving a call in progress without any notice (usually after a call barge in).

Barriers to Entry-Barriers to entry are requirements or factors that deter new people or companies from entering into a new business or marketplace.

Barring of All Incoming Calls (BAIC)-A system feature that restricts the delivery of all incoming calls to a phone.

Barring of All Outgoing Calls (BAOC)-A system feature that restricts the delivery of all outgoing calls from a phone.

Barring of Outgoing International Calls (BOIC)-Barring of outgoing international calls is a call-processing feature that allows a service provider to restrict the ability of users to make international telephone calls.

Barring Services-Barring services are operations or processes that can restrict the delivery or processing of requests or services. An example of barring services is the restriction of the delivery of incoming telephone calls from a list of barred telephone numbers.

Barter-A barter is an exchange of items, media or services that are used as value instead of the use of money.

Barter Agreement-A barter agreement is a document or understanding of terms for the exchange of products or services, which are used as value instead of the use of money.

Barter Syndication-A barter syndication is a company or organization that assists other companies in the exchange of goods and services using valuable assets other than money.

Barter Transaction-A barter transaction is the transfer of products, media or services that are used as value instead of the use of money. Barter transactions can range from simple one time use of media or broadcast to a mix of valuable services that would occur over a period of time (such as authorization to use equipment, ad insertions or branding rights).

Base Amount-A base amount is a reference or starting value. A base amount may be a target for a marketing campaign where the expected results are based on previous average performance.

Base Currency-Base currency is the reference unit that is used for a transaction.

Base Fee-A base fee (or base rate) is an amount charged for a service per a specific time period. Base fees are usually associated with a service that has a fixed charge (for a basic amount or type of service) and variable charge (applied only when more than the basic services are used).

Base Rate-A fixed amount charged for a service per a specific time period. Normally this term is associated with a service that has a fixed charge (for a basic amount or type of service) and variable charge (applied only when more than the basic services are used).

Base Rate Area-A base rate area is a geographic region where base rate tariffs apply.

Base Station (BS)-The radio part of a mobile radio transmission site (cell site). A single base station usually contains several radio transmitters, receivers, control sections and power supplies. Base stations are sometimes called a land station or cell site.

A base station contains amplifiers, radio transceivers, RF combiners, control sections, communications links, a scanning receiver, backup power supplies, and an antenna assembly. The transceiver sections are similar to the mobile telephone transceiver as they convert audio to RF signals and RF to audio signals. The transmitter output side of these radio transceivers is supplied to a high power RF amplifier (typically 10 to 50 Watts). The RF combiner allows

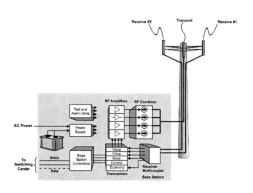

Base Station (BS) Functional

separate radio channels to be combined onto one or several antenna assemblies without interfering with each other. This combined RF signal is routed to the transmitter antenna on top of the radio tower via low energy-loss coaxial cable.

Base Traffic Load-In trunk forecasting, the average load offered on the first route available between two identified areas. Base load is found by averaging the traffic measured during the same 1-hour period each day over a period of several days. Base loads often are used to forecast future loads.

Base Year-Any 12 consecutive months for which data is collected for determining base loads.

Baseband Connection-A baseband connection is a transmission path between two or more points that uses the fundamental frequency or transmission components of an communication system.

Baseline-The process of determining and documenting network throughput and other performance information when the network is operating under what is considered a normal load. Measured performance characteristics might include error-rate and data-transfer information, along with information about the most active users and their applications.

Basic 911 (B 911)-Basic 911 is an emergency service call code that allows users to contact an emergency service operators. B 911 is the

early version of emergency call service, as it does not provide location information of the caller to the emergency operator.

Basic Budget Service-Basic budget service is a local phone service rate plan that is offered to people that have limited incomes. Basic budget services may include a reduced number of service features such as a limited number of outgoing and incoming calls.

Basic Cable Service-Basic cable service is the retransmission of a limited set (non-premium) of television broadcast signals.

Basic Exchange Telecommunications Radio Service (BETRS)-A wireless extension to a local telephone system to allow telephone service in rural areas. The BETRS system is a digital radio system that operates at frequencies near 150, 450, and 850 MHz. The system is used in primarily in areas where wired telephone service is not economically viable. BETRS is sometimes referred to as Basic Exchange Radio (BEXR) Service.

Basic Input/Output System (BIOS)-The software or processes that controls the input and output of information to devices connected to a computer. The BIOS contains the set of instructions that sets up and tests the hardware when the computer is first turned on (booted). It starts the loading of the operating system and coordinates the operation of computing devices, such as floppy drives, hard disks, CD ROMs, video cards, mouse, and keyboards. The BIOS program is stored in non-volatile memory and is rarely changed (if ever). It is pronounced "bye + Ose."

Basic Rate (BR)-(1-general) Basic rate is the default or reference transmission parameters for a device or system. (2-ISDN) ISDN basic rate is an access channel transmission rate of 144 kbps that is subdivided into two B channels of 64 kbps and one D channel of 16 kbps. (3-Bluetooth) Bluetooth basic rate is the physical transmission channel of 1 Mbps that is produced using GMSK modulation.

Basic Rate Access-Basic rate access is access to an Integrated Services Digital Network (ISDN) that provides data transmission at rates to 144 kbps. The information stream is divided into two B channels operating at 64 kbps and one D channel at 16 kbps. The B channels are used for voice or data. The D channel carries control and signaling information to set up and take down the voice and data calls, and also can carry X.25 data at rates up to 9.6 kbps. Basic rate access also is referred to as basic rate interface.

Basic Rate Interface (BRI)-Basic rate interface is a data interface that is used in the integrated services digital network (ISDN) system for end user devices or low speed connections. The BRI interface provides up to 144 kbps of information that is divided into two 64 kbps channels (voice or data) and one 16 kbps control channel (data). The 64 kbps channels are referred to as the B channels and the 16 kbps channel is called the D channel.

Basic Service Element (BSE)-A basic service element is a basic service "building block" that may be offered by a carrier.

Bastion Host-A bastion host is a computer server that acts as the main connection to a data network (such as the Internet) for users of a local area network (LAN).

Batch-A batch is the processing of information or data as a group operation. Information for batch processing is gathered for a period of time and then the information is processed as a single process.

Batch Costing-Batch costing is the measuring of the production or creation value of goods or services in quantity segments or groups.

Batch File-A batch file is a data file that contains operating system commands that can be performed by a batch processor. Batch files enable a group of commands to be performed at the same time.

Batch Processing-Batch processing is a set of tasks or instructions that do not occur in a real time mode. In a batch process, events are collected and then forwarded in batches to the processor which is "idle" until required to begin.

Batch Rating Engine-A batch rating engine is a function within a billing system that assigns the charging rates to a usage event or usage record in a group (batch) process.

Battery-A battery is a device that stores electrical energy for use at a later time. Some batteries are designed and used for a single use (primary cell) and other batteries are designed for repetitive use (rechargeable).

Battery Backup-Battery backup is the use of batteries to power devices or systems.

Battery Life-Battery life is the amount of time that a device will operate from a battery power supply when operating in a specific mode of operation.

Baud (Bd)-Baud is a unit measure of data symbols per second. This name is taken from the name of the 19th century French teletypewriter machine innovator Emiel Baudot. For a method of modulation or encoding in which there is a choice of only two symbol values per symbol interval, or one bit per symbol (such as two-level pulse voltages) the baud rate is equal to the bit rate (bits per second). For a method of modulation or encoding in which there are more than two symbol values per symbol interval (and thus 2 or more bits per symbol) the bit rate is higher than the baud rate. For example, QPSK phase modulation and 2B4Q pulse coding both have 4 symbol values per symbol interval and thus the bit rate (bits per second) is twice the symbol (baud) rate. (Please do not make the error of writing "baud per second.")

Baud Rate-The number of signaling elements (symbols) per second on a transmission medium. For some line codes, such as bipolar, baud rate is the same as bit rate. However, in many applications, the baud rate is below the bit rate. For example, in 2B1Q coding, each quaternary signaling element conveys 2 bits of information, so the baud rate is one-half the bit rate. The spectral characteristics of a line signal depend on the baud rate, not the bit rate. For high-speed digital communications systems, one state change can be made to represent more than one data bit.

This diagram shows that the baud rate is not always the same as the bit rate as each baud (symbol) can have several states that represent multiple binary bits.

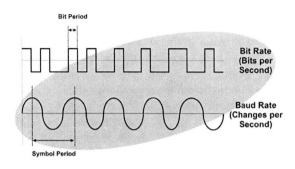

Baud Rate

Bay-An equipment casing that can hold electronic assemblies such as transmission or call processing assemblies. Commonly called an "Equipment Bay."

BBTV-Broadband TV

BC-Billing Center

BCC-Billing and Customer Care

BCC-Blind Carbon Copy

BCI-Billing Correlation Identifier

BCMCS-Broadcast and Multicast Services

BCP-Business Continuity Planning

Bd-Baud

BD-ROM-BD-ROM is a storage format specification for Bluray technology.

BE-Best Effort Service

Bearer-A bearer is transmission channel or medium that is used to transfer data or media .For integrated services digital networks

(ISDN), a basic rate interface is constructed of two 64 kbps bearer user data channels and one 16 kbps control channel.

Bearer Services (BS)-Bearer services are telecommunication services that are used to transfer user data and control signals between two pieces of equipment. Bearer services can range from the transfer of low speed messages (300 bps) to very high-speed data signals (10+ Gbps).

Bearer services are typically categorized by their information transfer characteristics, methods of accessing the service, inter-working requirements (to other networks) and other general attributes. Information characteristics include data transfer rate, direction(s) of data flow, type of data transfer (circuit or packet) and other physical characteristics. The access methods determine what parts of the system control could be affected by the bearer service. Some bearer services must cross different types of networks (e.g. wireless and wired) and the data and control information may need to be adjusted depending on the type of network. Other general attributes might specify a minimum quality level for the service or special conditional procedures such as automatic re-establishment of a bearer service after the service has been disconnected due to interference. Some categories of bearer services available via the telephone system include synchronous and asynchronous data, packet data and alternate speech and data.

Beginning of Message (BOM)-Beginning of message is a code or frame of data that indicates the beginning of a message data segment.

Behavioral Actions-Behavioral actions are processes or events caused by people or systems that result from exposure to information or other stimulus (such as viewing web pages).

Behavioral Data-Behavioral data is information about the activity of visitors or users of a system or service.

B

Behavioral Score-A measure of a customer's credit worthiness that is based on past payment history. Sometimes termed "Internal Score", or "Payment Behavior Score".

Behavioral Targeting-Behavioral targeting is a process that uses information gathered about the behavioral characteristics of a person (such as which objects or links on a web site that a visitor has interacted with) to change the promotions or offers with the objective of increasing the effectiveness of the marketing campaigns.

Bell Customer Code-A bell customer code is a three-digit number that is added to the main telephone number that is used to identify a customer or account.

Bellcore-A telecommunications research and development consortium that was developed by the seven regional companies that resulted from AT&T's divestiture of the Bell Operating Companies. These were Ameritech, Bell Atlantic Corporation, BellSouth, NYNEX, Pacific Telesis, Southwestern Bell Corporation, and US West. Bellcore was sold in the late 1990s to Science Associates, and is now called Telcordia.

Bellcore Automatic Message Accounting Format (BAF)-Bellcore automatic message accounting format is a standard billing record structure that was defined in Bellcore TR#030#NWT-001100.

Benchmark-Benchmarking is a process of establishing targets or objectives as compared to (benchmark to) existing systems or performance standards.

BER-Bit Error Rate

Best Effort Service (BE)-Best effort is a level of service in a communications system that doesn't have a guaranteed level of quality of service (QoS).

Beta-(1-general) Second letter of the Greek alphabet. Used as a mathematical symbol for the current amplification factor of a junction transistor. (2- testing) A second stage of product testing (following alpha testing), typically allowing selected potential consumers use the product in the field. (3- video) A now-obsolescent videotape format for consumer magnetic tape video recording, and a shortened form of a broadcast quality video recording system, named Betacam, that uses metallic particle video recording tape in a cassette having the same dimensions as the consumer Beta product. Betacam is used by several television network news broadcasters.

Beta Application-A beta application is a near-final version of a software product that is provided to a group of users (typically friendly users) who agree to use the product prior to its final release.

Beta Testing-Beta testing is the field testing process used to confirm the operation and performance of new hardware or software products before a product is officially released. Beta testing is the second stage for testing a new hardware or software product, usually performed by friendly customers or affiliates of the manufacturer or developer. The key purpose of Beta testing is to identify problems and the reliability of operation during normal field operating conditions. The typical number of Beta test participants is 50 to several hundred.

BETRS-Basic Exchange Telecommunications Radio Service

BG-Billing Gateway

BHCA-Busy Hour Call Attempts

BHCC-Busy Hour Call Completion

BHM-Busy Hour Minutes

BHMC-Busy Hour Minutes of Capacity

BHT-Busy Hour Traffic

Bias-(1-mathematic) A deviation from a reference value. (2-electrical) A force (electrical, mechanical, or magnetic) that is used to establish a level that is necessary to operate a device or assembly.

Bid Attacks-Bid attacks are rapid changes in bid prices, which are designed to obtain a favorable status (such as increased ad positioning) or to create a disadvantage for other bidders.

Bid Gap-Bid gap is the difference in bid amounts between two advertisers who are competing for a position (typically the top spot) on a pay-for-placement search engine.

Bid Management-Bid management is the process of monitoring and adjusting the bid amounts for pay for placement sponsored ad listings.

Bidding Tools-Bidding tools are programs or services that track and set bids for keywords and other marketing programs that rely on bidding methods.

Bidding War-A bidding war is a series of events which pushes the bid price higher. A bidding war may be performed to maintain a specific ad position or it may be a competitive tactic used to push the cost of an ad word program higher for a competitor.

Bi-Directional-Bi-directional communication is the ability to send and receive information on a communication channel. While bi-directional connections may allow two-way connectivity, the connections may not transmit and receive at the same time.

Bidirectional (b)-Bidirectional transmission is the transfer of information in two directions.

Bidirectional Carrier-A telephone company (telco) carrier system that utilizes 1 cable runs. This cable run uses 100 pair separation between transmit and receive pairs.

Bi-directional Services-Bi-directional services are communication connections that allow the operation of applications require two-way communication. An example of bi-directional services include video conferencing, chat rooms, and video mail.

Bill-A bill is a list or charges or fees that are associated with a product or service that is assigned to a customer or account.

Bill Calc-Bill Calculation

Bill Calculation (Bill Calc)-Bill calculation involves identifying the account that will pay for the service usage (guiding), determining the appropriate rate plan for the service and calculating the charge (rating) the cost for the usage event.

Bill Cycle Code-An identifier assigned to all customers who are to be billed on the same run of the billing process.

Bill Dispute-A bill dispute is a claim submitted or recorded from a person or company that asserts that a billing charge is invalid or incorrect.

Bill Generation-Bill generation is the process of gathering, sorting and combing billing records and adding service plan charges (such as monthly service charges) to produce bills.

Bill Inquiry-A bill inquiry is a request from a person or device to provide additional information or details about an invoice or a billed item.

Bill Insert-A bill insert is an item (such as a brochure) that is inserted into a bill that is sent to a customer.

Bill of Entry-Bill of entry is a document that defines the value of the consignment of goods that are to be transported. Customs officials use a bill of entry when products enter into a country.

Bill of Lading-A bill of lading is a document that is issued on behalf of the shipping carrier that defines the receipt for goods at a specific location.

Bill Of Materials (BOM)-A list of parts (components and other assemblies) that are the materials that are used to produce a product, quantity of product, or an assembly. The listing of parts in a BOM is often assigned a cost to estimate the construction cost of the product or assembly.

Bill of Quantity (BOQ)-A bill of quantity is a listing of materials that are anticipated for use in a project or job.

Bill on Demand-Bill on demand is the creation of a bill immediately or shortly after when the bill is requested.

Bill Payment-Bill payment is the process of transferring value to pay a bill or account.

Bill Period-The period for which the recurring charges apply. In monthly billing for instance, the Bill Period typically corresponds to the period between the current bill and the next bill for charges billed in advance, and between the previous bill and the current bill for charges billed in arrears (e.g. April 5th to May 4th).

Bill Pool-A bill pool is a group of billing records that have been processed by billing system to include the necessary charging rate information. The bill pool usually contains records that are ready for the final stage of bill processing.

Bill Posting-Bill posting is the process of recording an invoice payment into a billing system.

Bill Print-Bill print is the process of converting billing records (invoices) into a format that can be sent or displayed to a person (e.g. a customer).

Bill Processing-Bill processing is the set of steps and functions that are used to identify, gather, sort and combine billing records to produce bills.

Bill Processor-A bill processor is a company or system that can receive and process billing transactions to produce invoices, statements or other billing services.

Bill Rate-A bill rate is a list of charges that are associated with products or services that are offered. A bill rate can vary based on the type of customer such as a wholesaler, preferred vendor or direct customer.

Bill Rendering-Bill rendering is the process of converting a bill, invoice or related billing information into a form that can be displayed or communicated to users (e.g. customers).

This figure shows some of the different options available for bill rendering and distribution. This example shows that bills can be printed and mailed, created in text form and emailed, converted to stored media such as CDROM or tape, can be formatted for web viewing or converted to a standard EDI format to be directly sent to a customer's computer server.

Bill Rendering Fees-Bill rendering fees are charges for the production of billing records in specific formats such as paper or other forms of stored media (e.g. CD ROM). The fees for bill production (bill rendering) may vary based on

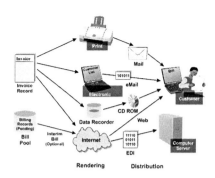

Bill Rendering Options

the detail level that is included in the bill format.

Bill Run-The process of triggering the start of the billing process that gathers and processes billing charges associated with specific customers or accounts. Also sometimes referred to as "Bill Round", or "Billing", or "Month End", etc. Typically, one or more cycles are assigned to a Bill Run; i.e. all customers who have been assigned those cycles will be processed during this bill run.

Bill Stuffer-Bill stuffer is an item (such as a brochure) that is inserted into a bill that is sent to a customer.

Bill to Cash Process-A bill to cash process is the capturing of the flow of events as well as the interaction points between the point where the bill for the service is generated or dispatched to the customer and the point where the service provider receives the payment for the service.

Billed Minutes-The actual time, in reported minutes, for messages of a duration equal to or greater than an initial period. When the reported minutes are less than the initial period, the initial period minutes are shown as billed minutes. For example, a one-minute conversation on a connection having an initial period of three minutes would be billed as three minutes.

Biller-A biller is a system or company that performs billing functions such as rating and invoicing.

Billing-The process of grouping service or product usage information for specific accounts or customers, producing and sending invoices, and recording (posting) payments made to customer accounts.

Billing Account-A billing account is a unique identifier that designates a customer or company that is used to associate billing charges for products and services. A single customer may have multiple billing accounts.

Billing Account Number (BAN)-A billing account number is a unique identifier used by communication service provider companies to designate a customer or customer location that will be billed for products and services. A single customer may have multiple billing accounts and BANs.

Billing Advice-Billing advice is a message or presentation of information on a bill to alert the bill recipient that an event may occur or that an action should be taken. An example of bill advice is including a message that states "intention to disconnect services if the past due amount is not paid".

Billing Aggregator-A billing aggregator is a company that gathers billing records from one or more companies and posts them to another billing system. An example of a billing aggregator is the gathering of billing information from many companies that provide information or value added services (e.g. news or messaging services) and transferring these to another basic services carrier for direct billing to a customer. The customer pays the carrier, the carrier pays the aggregator, and the aggregator pays the value added service provider company. The billing aggregator performs services similar to a clearing house.

Billing and Accounting for Roaming Group (BARG)-Billing and accounting for roaming group is a core working group of the GSM Association that focuses on the intercarrier business/wholesale charging framework, administrative, and procedural issues related to roaming (visiting) customer network usage.

Billing and Customer Care (BCC)-A set of functions and processes related to generating customer invoices. These generally include Event (network usage) Rating, Invoicing, and tools to establish and maintain the customer profile.

Billing and custom care systems convert the transfer of bits and bytes of digital information within the network into the money that will be received by the service provider. To accomplish this, billing and customer care systems provide account activation and tracking, service feature selection, selection of billing rates for specific calls, invoice creation, payment entry and management of communication with the customer.

Billing and customer care systems are the link between end users and the telecommunications network equipment. Telecommunications service providers manage networks, setup the networks to allow customers to transfer information (provisioning), and bill end users for their use of the system. Customers who need telecommunication services select carriers by evaluating service and equipment costs, reviewing the reliability of the network, and comparing how specific services (features) match their communication needs. Because most network operations have access to systems with the same technology, the billing and customer care system is one of the key methods used to differentiate one service provider from another.

This figure shows an overview of a billing and customer care system that is used for communication services. This diagram shows the key steps for billing systems. First, the network records events that contain usage information (for example, connection time) that is related to a specific call. Next, these events are combined and reformatted into a single call detail record (CDR). Because these events only contain network usage information, the identity of the user must be matched (guided) to the call detail record and the charging rate for the call must be determined. After the total charge for the call is calculated using the charging rate, the billing record is updated and is sent to a bill pool (list of ready-to-bill call records). Periodically, a bill is produced for the customer and as payments are received, they are recorded (posted) to the customer's account.

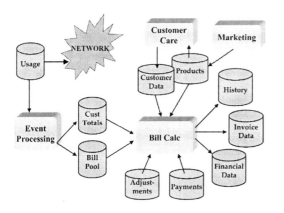

Billing and Customer Care (BCC) System

Billing Block-A billing block is the listing of credits (in a block format) for a program in public media such as newspaper ads, billboards, etc.

Billing Bureau-Billing bureaus are companies that provide the billing solutions as a completely outsourced 'service' to service providers.

Billing Call Type-A billing call type is a parameter used in rating a call. It is typically derived from a set of network elements recorded by the network (Ingress, Egress, Service Type, Routing, Special Features, etc.) Examples of a Call Type that are rated differently: Operator Assisted, International, Roaming, etc.

Billing Center (BC)-A billing center is a facility and is the associated equipment that is responsible for the billing of communication and/or information services.

Billing Company-A company that bills customers for services such as collect calls or long distance changes. The billing company may or may not be the same as the company that provides the service.

Billing Correlation Identifier (BCI)-A code (16-octets) that is used to identify a particular call that is made within the PacketCable system.

Billing Cycle-To distribute the bill processing requirements, billing systems divide customers into groups that allows their bills to be processed in specific cycles (or "billing cycles.") The billing cycles are different for groups of customers. This allows the billing system to only batch a portion of the billing records each time. These billing records must be forwarded for delivery (to a bill printer or for electronic distribution).

This diagram shows how a billing system can divide the invoicing process into billing cycles. In this example, each month is divided into 4 groups of billing cycles. At the end of each billing cycle, the billing records for each cus-

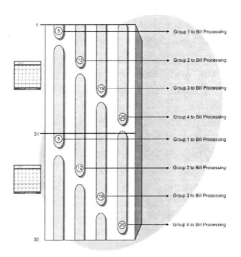

Billing Cycle Operation

tomer in a particular billing group are gathered, processed, and converted into invoices.

Billing Data-Billing data is information that relates to the purchase of products or the usage of services. For communication services, billing data usually includes call time, originating port number, calling telephone number, call class, dialed number, connection indicator, home/roam indicator (wireless), answer time, disconnect time, timeout, and other billing related information.

Billing Due Date-A billing due date is the date by which payment due must be received before the collections process is triggered. This date is typically dependent upon the invoice date, and is typically different for each cycle.

Billing Event-A measurable condition in a network that represents the usage of a network resource by a customer. Several billing events usually occur for each communications session (e.g. telephone call). Billing events are often supplied to a mediation device that combines the billing events into a single call detail record (CDR).

Billing Flow-Billing flow is a stream of billing related events and records that are combined to form billing records and invoices.

Billing Gateway (BG)-A billing gateway is a device or assembly that can process billing related data from one network, device or system into a form that that can be used by a billing system.

Billing Identifier (BillingID)-A billing identifier is a code or unique label that identifies data or information that relates to a telephone call or communication session.

Billing Increment-The smallest amount of time or resource that can be billed or charged. Billing increments include units of time (e.g. 1 minute or 6 second increments), number of messages, thousand bytes of data (kBytes), or other amount that can be calculated into a charge or criteria for a billing system.

Billing Information-Billing information is the data that is used to bill a customer. Billing information typically includes contact information, payment type and details and selected payment process.

Billing Interconnection Percentage (BIP)-The percentage of interconnection charge that is calculated based on the percentage of a communication route or amount of resources used when there are multiple carriers that are providing a service.

Billing Media Converter (BMC)-A device that transfers automatic message accounting (AMA) data from end offices to regional accounting offices using magnetic tape.

Billing Model-Billing models are the relationship of systems or processes that are used to gather usage data, service fees and payment information for users or accounts.

Billing Module-A billing module is a software program that is a self-contained process that can perform billing related functions such as record guiding and usage rating.

Billing Multiplier-A billing multiplier is a value that is used in determining a billed partial charge. For instance, if an average calculation is used, every month has an average of 30.417 days (365/12); therefore a customer who subscribes to a service on the 8th of the month will be charged (Y/30.417) x8 (where Y is the total charge for the month).

Billing Name and Address-The name and address provided to carrier (such as a local exchange company) by each of its customers to which the company directs bills for its services.

Billing Operations-Billing operations are the systems and processes that oversee the gathering, organizing, processing and billing related information.

Billing Options-Billing options are the charging or payment options for the sale of product or services. Some of the billing options for services include subscription accounts, metered (usage) billing and transactional (per event) billing.

Billing Partitions-Billing partitions are boundaries in a billing system or billing information that isolate the data or functions from other users or systems.

Billing Protocols-Billing protocols are sets of commands and procedures that are used for authentication, authorization and accounting for services that are provided on communication systems.

Billing Service Bureau-A billing service bureau is a company that provides billing services to other companies. The services provided can range from billing consolidation to complete billing operations that include gathering billing records, processing invoices, mailing or issuing the invoices, and posting payments.

Billing Stream-A billing stream is a sequence of billing records. A billing stream is sent to a billing process such as record guiding or rating.

Billing Support System (BSS)-Billing support systems are combinations of equipment and software that are used to allow companies to perform the service usage and product sales information. BSS systems can group this information for specific accounts for customers, producing invoices, creating reports for management, and recording (posting) payments made to customer accounts.

Billing System-A billing system is a combination of software and hardware that receives call detail and service usage information, grouping this information for specific accounts or customers, produces invoices, creating reports for management, and recording (posting) payments made to customer accounts.

This figure shows a standard billing process. In this diagram, the customer calls customer care or works with an activation agent to establish a new wireless account. The agent (customer care) enters the customer's service preferences into the system, checks for credit worthiness, and provides the customer with a phone number so that the customer may make and receive calls through the telephone network. As the customer makes calls, the connections made by the network (such as switches) create records of their activities. These

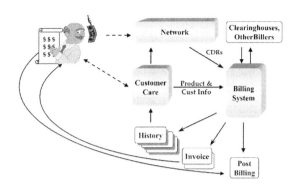

Standard Billing Process Operation

records include the identification of the customer and other relevant information that are passed onto the billing system. The billing system also receives records from other carriers (such as a long distance service provider, or a roaming partner). The billing system now guides and updates these call detail records (CDRs) to their correct customer and rating information. As information about the customer is discovered (e.g. rate plan), the updated billing records are placed in a billing pool so that they may be combined into a single invoice that is sent to the customer. The customer then sends his payment to the telecom service provider. Payments are recorded in the billing system. History files are then updated for the use of customer service representatives (CSRs) and auditing managers.

This figure shows standard processes used in a billing system. In this diagram, the customer calls customer care or a sales department and works with an activation agent to establish a new account. The agent (customer care) enters the customer's service preferences into the system, checks for credit worthiness, and provides the customer with a phone number so that the customer may make and receive calls through

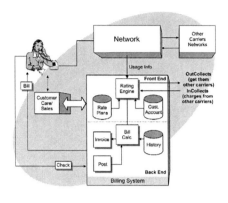

Billing System Operation

the telephone network. This diagram shows the billing system is divided into two parts; a front end (near real time processing) and a back end (periodic bill processing). The front end gathers call information as connections made through the network (such as switches) and uses these events create call detail records (CDRs) of the customer's communication activities. Each CDR includes the identification of the customer and other relevant information that are passed onto the billing system. The billing system also receives CDRs from other carriers (such as a long distance service provider). The billing system then guides the billing records to the correct account and rates the call using the rate tables. These updated billing records are placed in a billing pool. The back end of the billing system periodically combines records from the bill pool to create a single invoice that is sent to the customer. The customer then sends the payment to the telecom service provider. The payments are posted (recorded) in the billing system. History files are then updated for the use of customer service representatives (CSRs) and auditing managers.

Billing Tape-Monthly call detail records that are generated by the carrier. Resellers often use the billing tape to generate their own bills. (see Call Report).

Billing Telephone Number (BTN)-A number recorded by the switch on a Call Detail Record (CDR) identifying the party to be billed for the call. This number is not always necessarily a telephone number, it could be a Calling Card Number or any other number used to identify the party responsible for payment.

Billing Terms-Billing terms are the rules or conditions that are applied to the purchase of products or services.

Billing Unit-A billing unit is a value or characteristic that is used in the measurement of service or product usage. Examples of a billing unit include minutes, megabytes transferred or program viewings.

Billing Validation-Billing validation is a process that is used to ensure that billing records or other associated information is correct.

Billing Validation Application (BVA)-A billing validation application is a software program that is designed to perform processes that can validate billing records and service requests.

BillingID-Billing Identifier

Billings-Billings are the charges generated or received for products or services.

Billion Instructions Per Second (BIPS)-BIPS is a unit of measure for the billions of processing steps that can be accomplished by a computer-processing device in one second.

Binary-A numbering system based upon the powers of 2 used to represent data in digital computer systems. Binary data consists of a positional sequence of 1s and 0s. Each position is used to represent a specific power of 2. For example, the decimal number 19 is written in binary as 10011. This corresponds to (2 to the 4th power) + (2 to the 1st) + 2 to the 0th) which is 16 + 2 + 1 = 19.

Binary File-A binary file is a data segment where its information elements are in digital (binary) form. The term binary file is commonly associated with media files that require a player or decoder to convert the binary format that can be used or perceived by people or devices.

Binary Format-Binary format is the conversion of commands, data and/or information from one structure (such as text and/or other parameters) into a sequence of binary codes. Because these binary codes usually represent much longer sequences of information, binary formats can have much lower data transmission rates.

Binary/Hexidecimal (BinHex)-Binary/hexidecimal is an encoding structure that allows binary files to be sent in text file format.

BinHex-Binary/Hexidecimal

Biometric Access Control-Biometric access control is the process of authenticating access through the use of human measurements. Examples of biometric access control include voice printing, fingerprinting, handwriting analysis, and eye retinal scanning.

Biometric Device-A biometric device is a transducer (bio converter) that can sense and analyze physical characteristic of a human. An example of a biometric device is a fingerprint or retinal scanner.

Biometrics-Biometrics is the measurement of identifying life characteristics or behavioral traits such as fingerprints, retina patterns or voice prints

BIOS-Basic Input/Output System

BIP-Billing Interconnection Percentage

BIP-Borderline Interconnection Percent

BIPS-Billion Instructions Per Second

Biscuit-A term sometimes used for external wall mounted RJ-45/RJ-48 jack.

B-ISDN-Broadband Integrated Services Digital Network

Bit (b)-A bit is the most fundamental and widely used form of digital signals are binary signals in which one amplitude condition represents a binary digit 1 and another amplitude condition represent a binary digit 0. Thus a binary digit or bit is one of the members of a set of two in a numeration system that is based on two and only two possible values or states.

Bit Error Rate (BER)-BER is calculated by dividing the number of bits received in error by the total number of bits transmitted. It is generally used to denote the quality of a digital transmission channel.

BIT Rate-A measurement of the transfer rate of digital signals through a channel. The bit rate is the number of bits transmitted in a specified amount of time. Bit rate is usually expressed as bits per second (bps).

Bit Torrent (BitTorrent)-A bit torrent is a rapid file transfer that occurs when multiple providers of information can combine their digital data (bits) transfer into a single stream (a torrent) of file information to the receiving computer.

Bitrate-Bit rate is the number of bits that pass (transfer) between corresponding equipment in a data transmission network.

Bits Per Inch (BPI)-The number of bits that a magnetic tape or computer tape cartridge can store per linear inch of length.

Bits Per Second (BPS)-The common measurement for data transmission that indicates the number of bits that can be transferred to or from a communications device in one second.

Bitstream-The stream of compressed data that is the input to a decoder, especially for compressed audio or video.

BitTorrent-Bit Torrent

Black Box-A black box is a functional unit, assembly or software module that performs a defined process. Because the process of the black box is defined, a designer of a system or product may not need to know the inner workings of the black box to use it in their design.

B

Black Hole-A black hole is a system or device where signals or data may enter but they do not leave.

Black Market-A black market is a trading environment or place that is used to obtain goods that are not available or unauthorized for sale.

Blackout Area-A blackout area is a geographic region that does not or is not allowed to provide communication signals. Examples of blackout areas are cities that do not broadcast local sports games to encourage residents to attend the games rather than watching the games on television.

Blade-A blade is a card or module that can be placed into a backplane of a communication system.

Blanket Authorization Agreement (BAA)-A blanket authorization agreement is a document that defines people and/or companies who agree to perform a broad set of actions or provide authorizations (terms) and the value(s) that are exchanged in return for these terms.

Blanket Order-A blanket order is an authorization to perform an action such as to send products as they become available or when they are needed and to invoice for the contents of the order. A blanket order contains details of who ordered, what was ordered, the method of payment, and where it is shipped. The details of the blanket order are added as the products and services are shipped.

BLC-Broadband Loop Carrier

BLEC-Broadband Local Exchange Carrier

Blended Agent-A blended agent is a telephone person working in a call center who can receive calls from outside customers in addition to making outgoing calls.

Blended Payment-Blended payment is the combination of two or more payment types to complete a financial transaction.

Blended Services-Blended services are the mixing of services that have different fulfillment and operational requirements.

Blending-Blending is the process of routing outbound and inbound phone calls to the same agents in call centers.

BLES-Broadband Loop Emulation Service

Blind Carbon Copy (BCC)-Blind carbon copy is the address or addresses of recipients of a message (such as an email message) who will receive a copy of the message without informing others on the distribution list that they received it (they are "blind" to the copy).

Blind Dialing-Blind dialing is the process of dialing a telephone number without verifying that a dial tone is present on the line.

Blind Transfer (BT)-The process of transferring a call to another extension or phone number without telling the person who's calling that they are being transferred. Blind transfers are sometimes called cold transfer or unsupervised transfer.

Block Call-Block call is a call processing feature that allows a telephone user to block incoming calls from specific telephone numbers.

Block Diagram-A high level abstraction of the various functions or components of a network, system or circuit depicted using simple geometric figures. Lines and arrows between the geometric figures denote an interaction of some type between the connected blocks. Block diagrams are typically used to simplify the understanding of a complex system.

Block Sequence Number (BSN)-A block sequence number is a value or code that identifies a specific range of data elements (a block) in a data file or data stream.

Blockage-The temporary lack of access because of high traffic in a switching system or in a subscriber line concentrator.

Blocked Attempt-An attempted call that cannot be further advanced toward its destination because of an equipment shortage or failure in a network.

Blocked Currency-A blocked currency is a monetary unit that cannot be transferred or used in certain areas (such as outside the country which controls the currency).

B

Blocking-Blocking is a condition that occurs in a communication network when all permitted trunk paths or circuits are in use or busy that prevents a user from accessing services. In the case of telephone call blocking, a message may be returned to the customer that the system is unavailable.

Blocking IPR-An intellectual property right (IPR), patent, copyright or other right proprietary to an individual, group or company, which precludes someone else from making, using or selling that invention.

Blocking Probability-The percentage of calls that cannot be completed within a one hour period due to capacity limitations. For example, if within one hour 100 users attempt accessing the system, and 10 attempts fail, the blocking probability is ten percent.

Blocking Ratio-For a group of servers, the ratio of unsuccessful access attempts compared to total access attempts within a specified time interval.

Blog Community-A blog community is a group of people who participate in blogging (message logging) activities related to specific subjects (blogs).

Blogging Platform-A blogging platform is the combination of hardware and software that is used to receive, process and provide message logs (blogs).

Blue Screen Of Death (BSOD)-A blue screen of death is a graphic display that indicates when the Microsoft Windows operating system has encountered errors so serious that the operating system could not continue to operate.

Blue Sky Laws-Blue sky laws are rules and regulations that are designed to protect investors from excessive valuation or overstated claims of future compensation that may result from investments.

Bluetooth-Bluetooth is a standardized technology that is used to create temporary (ad-hoc) short-range wireless communication systems. These Bluetooth wireless personal area networks (WPAN) are used to connect personal accessories such as headsets, keyboards and portable devices to communications equipment and networks.

The Bluetooth system can dynamically discover and connect to other nearby devices (if the devices are setup to be discoverable). To setup a Bluetooth system, one of the devices is designated as a master (controlling) and the other devices are setup as slaves (responding to the commands of the master). When a small Bluetooth system has been setup, it is called a Piconet.

This figure shows the typical service discovery operation of a PDA device that desires to print to a nearby printer. In this example, a PDA unit (acting as a master unit) sends out an inquiry message to a printer (a slave unit). The printer (and possibly other devices) responds with its' Bluetooth device address. The PDA then sends a connection request using the printer's Bluetooth address along with a request for the capabilities of the printer. The printer then returns the capabilities requested (if available) from its SDP database and these capabilities are temporarily stored in the SDP database of the PDA. The PDA can now display the availability of the printer to the user. This allows the user to select the printer of choice that has specific printing capabilities (such as laser or color printing).

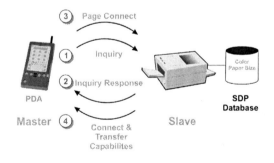

Bluetooth Discovery

This diagram shows how a Bluetooth system can connect multiple devices on a single radio link. This diagram shows that a laptop computer has requested a data file from a desktop computer. When this laptop computer first requests the data file, it accessed the Bluetooth radio through a serial data communication port. The serial data port was adapted to Bluetooth protocol (RFComm) and a physical radio channel was requested from the local device (master) to the remote computing device (slave). The link manager of the master Bluetooth device requests a physical link to the remote Bluetooth radio. After the physical link is created, the logical link controller sends a message to the remote device requesting a logical channel be connected between the laptop computer and the remote computer. The logical link continually transmits data between the devices. In this diagram, the user then requests that a CD Player send digital audio to a headset at the remote computer. Because a physical channel is already established, the logical link controller only needs to setup a 2nd logical link between the master Bluetooth device and the remote Bluetooth device. Now data from the CD ROM will be routed over the same physical link between the two Bluetooth devices.

This diagram shows the basic radio transmission process used in the Bluetooth system. This diagram shows that the frequency range of the Bluetooth system ranges from 2.4 GHz to 2.483 GHz and that the basic radio transmission packet time slot is 625 usec. It also shows that one device in a Bluetooth piconet is the master (controller) and other devices are slaves to the master. Each radio packet contains a local area piconet ID, device ID, and logical channel identifier. This diagram also shows that the hopping sequence is normally determined by the master's Bluetooth device address. However, when a device is not under control of the master, it does not know what hopping sequence to use so it listens for inquiries on a standard hopping sequence and then listens for pages using its own Bluetooth device address.

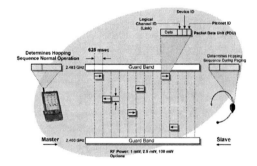

Bluetooth Radio

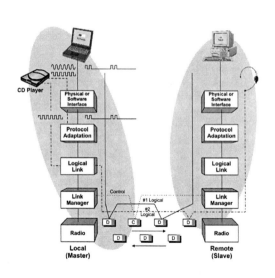

Bluetooth System Operation

BMC-Billing Media Converter
BoD-Bandwidth On Demand
BOF-Business Operations Framework
BOIC-Barring of Outgoing International Calls

Boilerplate-Boilerplate is a sample structure or document that is used to assist in the creation and formatting of other documents.

Bold Text-Bold text is a text feature that darkens and/or expands text to add emphasis. Bold text may be a markup attribute tag that is inserted in a document.

BOM-Beginning of Message

BOM-Bill Of Materials

Bonded DSL-A bonded DSL system combines (bonds) two or more copper lines to provide for higher-data transmission rates. Bonded ADSL systems can combine up to 32 individual communication lines to provide extremely high-speed data communication

Bong-A unique tone that telephone carriers add to an audio signal while a call is in progress to alert the user that another action on the users part is required (such as to dial an access code number for a calling card).

Booked Space-Booked space is the amount of ad views that are already sold.

Bookings-Bookings are reservations for resources that will be required to provide services (such as broadcasting television programs) in the future.

Bookmark-A bookmark is a code or label that identifies (tags) a specific location or entry point within a document or media file.

Boolean-Boolean is a logical mathematical operation. Boolean operators include logical conjunction (AND), logical inclusion (OR), exclusive or (XOR), and logical negation (NOR).

Boot-(starting a computer and/or loading an operating system into memory. (2-cable) A boot is a protective case that covers the physical attachment of a connector to a cable. A boot may only serve a cosmetic (appearance) or it may provide some mechanical isolation of cable to the connector minimizing the effects of pulling or wiggling of the cable.

Booting-Booting is the process of initially loading the operating instructions for a computing device. These initial instructions are loaded from long term memory modules in the computing device by a boot loader program.

Bootstraping-Bootstraping is a process of starting a project or company with a minimal amount of resources.

BOQ-Bill of Quantity

Border-A perimeter of a system or network. Borders can be physical or logical boundaries that can be crossed at locations known as border gateways.

Borderline Interconnection Percent (BIP)-Borderline interconnection percent is the ratio of the billable portion of jointly provided service (interoffice mileage).

Bottleneck-A point in a data communications path or computer processing flow that limits overall throughput or performance.

Bottom Line-The bottom line is a final measure of profit or cost (the bottom line of a financial statement). The phrase bottom line is commonly used to indicate the net profit of a company or a program.

Bouncing Busy Hour-A busy hour that, over several days, doesn't occur at the same time each day.

Boutique-A boutique is a company or group of products or services that have a small by related group of products or services.

BPI-Bits Per Inch

BPL-Broadband Over Powerline

BPO-Business Process Outsourcing

BPON-Broadband Passive Optical Network

BPS-Bits Per Second

BR-Basic Rate

Brain Drain-Brain drain is the process of hiring smart people from other companies resulting in the reduction in brainpower for that company.

Brainstorming Session-Brainstorming is an activity that involves the exchange of related ideas, terms, or images in a session to help stimulation thoughts that will lead to the ability to solve a problem or achieve an objective.

BRAN-Broadband Radio Access Networks

Brand-The overall image or concept of the meaning or value of a word, name, or address (such as a web address) by consumers. Consumers use brand concepts to identify the value of products or services.

Brand Equity -Brand equity is the term used to describe the extra value that a company or product can command in the marketplace because of the branding activity associated with it.

Brand Name-A brand name is a word or symbol that identifies a product or services.

Brand Recognition-Brand recognition is ability to recognize a name or identifying trade logos promoted by a company or person.

Brand Value-Brand value is a measure of the overall image or concept of the meaning or value of a word, name, or address (such as a web address) by consumers.

Brandable-Brandable is the ability to create an awareness of a company (corporate brand), product (product brand), or service (service brand) to people or companies.

Branding-Branding is the process of creating an awareness of a company (corporate brand), product (product brand), or service (service brand) to the purchasers, users, or influencers of products or services.

Breach-A breach is loss of security or operational functions that result of an attack or other unintended process that disables or penetrates a system or security system.

Breadcrumb Trails-Breadcrumb trails are a sequence of data that is left behind by web site visitors (possibly as a result of specific actions such as moving a mouse over an icon) that can be used by marketing professionals to help them understand the shopping and buying patterns of customers.

Break Even Analysis-A break even analysis is the calculation that uses the profit contribution of product sales (receipts less cost to product) to determine the approximate number of product sales that are necessary to recover the development, production and distribution costs.

Break Event Point (Breakeven)-Breakeven is a point or a number where the cost of developing and selling a product or service has been recovered from the profits generated by the sale of the product or service.

Breakage-A final balance on a prepaid service that is never used.

Breakeven-Break Event Point

Break-In-Busy Override

BRI-Basic Rate Interface

Brick and Mortar-Brick and mortar is a term used to describe a business that has one or more physical locations that sell products and services.

Bridge-A bridge is a data communication device that connects two or more segments of data communication networks by forwarding packets between them. Bridges extend the reach of the LAN from one segment to another. Bridge devices operate at layer 2 of the OSI reference model, used to connect two or more LAN segments. Bridges provide more intelligence and fault isolation than repeater, which operate at Layer 1. However, unlike routers, which operate at Layer 3, bridges do not provide broadcast domains.

Bridges operate by inspecting each packets MAC header and forwarding the packet based only on this information. Each bridge keeps a table which enables it to determine the egress port to which each packet should be forwarded. In general, there are two classes of bridges. Transparent bridges are commonly found in Ethernet-like networks and make use of only MAC addresses. Source Route bridges are commonly found in Token-Ring networks and utilizes a Routing Information Field in the MAC header to forward packets.

The terms Routing Information Field and Source Routing are slightly misleading. While these terms have the word "route" in them, these are purely Layer 2 protocols and should not be confused with Layer 3 routing protocols such as RIP or OSPF.

This figure shows the basic operation of a bridge that is connecting 3 segments of a LAN network. Segment 1 of the LAN has addresses 101 through 103, segment 2 of the LAN has addresses 201 through 203, and segment 3 of the LAN has addresses 301 through 303. The table contained in the bridge indicates the address ranges that should be forwarded to specific ports. This diagram shows a packet that is received from LAN segment 3 that contains the address 102 will be forwarded to LAN segment 1. When a data packet from computer 303 contains the address 301, the bridge will receive the packet but the bridge will ignore (not forward) the packet.

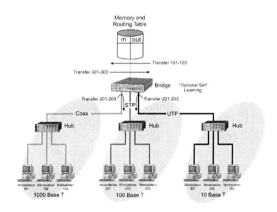

Data Bridge Operation

British Telecommunications Act-The British telecommunications act is a set of laws that was enacted in 1981 that separated telecommunications operation from the post office and created British Telecommunications (BT).

Broadband-(1-data transfer) A term that is commonly associated with high-speed data transfer connections. When applied to consumer access networks, broadband often refers to data transmission rates of 1 Mbps or higher. When referred to LANs, MANs, or WANs, broadband data transmission rates are 45 Mbps or higher. (2-radio bandwidth) A frequency bandwidth that is much larger than the required bandwidth to transfer the information signal. For example, using a 1 MHz wide radio channel to transmit a 4 kHz limited audio signal.

Broadband Access System-A broadband access system is a portion of a high-speed communication system (over 1 Mbps) that coordinates requests for services and enables the transfer of information when authorized.

Broadband Communications-Broadband communication service is the transfer of digital audio (voice), data, and/or video communications at rates greater than wideband communications rates (above 1 Mbps).

Broadband Integrated Services Digital Network (B-ISDN)-Usually refers to the portions of a digital network operating at data transfer rates in excess of 1.544 or 2.048 Mbps. The B-ISDN network often uses ATM to enables transport and switching of voice, data, image, and video over the same network equipment.

Broadband Local Exchange Carrier (BLEC)-A service carrier (provider) that offers broadband services locally. Broadband services are usually defined as the ability to transmit a large amount of information, including voice, data, and video, that has a combined data transmission rate that exceeds 1 Mbps.

Broadband Loop Carrier (BLC)-Broadband loop carrier is a company or system that provides broadband services to customers.

Broadband Loop Emulation Service (BLES)-Broadband loop emulation service is the end-to-end architecture that allows DSL systems to provide voice and data over frame-based DSL lines.

Broadband Network-A network that is capable of transmitting a large amount of information, including voice, data, and video that have a combined data transmission rate that exceeds 1 Mbps. Sometimes called wideband transmission, it is based on the same technology used by cable television.

Broadband Over Powerline (BPL)-Broadband over powerline is the transmission of broadband data signals (over 1 Mbps) over power lines. BPL can refer to sending broadband over high voltage or low voltage power lines.

Broadband Passive Optical Network (BPON)-A broadband passive optical network (BPON) is an ATM based broadband optical network that combines, routes, and separates optical signals through the use of passive optical filters that separate and combine channels of different optical wavelengths (different colors).

Broadband Radio Access Networks (BRAN)-A broadband radio access network is a portion of a communication network that allows individual subscribers or devices to connect to the network via radio signals and obtain broadband services (greater than 1 Mbps).

Broadband Router (Brouter)-A broadband router is a packet switching assembly that is able to receive and transfer packets in a broadband network (1 Mbps+).

Broadband Signal-A term that is used to describe high-speed data communication signals. For consumer data communication, broadband signals have a data transmission rate above 1 Mbps are usually considered Broadband signals. For telecommunication or data networks, data transmission rates above 45 Mbps (OC1) are considered broadband.

Broadband Telephony-The use of a broadband data connection (such as high-speed DSL or cable modem Internet connection) to make telephone calls.

Broadband Transmission-Broadband transmission is the transport of information on a broadband communication channel.

Broadband TV (BBTV)-Broadband television is the delivery of digital television services over broadband data connections. Broadband television systems may be able to control and guarantee the quality of television services if the underlying broadband connections have enough bandwidth.

Broadband Video-Broadband video is the transfer of digital video information through a broadband communication system. Broadband video is typically delivered at a data rate above 1 Mbps.

Broadband War-The broadband war is the conflict that is occurring between multiple types of communication service provider companies who can each provide broadband (1 Mbps+) communication services. Key tactics in the broadband war include increasing the performance of broadband connections (higher data transmission rates) and declining costs of broadband service. Broadband competitors include telephone companies (DSL), cable television (Cable Modems), mobile telephone (EVDO, HSPDA), fiber systems (FTTH or FTTC) and fixed wireless (WiMax).

Broadband Wireless-Broadband wireless is the transfer of high-speed data communications via a wireless connection. Broadband wireless often refers to data transmission rates of 1 Mbps or higher.

Broadband Wireless Access (BWA)-Broadband wireless access is a term that is commonly associated with wireless high-speed data transfer connections. When applied to consumer access networks, broadband often refers to data transmission rates of 1 Mbps or higher.

Broadcast-Broadcasting is the transmission of an information signal to a specified geographic area or network. This allows the same information to be received by all customers in that geographic area that can successfully receive (demodulate) and decode the information.

This diagram shows two broadcast examples: radio broadcast and network broadcast. Part (a) shows a radio broadcast tower that is sending an audio broadcast to all radios that are within its radio signal coverage area. Part (b) shows a network broadcast system that sends a data message that is coded to indicate the message is a broadcast message. This message contains an address that indicates it is a broadcast message. When routers or other data distribution devices receive this message, each distribution device forwards the data broadcast message to the other network parts for which it is connected to. All communication devices that are connected to the network can receive the broadcast message.

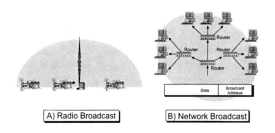

Broadcast Communication Operation

Broadcast Advertising-Broadcast advertising is the sending of the same advertising message to all the recipients (such as television viewers) who can receive the broadcast signal.

Broadcast and Multicast Services (BCMCS)-Broadcast and multicast services are broadcasting services that can be provided over (overlayed) existing CDMA mobile communication systems. The BCMCS system is a two-way broadcasting system that allows the user to interact with the broadcast network. The BCMCS system can provide bearer services (multicasting and broadcasting media) and user services (controlled media streaming).

Broadcast Distributor-A broadcast distributor is a system or company that distributes information through a broadcast (e.g. television radio signals) transmission system.

Broadcast Media-Broadcast media is the continuous transfer of media information that is sent through a communication system. Broadcast media commonly refers to audio and/or video images that are sent through a wireless network (such as a radio or television broadcast system).

Broadcaster-A broadcaster is a company that transmits or provides information to users that are connected or able to access signals on the broadcast network.

Broadcasting-Broadcasting is a process that sends voice, data, or video signals simultaneously to group of people or companies in a specific geographic area or who are connected to the broadcast network system (e.g. satellite or cable television system). Broadcasting is typically associated with radio or television radio transmission systems that send the same radio signal to many receivers in a geographic area. Broadcasting can also be applied to distribution or point to point networks where all users that are connected to the network can receive the same information signal.

Broadcasting Service-Broadcasting service is the transmission of information signals (audio, video and/or data) that is intended for direct reception by multiple receivers (e.g. consumers).

Broadcatching-Broadcatching is the process of consumers selecting, capturing and consuming the media they want from an endless array (broad selection) of choices.

Brochureware-Brochureware are images or brochures that are presented on a web site.

Broken Link-A broken link is a hyperlink that has a path that can no longer connect to a file or other resource.

Brouter-Broadband Router

Brownfield-Brownfield is the existing material such as equipment, architectures and procedures that is already used in a system or network.

Brownout-A reduction in the servicing of customers as a result of the demand for service exceeding the service processing capability of the service provider's equipment or staff. Brownout usually occurs during a peak period. Because service access attempts by customers increase during a brownout period (customers repeatedly attempt to get service), service providers may discontinue services to groups of customers during brownout.

Browser-A software program or module (called a client) that is used to convert information that is available on the Web portion of the Internet into forms usable by a person (text, graphics and sound). Also called a web browser.

Browser Plug In-A browser plug-in is a software program that can installed or linked to a web browser application to enhance its capabilities. An example of a browser plug-in is a media player plug-in. The media player decodes and reformats the incoming media so it can be displayed on the web browser.

Brute Force Attack-A brute force attack is the process of attempting to gain access to a system or to decode encrypted information through the repeated trying of potential keys to identify the correct key.

BS-Base Station

BS-Bearer Services

BSCS-Business Support And Control System

BSE-Basic Service Element

BSN-Block Sequence Number

BSOD-Blue Screen Of Death

BSP-Backbone Service Provider

BSS-Billing Support System

BT-Blind Transfer

BTA Authorization Holder-A basic trading area authorization holder is the individual or company that is authorized to provide radio service to the population of a BTA.

BTA Service Area-The basic trading area service area is the geographic area within the boundaries of BTA to which a radio license holder may provide radio services (e.g. MMDS service).

BTN-Billing Telephone Number

Buckslip-A buckslip is a financial bill (usually a small amount) that is inserted into a mailing item that helps to motive the recipient to respond to the offer or request.

Budget Management Discipline -A process utilized by telephone company (telco) management personnel to get departments and project managers to adhere more tightly to standardized proposal, funding and follow-up procedures for budgeted projects and organizational units.

Buffer-(1-memory) An allocation of memory storage that set aside for temporary storage of data. The use of memory buffers allow data transfer rates to vary so that differences in communication speeds between devices do not interrupt normal processing operations. (2-optical) A layer placed between a fiber and its jacket to provide additional protection to the fiber. The layer is often made of a thermoplastic material.

Bug-(1-problem) A problem that occurs in software or hardware that can be fixed by changes to the software or design changes to the hardware. The term seems to have started from a problem that occurred with an early model computer system when a moth got caught inside the machinery resulting in a bug problem. (2-audio) A microphone or listening device that is concealed for audio surveillance.

Bug Rate-Bug rate is a measure of how many new software errors or process challenges are discovered (and potentially corrected) over a period of time.

Bulk Basis-Bulk basis is a billing system that is based on the billing name and address information for all the local exchange service subscribers of local exchange carrier.

Bulk Billing-A method of billing telephone customers in which the charges for all messages of a given type, chargeable to a particular account for a billing period, are combined and billed as a single amount. Message-unit billing is the most frequent application of the bulk-billing procedure.

Bulk Buying-Bulk buying is the process of buying products or services in large quantities to obtain them at a lower average cost.

Bulk Mail-Bulk mail is a category of mail services that processes mailing items in group (bulk) form. Bulk mailing typically requires specific markings and mail items that are similar.

Bulk Orders-Bulk orders are the submission, processing or creation of multiple orders in a standard or predefined format. Bulk orders may be used for enterprise to assist in the entry for orders that have similar data or formats such as service orders for new employees.

Bulk Postage Rate (Bulk Rate)-Bulk postage rate is a reduced level postage fee that is provided in return for the use of bulk mailing.

Bulk Rate-Bulk Postage Rate

Bullet Proof (Bulletproof)-Bullet proof is the ability of a system or process to deter or remain unchanged to the effects of unauthorized users or processes.

Bulletproof-Bullet Proof

Bullwhip Effect-A bullwhip effect is a business process where a change in a portion of a value chain affects other parts of the processes.

Bundled-Bundled is the combination of one or more products or services to create a new combined product or service. Combining several products or services may offer the customer or user a better pricing option.

Bundled Offer-A bundled offer is a promotion of a set of product or services. Bundled offers usually offer a benefit (such as price or shipping cost reduction) of purchasing two or more products together.

Bundled Rates-Bundled rates are the cost assignments for products or services that are composed of two or more underlying products or services.

Bundling-Bundling is the process of combining different products and services into a "package" offer, and then offering it to a customer at a separate, combined price.

Bundling Services-The combining of different services into one service offering so the customer can communicate with one company for several different services. An example of bundling is the combination of cellular, PCS, local and long distance services as one service package.

Burn In-Burn in is the process of operating newly produced devices to order ensure they operate as expected. Burn in is commonly performed to discover products with early lifecycle failures. Burn in may be performed in harsh environmental conditions such as high-heat or high-humidity.

Burn Out-Burn out is a condition where stress causes people or workers to become apathetic or less responsive.

Burn Rate-Burn rate is the speed at which a company uses up its available cash when developing or launching new products.

Business Applications-Business applications are processing systems (such as software programs) that are used to perform of business related functions.

Business Continuity Planning (BCP)-Business continuity planning is the processes of planning for changes in operations that may be necessary to maintain business functions during major disruptions (such as an Earthquake, fire at data centers or operations centers or a terrorist attack). BCP generally involves moving the use traffic and services to a geographically distant location or to partner's resources.

Business Drivers-Business drivers are the characteristics of a business or external factors that influence a business that can be used to characterize or classify its performance or likely future market potential. Examples of business drivers include operations efficiency, patents, new products or services and the consumer awareness.

Business Group-A collection of lines that serve a single business location, share a common number dialing plan and are assigned other business features. It is commonly called a Centrex group.

Business Logic-Business logic is a set of rules and processes that are based on discrete values (such as if or when events occur) which are used to determine or define business actions.

Business Metrics-Business metrics are the Key Performance Indicators (KPIs) of the business functions. Key metrics for telcos are ARPU (Average Revenue Per User), Cycle time (mean time for order to activation, order to bill, and lead to cash), first time right, CAPEX (Capital Expenditure), OPEX (Operational Expenditure). The TMForum GB935 covers the business metrics framework.

Business Model-A business model is a representation of the processes of a business where the representations may be in mathematical formulas or as functional descriptions of processes that are used to define how a business will generate revenues or profits. Business models are used to simulate what a business will accomplish when it is provided with various conditions or opportunities.

Business Operations Framework (BOF)-Business operations framework is a hypothetical or desired structure that is used to plan or describe the business aspects of a system as they relate to the operational components of the business.

Business Plan-A business plan is a document that defines the objectives of a business, what resources it needs and how it expects to use these resources to achieve its objectives.

Business Process-A business process is a set of steps, conditions and functions that are used to enable or perform business operations. Key business processes in the telecommunications industry include lead to cash (L2C), trouble to resolve (T2R)) and concept to market (C2M). The Tele Management forum NGOSS - eTOM framework defines the processes under strategy, infrastructure, product (SIP), operations (readiness, fulfillment, assurance and billing) and enterprise.

Business Process Management-Business Process Management is a management model that allows the organizations to manage their processes as any other assets improve and manage them over the period of time. The activities which constitute business process management can be grouped into five categories including design, modeling, execution, monitoring, and optimization.

Business Process Outsourcing (BPO)-Business process outsourcing is the contracting of a specific business tasks to a third-party service provider. Usually, these are the tasks that a company requires but does not depend upon to maintain their position in the marketplace. BPO is often divided into two categories that include back office outsourcing, which includes internal business functions such as payroll, billing or purchasing, and front office outsourcing, which includes customer-related services such as marketing or tech support.

Business Rating-Business rating is the evaluation of the characteristics of a company by another person or company to provide a rate value that indicates the ability or likelihood of a company to perform actions (such as the ability to pay bills or produce products).

Business Rules-Business rules are the requirements, processes and actions that are created to manage business operations.

Business Strategy-The objectives and planned actions that will contribute to the successful operation of a business.

Business Summary-A business summary is a short description (typically 1 to 2 pages) of a business. A business summary typically includes the key problems or needs the business satisfies, how the business solves these needs and why the business is likely to succeed.

Business Support And Control System (BSCS)-The billing system used in the GSM network.

Business to Business (B2B)-Business to business is the operations and communication that facilitate business between a business and other businesses.

Business to Business Advertising (B2B Advertising)-Business to business advertising is the process of directing communication messages to businesses that buy products or services.

Business to Consumer (B2C)-Business to consumer is the operations and communication that facilitate business between a business and consumers.

Business to Distributor (B2D)-Business to distributor is the operations and communication that facilitate business between a business and distributors.

Business Web Transaction Processing (BWTP)-Business web transaction processing is the set of tasks and processes that are used to interact and coordinate business transactions over the Web.

Busy Hour-The hour in a day when the total usage of the network, trunk connection, or the switching system is greater than at any other hour during the day. Telephone systems and networks are typically designed to meet a specific quality level that can be provided during the busy hour (e.g. maximum number of blocked calls).

Busy Hour Call Attempts (BHCA)-Busy hour call attempts is the number of call or session setup attempts that occur during the hour of the day that has the highest amount of service usage.

Busy Hour Call Completion (BHCC)-Busy hour call completion is a measurement of the number of successful call completions that occur during a busiest hour of the day.

Busy Hour Minutes (BHM)-The number of minutes during "busy hours" ordered by a customer. Other terms: BHMOT represents originating usage, BHMTT represents terminating usage.

Busy Hour Minutes of Capacity (BHMC)-The maximum amount of access minutes that an interconnecting carrier (such as an IXC) expects to route through end office switch at peak activity.

Busy Hour Traffic (BHT)-Busy hour traffic is the amount of call traffic in hours that occur during the busiest time of the day.

Busy Hour Usage Profile-Busy hour usage profile is the characteristics of how a system or service that occurs during the hour during the day that has the highest service usage.

Busy Override (Break-In)-A feature that allows access to a busy number for an emergency.

Busy Tone-An audible tone or signal that indicates to a caller that their call cannot be received because a called line is unavailable (in use).

Butt Set-A butt set is a telephone set that used as a test instrument by telephone technicians. The butt set usually has a set of alligator clips or test probes instead of a standard telephone plug to permit the direct connection to the wires test terminals.

Button Ad-A button ad is a selectable graphic or image, which is located on a web site which appears as a button.

Buy Data-Buy data is information that defines purchases of products or services.

Buyer's Market-A buyer's market is a condition of an industry or geographic region that the demand for products or services is low. This provides buyers with additional negotiating value.

Buying Cycle-A buying cycle is the processes and tasks that are used by a person or company when they are buying or replacing a product or service.

Buying Decision-A buying decision is the facts and/or emotional elements that are considered when making a decision to buy or not buy a product or service.

Buying History-Buying history is the purchase records of a person of company over a period of time. The buying history can be used to evaluate the types of products or services that a customer may be interested in purchasing more of.

Buzzword-A buzzword is a term that is invented for a specific use or technology to make it seem more exciting, unique, useful or proprietary

BVA-Billing Validation Application

BWA-Broadband Wireless Access

BWTP-Business Web Transaction Processing

Bypass-Bypass service is the routing of calls or communication sessions around any other networks facilities to avoid toll charges. Bypass is commonly discussed with the ability to bypass local exchange carriers (LEC) to save on interconnection charges.

Bypass LAN-The capability of a station to be electronically or optically isolated from a network while maintaining the integrity of the network ring. This is found in the newer hub-based token ring networks.

Byte (B)-A group of bits, typically eight, used to represent data values from 0 to 255. These values may represent alphabetic or control characters or numbers less than 255. A single byte is typically the smallest data value manipulated by a computer. In order to represent larger values, multiple bytes are grouped together into words that are 16, 32, 64 or 128 bits in length.

C

C+ Or C++ Programming Language-A high-level programming language that is commonly used to create commercial software programs.

CA-Call Agent

CA-Certificate Authority

CA-Certification Authority

CA-Conditional Access

Cabinet-In communication systems, a cabinet is an enclosure that is used to hold equipment or electronic assemblies.

Cable-A cable is a grouping of one or more transmission lines (such as copper wire, coax, or optical fibers) into a single sheathed (coated) line.

Cable Act of 1984-The cable act of 1984 defined new regulations for the cable industry, which included the requirements for renewing franchise agreements.

Cable Assembly-A cable assembly is composed of one or more cables (multiple transmission lines) with connection points (usually connectors or terminations).

Cable Box-(1-optical or electrical) A cable box is a container assembly that holds wires, cables and devices. (2-television) An electronic device that adapts a communications medium to a format that is accessible by the end user. Set top boxes are commonly located in a customer's home to allow the reception of video signals on a television or computer.

Cable Budget-A cable budget is the maximum length of cable that is allowed between a users equipment and the modem or concentrator in a system that will provide signals to the equipment.

Cable Cap-A cable cap or water seal is a covering that is applied to the end of a cable to seal the end of the cable from the entry of water and other unwanted substances.

Cable Card (CableCARD)-A cable card is a portable credit card size device that can store and process information that is unique to the owner or manager of the cable card. When the card is inserted into a cable card socket (such as into a set top box), electrical pads on the card connect it to transfer information between the electronic device and the card. Cable cards are used to identify and validate the user or a service and may contain decryption keys that allow for the descrambling of cable transport channels.

Cable Converter-Cable converters, commonly called a "set top" box are electronic devices that convert an incoming cable television signal into a form that can be displayed on a video device typically a television or computer. The set-top box is typically located in a customer's home to enable the reception and/or interaction with services on the customer's television or computer.

Cable Distributor-A cable distributor is a system or company that distributes information through a cable television (CATV) transmission system.

Cable Drop-A cable drop is a segment of cable or communication line that runs from a communication network to a home or business.

Cable Duct-Cable ducts are pipes or conduits that are installed underground or in a building, whose purpose is to protect the cables installed within them.

This figure shows cable ducts that are inside a duct. These cable ducts are hollow and the allow cables to be pulled through the cable after the duct has been installed.

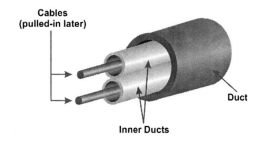

Cable Ducts

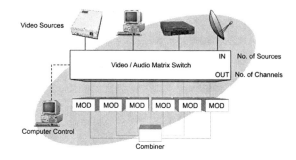

Head End System

Cable Enclosure-A cable enclosure is a plastic or metal container that is used to cover and protect wires or cables. Cable enclosures may be used to protect the ends of cable or to protect splices. Cable enclosures may contain multiple container shells. An outer shell may be used to provide mechanical and environmental protection and the inner shell may be used to hold the cables or a splice tray.

Cable End Box-A cable end box is a container assembly that is used to provide access for pulling cables and wires through conduits that are connected to the cable end box.

Cable Head End-The portion of a cable television communication system that receives, formats, and transmits the carrier signals to the distribution system. Because cable television companies have started to offer two-way telecommunication services such as Internet and telephone service, the head end equipment has been expanded to include voice and data gateways.

This figure shows a diagram of a simple head-end system. This diagram shows that the head-end allows the selection of multiple video sources. Some of these video sources are scrambled to prevent unauthorized viewing before being sent to the cable distribution system. The video signals are supplied to video modulators that convert the low frequency video signals into their radio frequency televi-

sion channel. The output of each modulator is combined and connected to the distribution trunk.

Cable Home Wiring-Cable home wiring is the internal wiring that is contained within the premises of a customer's home. The cable home wiring begins at the demarcation point and it does not include converter boxes, decoders, amplifiers, or remote control units.

Cable Installation-Cable installation is the process of identifying and/or creating routing paths for cable lines, handling cable rolls and pulling through conduits or attaching cable to supporting structures.

Cable Lasher-Cable lasher is a machine that dispenses lashing wire around a cable and messenger wire as it travels along the length of the messenger cable.

Cable License-A cable license is authorization for a company to provide media transmission services in a geographic area using a cable distribution network.

Cable Locating Equipment-Cable locating equipment is a test instrument that can be used to locate buried or covered cables, lines, or pipes. Cable locating equipment typically operates by injecting a radio frequency signal onto the cable, lines, or pipes to allow a receiver to detect the RF signal located within or near the cable, lines, or pipes.

Cable Management System-A cable management system is the process used to plan and organize the routes and grouping of cables within a building or facility. Cable management systems commonly include trays, ducts, risers and other types of channels to hold cables. Cable management systems may include diagrams and maps to help contractors and technicians install and service cables.

Cable Modem (CM)-A communication device that Modulates and demodulates (MoDem) data signals to and from a cable television system. Cable modems select and decode high data-rate signals on the cable television system (CATV) into digital signals that are designated for a specific user.

There are two generations of cable modems; First Generation one-way cable modems transmit high speed data to all the users into a portion of a cable network and return low speed data through telephone lines or via a shared channel on the CATV system. First generation cable modems used asymmetrical data transmission where the data transfer rate in the downstream direction was typically much higher than the data transfer in the upstream direction. The typical gross (system) downstream data rates ranged up to 30 Mbps and gross upstream data rates typically range up to 2 Mbps. Because 500 to 2000 users typically share the gross data transfer rate on a cable system, cable modems also have the requirement to divide the high-speed digital signals into low speed connections for each user. The average data rates for a first generation cable modem rage up to 720 kbps. Until the late 1990's, most cable modems used first generation technology.

Second generation cable modems offered much data transmission rates in both downstream and upstream directions. Second generation cable television systems use two-way fiber optic cable for the head end and feeder distribution systems. This allows a much higher

data transmission rate and many more channels available for each cable modem. As of the year 2000, approximately 35% of the total cable lines in the United States had already been converted to HFC technology.

This figure shows a basic cable modem system that consists of a head end (television receivers and cable modem system), distribution lines with amplifiers, and cable modems that connect to customers' computers. This diagram shows that the cable television operator's head end system contains both analog and digital television channel transmitters that are connected to customers through the distribution lines. The distribution lines (fiber and/or coaxial cable) carry over 100 television RF channels. Some of the upper television RF channels are used for digital broadcast channels that transmit data to customers and the lower frequency channels are used to transmit digital information from the customer to the cable operator. Each of the upper digital channels can transfer 30 to 40 Mbps and each of the lower digital channels can transfer data at approximately 2 to 10 Mbps. The cable operator has replaced its one-way distribution amplifiers with precision (linear) high frequency bi-directional (two-way) amplifiers. Each high-speed Internet customer has a cable

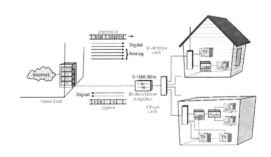

Cable Modem Overview

modem that can communicate with the cable modem termination system (CMTS) modem at the head end of the system where the CMTS system is connected to the Internet.

Cable Modem Termination System (CMTS)-A cable modem termination system is a system located in the headend of the cable television system that coordinates the overall operation of the cable modem system. The CMTS controls the gateways (Internet to data) and end user cable modems. The CMTS not only manages the data paths to allow end users to connect to the Internet, but also provides cable modem authentication, IP address assignment, billing functions and is responsible for the majority of Media Access Control (MAC) functionality in a cable modem network. A single CMTS typically controls hundreds or even thousands of end-user cable modems.

Cable Operator-A cable operator is a company or person who operates and/or provides broadcast services over cable systems. A cable operator may include affiliates who directly or indirectly provide and/or control the operations of the cable system.

Cable Plant-Cable plant is the physical property and facilities of a telephone, cable or company that uses a cable based transmission system. Cable plant property includes the transmission cables, amplifiers, and channel multiplexing equipment.

Cable Programming Service-Cable programming service is the providing of information content (programs) by a cable television system operator.

Cable Ready Television-Cable ready television is a video display device (a television) that is capable of receiving and displaying channels from a cable television system without the need for external adapters or devices.

Cable Run-A cable run is the physical path that is used by a cable between its end connections.

Cable Service-Cable service is the transmission of video programming and other media services to customers (subscribers) via a video and high-speed media distribution system (typically over fiber and/or cables).

Cable System Operator-A cable system operator is a company or organization who provides cable service (video and other media distribution) over a high-speed communication system who performs the operation the distribution system and controls access to the media that is distributed on the system.

Cable Tag-A cable tag is an identifying plate, sticker or label that is attached to a cable or located somewhere near the cable (such as on a telephone pole).

Cable Telephone Adapters-Cable telephone adapters are devices that convert telephone signals into another format (such as digital Internet protocol) that can be transferred on a cable television system. These adapter boxes may provide a single function such as providing digital telephone service or they may convert digital signals into several different forms such as audio, data, and video. When adapter boxes convert into multiple information forms, they may be called multimedia terminal adapters (MTAs) or integrated access devices (IADs).

Cable telephone adapters must convert both the audio signals (voice) and control signals (such as touch-tone or hold requests) into forms that can be sent and received via the cable television network.

Cable Telephony-A process of providing telecommunications services through the use of community access television (CATV) systems. Cable telephony services usually combine voice telephone, Internet access, digital cable television (TV), and analog cable TV.

This diagram shows a CATV system that offers cable telephony services. This diagram shows that a two-way digital CATV system can be enhanced to offer cable telephony services by adding voice gateways to the cable network's head-end CMTS system and media terminal adapters (MTAs) at the residence or business. The voice gateway connects and converts signals from the public telephone network into data signals that can be transported on the cable modem system. The CMTS system uses a portion of the cable modem signal (data channel) to communicate with the MTA. The MTA converts the telephony data signal to its analog audio component for connection to standard telephones. MTAs are sometimes called integrated access devices (IADs).

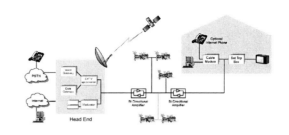

C

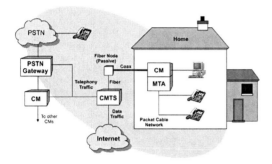

Cable Telephony System

Two-Way Cable Television System

This figure shows a cable television network that has both television distribution and cable modem transmission capability. This diagram shows that the head-end of a cable television (CATV) system is the initial distribution center for the CATV system. The head end is where incoming video and television signal sources (e.g., video tape, satellites, and local studios) are received, amplified, and modulated onto TV carrier channels for transmission on the CATV cabling system. The cable distri-

Cable Television (CATV)-Cable television is a distribution system that uses a network of cables to deliver multiple video and audio channels. CATV systems typically have 50 or more video channels. In the late 1990's, many cable systems started converting to digital transmission using fiber optic cable and digital signal compression.

This figure shows a two-way cable television system. This diagram shows that the two-way cable television system adds a cable modem termination system (CMTS) at the head-end and a cable modem (CM) at the customer's location. The CMTS also provides an interface to other networks such as the Internet.

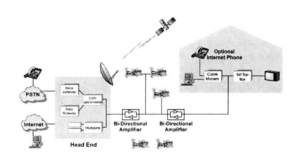

Cable Television Network

bution system is a cable (fiber or coax) that is used to transfer signals from the head end to the end-users. The cable is attached to the television through a set-top box. The set-top box is an electronic device that adapts a communications medium to a format that is accessible by the end-user.

Cable Transfer-Cable transfer occurs when working cables are spliced and the cable count is changed per engineering documents.

Cable Tuner-A cable tuner is a device that can select and pass through television channels that operate on cable television systems.

CableCARD-Cable Card

Cablecasting-Cablecasting is the providing (transmitting) of television programs or other media on a cable broadcast system.

Cabling Safety-Cabling safety are the procedures used to ensure the safety of installers and users of systems that use cables.

CABS-Carrier Access Billing System

CABS End Office (CEO)-CABS end office is the system that processes service orders, allocates resource usage to end office switching system and calculates the distance between EO and the customer's POP to assist in the billing process.

CAC-Call Admission Control

CAC-Customer Acquisition Cost

Cache-A memory storage area that temporarily stores information that is repeatedly accessed several times. The cache memory can be accessed more quickly than other memory storage areas such as a CD-ROM or hard disk. This allows computers to process information faster. A cache is also used to temporarily store files from other locations (such as Internet Web pages) so it is possible to return to this information without having to transfer the information again.

Cache Hit-A cache hit is a request for a web page file transfer for a web site that is provided from a cache sever that has previously stored the web page. Cache hits result in a lower hit count on the web servers as the web page is actually provided from the cache server.

Cache Memory-A cache memory is an area of memory that can be used by a computing system to assist in the processing of instructions or data. Cache memory can be used as a temporary scratchpad for data and typically has faster memory than other memory segments in the system.

Cache Server-A cache server is a computing device that is used to store data that is transferred through a network (such as through the Internet) so it can provided the data from the cache service to a user so it will not be necessary to transfer the same data again. The use of cache servers can reduce the number of web page views or visits counted on a web site as these pages may be provided by another computer.

Caching-Caching is a process by which information is moved to a temporary storage area to assist in the processing or future transfer of information to other parts of a processor or system.

Caching Rights-Caching rights are the authorizations to temporarily copy and store information for the potential need to refresh the current or recently rendered version of content.

Caching Server-A caching server is a computer that can receive, process, and respond to an end user's (client's) request information or information processing and temporarily store (cache) information that may be used again during a server's communication session.

CAD-Computer Aided Design

CAD-Computer Aided Dispatch

Cadence-A cadence is a signal pattern such as a ringing signal that has one or more repeated tones followed by a silence period.

CALEA-Communications Assistance For Law Enforcement

Calendar Routing-Calendar routing is the process of selecting routing paths for calls to call centers based to the calendar day and potentially the time of day.

Call Accounting-Call accounting is the processing of call routing information to track the services used and their amount of use associated with the call. Call accounting detects, captures and organizes records so they can be monetized.

Call Accounting System (CAS)-A call accounting system is the hardware and software services unit that can be integrated with telephone systems (such as a PBX system) or it can be an independent system (such as a computer). A call accounting system is sometimes called a station message detail recording (SMDR) system.

Call Admission Control (CAC)-(1-ATM) A traffic management feature of ATM networks that ensures that virtual channel connections are not offered unless enough bandwidth is available on its network. (2-voice) Call Admission Control (CAC) is a concept that applies to voice traffic. CAC is a deterministic and informed decision that is made before a voice call is established and is based on whether the required network resources are available to provide suitable QoS for the new call.

Call Agent (CA)-In a communication network, the call agent is responsible for call processing functions related to setup, maintenance and teardown communication sessions (e.g. voice calls).

Call Back-Call back is a process of return calls as part of a service. Call back can be part of a security procedure that ensures the person or device that called in to a system or service is who they claim to be as if they are not the authorized callers, the call back will not go to the fraudulent caller.

Call Barring-A telecommunications network feature which restricts the origination or delivery of calls to a device operating in its system. Call barring can restrict some or all parts of outgoing or incoming calls. When all calls are restricted (may be activated when the customer does not pay their bill), no calls may be made from or answered by the telephone device. Outgoing call restriction may limit the calls to emergency only, no international calls or local calls only. Restrictions on incoming calls may allow calls from specific people in a closed user group (CUG) or non-toll calls (such as collect calls).

Call Blocking-(1-system unavailable) The limitation on the ability to originate or deliver a call. Call blocking can occur on origination if telephone system resources are not available (such as when all the channels are occupied on a cellular system). Call blocking can also be a system feature where the delivery of calls can be blocked. (2-service) A service that permits a customer to restrict the delivery of incoming calls from numbers on a pre-defined (blocked number) list.

Call Center (Call Centre)-A call center is a place where calls are answered and originated, typically between a company and a customer. Call centers assist customers with requests for new service activation and help with product features and services. A call center usually has many stations for call center agents that communicate with customers. When call agents assist customers, they are typically called customer service representatives (CSRs).

Call centers use telephone systems that usually include sophisticated automatic call distribution (ACD) systems and computer telephone integration (CTI) systems. ACD systems route the incoming calls to the correct (qualified) customer service representative (CSR). CTI systems link the telephone calls to the accounting databases to allow the CSR to see the account history (usually producing a "screen-pop" of information).

Call Centre-Call Center

Call Clearing-The process used (e.g. signaling messages) to clear a call connection. Call clearing involves informing the network elements of the release of connection ports and call processing (e.g. echo canceling) activities.

Call Completion Rate-Call completion rate is the ratio of successfully completed calls to the total number of attempted calls. This ratio is typically expressed as either a percentage or a decimal fraction.

Call Detail Record (CDR)-A call detail record that holds information related to a telephone call or communication session. This information usually contains the origination and destination address of the call, time of day the call was connected, added toll charges through other networks, and duration of the call.

This figure shows the basic structure of a call detail record (CDR). This diagram shows that a usage data report (UDR) contains a unique identification number, the originator of the call, the called number, the start and end time of the call. This diagram also shows an additional charge for operator assistance and that a UDR dynamically grows as more relevant information becomes available.

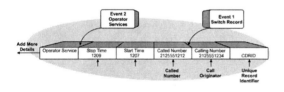

Call Detail Record (CDR) Structure

Call Detail Record (CDR) Processing-The process of creating a call billing record through the gathering and manipulating of billing event and service record information related to the call. Call record processing involves receiving the call record, guiding the record to the specific billing account, rating the record,

and routing the record to the appropriate database for collections or account settlement. This figure shows the general process that is used to identify and rate (bill) a call. This diagram shows that a call detail record evolves as it passes through the rating process. In the first step, a rate band is determined. Then, the identification information on the CDR is used to identify a specific customer's account (guide the record). The customer's rate plan is discovered and the unit (usage) and fixed (per event) charging rates are gathered and calculated. The new information (rate band, call charge amount) is added to the call detail record and it is moved to the bill pool, as it is ready to be billed.

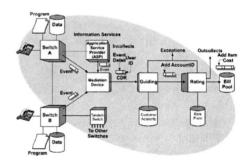

Call Detail Record (CDR) Processing
Operation

Call Detail Record Exclude Table (CDR Exclude Table)-A call detail record exclude table is a group of codes that list local central offices that are not to be monitored (to be ignored) by a billing system.

Call Detail Record Server (CDR Server)-A call detail record server is a computing device that collects and processes call usage information for the creation of call detail record (CDR) accounting and billing information.

Call Diversion-The automatic forwarding of telephone calls to a programmable number if the called telephone number is non-operational. This number is chosen and changed by the user.

Call Duration-Call duration is the amount of time between when a communication connection is setup or started to when the communication connection ends or the connection is terminated.

Call Forward Busy-The process of forwarding a call to another extension or telephone number when the selected extension or telephone number is in use (busy).

Call Forwarding-A call processing feature allows a user to have telephones calls automatically redirected to another telephone number or device (such as a voice mail system). There can be conditional or unconditional reasons for call forwarding. If the user selects that all calls are forwarded to another telephone device (such as a telephone number or voice mailbox), this is unconditional. Conditional reasons for call forwarding include if the user is busy, does not answer or is not reachable (such as when a mobile phone is out of service area).

This diagram shows how call forwarding can be used to automatically redirect telephone calls based on specific conditions. This example shows that a call may be redirected by the switching system to one extension if the user is busy and to a different extension if the user

has programmed the extension as "Do Not Disturb" or if it is busy. When the call is received by the system destined for extension 1001, the call processing system uses the indication of busy along with a redirection table to determine the call must be automatically transferred to extension 1003.

Call Forwarding No Response (CFNR)-Call forwarding no response is a call routing process that redirects incoming calls addressed to a called subscriber's telephone number to another telephone number (forwarded to number) when there is no answer on a called telephone number after a predefined period of time. CFNR is commonly used to redirect an incoming call to a voice mailbox.

Call Forwarding on Mobile Subscriber Busy (CFB)-Call forwarding on mobile subscriber busy is a call processing service that redirects a call path to an alternative destination when the destination mobile telephone is in use.

Call Forwarding-Busy (CFB)-A call routing process that redirects incoming calls addressed to a called subscriber's telephone number to another telephone number (forwarded to number) when the telephone number is busy processing a call. CFB is commonly used to redirect an incoming call to a voice mailbox.

Call Gapping-A network management control that limits the rate of traffic flow to a specific destination code or station address.

Call Hold-A feature that allows a user to temporarily hold and incoming call, typically to use other features such as transfer or to originate a 3rd party call. During the call hold period, the caller may hear silence or music depending on the network or telephone feature.

This diagram shows how a call can be temporarily placed on hold so the call can stay connected without the user having to continue conversation with the caller. During hold, the audio from the user is muted. For an analog

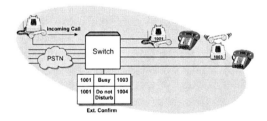

Call Forwarding Operation

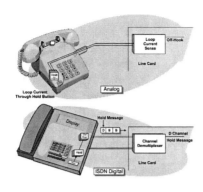

Call Hold Operation

line, the call hold feature involves placing a load (connection) across the line so that current may continue to flow through the circuit. For digital systems, the call hold feature may send a call hold message back to the system (such as a signaling message on the signaling channel) so the system can know that the status of the telephone station has changed to "hold."

Call Intrusion-The unannounced or unrequested connection of another caller to a call that is in progress.

Call Log-(1-communication) The recording of information that is associated with a communication device or service. (2-Sales) A call log is the recording of information about a telephone or communication event between a salesperson and a customer or prospective customer.

Call Logging-Call logging is the processing of call routing information to track the services used and their amount of use associated with the call. Call logging detects, captures, and organizes records so they can be monetized.

Call Management Record (CMR)-Records relating to a call that was processed by a call manager in an Enterprise communication system (such as an iPBX system that uses AVVID protocols).

Call Me Later-Call me later is a service that allows an Internet user or television viewer to select an option that requests that someone from a company or offered service call them later.

Call Me Now-Call me now is a service that allows an Internet user or television viewer to select an option that requests that someone from a company or offered service call them immediately.

Call Mix-Call mix is the division of call types that are received or originated in a phone system or call center.

Call Observing-The ability to monitor a call while it is in progress. Call observing may be performed by managers who want to monitor the communication (typically sales or customer service) performance of employees.

Call Overflow-Call overflow is a processing state of a telephone system (such as a PBX) that occurs when more calls are received than the call center can process. During call overflow conditions, calls may be redirected to other call centers or they may be transferred to a voice mail system to allow the caller to leave a message.

Call Park-A call hold feature that allows a telephone station (usually a PBX telephone) to initiate other calls or services while a call is parked.

Call Pickup-A telephone call processing (switch control) feature that enables a telephone user to answer a telephone from another telephone station. When a call is received, a key sequence is entered from specific groups or any telephone (dependent on how the system is setup) and the call is redirected to the extension or line that has picked up (entered the code) the line.

This diagram shows the operation of a telephone system that has a call pickup group feature. In this example, an incoming call is received and is directed toward the main extension 1001. When the call is received for extension 1001, the system uses the call pickup group list to determine that extensions 1001, 1002, 1003, and 1004 are programmed for call pickup. This allows any of these extensions to answer the call by pressing a key sequence (key "3" in this example).

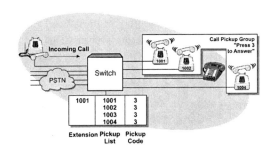

Call Pickup Operation

Call Processing-Call processing is the steps that occur during the setup and management of a telephone call or communication session. The call setup steps are typically associated with the authorization and routing of the call connection and the call management is associated with the feature control that occurs during the call (e.g. 3 way calling). When used as part of a billing system, call process involves gathering messages from various sources (event records), reformatting and edit the messages, calculate a charge for the message, assign a customer to the message, and getting the message ready for billing.

This figure shows the basic functions of the call processing section of a billing system. This diagram shows how different event sources are received by the call processing system. These event sources may be from the network elements or from other companies that have provided services to your customers. These records are reformatted to a common CDR format and duplicate CDRs are eliminated. Identification information in each call detail record is used to guide (match) the record to an account in the customer database. The customer's information determines the rate plan to use in charge calculation. The rating database uses rate tables, the customers selected rate plan, and possibly other information (e.g. distance, time of day) to calculate the actual charge for each call. All of the information is added to the CDR and it is either placed in the bill pool (ready for billing), or it is sent to another company to be billed if the customer identification is not part of this network's customer database. If there are any problems with call processing, the call detail records are sent to message investigation for further analysis.

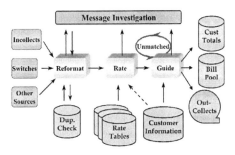

Billing Record Call Processing Operation

Call Processing Language (CPL)-A language that is based on extended markup language (XML) that is used to describe and control Internet (VoIP) telephony services. CPL is used by end users to create telephone services that integrate VoIP with existing Email, WEB, and other applications. It can be used with other VoIP protocols such as session initiation protocol (SIP).

Call Progress Tones-Signaling control tones that are sent to inform the device or system of the progress of a call. Examples of call progress tones include dial tone, ringback and busy tone.

Call Rate (CR)-The number of calls that are received over a defined period of time (such as during an hour or day.) The call rate may be characterized by a specific period of time such as the busy hour (BH) so its usage often include enough qualifying information to assure that it will be properly understood.

Call Rating-A process of assigning a value or cost to a telephone call or communication session.

This figure shows the basic call rating process used in a billing system. This diagram shows that a call detail record evolves as it passes through the rating process. In the first step, a rate band is determined. Then, the identification information on the CDR is used to identify a specific customer's account (guide the record). The customer's rate plan is discovered and the unit (usage) and fixed (per event) charging rates is gathered and calculated. The new information (rate band, call charge amount) is added to the call detail record and it is moved to the bill pool as it is ready to be billed.

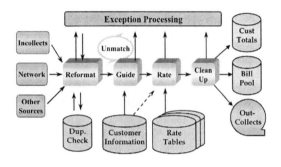

Billing System Call Rating

Call Reorigination-Caller reorigination is a prepaid calling card process that allows a caller who has dialed into a phone system using a calling card to originate new calls without redialing and entering account numbers or pin codes.

Call Report-A cellular telephone record stored on a data acquisition system (DAS) tape containing the overall timing information, mobile number, dialed digits, and appropriate indicators to ticket a call for every call completed or attempted through the system. (See billing Tape).

Call Restrictions-In call-processing feature that restricts telephone call origination to spe-cific groups of authorized numbers (typically local phone numbers or no-international calls).

Call Routing-(1-circuit switching) The process of determining the path of a call from point of origination to point of destination. (2-packet switching) The steps taken to ensure a connection can be made from an origination point to its destination point. The packets that transfer through this connection may actually take different paths.

Call Routing Table-A call routing table is a list of call connection connection paths (routes) that are associated with dial plans that will be selected based on the dialed digits received from the caller. Call routing tables determine if the call is allowed from the caller (international calls may be restricted) and which network will handle the call (local telephone line or long-distance network).

Call Screening-A process that allows a call recipient to review the incoming telephone number (and name if available) to determine if they desire to answer the incoming call or to transfer the call to another person.

Call Server-A call server is a particular form of application server that manages the setup or connection of telephone calls. The call server will receive call setup request messages, determine the status of destination devices, check the authorization of users to originate and/or receive calls, and create and send the necessary messages to process the call requests.

Call Setup-Call setup are the call processing steps (events) that occur during the time a call is being established, but not yet connected.

Call Sheet-A call sheet is a data gathering form that is used to record information that results from a call or communication session with a person or company.

Call Supervision-The process of monitoring a communication line or trunk for changes in call status. These changes can include start-dial (off-hook), feature change requests, or call termination (on-hook).

Call Termination-The process of terminating a call or communication session. Call termination can be initiated by any of the users of the call or the system that is connecting the call.

Call Timers-Clock timers that are used to count time or events that occur with a telecommunications device. This may be the airtime or usage of a telecommunications service or how many calls were made during a particular time period.

Call Trace-Call trace allows a subscriber to initiate a call trace request message that allows the dialed digits of a caller to be stored for investigation. The activation of the call trace service alerts the telephone service operator to "tag" the originator's number to allow authorities to investigate the originator of the unwanted or unauthorized call. Some of the call trace activation codes include *57 on a touchtone phone or the dialing of 1157 on a rotary (pulse) phone. If the call trace of the last call was completed successfully, an announcement should be heard. The service operator will usually release the call trace information to law enforcement agencies and a signed authorization from the subscriber may be required.

This diagram shows the call trace operation. The call trace operation starts with the reception of an unwanted call. When the call is received, an event record is created in the switch that records the connection through the switch (resources used) and this record usually contains the calling party number associated with the incoming call. This example shows that the customer dials a call trace feature code to inform the carrier to trace the last call. The local switch (recipients phone carrier) reviews the call detail information stored within the switch (or billing system) to determine if a calling number was provided. If the number was provided and is available, the information will be stored in the carriers system. The customer then calls the carrier and requests that the number be sent to local authorities for further action. This example shows that the customer completes a call trace release form that authorizes the transfer of the number to authorities.

Call Tracking Number (CTN)-A call tracking number is an identification code that is used to track the status of communication with a customer on a customer care issue.

Call Vectoring-Call vectoring is a call routing processes that redirects incoming calls to ports or additional locations based on a set of pre-defined commands or values. Call vectoring allows incoming calls to be directed differently depending on variety of factors including the time of day, incoming call location and how busy the system is along with other factors.

Call Waiting (CW)-A telephone call processing feature that notifies a telephone user that another incoming call is waiting to be answered. This is typically provided by a brief tone that is not heard by the other callers. Some advanced telephones (such as digital mobile telephones) are capable of displaying the incoming phone number of the waiting call.

After the service provides the subscriber with the notification of an incoming call while the subscriber's call, controlling subscriber can either answer or ignore the incoming call. If the controlling subscriber answers the second

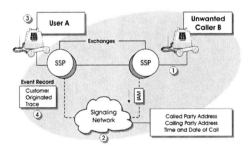

Call Trace Operation

call, it may alternate between the two calls. This diagram shows how call waiting service may be provided on an analog telephone line. In this example, a call is in process with caller 1. During the call, a second caller (caller 2) dials the telephone number of the user that has call waiting. The system discovers that the line or extension is busy on another call. The system also determines that this user has the call waiting service processing feature available so it sends a call waiting message tone to the user (only heard by the user). If the user desires to answer the call, the user sends a flash message (a momentary open on the line) that indicates to the telephone system to place the current call in progress on hold and switch to the other incoming call (caller 2). Each time a flash message is sent, the line alternates between each incoming caller.

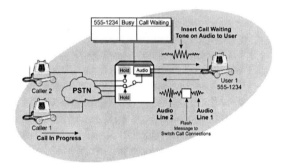

Call Waiting Operation

Call Waiting Identification (CWID)-A telephone service that provides a receiving telephone device with the phone number of a new originating caller (call to be received) while the callee is on the line with another caller.

Callback-A call processing service that reverses the connection of calls. This process is divided into the call setup (dial-in) and callback stages. The caller dials a number that provides access to the callback service. The callback gateway receives the call and prompts the caller to say or enter (e.g. by touch tone) the number they desire to be connected to and the number they want the callback service to connect to. The callback center then originates calls to both numbers and connects the two individuals to each other.

Call-Back Unit-A device that calls a user at a pre-programmed telephone number when a connection is requested to prevent unauthorized access from unknown location, or to establish low cost international calls between a nation with low originating costs, automatically directing call-backs into a target nation with higher originating costs.

Called Line Identification (CLI)-The identification information carried by an SS7 packet that provides the destination receiver to identify the source of the call.

Called Number Delivery (CND)-The calling number from the originator of the call.

Called Party-A called party is the person who receives a telephone call.

Called Subscriber Identification (CSI or CSID)-Called subscriber identification is an identifier whose coding format contains a number (such as the telephone number from a called fax machine).

Caller ID-An optional telephone service that provides a receiving telephone device with the phone number of the originating caller, which can be displayed to the destination person prior to receiving the call. The caller ID is transmitted as a data parameter in the SS7 Initial Address Message from the originating end switch to the destination end switch in the process of setting up the call. Some caller ID services can also provide directory name listing information derived separately from the LIDB database. Caller ID information is typically transferred as a type-202-modem-compatible data signal between the first two ringing cadence cycles of the alerting tone.

Caller ID Type II-The ability of the system to announce the phone number of the caller who is calling in while the line is in use (call waiting).

Calling Card-An identifying number or code unique to the individual, that is issued to the individual by a common carrier and enables the individual to be charged by means of a phone bill for charges incurred independent of where the call originates.

This figure shows how a calling card can be used to initiate calls through a telephone network. In this example, the customer uses the telephone number on the pre-paid calling card to initiate a call to a switching gateway. The gateway gathers the calling card account information by either prompting the user to enter information or by gathering information from the incoming call (e.g. prepaid wireless telephone number). The gateway sends the account information (dialed digits and account number) to the real time rating system. The real-time rating system identifies the correct rate table (e.g. peak time or off peak time) and inquires the account determine the balance of the account associated with the calling card. Using the rate information and balance available, the real time rating system determines the maximum available time for the call duration. This information is sent back to the gateway and the gateway completes (connects) the call. During the call progress, the gateway maintains a timer so the caller cannot exceed the maximum amount of time. After the call is complete (either caller hangs up), the gateway sends a message to the real time rating system that contains the actual amount of time that is

used. The real time rating system uses the time and rate information to calculate the actual charge for the call. The system then updates the account balance (decreases by the charge for the call).

Calling Line Identification (CLI)-A service which displays the calling number prior to answering the call that allows telephone customers to determine if they want to answer the call. The calling number may be used by the telephone device to look-up a name in memory (e.g. mom) and display the name along with the phone number.

This figure shows the calling number identification operation. Calling number identification operation starts with the reception of a call. When the call is received, the initial address message (IAM) contains the calling party number of the incoming call. The IAM may contain additional information such as the text name of the calling party. This example shows that the local switching system extracts this information and combines this information with the ring signal (using different frequencies and amplitudes) and sends it to the customer during the alerting (ringing) process. If the customer has the appropriate display equipment, the calling number information is display as the telephone rings.

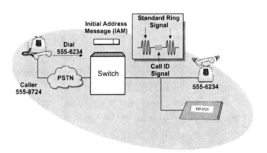

Calling Line Identification (CLID) Operation

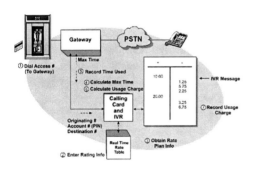

Calling Card Operation

Calling Line Identification Display (CLID)-Calling line identification display (CLID) identifies for display the originating or calling phone number.

Calling Line Identification Presentation (CLIP)-The transmission of a calling number to the receiver of a call. The calling number may originate from the caller's equipment or it may be created by the network.

Calling Line Identification Restriction (CLIR)-The ability of a caller to block the delivery of their telephone number from being displayed on a caller identification device. The CLIR may be on a per call basis or per line (continuous) block. CLIR provides the ability for the caller to remain anonymous to the called person.

Calling Name Delivery-A feature that enables a customer to view the name of the calling party as well as the date and time of the call on a display device.

Calling Number Delivery (CND)-Calling number delivery is the process of providing the calling number through telephone systems.

Calling Number Display Blocking-A feature that gives calling customers the option to change the caller ID status of their number on a per-call basis. When the calling directory number is private, the caller ID display will not be available to the called party.

Calling Number Identification (CNI)-Calling number (and/or name) identification is a service that provides the number and/or name of the incoming to the telephone user prior to the user's answering of the call. This allows the customer to view the telephone number of the person who is calling before deciding to accept the call.

Calling Party-A calling party is the person who originates a telephone call.

Calling Party Number (CPN)-The calling party number is the number of persons or devices that are originating a call.

Calling Party Pays (CPP)-A communication service that bills the calling party for the delivery of their call through a network (such as a mobile communication network or freephone service) or for the providing of information (such as a news service.)

CAM-Computer Aided Manufacturing

CAMA-Centralized Automatic Message Accounting

Campaign Management-The process usually utilized for direct marketing (i.e. direct mail or phone sales), to manage the conceptualization, prioritization, planning, execution, and post campaign measurement of activities.

Cancellation Fee-A cancellation fee is a charge or value that is assessed when a customer terminates or changes their service agreement.

Cancellation Policies-Cancellation policies are the rules that are used to determine what responsibilities each party in a contract has in the event the contract is terminated.

Cancellation Rate Card-A cancellation rate card is a fee schedule that is applied for services that are changed or cancelled.

Canonical-Canonical is the use of rules or logical functions to process information or data.

Canonical Address-A canonical address is a unique code that is composed of unique code segments that conform to a set of rules or logical functions.

Canonicalization-Canonicalization method is the process that is used to standardize the format or location of objects or information.

Capacity-(1-communication) The maximum information carrying ability of a communications facility or system. The unit of capacity measurement for the facility or system depends on the type of services or information content that are provided by the facility or system. (2-energy) The amount of electrical energy that a device such as a capacitor or battery can store. The unit of measure for capacity of a capacitor is the Farad.

Capacity Limit-A capacity limit is the maximum amount of service (such as data transmission rate) or number of customers that a system can provide services to at a defined level of service.

Capacity Management-Capacity management is the use of processes and system for the assignment and optimization of network bandwidth through the monitoring of performance metrics, workload analysis, application resources, and overall network demand.

Capacity Planning-Capacity planning applies to proactively assessing systems, whether network or computer based systems, and forecasting the needed future growth of the system. Capacity planning in network management can be a function of the Configuration or Performance Management level of the FCAPS model.

CapEx-Capital Expenses

Capital Cost Model-A financial cost model that recognizes the differences in regulatory accounting versus tax accounting for telecommunications networks and other systems.

Capital Cost per Subscriber-Capital cost per subscriber is the total cost of network investment required to provide services to subscribers (customers). Capital costs include the purchase of network equipment, installation costs, initial license fees, development costs, and other non-operational costs.

Capital Expenses (CapEx)-CapEx is the capital expenses that include the purchase of land, buildings and, most importantly, the build out of network capacity in a telecommunications system.

Capital Plan-A capital plan is an outline of the expected investment and timing for the acquisition and installation of equipment and systems.

CAPs-Competitive Access Providers

Captive Market-A captive market is a group of people or consumers that have few options among products or services they may purchase.

Capture-Capturing is the process of gathering information or media.

Card Rate-Card rate is the price that is determined from set media publication rates (from the rate card).

Card Security Code (CSC)-A card security code is an additional number located on a credit card that can be used to verify that a pur-

chaser has possession of the card.

Card Verification Value 2 Code (CVV2 Code)-A CVV2 code is an additional number located on a credit card that can be used to verify that a purchaser has possession of the card.

Carding Fraud-Carding fraud is the use of a fake billing issue announcement email to trick email recipients to disclose private security information.

CARE-Customer Account Record Exchange

CARE/ISI-Customer Account Record Exchange/Industry Standard Interface

CAROT-Centralized Automatic Reporting On Trunks

Carriage-The established procedures of a cable TV system regarding the carrying of signals of local TV stations on its channels.

Carrier-(1-signal) A sine wave that can be modulated in amplitude, frequency or phase for the purpose of carrying information. (2-company) A business organization providing telecommunications service, such as a radio common carrier or a telephone service provider. (3-frequency) The frequency of the radio carrier signal. (4-level) The radio energy (power) of a carrier signal, typically expressed in decibels in relation to some nominal (reference) level.

Carrier Access Billing System (CABS)-A system that is used by network access providers to bill carriers for their customer's access to the network facilities.

Carrier Access Tariff-A carrier access tariff is the rates that are charged to carriers (such as long distance voice providers) to access to local networks (such as the local telephone network).

Carrier Ethernet-Carrier Ethernet is a packet based transmission protocol based on 802.3 (Ethernet) that is used by service providers to provide data services to customers.

Carrier Identification Code (CIC)-A 3-digit code that uniquely identifies a telecommunications carrier within the North

C

American Numbering Plan (NANP). The CIC is indicated by an XXX in a Carrier Access Code where X can be any digit, 0 through 9. After an XXX code has been assigned to a carrier, the code is retained for use with either Feature Groups B (95OA-0XXX, 950-1XXX) or D (10XXX) throughout the area served by the NANP. (See also: Carrier Access Code, presubscription, primary interexchange carrier.

-A long-term plan for inter-connecting interexchange and international carriers to exchange carrier intraLATA networks. This plan also can be called the Exchange Access Plan. The term equal access commonly is used to refer to the features provided by this plan.

Carrier Neutral-Carrier neutral is the providing an unbiased carrier selection process when communication services are requested or provided.

Carrier's Carrier-A carrier's carrier is a company that provides telecommunications services to other carriers. Because a carrier's carrier does not provide service to the public, it may be subject to fewer regulations.

Carryover-Carryover is the transferring of previous units or amounts to another account or transaction.

Carterfone Decision-A Federal Communications Commission (FCC) decision in 1968 that allowed customers to directly connect their own customer-provided equipment (CPE) to telephone company networks. Prior to this decision, some telephone companies prohibited the attachment of such equipment to their networks. Equipment that is to be connected to telephone networks must conform to industry standards and FCC regulations to ensure the protection of the telephone network from potential damage that may result from defective CPE.

CAS-Call Accounting System

CAS-Centralized Attendant Service

CAS-Conditional Access System

Cascading Style Sheets (CSS)-A cascading style sheet is a data file that describes the layout (style) of a web document. A CSS can define the spacing, font types, colors and other characteristics within an electronic document.

Case Code-A case code is a unique number or label that identifies a case of (group of) product or materials.

Cash Basis-Cash basis is an accounting process that only records cash transactions. An example of a cash basis is the recording of a sale only when the cash was received from the sale.

Cash Cost Per User (CCPU)-Cash cost per user is the amount of cash that is required to provide services and/or equipment to a user.

Cash Cow-A cash cow is a product, service or business that is generating revenue exceeds the costs of producing, selling and distributing the products or services offered.

Cash Flow Statement-A cash flow statement is a summary list of cash receipts and cash deposits for a business and over a period of time.

Cash on Delivery (COD)-Cash on delivery is a payment option that allows a buyer to pay for goods or services when they are delivered.

Cashier Check-A cashier check is a written order from a bank to pay money where the funds have been verified as available and reserved from the payer's account.

Casual Billing-Casual billing is an agreement between companies (such as local and long distance service providers) where each service provider agrees to provide services to customers of the other company even though they do not manage an account for the user.

CAT5-Category 5

Catalog-A catalog is a collection of products or services and information related to them that are accessible by people or systems that desire to obtain information about the items. A catalog may be produced in various formats including printed, electronic, online and video.

Catalog Management-Catalog management is the process of identifying, organizing, assigning, and tracking products that are included in catalogs.

Catalog Ontology-A catalog ontology is a structure of categories that are used to organize the access or presentation of products or services that are included in a catalog.

Catalog Schema-A catalog schema is a hierarchical structure that describes the relationship of categories used in a catalog to organize or presentation of products or services that are included in a catalog.

Cataloging-Cataloging is the process of identifying media and selecting groups of items to form a catalog.

Catastrophe-A catastrophe is a disaster that is sudden and destructive.

Catch Up Television (CUTV)-Catch up television (CUTV) is a television delivery service that allows a user to begin watching a television program any time after it has been broadcasted.

Category 5 (CAT5)-Category 5 unshielded twisted-pair wiring commonly used for 10baseT and 100BaseT Ethernet networks and rated by the EIA/TIA

Category Management-Category management is the process of identifying, defining and assigning the relationship of categories for data, products or services.

Category Selection (Selects)-Category selections (selects) are the available characteristics that are associated with a contact or data element in a list or database.

CATV-Cable Television

CATV-Community Access Television

Caveat-A caveat is a notice, warning or the providing of information that could influence a persons decision.

CBD-Central Billing District

CBR-Constant Bit Rate

CBR Feed-Constant Bit Rate Feed

CBT-Computer Based Training

CC-Conference Calling

CCB-Common Carrier Bureau

CCB-Customer Care and Billing

CCBS-Customer Care And Billing System

CCC-Clear Channel Capability

CCITT-Consultative Committee for International Telephony And Telegraphy

CCPU-Cash Cost Per User

CCS-Centum Call Seconds

CCS-Common Channel Signaling

CCTV-Closed Circuit Television

CD-Commitment Date

CDA-Content Distribution Agreement

CDDD-Customer Desired Due Date

CDN-Content Delivery Network

CDP-Customer Demarcation Point

CDPD-Cellular Digital Packet Data

CDR-Call Detail Record

CDR Exclude Table-Call Detail Record Exclude Table

CDR Processing-Call Detail Record

CDR Server-Call Detail Record Server

CE-Consumer Electronics

Cease and Desist Letter-A letter that requests or demands that an activity or activities of a company or person stop to prevent legal action.

Cease and Desist Order-A Cease and Desist Order is a directive that is issued by a court or governmental agency that instructs a person or business to halt the sale or specific activities.

CEI-Customer Experience Indicator

Ceiling-The ceiling is the maximum or highest available price.

CEL-Contracts Expression Language

Cell Loss Rate (CLR)-Cell loss rate is a ratio of the number of data packet cells that have been lost in transmission compared to the total number of cell packets that have been transmitted.

Cellemetry-Cellemetry is a brand name for a telemetry (measurement) service that uses the cellular network to carry short data messages such as utility meter reading, vending machine status and vehicle or trailer tracking.

Cellular-Cellular radio is wireless telephone system that divides geographic areas into small radio areas (cells) that are interconnected with each other. Each cell coverage area has one or several transmitters and receivers that communicate with mobile telephones within its area.

Cellular Carrier-A radio common carrier that provides cellular telephone service.

Cellular Digital Packet Data (CDPD)-Cellular Digital Packet Data (CDPD) is a wide area data transmission technology developed for use on cellular phone frequencies. CDPD uses unused cellular channels to transmit data in packets. This technology offers data transfer rates of up to 19.2 Kbps, better error correction and quicker call set up than using modems on an analog cellular channel.

Cellular Intercarrier Billing Exchange Roamer Recrod[tm] (CIBER)-Cellular intercarrier billing exchange roamer is a set of proprietary protocols for the exchange of billing information among AMPS analog, CDMA and TDMA carriers wireless carriers, billing vendors, clearinghouses, and clearing banks. It is a defacto billing standard originally developed and maintained by the Cibernet subsidiary of the Cellular Telecommunications Industry Association (CTIA). MACH Cibernet now manages CIBER.

This figure shows some of the information (fields) contained in a type 22 CIBER record. This example shows that the type 22 Ciber record field structure has been updated from the previous type 20 record structure to include additional fields that allow for telephone number portability (enabling telephone number transfer between carriers). This list shows that fields in the Ciber record primarily include identification of airtime charges, taxes, and interconnection (toll) charges.

Cellular Operator-The owner and/or operator of a cellular network

Cellular Subscriber Station (CSS)-Another name for mobile phone. This is the preferred term to network providers when Mobile Stations may be fixed in location.

```
Type 22 Record - Sample of Fields

◆ Home Carrier SID/BID        ◆ Caller ID
◆ MIN/IMSI                     ◆ Called Number
◆ MSISDN/MDN                   ◆ LRN
◆ ESN/IMEI                     ◆ TLDN
◆ Serving Carrier SID/BID      ◆ Time Zone Indicator
◆ Total Charges and Taxes      ◆ Air Connect Time
◆ Total State/Province Tax     ◆ Air Chargeable Time
◆ Total Local Tax              ◆ Air Rate Period
◆ Call Date                    ◆ Toll Connect Time
◆ Call Direction               ◆ Toll Chargeable Time
◆ Call Completion Indicator    ◆ Toll Carrier ID
◆ Call Termination Indicator   ◆ Toll Rate Class
```

Cellular Intercarrier Billing Exchange Roamer (CIBER) Structure

Cellular System- A cellular system is a fully automatic, wide-area, high-capacity RF network made up of a group of coverage sites called cells. As a subscriber passes from cell to cell, a series of handoffs maintains smooth call continuity.

This figure shows a basic cellular system. The cellular system connects mobile radios (called mobile stations) via radio channels to base stations. Some of the radio channels (or portions of a digital radio channel) are used for control purposes (setup and disconnection of calls) and some are used to transfer voice or customer data signals. Each base station contains transmitters and receivers that convert the radio signals to electrical signals that can be sent to and from the mobile switching center (MSC). The MSC contains communication controllers that adapt signals from base stations into a form that can be connected (switched) between other base stations or to lines that connect to the public telephone network. The switching system is connected to databases that contain active customers (customers active in its system). The switching system in the MSC is coordinated by call processing software that receives requests for service and processes the steps to setup and maintain connections through the MSC to destination communication devices such as to other mobile telephones or to telephones that are connected to the public telephone network.

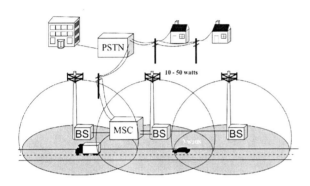

Basic Cellular System

Cellular Telecommunications and Internet Association (CTIA)-Formerly the Cellular Telecommunications Industry Association representing only cellular network operators, CTIA-The Wireless Association ®, as it is now known, is a trade organization representing service providers, manufacturers, and wireless data and Internet companies.

Cellular Video-Cellular video is the transferring of signals that carry moving picture information through the use of cellular (mobile telephone) radio channels. Cellular video may be sent over common traffic channels (e.g. GSM) or it may be sent over high-speed cellular packet data channels (such as EDGE, EVDO or HSPDA).

Central Billing District (CBD)-The billing district that is at the center of a city or network.

Central Exchange (Centrex)-Centrex is a service offered by a local telephone service provider that allows the customer to have features that are typically associated with a private branch exchange (PBX). These features include 3 or 4 digit dialing, intercom features, distinctive line ringing for inside and outside lines, voice mail waiting indication and others. Centrex services are provided by the central office switching facilities in the telephone network.

Central Office (CO)-The name commonly used in North America to identify the switch in a telephone network that connects local customers. The international name is "public exchange" or "telephone exchange." CO actual refers to the building that contains the switching and interconnection equipment.

Central Office Area-A geographic area served by an end office.

Central Office Equipment-Central office equipment is any assembly or system that is installed in a central office.

Central Processing Unit (CPU)-A central processing unit is the electronic components of a computer that interpret instructions, performs calculations, moves data in main computer storage, and controls input/output operations. Often this expression refers to a mainframe or midrange machine.

Central Subscriber Management-Central subscriber management is a process or system that coordinates the additions, changes, and terminations of subscribers of a service from a central location.

Centralized Architecture-Centralized architecture is a system that is designed to use a central intelligence to coordinate the overall operation of the system. The use of a centralized communication system allows one point to coordinate the operation of all other points and devices within the network.

Centralized Attendant Service (CAS)-A centralized group of customer service operators (attendants) who answer incoming calls for multiple telephone systems.

Centralized Automatic Message Accounting (CAMA)-A system for the recording of detailed billing information at a central location other than an end office, usually at a tandem.

Centralized Automatic Reporting On Trunks (CAROT)-A system that automatically and routinely reports on the performance of trunk lines to ensure high availability and to avoid potential failures. The reporting system may check for return loss, line noise, timing precision, transmission, and call processing performance criteria.

Centralized Control-A system that relies on a central intelligence to coordinate the overall operation of the system.

Centralized Message Data System (CMDS)-A system that allows interexchange of billing and usage data between telephone exchange networks. CMDS has four parts: (1) Centralized Message Data System (CMDSI) provides for collect services (collect, third party number, and calling card),(2) Carrier Access Billing System (CABS) that bills interexchange carriers (IXCs) for local exchange access services, (3) 800 Service Usage; and (4) Meet Point Billing which involves the billing of access services (via CABS) provided to two or more interexchange carriers.

Centralized Repository-A centralized repository is a storage system that is used to hold and/or organize data in a single location.

Centralized Security-Centralized security is the coordination of access control, authorization and encryption of data through a communication network from a centralized authority.

Centrex-Central Exchange

Centric-Centric is the central part or core focus of a process or system.

Centum Call Seconds (CCS)-A measurement of communication trunk usage that equals 100 seconds of continuous usage. Because the standard time interval for communication network engineering is based on activity over an hour, the system load is expressed in hundred call seconds (CCS) for one hour. A single hour has thirty-six hundred call seconds (equal to one erlang).

CEO-CABS End Office

CEPT-Conference Of European Postal And Telecommunications Administrations

Certificate Authority (CA)-A certificate authority is an authorized company that validates and provides secure socket layer (SSL) certificates that enable SSL authentication and encryption between communication devices. There are several certificate authority companies and are sometimes referred to as a "trusted authority."

Certificate Chain-A certificate chain is the linking of certificates that reference each other.

Certificate of Inspection-A certificate of inspection is a document or verification information from a third party (typically a recognized company) that declares the authenticity of an item and its associated characteristics.

Certificate Policy-A certificate policy is a set of rules that define how certificates and information within certificates will be used in security processes.

Certificate Repository-A certificate repository is a storage system that is used to hold certificate information. The certificate repository holds information that a security system can trust.

Certificate Revocation-Certificate revocation is the process of sending a command to a user or device that updates, disables or removes or changes the usage access rights for its associated media or data.

Certificate Revocation List (CRL)-A certificate revocation list is a group of users or devices that have been identified as having expired or invalid certificates.

Certificate Signing Request (CSR)-A certificate signing request is a message that is sent from a user (an applicant) to a certificate authority to obtain a digital identity certificate.

Certificate Store-A certificate store is a grouping (database) of security elements (e.g. keys or certificates). A certificate store holds and provides access to authenticated documents or electronic keys that are signed, typically by some higher authority, and that usually contains useful information in an encapsulated unit that provides a higher level of assurance of a request or service.

Certification-Certification is the process of testing people or products to ensure they conform to industry standards or specifications. Certification helps to ensure that products operate as expected and are interoperable with other products that are produced to the same standards or specifications.

Certification Authority (CA)-A certification authority is a company that can validate (certify) that specific identities, components or transactions are valid or are authentic.

CEV-Controlled Environmental Vault

CFB-Call Forwarding on Mobile Subscriber Busy

CFB-Call Forwarding-Busy

CFNR-Call Forwarding No Response

CFO-Chief Financial Officer

CFR-Code of Federal Regulations

CFR-Cost and Freight

CG-Charging Gateway

CGF-Charging Gateway Function

CGI-Common Gateway Interface

Chain of Custody-Chain of custody is the sequence of people and/or companies that have had control of an asset.

Chain of Title-A chain of ownership based on history of a literary property, which can be traced back to the time of creation.

Challenge Handshake Authentication Protocol (CHAP)-Challenge handshake authentication protocol is a security protocol that is used to authenticate (validate the identity) a user or device that is requesting a service before they are given access to service. CHAP protocol uses previously known information to calculate an authenticated response and only the calculated response is transferred during the authentication process rather than the key or other secret data.

Challenge Phrase-A challenge phrase is an authentication process that requires a user or device to respond correctly to a challenge message. The challenge phrase usually contains information only the recipient knows or can respond to.

Challenge Response-A challenge response is an authentication process that requires a user or device to respond correctly to a challenge message. The challenge message usually requests information only the recipient knows or can respond to.

Chamber of Commerce-A chamber of commerce is a group of people or companies that work together to assist in the economic development of a community.

Change Agent-A change agent is a person or influence on a process or system that assists or has the potential to modify the process or system.

Change In Rate Structure-Change in rate structure is a modification of the rate components for existing products or services.

Change Log-A change log is a listing of the changes (edits) that have been applied to a media program.

Change Orders-Change orders are instructions that indicate alterations or modifications to a specification or plan.

Channel-(1-general) In general, a stream of information transmitted as part of a distinct communication, or conversation, or for a particular purpose or end use. One channel may be distinguished from other channels by the time of occurrence of the transmission, by the format or organization of its content, by the frequency of a carrier signal used to transmit it, or by some secondary property such as the type of error detecting code used for it, or by other properties. Due to the advance of technology, a single channel may at some time historically be modified so that it carries multiple channels. For example, when the only distinguishing feature of two radio signals was their carrier frequency and each one carried only one conversation, each one was described as a channel. Different carrier frequencies were designated by distinct channel numbers. At a later date, time division multiplexing was used with radio signals to distinguish 3 or 8 distinct conversations on one modulated radio carrier frequency. Confusion arose because the 3 or 8 distinct conversations, each using a designated time slot in time division multiplexing, were described as separate channels. At the same time, the entire signal (comprising 3 or 8 channels) was also described by some as a channel. In some cases the reader must read carefully, with awareness of the changing historical meanings of some terms. Regarding this particular example, the entire conglomeration of 3 or 8 time slots is described best as a "modulated carrier" and not as a channel. (2-broadcasting) A portion of the radio frequency spectrum assigned to a particular broadcasting station. (3- signal path) A transmission path between two or more termination points. The term channel can refer to a 1-way or 2-way path. (4-video effects) A digital effects processing path for video. (5-programming) Channels are collections of content that can be

associated with a brand or media access point. This diagram shows the difference between physical and logical channels. In this example, a physical channel transports information between two points using electrical signals. The physical channel is divided into frames that contain various fields (groups of information within the frame). This diagram shows that the frames on the physical channel are divided into 4 logical channels; 3 logical channels for data and one logical channel for control. The exact relationship between the frame structure and the logical channel is called mapping. This example shows two different mapping examples. In the first example, the bits in the information portion of each frame are equally divided. In the second mapping example, more bits are proportioned to channel 2. This results in a lower data transfer rate for channels 1 and 3 while channel 2 has a higher data transfer rate.

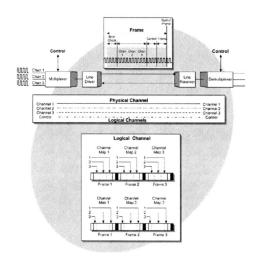

Channel Systems

Channel Assets-Channel assets are the media programs that are associated with a channel. Channel assets can be branded collections of content or other program media that is available through or associated with a channel.

Channel Bandwidth-(1-data) Channel bandwidth the maximum or fixed data transmission rate of a communication channel. (2-radio) The difference between the upper frequency limit and lower frequency limit of allowable radio transmission energy for a channel.

Channel Block-Channel blocks are groups of channels that are assigned or used together.

Channel Busy Tone-An audible signal indicating that a call cannot be completed because all switching paths or toll trunks are busy, or that equipment is blocked. The tone is applied 120 times per minute. The channel busy tone also is called fast busy, all trunks busy, or reorder tone.

Channel Capacity-The amount of data or channel transmission capability of a communication channel.

Channel Change Time-Channel change time is the duration between when a channel change request is initiated (e.g. pressing a channel button) and when the display of the selected channel video begins.

Channel Conflict-Channel conflict is the overlap of marketing objectives or interference that occurs when multiple channels are used to distribute products or services.

Channel Count-Channel count is the number of communication channels that are or can be sent on a communication system. For cable television systems, the channel count is the number of television channels that can be simultaneously sent on a cable distribution system.

Channel Efficiency-(1-communication) Channel efficiency is a measure of the amount of user data or information that is sent on a communication channel as compared to the total transmission data or information that is transmitted. (2-business) Channel efficiency is the amount of cost that is required to support a distribution channel as compared to the gross sales that is generated by the distribution channel.

Channel Equipment-Channel equipment is an interface assembly that allows end user devices to send and receive control signals. For telephone line channel equipment, these control signals typically include tip and ring along with supervisory signaling leads.

Channel Extension-A system that enables peripheral equipment, such as high-speed tape spoolers, printers, and terminals, to be connected to mainframe computers many miles away.

Channel Loading-Channel loading is a ratio of the number of users authorized to operate on a particular channel or system compared to the number of users that actively transmit on a system. An example of channel loading is a private telephone system that may have 5 extensions for every telephone line (loading of 5:1) because the average business telephone user only talks for 1-2 hours per day.

Channel Management-Formal process utilized to manage the creation, staffing, tasking, and measurement of sales and customer support channels.

Channel Manager-A channel manager is an assembly or process that is responsible for creating, managing and ending communication channels.

Channel Mapping-A relationship that allows multiple communication channels to be assigned to other transmission channel formats. An example is mapping multiple DS3 channels to the frames within a synchronous optical network (SONET) transmission channel.

This diagram shows how multiple communication channels can be mapped (related to) specific portions of another communication channel (payload). Mapping allows several communication channels to share the same physical or logical data transmission medium.

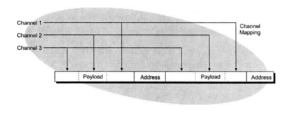

Channel Mapping Operation

Channel Pair-Channel pairs is the grouping of channels together to form a bi-directional (two-way) communication circuit.

Channel Rate-The data rate at which information is transmitted through the channel or communications media, typically stated in bits per second (BPS).

Channel Reliability-The percentage of time a channel is available for use in a specific direction during a specified period.

Channel Structure-Channel structure is the division and coordination of a communication channel (information transfer) into logical channels, frames (groups) of data, and fields within the frames that hold specific types of information.

Channel Tier-Channel tier is the grouping (partitioning) of channels into specific category levels that have similar service or content characteristics.

Channel Type-A channel type is a set of characteristics associated with how products are transferred from the manufacturer or source of the products to the customers who use or consume them.

Channels of Distribution-A channel of distribution is the routes or sales path that a product service uses to get from the original manufacturer or supplier to the customer or end user.

CHAP-Challenge Handshake Authentication Protocol

Character Set-A set of different characters such as letters, numbers, and symbols that are agreed upon to represent characters in a computing system.

Charge-(1-electric) The product of electric current and time. The SI unit of electric charge is the coulomb, named for a 19th century French physicist. One coulomb is the product of one ampere of current and one second of time. There are approximately 6,250,000,000,000,000,000 electrons in one coulomb of electric charge. To state that in another way, the electric charge of one electron is negative and its magnitude is 1.6 E-19 coulomb.(2-verb) The process of replenishing or replacing the electric charge in a capacitor, secondary cell or storage battery. (3-cost) Charge is the name for a price or fee or cost of some goods or services. The verb form of this word is the action of computing or applying a cost.

Charge Account-A charge account is an transaction recording area within a financial system that can store charges and payments for products and services by a company or person. A charge account may have limits associated with it such as maximum allowable balance (credit limit) and who can authorize transactions.

Charge Initiation-Charge initiation is an event that produces or initiates a usage record.

Charge Number-A charge number is an identification code that is used to identify specific accounts for billing purposes.

Charge Rate-A charge rate is the service fee or charge unit for a specific service or service category.

Charge Setting-Charge setting is the initialization, adjustment or posting of a charge amount.

Charge Waiver-Charge waivers are credits that are applied to an account or invoice to reverse a previously charged item or fee.

Chargebacks-Chargeback are the reversal of

a deposit or payment. A chargeback is commonly associated with credit card transactions that are disputed and the credit card processor (merchant processor) taking the money from a credit card sale and refunding the charge to the credit card user.

Charges-Charges are the rates or costs associated with a usage of a service or product.

Charging Attributes-Charging attributes are the characteristics of a service object (e.g. data transfer rate) that can be used to identify, qualify, or assist in the billing of that service.

Charging Categories-Charging categories are classifications that are used to group or relate charge events with each other. Examples of charging categories include reoccurring monthly fees, data usage, and voice service.

Charging Control-Charging control is the interaction of billing systems to receive information and interact with the provisioning and operation services.

Charging Correlation-Charging correlation is the process of identifying and associating accounting related events and services to a user, device or account for which they should be associated.

Charging Gateway (CG)-A charging gateway is a device or processing system that combines, processes, and reformats billing detail records (CDRs, IPDRs, etc) into a format that can be used by a billing system.

Charging Gateway Function (CGF)-A charging gateway function is a device or processing system within a GSM, GPRS, or WCDMA system that combines, processes, and call detail records (CDRs) into a format that can be processed by a billing centre.

Charging Granularity-Charging granularity is the steps of billing a unit that is used to measure service or resource usage. An example of charging granularity is the billing of time usage in 6-second increments.

Check-A check is written order directing a bank to pay money.

Check Authorization-Check authorization is the process of validating funds that are available for a check that has been presented as a form of payment.

Check Clearing-Check clearing is the process of receiving checks, submitting them to banks for payment and receiving the money from the bank that the checks are drawn on.

Checkout Counter-A checkout counter is the location where a sales transaction occurs.

Checkout Manager-A checkout manager is a program or system that coordinates the setup, selection and sequencing of screens or web pages for an online store order processing system.

Checksum-A number that is calculated from a block of data or sequence of data bits that is sent along with the data and used by the receiver to detect if transmission errors occurred. The receiver will use the same calculation process on the received data to calculate an compare the appended check sum with the value it has calculated. In some cases, the checksum can be used to correct some of the bit errors.

Chief Financial Officer (CFO)-A chief financial officer is a person who is responsible for the planning, selection and implementation of financial systems.

Chief Information Officer (CIO)-A chief information officer is a person who is responsible for the planning, selection and implementation of information systems.

Chief Operations Office (COO)-A chief Operations officer is a person who is responsible for the planning, selection and implementation of operations systems.

Chief Technology Officer (CTO)-A chief technology officer is a person who is responsible for the acquisition, research and development of technologies.

Chip-(1- component) A small slice of material, usually composed of a semi-conductor such as Silicon, on which a electronic assembly is built (such as a microprocessor). The name "chip" is commonly used for an "integrated circuit." (2-spread spectrum) The smallest information element for the transmission of a spread spectrum signal. Each bit in a spread spectrum system is composed of several chips.

Churn-The process of customers disconnecting from one telecommunications service provider. Churn can be a natural process of customer geographic relocation or may be the result of customers selecting a new service provider in their local area.

Churn Posture -The churn posture defines the default strategic position that upper management desires to communicate to others (within the company and to outsiders such as investors) regarding their values related to customer churn. Examples of churn posture include low or high value of churn management (customer retention).

Churn Rate-Churn rate is the ratio of the net number of subscribers who disconnect over a period (adds-disconnects) of time as compared to the number of active subscribers in the system over a period of time. For communication systems, the churn rate is typically calculated on a monthly basis and expressed in percentage form (e.g. 1.2% churn rate means that 1.2 customers disconnected for each 100 customers who subscriber to system services).

CIBER-Cellular Intercarrier Billing Exchange Roamer

CIC-Carrier Identification Code

CID-Conference Identifier

CID-Connection Identifier

CIF-Common Intermediate Format

CIF-Cost, Insurance and Freight

CIM-Common Information Model

CIM-Customer Interface Management

CIO-Chief Information Officer

Cipher-An algorithmic transformation of data typically used to prevent unauthorized viewing or alteration of the original data. There are many different types of ciphers. One of the most common is based upon the Digital Encryption Standard, DES, algorithm.

CIR-Committed Information Rate

Circuit-(1-communication) Any communication path through which any information can be transferred. (2-electronics) A combination of electrical processing components that perform a process (such as signal amplification) or function (clock display processor).

Circuit Assignment-A process that identifies, reserves, or designates a partially or wholly inventoried equipment item to a circuit.

Circuit Bonding-The combining of multiple circuits to one or more communication channels to increase reliability through circuit redundancy or to share data services.

Circuit Charges-Circuit charges are fees or assessments for the allocation or dedicated of communication paths between access points in a communication network.

Circuit Layout Record (CLR)-A circuit layout record is a set of information associated with a circuit connection that defines the types of services the circuit has.

Circuit Layout Record Card-The card containing a circuit layout record and related information.

Circuit Miles-The route miles of circuits in service, determined by measuring the length in miles of an actual path followed by a transmission medium.

Circuit Provision-The process used by carriers to determine the need for trunks and special-service circuits.

Circuit Provision Center-A center that assigns facilities and equipment for message trunk circuits and designs special-service circuits and carrier systems. A circuit provision center also generates and maintains circuit records and inventory and assignment records for all interoffice facilities and equipment.

Circuit Provisioning-Circuit provisioning is the process used by a service provider (network operator) that provides a circuit connection between two (or more) points.

Circuit Reliability-The percentage of time a circuit is available to the user during a specified period of scheduled availability.

Circuit Switch-A circuit switch is a device or assembly that receives signals on input ports and provides a continuous connection (may be a physical or logical connection) to output ports.

Circuit Switched Data-Circuit switched data is a data communication method that maintains a dedicated communications path between two communication devices regardless of the amount of data that is sent between the devices. This gives to communications equipment the exclusive use of the circuit that connects them, even when the circuit is momentarily idle.

To establish a circuit-switched data connection, the address is sent first and a connection (possibly a virtual non-physical connection) path is established. After this path is setup, data is continually transferred using this path until the path is disconnected by request from the sender or receiver of data.

This figure shows the basic operation that uses circuit-switched data. In this example, a laptop computer is sending a file to a company's computer that is connected to the public switched telephone network (PSTN). The laptop computer data communication software requests the destination phone number from the user to connect to the remote computer.

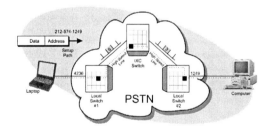

PSTN Circuit Switched Data

This telephone number (the address) is used to connect a path through the PSTN switches until the call reaches the destination computer. The dialed number is first connected through local switch #1, port number 4236. This port number is assigned to a memory location in the switch that routes the data connection through a high-speed line, time slot 6 to an Inter-eXchange Carrier's (IXC) switch. The IXC switch then assigns a memory location in its switch to a high-speed line, time slot 3 that connects to local switch #2. Local switch #2 assigns a memory location in its switch to port number 1249. This port connects to the remote computer. Once this path through the network is setup, it remains constant throughout the data communications session regardless of how much data is transferred between the laptop computer and the company's computer.

Circuit Switched Voice Service-Circuit switched voice service is a communication method that maintains a dedicated communications path between two communication devices regardless of the amount of information that is sent between the devices. This gives to communications equipment the exclusive use of the circuit that connects them, even when the circuit is momentarily idle.

Circuit Switching-A process of connecting two points in a communications network where the path (switching points) through the network remains fixed during the operation of a communications circuit. While a circuit switched connection is in operation, the capacity of the circuit remains constant regardless of the amount of content (e.g. voice or data signal) that is transferred during the circuit connection.

This figure shows how circuit switching is used for voice communication. In this example, a telephone is dialing a telephone that is connected to a distant switch. When the user dials the telephone, the dialed digits are captured and used to program the circuit switches between the two telephones. Each switch

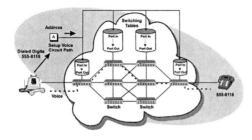

Circuit Switching Operation

then has assigned input ports and output ports and each switch only adds a small amount of transfer time between ports. After all the switching connections are made, an audio path can be connected between. Throughout the connection, this path will be maintained through the initial path (the same switch ports) without any changes.

Circuit Work Location-Any work location associated with a given circuit, such as the service address for a customer termination, for the central office or intermediate locations and for remaining terminations.

Circulation-Circulation is the number of copies of a media item (such as magazines) are distributed.

Clarification Requests-Clarification requests are questions that ask for details or an expanded description of terms in RFPs or other documents.

Clarification Responses-Clarification responses are details or expanded descriptions that are provided to people or companies that sent clarification requests for RFPs or other documents.

CLASS-Custom Local Access Signaling Services

CLASS (TM) Services-A group of telecommunication services that provide selective-call screening, alerting, and calling-identification delivery functions. Bellcore owns the trademark for CLASS (TM) services.

Class 3 Switch-The class 3 switch is the hierarchical interconnection for Class 4 and optional Class 5 switches. Class 3 switches are commonly used for Interexchange Carrier (IXC) services and are also known as primary switching offices.

Class 4 Switch-A voice communications switching system used to interconnect local telephone switching systems. The class 4 switching system was one level above the class 5 end office switching system. Class 4 switches are also known as "Tandem Switches."

Class 5 Electronic Switching System-A classification of a switching system that is used by local telephone service providers. A class 5 switch is the last point in the network prior to the customer. Class 5 switches usually can handle anywhere from 10,000 to 100,000 customers.

Class Of Device(CoD)-Class of device is a parameter or code that is used to identify the capabilities of a device. For the Bluetooth system, the class of device is obtained during the device discovery procedure. This parameter is conveyed from the remote device, indicating the type of device it is and types of services it supports.

Class Of Service (COS)-(1-multimedia) Class of service is the communication parameters that are assigned or associated with a particular application or communication session. The class of service usually requires a specific quality of service (QoS) level. (2-telecommunications) Categories of services that are provided by tariff for charging customers for the particular service they select. Examples of class services include flat rate, coin, toll free (800) service, and PBX. (3-SS7) Services provided by the Signaling Connection Control Pant (SCCP) to its users.

Classified Advertising-Classified advertising is the promotion of advertising messages in a specific area that allows the advertising messages to be grouped into specific classifications.

CLD-Competitive Long Distance Carrier

Clean Bill of Lading-A clean bill of lading is a document that is issued on behalf of the shipping carrier that defines the receipt for goods at a specific location that does not have any modifications or changes to the goods, terms or conditions.

Clear Channel-(1-radio) A carrier frequency that is licensed to only one transmitter in the entire nation, thus ensuring no radio interference. (2-digital telephone systems) A digital channel, typically 64 kb/s, for the subscriber that can carry any binary bit stream without restriction, including a string of binary zeros.

Clear Channel Capability (CCC)-A characteristic of DS1 transmission in which the 192 information bits in a frame can carry any combination of zeros and ones.

Clear Defective Pair-A facility modification that requires the repair of a defective wire pair for a specific service order or related line and station transfer.

Clearing House (Clearinghouse)-A clearinghouse is a company or association that transfers billing records and/or performs financial clearing functions between carriers that allow their customers to use each other's networks. The clearinghouse receives, validates and accounts for telephone bills for several telephone service providers. Clearinghouses are particularly important for international billing because they convert different data record formats that may be used by some service providers and convert for the currency exchange rate.

Clearinghouses provide a variety of services including processing proprietary records (e.g. switch records) into formats understandable by the member carriers' billing systems, validate charges from carriers with intersystem agreements, and extract unauthorized or unbillable billing records. Clearinghouses transfer messages in a standard format such as exchange message record (EMR), cellular inter-carrier billing exchange roamer (CIBER), or transferred account process (TAP)

format. The EMR format is often used for billing records in traditional wired telecom networks and the CIBER and TAP formats are used for wireless networks. The records may be exchanged by magnetic tape or by other medium such as electronic transfer or CD-ROM.

Clearing of the Boards-Clearing of the boards is the removal of all the scheduled programming time slots so that a new schedule can be created from a blank schedule. Clearing of the boards may be performed periodically in a television broadcast system to re-prioritize programming schedules.

Clearinghouse-Clearing House

Cleartext-Cleartext is data or information that has not been encrypted. The headers of data packets may be sent in cleartext to allow the communication device to setup and receive data packets.

CLEC-Competitive Local Exchange Carrier

CLI-Called Line Identification

CLI-Calling Line Identification

Click And Mortar-Click and mortar is a term used to describe a business that has one or more physical locations that sell products and services and has an online storefront.

Click Counter-Click counter is device or software program that counts the number of clicks that occur on a link or other selectable item (such as an image) over a period of time.

Click Fraud-Click fraud is the selection of links ("clicks") by users who do not have an interest in using the link for its intended purpose. Click fraud may result from users who receive compensation for link clicks on their web sites or web sites they manage from advertisers.

Click Fraud Detection-Click fraud detection programs or services operate by identifying the visitor and storing the information about the visitor to allow comparison of click selections that occur at another time.

Click Fraud Detection Message-A click fraud detection message is a text or image display that informs a web user that they have been observed to repeatedly click on a link which is not authorized. A click fraud detection message may indicate what actions may be taken if the user continues to click on the link.

Click Fraud Monitoring-Click fraud monitoring is the identification of link selections (clicks) that are performed by people or systems for the purpose of gaining value (e.g. affiliate advertising revenue) or harming the owner of the link.

Click Fraud Tools-Click fraud tools are programs, systems or processes that can be used to identify and analyze the sources of click fraud.

Click Rate-Click rate is the ratio of how many times an item is clicked (selected) as compared to the number of time it is viewed.

Click Through Rate (CTR)-Click through rate (usually in percentage form) is a ratio of how many clicks a link or advertising message receives from visitors compared to the number of times the link or advertising message is displayed. An example of click through rate is a link that is clicked 5 times out of 100 displays to visitors is 5%.

Click Through Tracking-Click through tracking is gathering of click selection information for web pages and grouping them with the identity or IP address of the visitor.

Click to Call (CTC)-Click to call is a combined voice and data service that allows a user who is viewing a web page or IP television program to click a link or image on the display to initiate a voice call. The link contains an embedded address (URL or IP address) that connects to a call server along with the necessary software (such as SIP) that allows for the setup and connection of the call. Click to dial service is similar in concept to the 'mailto:' link that can launch a user's email software when selected.

Click to Dial or Click to Call-Click to Dial is a combined voice and data service that allows a user who is viewing a web page to click a link on that web page to initiate a voice

over Internet call. The link contains an embedded address (URL or IP address) that connects to a call server along with the necessary software (such as SIP) that allows for the setup and connection of the call. Click to dial service is similar in concept to the 'mailto:' link that can launch a user's email software when selected.

Clickstream Analysis-Clickstream analysis is a review of visitor link selection (click) information during communication sessions (a web site visit) to determine how visitors are reacting to promotions, offers or other factors related to Internet marketing.

Clickthrough-A clickthrough is the selection of a link by a web site visitor or email recipient.

CLID-Calling Line Identification Display

Client-A client is a computer, hardware device or software program that is configured to request services from a network. Client also may refer to the codec or terminating device located at one end of a network node.

Client Authentication-Client authentication is a process of exchanging information between a user of services (a person and/or a device) and the provider of services (such as a communication network).

Client Device-A client device is a hardware assembly that is configured to request services from a system or network.

Client Integration-Client integration is the process of installing, testing and validating that a software program or device that can request services from another device or system performs its desired operations.

Client Side-Client side is the requesting (user) side of a system or service.

Client Side Tracking-Client side tracking is the process that uses the requestor or user access device to monitor (track) the usage of a product or service.

Client Software-Software that is controlled by a user that requests processing services from another computer or system. An example of client software is an Internet browser that requests information from a web site.

Client/Server Architecture-A form of distributed computing in which each application is viewed as a series of inter-dependent tasks accomplished by several different computers linked on a network, for example, a personal computer can be a client while it is linked to a remote processor acting as a server. The various tasks performed by the client is considered to be front end operations and those performed by the server are backend operations

This diagram shows the configuration of a client/server network. A client/server network is a form of distributed computing in which each application is viewed as a series of inter-dependent tasks being accomplished by several different computers linked on a network. For example; a personal computer can be a client while it is linked to a remote processor while it is acting as a server for another computer. The various tasks performed by the client are considered to be front end operations and those performed by the server are back-end operations.

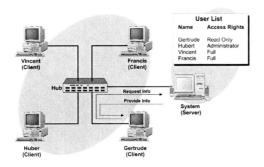

Client/Server Network

CLIP-Calling Line Identification Presentation
Clip Art-Clip art is an existing set of diagrams, images or animation segments that can be used or inserted into media materials.
CLIR-Calling Line Identification Restriction

CLLI-Common Language Location Identification
CLLI Code-Code Common Language Location Identifier Code
Clock-A clock is a reference source of timing information that is used to coordinate the operation of equipment, machines or systems. A clock can be an electronic circuit that produces timing pulses that is used or shared by several components, circuits or assemblies.
Clock Synchronization-Clock synchronization is a process that adjusts the timing of a signal to match or become relative to a clock signal.
Clone-A clone is a device or assembly (possibly a mobile telephone) that contains identification information that is acquired from another device or assembly. This information may come from a stolen device or they may be duplicated from identification numbers.
Cloning-Duplicating certain information necessary for the connection of communication (such as wireless telephones) devices in a fraudulent manner. Cloning the ESN's of existing customers is an example of cloning.
Closed Circuit Television (CCTV)-Closed circuit television is a private (closed circuit) network video transmission system that can display images on one or more television (video) monitors.
Closed Network-A closed network is a communication system that is accessible only by private users.
Closed Order-A closed order is a purchase authorization that has been completely fulfilled.
Closed Pricing-Closed pricing is the restriction of price lists and intended price changes to a limited number of people or companies who want to see them.
Closed User Group (CUG)-(1-access restriction) A group of directory numbers sharing an access restriction such that any directory number can reach others in the group but cannot access outside numbers. (2- cellular system) Advanced features such as 4-digit dialing

authorized for a closed group of users of the service. (3-X25 protocol) In the X.25 packet-switching protocol, a facility indicating a virtual grouping of terminals that can communicate only with other members of that group. The feature can be extended to a closed user group with outgoing access, or a closed user group with incoming access.

Closing Submission Date-A closing submission date is the last date possible to submit a document or proposal (such as an RFP).

Closing the Sale-Closing the sale is the process of completing an agreement (getting authorization for an order) that defines terms that are agreeable to a sales prospect for the acquisition of your product or service.

CLR-Cell Loss Rate

CLR-Circuit Layout Record

Cluster-(1-data communications) A group of storage devices, processors, or computers that share a common bus or network and function as a single system or sub-system within a larger network. These devices are often coordinated by a cluster controller. (2-memory device) The amount of memory storage capability within a sector on a diskette or hard disk. (3-SS7) In the Signaling System 7 protocol, a set of signaling points that are identifiable as a group within the signaling-point code-address spare.

Clustering-Clustering is the grouping of resources or servers within a network to increase system reliability and/or to distribute the processing requirements of the system.

CM-Cable Modem

CM-Configuration Management

CMDS-Centralized Message Data System

CMLA-Content Manager License Administrator

CMR-Call Management Record

CMRA-Commercial Mail Receiving Agency

CMS-Content Management System

CMTS-Cable Modem Termination System

CND-Called Number Delivery

CND-Calling Number Delivery

CNG-Comfort Noise Generator

CNI-Calling Number Identification

CO-Central Office

Coax-Coaxial Cable

Coaxial Cable (Coax)-A coaxial cable (coax) is a multi conductor cable comprising a central wire conductor surrounded by a hollow cylindrical insulating space of air, or solid insulation, or mostly air with spaced insulating disks, finally surrounded by a hollow cylindrical outer conductor. Invented by Lloyd Espenschied in the 1930s for transmission of wideband television and radar signals, it is also used for other purposes. The hyphenated form of the name is preferable to avoid confusion with the English word "coax." Versions using more than one internal conductor are called "tri-ax," "quad-ax" for three total conductors or four total conductors respectively. Co-ax exhibits lower radiated electromagnetic power losses because the electromagnetic field is confined inside the outer conductor. The form with the hyphen is preferable to avoid confusion with the English word coax.

This figure shows a cross sectional view of a coaxial cable. This diagram shows a center

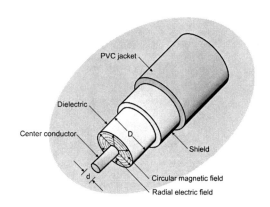

Coaxial Cable Diagram

conductor that is surrounded by an insulator (dielectric). The insulator is surrounded by the shield. This diagram shows that during transmission, electric fields extend perpendicular from the center conductor to the shield and magnetic fields form a circular pattern around the center conductor.

COD-Cash on Delivery

CoD-Class Of Device

COD-Connection Oriented Data

Code-(1-processing) An instruction in computer language. (2-software) Software instructions that are used within a computer program. (2-identifier) A unique pattern or label that is used to identify an item or object. (4-morse code) A code is the system of dots and dashes used to represent the letters of the alphabet, numerals, punctuation and other symbols.

Code Common Language Location Identifier Code (CLLI Code)-CLLI codes are used to provide unique identification of facilities (equipment and cables) between any two interconnected CLLI coded locations. The CLLI code is mnemonic code that can be a maximum of 38 characters. An example of a CLLI code is 115T3NYCNY20DALTX. This example shows a T-3 carrier is connected between New York City, New York and Dallas, Texas.

Code of Conduct-A code of conduct is a set of rules or guidelines that define how people or companies should act. A code of conduct may be specified as part of an agreement to help ensure that employees or contractors behave in a consistent and acceptable way as part of the terms of their agreement.

Code of Federal Regulations (CFR)-Code of federal regulation is a set of US government rules that represent broad areas subject to federal regulation. There are different parts of CFR that relate to different industries such as title 47 that pertains to telecommunications.

Code of Practice #3 (COP3)-Code of practice number 3 specifies working guidelines for the deployment of television services over tele-com lines, notably over digital subscriber line (DSL) and fiber transmission systems. The most significant attribute of COP3 is the use of forward error correction (FEC) on the packet stream to compensate for the high transmission error rate found on DSL circuits.

Code Signing-Code signing is a process that adds an identifying signature to software code to help ensure that it is not malicious or that it has not been altered from its original source.

Codec-Coder/Decoder

Coded Speech-A speech signal that has been changed into a standard digital code form.

Coded Trunks-Trunks that are coded for identification so they can be identified as they hunt (search) for connections.

Coder/Decoder (Codec)-Codecs are devices or software that are used to compress (code) or expand (decode) information to a fewer number of bits for more efficient transmission and storage. Normally the term codec applies only to compression of human-perceived signals such as speech, audio, images, or video.

Codifying-Codifying is the process of organizing and arranging a system or process.

Coding-(1-digital) A process of changing digital bits to include error protection bits and/or signaling bits prior to the sending or storing of the information. (2-software) The process of writing instructions or commands for software programs.

Coexistence-Coexistence is the ability of two or more systems to operate independently with each other with no or reduced performance levels.

Coin Collect Message Type-A message charged to the coin station at which it terminates.

Coin Paid - Message Type-A message sent and paid from a coin station.

Coin Supervisory Trunk Group-A communication trunk that permits call processing features so an operator can check the operation of a pay telephone (e.g. stuck coins) and to allow the recording (coin entry) of overtime charges.

Coin Telephone-A telephone that allows the collection of coins. William Gray invented the coin telephone.

Coin Zone Trunk Group-A trunk group that enables cord switchboard operators to supervise charge arrangements for coin dialing to destinations beyond the basic coin-rate zone.

Coinless Public Telephone Service-A service for use of a coinless public telephone for originating calls only. The phone is served by a single-party, loop-start line with no extensions and no custom calling features.

Cold Boot-The startup process that begins when equipment (such as a computer) or device is first powered on. A cold boot is used to start an operating system from a known good point (before other programs are activated).

Collaboration-The process of two or more people working together on a shared medium (such as a word processor file or sketch whiteboard). Collaboration usually enables a group of people to work together in real-time to share screen displays, documents, and video images.

Collaborative Phase-A collaborative phase is a portion of a development process when people or companies to work together to develop or improve standards, products or systems.

Collaborative Planning, Forecasting and Replenishment (CPFR)-Collaborative planning, forecasting and replenishment is a set of processes that are used by several participants who can interact with each other to identify, forecast and schedule the acquisition or development of resources that are used to produce products or services in the future for a business or company.

Collar-A collar is a financial guarantee.

Collateral Materials-Collateral materials are items or tools that are used to assist in the selling of products or services.

Collect Call-A collect call is a telephone call where the call recipient pays for the call. The person who initiates the call requests that call be made as a collect call. Collect calls usually require the obtaining of authorization from the call recipient to accept the charges.

Collection-The methods and procedures used by the service provider to receive payments due from a customer (also known in Europe as "chasing up".) Sometimes also refers to the methods used to minimize the risk that a customer will end up in arrears.

Collection Agency-A collection agency is a company that performs bill collection activities.

Collections-Collections are activities that a service provider performs to receive money from their customers. Ideally, all customers will receive their bills and pay promptly. Unfortunately, not all customers pay their bills and service providers must have a progressive collection process in the event a customer does not pay their bill.

When customers are first added to a system, they are rated on the probability that they will pay their bills. This is accomplished by using information on their application and reviewing the credit history as provided by an independent credit reporting agency.

The collection process for delinquent customers usually starts by sending a reminder messages to the customer be mail or recorded audio message. If initial attempts to collect are unsuccessful, more aggressive collection activities will progress that include restricted calling, service disconnection and sending or selling the uncollected invoice to a collection service.

Collective Action-Collective action is a process or activity that is performed by or for the benefit of a group of companies and/or people who share a common interest.

Collective Licensing System-A collective licensing system is a process that allows a collective group of technologies or intellectual property to be licensed as a complete group instead of identifying and negotiating licenses for each part separately.

Co-location-The location of equipment from systems or multiple carriers at the same facility. Co-location commonly refers to the location of a competing telephone service provider's equipment at a local telephone company's switching facilities. This enables providers of interstate or competing telecommunications service providers to connect their facilities directly to those of a local exchange carrier (LEC).

Co-location-Colocation

Colocation (Co-location)-Collocation (also co-location) is the placement of equipment (such as multiple antennas) at a common physical site to reduce environmental impact and real estate costs and speed zoning approvals and network deployment. Co-location can be affected by competitive and interference factors. Some co-location services are performed by antenna leasing companies who act as brokers arranging for sites and coordinating several carriers' antennas at a single site.

COM-Common Object Model

Co-marketing-Co-marketing is the promoting of products or services by more than one company. Co-marketing of products or services commonly involves the manufacturer or service provider giving marketing allowances or incentives to retailers or other sales focused companies for specific marketing promotions in their market areas.

Combination Offer-A combination offer is a promotion of more than one product or service in the same brochure or mailing.

Combination Rate-A combination rate is a price that is offered to a buyer (such as an advertiser) when they buy multiple products or services (such as advertising space or time in more than one venue).

Combination Trunk-A trunk line that can operate as a direct inward dial (DID) or direct outward dial (DOD) trunk. This allows them to receive (ring-in) and send calls.

Comfort Noise Generator (CNG)-A circuit or system that creates a noise signal that is inserted as background noise on a communication line to help the call participants sense that the line is still active when none of the participants are talking.

Comma Delimited File-A comma delimited file is a data file format that divides the fields in each record by commas.

Comma Separated Value (CSV)-Comma separated value is a record structure that is used import data by separating the information into records (lines) and items (fields) where the fields are separated by commas.

Command Set-A command set is a set of instructions that are used to initiate and control processes or devices.

Commerce Extensible Markup Language (cXML)-Commerce extensible markup language is a set of commands and processes that define exchangeable elements for electronic business transactions. More information on cXML can be found at www.cXML.org.

Commerce Revenue-Commerce revenue is a source of money that is created through the produce sales activity of customers. Commerce revenue is different than service revenue. An example of commerce revenue is the purchasing of a pizza through an interactive television advertisement message. The commerce revenue is the money generated from the pizza sale and not the money that is generated from the television service.

Commerce Server-A commerce server is data processing device that receive, process and respond to commercial transaction requests. Commerce servers may be able to maintain inventory databases, manage shopping carts and process payments.

Commerce Service Provider (CSP)-A commerce service provider is a company that provides companies with tools or services that are used to perform business transactions over the Internet. These services may include order processing, inventory management, distribution control, web marketing and other supporting services.

Commercial Advertising-Commercial advertising is the sending of promotional messages when there is a form of compensation that results from the sending of the messages.

Commercial Contractor-A commercial contractor is a company to whom a government agency awards a contract.

Commercial Loan-A commercial loan is the temporary providing of cash or assets where the loans are repaid or secured by a business or organization.

Commercial Mail Receiving Agency (CMRA)-A commercial mail-receiving agency is a company that is authorized to receive mail on behalf of other companies.

Commercial Service (CS)-Commercial service is the providing of communication and/or processing of data or media where a usage charge or other value transfers occur to obtain access and/or use of the service.

Commercial Use-The providing of communication and/or media transmission on a fee basis.

Commission Import-Commission importing is the recognizing of commission costs or assignments and the capturing of these costs into financial systems.

Commission Management-Commission management is the process that records, coordinates and processes information that is used for identification, entry and tracking of payment for commissions that are paid for the sale of products or services.

Commission Model-A commission model is a representation of the commission structure and how they will be used to define how a business will generate revenues or profits.

Commission Schedule-A commission schedule is an itemized list or table that provides commission levels that are paid for the sales of products or services that occur in specific ranges or categories.

Commission Structure-A commission structure is a group of sales commission rules that define how people and companies will be com-

pensated on a commission basis for their sales efforts.

Commitment-Commitment is the reservation of resources (such as money) that are assigned for specific projects or acquisitions.

Commitment Costs-Commitment costs are the fees and assessments that are charged for the installation of cable or optical fiber pairs.

Commitment Date (CD)-A commitment date is a date that is agreed by two or more parties on which work activities will be completed or services will commence.

Committed Information Rate (CIR)-Committed information rate is a guaranteed minimum data transmission rate of service that will be available to the user through a network. Applications that use CIR services include voice and real time data applications. CIR can be measured in bits per second, burst size, and burst interval.

Some service providers allow users to transmit data above the CIR level. However, when data is transmitted above the CIR level, some of the data may be selectively discarded if the network becomes congested.

Commodity-A commodity is a product or service that can be bartered or sold.

Commodity Code-A commodity code is a unique identifier that is assigned to a product or service that can be bartered or sold.

Common Carrier-(1-general) A company that carries goods, services, or people from one point to another for the public. (2-telecommunications) A company that provides communications services and typically is subject to a regulatory agency.

Common Carrier Bureau (CCB)-Common carrier bureau is a division of the FCC that regulates interstate telephone systems.

Common Channel Signaling (CCS)-Common channel signaling is a network control process in which signaling information relating to a multiplicity of circuits (trunks), is conveyed over a separate single channel by addressed messages. Common channel signaling system #7 ("SS7") is the primary system used for interconnection of telephone systems. SS7 sends packets of control information between switching systems.

The SS7 network is composed of its own data packet switches, and these switching facilities are called signal transfer points (STPs). In some cases, when advanced intelligent network services are provided, STPs may communicate with signal control points (SCPs) to process advanced telephone services. STPs are the telephone network switching point that route control messages to other switching points. SCPs are databases that allow messages to be processed as they pass through the network (such as calling card information or call forwarding information).

This diagram shows the basic structure of the SS7 control signaling system. The SS7 network is composed of its own data packet switches, and these switching facilities are called signal transfer points (STPs). In some cases, when advanced intelligent network services are provided, STPs may communicate with signal control points (SCPs) to process advanced telephone services. STPs are the telephone network switching point that route control messages to other switching points. SCPs are databases that allow messages to be processed as they pass through the network (such as calling card information or call forwarding information). Messages originate and

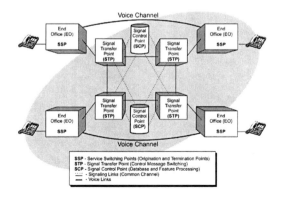

Common Channel Signaling System

terminate at a service switching point (SSP.) A SSP is a part of the end office (EO) switching system. End offices are sometimes called central offices (COs).

Common Channel Signaling Network-The separate control signaling network that sets up, maintains and disconnects communication paths that are used to transfer customer voice and data information. The control signals are routed via specialized network connection points (nodes) that do not interact directly with the customer voice or data information.

Common Control-A system in which items of control equipment are commonly shared (such as network control signaling).

Common Equipment-Any apparatus used by more than one unit of equipment, or any equipment used by more than one channel; that is, equipment common to two or more channels.

Common Gateway Interface (CGI)-A standard software program interface that allows Web servers to interact with user specified applications such e-commerce programs and databases. CGI describes and uses variable control tags that are inside an HTML file to allow information to transfer from user applications (programs and web pages) to the Web server host program. Because the CGI defines specific information variables and how these variables (information elements) are processed, CGI scripts can be created and used on any web based operating system.

Common Information Model (CIM)- Common information model is the use of a business objects or entities to identify, describe and manage common information that apply to multiple applications or systems. The TM Forum shared information data model (SID) provides guidelines for the information modeling that can be used in the communications industry.

Common Intermediate Format (CIF)-A video format, adopted by the Specialists Group of the International Telegraph and Telephone Consultative Committee (CCFIT), for transmission of video signals on Integrated Services Digital Networks (ISDNs). The standard CIF has 288 lines of 360 pixels each for luminance and half as many lines and pixels for each of two chrominance, or color, components. A second version, 1/4 CIF, has one-fourth that resolution.

Common Language Location Identification (CLLI)-A standard code used by telecommunications systems to identify specific locations of a switching office or network element. A CLLI code is composed of 11 alphanumeric characters. The first four characters of the CLLI code are an abbreviated place name. Characters 5 and 6 are state abbreviations. Positions 7 and 8 identify a specific building and 9,10 and 11 represent a particular piece of equipment.

Common Object Model (COM)-Common object model is a specification developed by Microsoft and DEC to allow objects to operate and interact across multiple platforms.

Common Object Request Broker Architecture (CORBA)-Common object request broker architecture is a set of defined software programs and processes that operate between (middleware) software application objects. CORBA allows different programs from different companies and different systems to interact with each other. CORBA is managed by the open management group (OMG). More information about CORBA can be found at www.OMG.org.

Common Open Policy Service (COPS)-The COPS protocol allows a system to implement policy decisions by allowing a client to obtain system configuration and parameter information from a policy server. COPS is defined in RFC 2748.

Common System-A system that shares power, interconnections, or environmental support for network elements associated with transmission and switching.

Common Trunk-Trunks within a telephone systems that are accessible to all groups of service (trunk) grades.

Communication Center-A facility responsible for the reception, transmission, and delivery of information.

Communication Channel-A communication channel is a medium that transfers information from a source to a destination. A communication channel may transport one or many communication circuits.

Communication Device-A communication device is a product, assembly, network unit that can receive and/or transmit information with one or more communication devices.

Communication Link-A communication link is a transmission system associated switch that allow information to pass between two points.

Communication Procedures-Communication procedures are the rules and processes that are used to communicate with other people or companies.

Communication Server-A communication server is a computing system that can receive, process, and respond to an end user's (client's) request for communication services.

This figure shows the different types of servers used in some SIP based communication systems. This example shows that a call manager (proxy server) receives and processes call requests from communication units (IP tele-

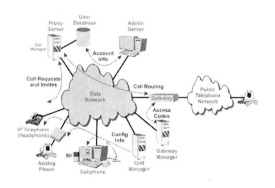

SIP Communication Servers

phones). The administrator server coordinates accounts to the system. A unit manager (location server) functions as a location server by tracking the IP address assigned to the communication units. The gateway manager identifies and coordinates communication through the available gateways. The system manager coordinates the communication between the different servers and programs available on the system.

Communication Service Provider (CSP)- A communication service provider is a company that provides information and/or performs information transfer actions (services) to customers.

Communication Services-Communication services are the processes that transfer information between two or more points. Communication services may involve the transfer of one type of signal or a mix of voice, data, or video signals. When communication services only involve the transport of information, they are called bearer services. When communication services involve additional processing of information during transfer (such as store and forward), they are known as teleservices.

Communication Session-A communication session is the setup, transfer and disconnection of information flow between two or more communications points or devices.

Communication Workers of America (CWA)-Communication Workers of America is a national union of communication industry employees.

Communications-The conveyance of information, including voice, images, and/or data, through a transmission channel without alteration of the original message.

Communications Assistance For Law Enforcement (CALEA)-A statute that was enacted by the U.S. congress in 1994 to define requirements of telephone service providers to provide wiretap capabilities to law enforcement agencies. To attach a listening device to a communication line, the law enforcement agency must have a surveillance order from a court of competent jurisdiction.

Communications Channel-The medium and Physical layer devices that convey signals among communicating stations.

Communications System-A collection of individual communications networks, transmission systems, relay stations, tributary stations, and terminal equipment capable of interconnection and interoperation to form an integral whole. The individual components must serve a common purpose, be technically compatible, employ common procedures, respond to some form of control, and, in general, operate in unison.

Community-(1-members) A community is a set of people who share in common interests or activities. (2-network) A network community is a set of communication and/or processing entities that can share and exchange information.

Community Access Television (CATV)-A system or process of delivering quality television reception by taking signals from a well-situated central antenna and delivering them to people's homes by means of a coaxial cable network.

Company Account-A company account is a customer account that is directly managed by the company that produces or provides the products or services.

Company Background-Company background is a description of the key attributes and objectives of a company. A company background may include product types and industry categories a company participates in along with the core values of the company.

Company Credit Card-A company credit card is a financial instrument that can be processed (charged) by companies for the payment for products or services and the charges accumulated on the card are guaranteed to be paid for by the company.

Company Logo-A company logo is an image or shape that is used to help identify a company or business,

Company Policies-Company policies are a set of rules or guidelines that define acceptable and/or unacceptable types of activities and actions that may be performed by employees, contractors or other people who work with the company.

Company Slogan-A company slogan is a short phrase that represents a company image or objectives, is recognizable, and relatively easy to remember.

Company Television Programming (Company TV)-Company television programming (corporate TV) is media that is created and managed for or by a company. Company television may be produced for the public and/or for internal communication purposes. Public company television channels may provide information about products, services or applications of the products or services that are of interest to the public. Internal ("in-house") company television programs may be used to provide employees with educational and company specific information (such as the location of a company meeting or party). Employees, vendors or others who are provided with access may distribute company television programs to monitors within company buildings or for distribution to multimedia computers that are only accessible by company employees.

Company TV-Company Television Programming

Compartmentation-Compartmentation is a process or method that is used to separate information or resources from each other. Compartmentation may be used to coordinate security access to different types or levels of information.

Compatibility-The ability of different systems to exchange information in usable form.

Compensation-(1-signal) The process of passing a signal through an element or circuit with characteristics the reverse of those in the transmission line, so that the net effect is a received signal with an acceptable level/frequency characteristic. (2-financial) Compensation is the providing of financial or other rewards for performance of services or products.

Competition Oriented Pricing-Competition oriented pricing is a setting or assigning of product or service prices that are based or related to the prices charged by other people or companies.

Competitive Access Providers (CAPs)-A telecommunications service provider that offers competing services to an established (incumbent) telephone service provider. CAPs typically compete with a local exchange carrier (LEC). CAPs can provide service by reselling local service from the LEC.

Competitive Advantage-A competitive advantage is an asset and/or processes that are owned or used by a person or company that provides advantages as compared to other businesses.

Competitive Analysis-Competitive analysis is the identification of key competitive components, researching of information of these components that are related to competing companies and analyzing this information to determine strengths and weaknesses of competing companies.

Competitive Intelligence-Competitive intelligence is information or the process of gathering information about a competitor's business operations, products, services, customers, or other business related data to help make business decisions.

Competitive Local Exchange Carrier (CLEC)-A telephone service company that provides local telephone service that competes with the incumbent local exchange carrier (ILEC).

Competitive Long Distance Carrier (CLD)-A competitive long distance carrier is a communication system that can connect local telephone systems with each other. CLDs are also known as long distance carriers.

Competitive Phase-The competitive phase is a portion of a development process that is used to obtain and evaluate competitive proposals for a project.

Competitive Service Provider (CSP)-A competitive service provider is a company or organization that competes with one or more established service providers (such as established telephone companies).

Competitor Click Fraud-Competitor click fraud is the selection of links ("clicks") by companies that compete with the link sponsor who have an interest in using the link for malicious or a non-intended purpose. These intentions may include depleting the PPC advertising funds or distorting the marketing results of link marketing programs.

Compile-To compile is the process of converting a program or data from one language (usually a high-level language) or format into a language (machine language) or format that can be used by a computer.

Compiled List-A compiled list is group of contacts or other information items that has been gathered from a variety of sources which have not requested to be part of the list.

Compiler-A compiler is an executable program that converts a instructions of a computer programming language into the necessary machine level instructions needed to carry out the operations specified in the statements on a specific type of microprocessor. These grammars are typically called high-level programming languages and the compiler will insure that the language constructs obey the rules specified by the particular grammar. Compilers are also capable of linking together multiple files containing machine level instructions into a single executable program. Most compilers do much more than this, however; they translate the entire program into machine language, while at the same time,

they check your source code syntax for errors and then post error messages or warnings as appropriate.

Completion Rate-Completion rate is a ratio of the number of calls or service attempts that are completed as compared to the total number of attempts that were performed.

Compliance-Compliance is the process of following rules or operating within established parameters.

Compliance Monitoring-Compliance monitoring is the process of capturing and evaluating media that is broadcast or transferred so that it can be determined if the media was authorized for use or if it was broadcast when and where it was scheduled to be broadcast.

Compliance Program-A compliance program is a system or process that is used to ensure services (such as billing) are correctly recorded and collected.

Compliance Recording-Compliance recording is the process of capturing and storing media that is broadcasted or transferred so that it can be accessed and reviewed at a later time to determine if the content conformed to rules and regulatory requirements.

Compliance Rules-Compliance rules are the properties and processes that must be followed by a system or service.

Compliant Device-A compliant device is a component or assembly that is validated to conform to a specification.

Component-(1-general) An assembly, or part thereof, that is essential to the operation of some larger circuit or system. A component is an immediate subdivision of the assembly to which it belongs. (2-SS7) in the Signaling System 7 protocol, the portion of the Transaction Capabilities Application Part that identifies a component type, provides correlation between components, specifies operations to be performed, and contains the parameters relevant to that operation.

Component Video-Component video consists of three separate primary color signals: red, green, and blue (RGB). The combination of component video can produce any color and intensity of picture information. Some component video systems are converted into a luminance (brightness) signal and two color difference signals.

Composite-(1-signal) A composite signal is the combination of multiple signals. Composite signals may be created by using a signal multiplexer that adds or multiplies the information from the multiple input signals. (2-video) Refers to a type of video signal or color monitor where color signals are carried by a single input signal and electrically separated (processed) inside the monitor.

Compound Error-Two or more errors on a data element or record.

Compound Mailbox-A compound mailbox is a media storage system or service that holds multiple types of media and messages such as fax, voice mail, e-mail, and video mail.

Compounded Interest-Compounded interest is the loan fee that is calculated on the outstanding principle of the loan and on the interest that has previously been charged on the loan balance (accrued interest).

Compressed Video-Compressed video is a sequence of images or moving picture segment that has been processed using a variety of computer algorithms and other techniques to reduce the amount of data required to accurately represent the content.

Compression-(1-digital) The processing of digital information to a form that reduces the space required for storage. (2-distortion) The altering of signal quality caused by the non-linearly conversion (compression) process in audio and video compression systems.

Compression Ratio-Compression ratio is a comparison of data that has been compressed to the total amount of data before compression.. For example, a file compressed to 1/4th its original size can be expressed as 4:1. In telecommunications, compression ratio also refers to the amount of bandwidth-reduction achieved. For example, 4:1 compression of a 64 kbps channel is 16 kbps.

Compulsory Licensing-Compulsory licensing is the requirement imposed by a governing body that forces a holder of intellectual property (e.g. a patent) to allow others to use, make or sell a product, service or content. Compulsory licensing usually requires the user of the intellectual property (licensee) to pay the owner (licensor) a reasonable license fee along with non-discriminatory terms.

Computer-A device, circuit, or system that processes information according to a set of stored instructions. Typically, an electronic computer consists of data input and output circuits, a central processing unit, memory storage, and facilities for interaction with a user, such as a keyboard and a visual monitor or readout.

Computer Aided Design (CAD)-Computer aided design is a computerized information processing system that can process design information into other usable forms. CAD systems may be able to assist in the design data entry (design rules) and render (display) products or operations from designs and other information entered into the CAD system.

Computer Aided Dialing-The use of a computer system to assist in dialing telephone numbers. Computer aided dialing systems are often used for telemarketing services.

Computer Aided Dispatch (CAD)-A computerized communication system that can coordinate and/or track mobile vehicles. CAD systems can be automated messaging devices to complex computer systems that display maps and vehicle positions on a computer monitor.

Computer Aided Manufacturing (CAM)-Computer aided manufacturing is a computerized data processing system that can use product design and production information to assist in manufacturing processes.

Computer Based Training (CBT)-The use of computers to assist the providing of training. Early CBT systems used text based screens and CD-ROM technology to provide for self paced learning experiences. CBT systems commonly provided training in small sections followed by a test section to ensure student learning prior to allowing additional sections to be presented. CBT is sometimes called computer-assisted instruction (CAI).

Computer Fraud-Computer fraud is the misrepresentation, alteration or use of computer-based data or media to obtain something of value.

Computer Language-The words and rules of construction for phrases and sentences that direct the operation of a computer.

Computer Modeling-Computer modeling is the representation of processes using software programs.

Computer Nerd-A geek is a person who is focused on computers and computer technology. Computer nerds typically do not tend to conform to mainstream habits such as dressing for success and/or regular bathing.

Computer Peripheral-An auxiliary input, output, or storage device under the control of a computer.

Computer Supported Telephony (CST)-Computer supported telephony is the management of call processing within a telephone system through the use of computers.

Computer System For Mainframe Operations (COSMOS)-Software that assigns and inventories central-office equipment to provide effective short-jumper frame management, assignment of facilities, and load balance in a switching system.

Computer Telephony (CT)-Computer telephony (CT) systems are communication networks that merge computer intelligence with telecommunications devices and technologies. This figure shows a sample CTI system computer that contains a voice card. This voice card is connected to a multiple channel T1 line. The voice card connects digital PBX sta-

tions through the voice card to individual DS0 channels on the T1 line when calls are in progress. Several software programs are installed on this system that provide for call processing, IVR, ACD, voice mail, fax, and email broadcasting. The monitor shows a directory of extensions. The advanced call processing feature shows text names along with the individual extensions to allow callers to automatically search through a company's directory without the need to use an operator.

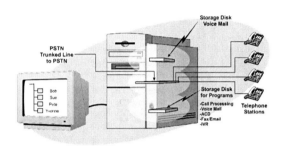

Computer Telephony (CT) System

Computer Telephony Integration (CTI)-CTI is the integration of computer processing systems with telephone technology. Computer telephony provides PBX functions along with advanced call processing and information access services. These services include, prepaid telephony access control, interactive voice response (IVR), call center management, and private branch exchanges (PBX).

This diagram shows how a telephone system can be integrated with a computer system to provide for advanced call processing and information services. This diagram shows the core of the system is a voice card or PBX switch that is controlled by call processing software. The call processing software is customized with information about the system equipment it is operating with. This diagram shows that

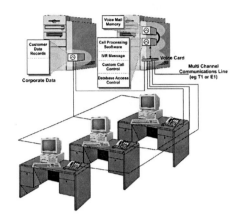

:Computer Telephony Integration (CTI)
System

the computer telephone integration (CTI) system also interfaces to a company database to allow the call center to receive and update information based on both telephony commands (automatic number identification and user selected options) and customer service representative (CSR) screen commands.

Computer Terminal-Computer terminals are input, output, and processing devices that interface human operators to data communication systems. Computer terminals frequently consist of a keyboard, video display monitor, central processor unit (CPU), and communication circuitry that can connect the computer terminal with a data communication network. The term "computer terminal" is often used to describe multiple types of devices including dedicated data "dumb" terminals, scientific workstations, and other types of computers that can communicate with other computers or a host computer.

Concatenate-Concatenation is the process of merging two or more components together to form a single item.

Concentration-The process of combining traffic from many low-usage communication lines to a lesser number of high-usage communication lines (trunks).

Concentrator-In a communication system, a concentrator combines multiple communication sources on to a higher capacity communication path. A concentrator usually provides more efficient communication capability between multiple low-speed channels to one or more high-speed channels.

Concession Schemes-Concession schemes are a structure that is used for providing discounts or offers to customers for retention or sales motivation purposes.

Concurrency Control-Processes or rules that coordinate the access of multiple users who access a database of information.

Conditional Access (CA)-Conditional access is a system or service access control process that is used in a communication system (such as a broadcast television system) to limit the ability of users to obtain or use media or services. Conditional access systems can use uniquely identifiable devices (sealed with serial numbers) and may use smart cards to store and access secret codes.

Conditional Access System (CAS)-A conditional access system is a security process that is used in a communication system (such as a broadcast television system) to limit the access of media to authorized users. Conditional access systems can use uniquely identifiable devices (sealed with serial numbers) and may use smart cards to store and access secret codes.

Conditions of Proposal-Conditions of proposal are the terms for the submission of a document or proposal. Examples of conditions of proposals include how much cost (if any) is authorized for the preparing of a proposal and how confidential or proprietary information will be handled.

Conference Bridge-(1-telephone) A telecommunications facility or service which permits callers from several diverse locations to be connected together for a conference call. The conference bridge contains electronics for amplifying and balancing the loudness of each speaker in a conference call so everyone can hear each other and speak to each other. Background noises are suppressed and typically only the current two or three loudest speakers' voices are retransmitted to other participants by the bridge, while a speaker's own voice audio is not sent back to that speaker to avoid audio feedback, echo or "squealing" self-oscillation. (2- text) A facility to receive the character codes from the keyboards of multiple participants and retransmit this text to the display or printer of all participants in the conference. First used with electromechanical teletypewriters for private networks of automobile parts suppliers. Modern implementations on the Internet are typically called a "chat room."

Conference Call-The connection of three or more telephones to a telephone conversation. This diagram shows how a conference call can use a conference bridge to allow several users to effectively communicate in a conference call (3 or more users). This example shows that this conference bridge uses audio level detectors to determine the level of the microphone audio level for each conference call participant that is talking. As a person begins to talk, the conference bridge increases the gain on the

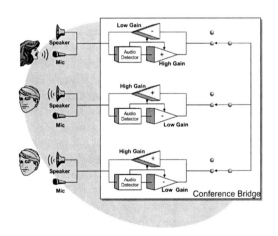

Conference Call Operation

microphone and decreases the gain on the speaker line. This process effectively dynamically reduces the background noise from non-participating members while providing good sound quality to participants that are talking.

Conference Calling (CC)-The process of connecting callers to a communication session that is shared between three or more callers (conferees).

Conference Identifier (CID)-A conference identifier is a unique code that is assigned to specific communication conference sessions.

Conference Of European Postal And Telecommunications Administrations (CEPT)-Original French language name "Conference Européenne des Administrations des postes et des télécommunications" abbreviated CEPT. A European standards body composed of national telecommunications administrators and official carriers. A standards organization for European telecommunications administrations, CEPT formerly issued technological standards. Today all relevant technological standards formerly issued by CEPT have been taken over by ETSI, and CEPT is devoted to tariff, legal and other nontechnical topics.

Confidentiality-Confidentiality is the protection or processing of data to ensure that an unauthorized user or receiver cannot use the information.

Confidentiality Agreement-A confidentiality agreement is a set of conditions or terms that defines how information or inventions can be discussed, shared or distributed.

Configurability-Configurability is the ability to change parameters or information that can control the operation of a device or service.

Configuration-(1-system) A relative arrangement of interconnected equipment or software in a system. (2-file) The information used to adapt an equipment or software program to its environment (configuration).

Configuration File-A configuration file is a set of configuration parameters that are used to adjust the operation or characteristics of a device or application. Application configuration files may have a filename extension of CFG or SET.

Configuration Management (CM)- Configuration management is the process or system that can identify, gather and set configuration parameters in hardware devices or software entities. Configuration management maintains an inventory of equipment and programs, which is typically updated regularly. Configuration management is one of the five functions defined in the FCAPS model for network management.

Configuration Manager-A configuration manager is a system or service that manages configuration information of devices and services.

Configuration Parameters-Configuration parameters are values or optional selections that can be applied to a device, system or service to alter its functions and/or operation.

Configuration Rights-Configuration rights are the authorizations to install, modify installation options or uninstall programs.

Configuration Settings-Configuration settings are values or options that are chosen to adjust the operation or performance of a product or service.

Confirmation-Confirmation is an indication that an event has occurred (such as the receipt of a product or service).

Confirmation Page-A confirmation page is a web page that is provided to a customer after they have performed a task (such as completing an order) that confirms their submission or action was successful.

Conflict Checking-Conflict checking is the review of assignments and resources to determine if overlaps in assignments (conflicts) have occurred.

Conflict of Interest-A conflict of interest is situation where a person or a company has responsibilities, agreements or personal interest to perform actions that could limit, reduce or harm their ability to perform actions for another person or company.

Conformance-(1-general) The ability or certification of a device or system to perform or act as defined to an agreed specification or process. (2-Bluetooth) When conformance to a Bluetooth profile is claimed by a vendor, all mandatory capabilities for that profile must be supported in the specified manner (process-mandatory). This also applies for all optional and conditional capabilities for which support is indicated. All mandatory, optional and conditional capabilities for which support is indicated are subject to verification as part of the Bluetooth certification program. See also Bluetooth Qualification Review Board.

Conformance Points-Conformance points are a combination of profiles and levels in a system (such as an MPEG system) where different products can interoperate (by conforming to that level and profile). An example of conformance points is if ability of a mobile video server to support creation and playback of a simple visual profile at level 0, any mobile phone that has these conformance points should be able to play a video with these profiles and levels.

Conformance Policing-Conformance policing is the monitoring and analyzing services or applications to ensure their assigned quality of service limits have been met.

Conformance Testing-Conformance testing is a group of operational and performance tests that help to determine if a product or system satisfies the requirements of an industry standard or product and/or service specification. Conformance testing helps to ensure (but does not guarantee) that a product will be interoperable with other products or systems of the same type. Conformance testing is often performed by an independent company or laboratory to ensure accurate and unbiased results.

Congestion-A condition that exists when the demands for service on a communications network exceed its capacity to deliver that service.

Congestion Time-The time or probability that a system is congested over any time period.

Connect Day-The day of the week on which a connection is made.

Connect Time-(1-general) The amount of time that is used between the requesting for a connection or service and when the connection or service is provided. (2-radio) That period of time a cellular phone is in radio contact with a cell site. The connect time is not the same as the duration of the conversation. (3-call processing) The local time at a calling party's location when a connection was made.

Connected Applications-Connected applications are software programs that are designed to perform operations using commands or information from users (such as a user at a keyboard), which also have the capability of communicating with other services or applications.

Connected Time-Connected time is the duration between when a path between two objects is initially established and when the connection path has ended.

Connection-(1-circuit) A connection is transmission path between two or more points. (2-communication) A connection is a 2-way communication session that can pass user information between either user.

Connection Identifier (CID)-A connection identifier is a unique name or number that is used to identify a specific logical connection path in a communication system. For the WiMax system, the connection identifier is a 16 bit code.

This diagram shows how a connection identifier (CID) is used in a data communications network to subdivide frames that are transmitted through a communication network so the portions of data can reach their specific destination channels. This example shows that the CID is used to de-multiplex the frames (divide channels) that pass through an access device. In this example, 4 telephone channels are multiplexed onto frames that are sent through

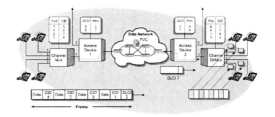

Connection Identifier (CID) Operation

access device 1. These frames are sent on a permanent virtual circuit (PVC) to access device 2. Access device 2 uses the data link connection identifier 7 to route the frames to channel demultiplexer. The channel demultiplexer uses the channel identifier (CID) code to route the data to each specific digital telephone device.

Connection Oriented Data (COD)-A communication connection that allows the sequential delivery of its component parts (packets or frames). COD assures that a supported application such as voice or video will receive data with a minimum amount of delay.

Connection Success Rate (CSR)-Connection success rate is a measure of how many connection request attempts have achieved their desired or allowable connection results. Connection success rates can be influenced by network activity or packet congestion resulting in the inability for devices or services to request and to get assigned a service connection.

Connectionless Network Service-A network service that transfers information between end users without establishing a logical connection or virtual circuit.

Connectionless Data Service-Service at a given layer of the OSI Reference Model in which there is no connection setup phase.

Connector-A connector is a device or assembly that mounted on a cable or device that permits the acoustic, electrical, or optical connection and disconnection to other cables or assemblies.

Consideration-Consideration is the item or value (such as money or services) that is provided for the exchange of assets or rights to the use assets (such as the right to view a movie).

Consignment-Consignment is a process where a person or company agrees to sell a product or service using their facilities when the owner or supplier agrees to receive payment after the products or services have been sold.

Console-The computer access device that allows an operator to communicate with or control a network. A console commonly consists of a display monitor and keyboard.

Consolidated Billing-Consolidated billing is the process of combining all of the customer's communication charges and credits for multiple types of services (e.g. wireless and data lines) on one bill or invoice.

Consolidated Carrier-A communications service provider that integrates several communication services. Also known as an integrated carrier.

Constant Bit Rate (CBR)-Constant bit-rate (CBR) service is a class of telecommunications service that provides an end-user with constant bit data transfer rate. CBR service is often used when real-time data transfer rate is required such as for voice service. An example of constant bit rate service is audio transmission.

Constant Bit Rate Feed (CBR Feed)-A constant bit rate feed is a media source that has a data transmission rate that does not vary over time. CBR feeds are commonly used for transmission systems that have a fixed data transmission rate such as DSL.

Consultant-A person with special skills. Also a person who borrows your watch, tells you what time it is, then charges you for the time of day. See also Guru.

Consultation Hold-A PBX feature that allows a telephone extension to place a call on hold while communicating on another line to receive a consultation before returning to process the call that is on hold. This consultation is often to determine the routing of the call (if the consulted person desires to receive the call).

Consultative Committee for International Telephony And Telegraphy (CCITT)-Original French language name "Comité Consultatif International Télégraphique et Téléphonique" abbreviated CCITT. A part of the United Nations Economic Scientific and Cultural Organization (UNESCO), based in Geneva, that develops worldwide telecommunications standards. Standards (called recommendations) were published every four years in complete sets. Each section of the standards appeared in a book having a letter code such as G for coding/decoding, Q for signaling, I for ISDN, V for modems and multiplexers, etc. Individual standards within each book are designated by a number following the letter, as in Q.931 or I.451 for ISDN call processing signals (appears in two different books for historical reasons), V.32 for certain modems and error correcting codes, X.25 packet networks, T.30 for facsimile, etc. In 1993, after a reorganization, the organization's name was changed to International Telecommunication Union-Telecommunications Sector (ITU-T or just ITU), and even though ITU now creates recommendations and standards, you will still hear the CCITT standards mentioned. ITU standards are still identified by letter-number codes of the same form, but they are no longer published in a full book, nor are they issued on a rigid four year schedule. Instead, each standard appears when a revision is deemed necessary. ITU standards can be purchased as a downloaded document via the ITU Internet wet site www.itu.int

Consulting-Consulting is the providing of information by skilled people or companies in an interactive format.

Consumable Materials-Consumable materials are objects or items that are used to support business operations.

Consumables-Consumables are items or materials that are used when operating a product, service or business.

Consumer-A consumer is a person or company who uses a product or service.

Consumer Acceptance-Consumer acceptance is the development of awareness or willingness by customers to purchase or use a product, service or tool.

Consumer Demand Curve -Economic model which defines the different level of service that an individual customer will demand at a different price/service level.

Consumer Electronics (CE)-Consumer electronics is the devices or products that are purchased or used by the general public (consumers).

Consumer Products-Consumer products are devices or items that are designed to appeal to a wide range of customer types. Consumer products may be sold in high-quantity to mass market outlets (such as retail store chains).

Consumerism-Consumerism is the process of advertising and the promotion of products and services to customers.

Contact-(1-electrical) A contact is the points that are meet to complete an electric circuit. (2-marketing) A contact is a person and associated information that can be used to communicate (contact) that person (such as a mailing address and phone number).

Contact Address-A contact address is the combination of a name, associated organization (if any) and a physical mailing address.

Contact Information-Contact information is the set of data that relates to a person or company.

Contact Life Cycle-Contact life cycle is the time period that a sequence of communication events occurs between a customer (a contact) and a company.

Contact Management-Contact management is a system that can add, modify and organize lists of people or companies along with activities associated with these contacts. Contact management applications can usually schedule events and keep track of all the communication that occurs between a salesperson and/or customer service representative with people and companies.

Content-Content is the information contained within a message, call, media program or web site display. Content is raw media (essence) that is combined with descriptive elements (metadata).

Content Acquisition-Content acquisition is the gathering of content from networks, aggregators and other sources. After content is acquired (or during the content transfer), content is ingested (adapted and stored) into the asset management system.

Content Actions-Content actions are processes that are performed on media programs or data objects. Content actions may include playing a media program, transferring it, or deletion.

Content Aggregation-Content aggregation is the process of combining multiple content sources for distribution through other communication channels.

Content Aggregator-A content aggregator obtains the rights from multiple content providers to resell and distribute through other communication channels. A content aggregator typically receives and reformats media content, stores or forwards the media content, controls and/or encodes the media for security purposes, accounts for the delivery of media and distributes the media to the systems that sell and provide the media to customers.

Content Asset-A content asset is a media file or streaming signal that can be used or monitized.

Content Authentication-Content authentication is a security process that verifies the identity and/or integrity of a media program or data item.

Content Consumption-Content consumption is the acquisition and usage of content by consumers.

Content Control-(1-service) Content control is the process of authorizing and provisioning of access of content to users. (2-interaction) Content control is the ability of a user to review, select and interact with content sources.

Content Conversion-Content conversion is the process of modifying or changing data or media (content) into a different form. An example of content conversion is the changing of an analog videotape format into a digital video disk (DVD) format.

Content Delivery-Content delivery is the transfer of content from one location to another location.

Content Delivery Method-Content delivery method is the transfer processes and/or systems that are used to transfer media from a content provider to a content user.

Content Delivery Network (CDN)-A content delivery network is a network of service that is designed or used for the distribution of content. CDNs may have redundant links and multiple streaming servers to ensure file downloads occur efficiently and reliably.

Content Delivery Platform-A content delivery platform is the combination of system hardware and software that is used to process and distribute content.

Content Delivery System-A content delivery system is the combination of equipment, protocols and transmission lines that are used to transfer content between addressable locations.

Content Developer-A content developer is a person or company who creates images or video.

Content Directory-A content directory is a listing of the content files that are stored or available on a system and/or network.

Content Distribution-Content distribution is the process of transferring content to one or more persons, companies or points.

Content Distribution Agreement (CDA)-A content distribution agreement is a document that defines the authorizations and terms between a content provider and a distributor.

Content Feed-A content feed is a media source that comes from a content provider or stored media system.

Content Ingestion-Content ingestion is the process of transferring media into a storage or content management system.

Content License-A content license is a contract that grants specific rights to use of content. An example of a content license is permission to distribute a television program to users in a cable television system.

Content Licensing-Content licensing is the defining, authorizing and compensating for the rights to develop, use or sell content (media or information).

Content Licensing Management-Content licensing management is the identification, entry, organizing and processing of rights to develop, use or sell content (media or information).

Content Lifecycle-Content lifecycle is the progression of content from its concept stage to its end of use or destruction. A typical content lifecycle for media includes the concept, development, production, packaging, distribution and repurposing.

Content Management System (CMS)-A content management system identifies, categorizes, and manages the storage and distribution of content.

Content Manager License Administrator (CMLA)-Content manager license administrator is a licensing and compliance organization that helps content owners, system operators and device manufacturers to develop and deploy systems that that enable secure delivery of digital media. The CMLA was formed to assist with the implementation of the open mobile alliance (OMA) digital rights management (DRM) specifications.

Content Monetization-Content monetization is any process that can be used to generate revenue from the use of content.

Content Monitoring-Content monitoring is the process of detecting the location, modification and status of content.

Content Network-A content network is a group of web pages that have similar content.

Content Owner-A content owner is a person or company that owns the rights to intellectual property (content).

Content Packager-A content packager is a program or system that is used to combine content (digital audio and/or video) and product information (e.g. Metadata) into a media format or data file that is sent from a content provider to a user or viewer of the content.

Content Partner-A content partner is a person, company or organization that participates in the creation, financial backing or other risk aspect of content production and distribution.

Content Producer-Content producers are companies or developers of media content. Content producers may directly provide distributors or network providers with access to content.

Content Protection (CP)-Content Protection is the end-to-end system preventing content from being pirated or tampered with in a communication network (such as in a television system). Content protection involves uniquely identifying content, assigning the usage rights, scrambling and encrypting the digital assets prior to play-out or storage (both in the network or end user devices) as well as the delivery of the accompanying rights to allow legal users to access the content.

Content Provider-A content provider is a person or company that provides content or intellectual property to distributors or users of content.

Content Rating-Content ratings are the qualification and assignment of codes (rating) to content that identifies characteristics of the content. These content ratings can be used to determine suitable uses and distribution channels for the content.

Content Rendering-Content rendering is the process of converting media into a form that a human can view, hear or sense. An example of content rendering is the conversion of a data file into an image that is displayed on a computer monitor.

Content Repository-A content repository is a storage system, which holds content. Content repositories are commonly located inside a network that is protected from unauthorized users.

Content Rights-Content rights are the authorized uses and the allowable distribution methods that can be used for content (typically data or media).

Content Rules-Content rules are the properties or access right assignments associated with files, directories, or communication services that define the ability of users to change, interact or access content.

Content Scheduler-A content scheduler is a media management system that is capable of accepting scheduled time periods and transferring media within an content management system when the transfer times occur.

Content Seeding-Content seeding is the process of providing an initial amount of content into a system to enable or initiate additional uses or distribution (e.g. peer to peer) of the content.

Content Segment-A content segment is a portion of content or media such as a television program, show or event.

Content Segment License (CSL)-A content segment license is a command or message that authorizes the decoding and usage rights (such as viewing) of a content segment (such as a television program).

Content Sensitive Help-Content sensitive help is adjusting of information presentation or customer support based on the level of assistance needed (such as to users or installers).

Content Server-A content server is a computer system that provides content or media to devices that are connected to a communication system (such as through the Internet). The content servers' many function is to process requests for media content, setup a connection to the requesting device and to manage media transfer during the communication session.

Content Source-A content source is the system or provider of content. Examples of content sources include satellite systems, video servers and live network feeds.

Content Tracking-Content tracking is the monitoring or following of the flow or usage of information content.

Content User-A content user is a person, company, or group that receives, processes or takes some form of action on information services or intellectual property (content).

Content Value Chain-A content value chain is the operational model that describes the core functions that are required to obtain and product media content that will be provided to the end viewers or users. The blocks in a typical content value chain include content producers, content aggregators, media servers, and distribution systems and media players.

Contest-A contest is an offering of a thing of value in return for registration or participation in some event.

Context-Context is the set of parameters a device or recipient currently has which can influence their interaction with content or services.

Context Rules-Context rules are conditional expressions that can use the meanings of words or elements to determine how rules will be applied instead of requiring keywords or data elements to exactly match to apply a rule.

Contextual Ecommerce-The grouping of products for sale via electronic connections (eg. Web shopping) based on the category of products rather than be specific attributes.

Contingency Plan-A contingency plan is a document or a proposed set of actions that should be performed if certain events occur.

Contingency Planning-Planning of a course of action such that the plan will be invoked only if the contingency materializes.

Contingent Compensation-Dependent on event(s) where as goals of a film are reached, the sum that is due to be paid, or that has been paid.

Continuing Property Record (CPR)-A continuing property record is a set of added work and resource details that identify the value of assets. CPRs for telephone companies are assigned by Telcordia.

Continuous Advertising-Continuous advertising is the process of presenting ads multiple times to provide repeated reminders of a message or brand.

Continuous Recognition-Continuous recognition is an analytical process (such as speech recognition) that can continuously operate when any amount of information is provided to it (such as when a person continues to talk).

Continuous Replenishment-Continuous replenishment is the process of continuously or periodically monitoring the amount of inventory or materials and ordering replenishment materials at predefined limits.

Contract-A contract is a set of terms and obligations that have been agreed to by two or more entities. Contracts can be active or passive. Active contracts are created when the parties actively in the creation and acceptance and passive contracts are entered into without the direct participation of the entities (parties) involved in the contract.

Contract Award-A contract award is a notification to a supplier that their contract proposal has been accepted.

Contract Based Tariff-A contract-based tariff is a rate structure that is based on a service contract between service providers.

Contract Law-Contract law is the set of rules and processes that are applied to the interpretation, application and enforcement of contracts.

Contract Manufacturing-Contract manufacturing is the use of a different company to produce an assembly or product.

Contract Model-A contract model is a representation of a binding agreement between two or more parties where the representations may be in rules or as functional descriptions of processes that are used to define how the terms of the agreement may be exercised or fulfilled.

Contract Modeling-Contract modeling is the conversion of the elements, terms and conditions of a contract into a form that can be used to define, determine or control actions or events that may or should occur as a result of the contact.

Contract Negotiation Date-The contract negotiation date is a day that negotiations begin to define the details of a project or contract.

Contract Terms-Contract terms are expressions of authorizations, requirements or conditions that are part of an agreement between two or more entities.

Contracting Authority-A contracting authority is a person or organization within the company that has the authority to sign a contract.

Contracting Officer-A contracting officer is a person who has been authorized by a company to initiate, negotiate and authorize contracts for products or services.

Contractor-A contractor is a company or person who provides products or services under the terms of a business agreement (a contract).

Contracts Expression Language (CEL)-Contract expression language is a set of commands and procedures that are used to express the terms of contractual agreements in a text based extensible markup language (XML) format. Because CEL expressions are text based, they are understandable and usable by both humans and machines.

Contracts Management-Contract management is the identifying, reviewing and assignment of terms to the appropriate systems, departments or people.

Contribution Margins-Contribution margin is the amount (portion) of revenue that is obtained from the sale of products or services that is available to be used to cover the fixed cost associated with the product or service.

Control-(1-general) The supervision that an operator or device exercises over a circuit or system. (2-comparison) A comparison of record counts or attribute totals against invoice-type data.

Control Center-(1-facility) A control center is a centralized work location from which managers and support staff administer the bulk of central office daily work requests. Examples include network terminal equipment centers, switching control centers, and frame control centers. (2-interface) A control center is a web page or interface device that allows administrators to change parameters within a system or service. An example of a control center is a display area that allows a web site manager to change the services and billing information for their hosted application.

Control Character-A character in a computer program whose occurrence in a particular context initiates, modifies, or stops an action that affects the recording, processing, or interpretation of data.

Control Console-A control console is an access device (such as a desktop computer) that allows an operator to communicate with or control a network or its services. A console commonly consists of a display monitor and keyboard.

Control Envelope-A control envelope is data encapsulation process used in an electronic data interchange (EDI) financial transaction. The control envelope is used to hold and validate that a transaction is authentic.

Control Office-An exchange carrier center or office responsible for the installation and maintenance of a given access service furnished to an access customer.

Control Plane-A control plane is the portions of a network system that are involved in the setup, management and termination of communication sessions and services.

Control Structure-A control structure is the defined division (such as fields, packets and frames) and process used for commands and data that are sent between devices or systems.

Controlled Environmental Vault (CEV)-A controlled environmental vault (CEV) is an underground room that is used to contain electronic and/or optical equipment under controlled thermal and humidity conditions.

Controlled Load Service-Controlled-load service provides a variable bandwidth for each communication session that varies based on factors including the amount of network activity (e.g. heavy traffic) and quality of service requirements (e.g. real-time compared to non-real time communication application).

Convergence Billing-An all inclusive bill that combines charges for: local, long distance, data, Internet, and possibly utilities (water, gas, electric) and cable TV. Convergent billing generally implies one all-encompassing view of the Customer for all subscribed services, and a unified view of the product portfolio to enable cross-product packages.

Convergence Sublayer (CS)-A convergence sublayer is a functional process within a communication device or system that adapts one or more transmission mediums (such as radio packet or circuit data transmission) to one or more alternative transmission formats (such as ATM or IP data transmission). The use of a convergence sublayer in a communication system typically allows for the transparent flow of

commands and media regardless of the media and type of transmission systems.

Convergent Billing-Convergent billing is the combining of billing information for multiple types of services such as television, telephone service and data communication services. Convergent billing or systems may be tightly integrated or loosely tied together (sometimes called "stapled") so allow a customer to have a single billing and customer care access point.

Convergent Network-A network incorporating wireless, optical and copper transmission media and multiple protocols. In the past, networks have often been built for a specific purpose and/or based on a specific technology. In the future, more and more interconnections among networks are anticipated to maximize communications options.

Conversation Class-Conversation class is the providing of communication service (typically voice) through a network with minimal delay in two directions. While conversation has stringent maximum time delay limits (typically tens of milliseconds), it is typically acceptable to loose some data during transmission due to errors or discarding packets during system overcapacity.

Conversation Minute Miles-The product of the total number of message minutes carried on a trunk or circuit group and the average route miles of the trunk or circuit group.

Conversation Time-Conversation time is the duration between the initiation of a conversation to the ending of the conversation.

Conversion Rate (CR)-(1-Internet Marketing) Conversion rate is a measure of the people who log on to a web page and select a process or purchase via that web site. A high conversion rate percentage usually indicates how much more valuable a web site is to its visitors.

Conversion Table-A conversion table is a group of structured information that can be used to convert or filter data from one source into another form. An example of a conversion table is a set of currency conversion rates that are used to convert prices or sales amounts from multiple currencies into a form that uses a single currency.

COO-Chief Operations Office

Cookie-A cookie is a small amount of information that is stored on a web user's computer (a client) that is used by a web site (web server) to help control the content and format of information to the user during future visits to the web site.

Co-Op-Co-Operative Advertising

Co-op-Cooperative Advertising

Cooperative Advertising (Co-op)-Cooperative advertising is the sending of promotional messages where the costs and benefits are shared between two or more people or companies.

Co-Operative Advertising (Co-Op)-An amount of funds, percentage of sales, or marketing allowance that is provided to a distributor or retailer for their advertising of specific products or services. To receive the co-operative advertising funds, the distributor or retailer may be required to provide proof that the advertising was performed and paid for.

Cooperative Agreement-A cooperative agreement is a set of terms and obligations that have been agreed to by two or more entities that define how they will work together to achieve a common goal. Cooperative agreements may be setup between governments and companies to stimulate business development investment and activities.

Cooperative Multitasking-Cooperative multitasking is the sharing of processing resources by multiple applications where control is transferred between each application to ensure the same processes or programs cannot access the same resources at the same time.

Cooperative Processing-Cooperative processing is the sharing of computing resources (such as sharing a mainframe computer)

where the instructions from software applications share the computer processor.

Coordinated Universal Time (UTC)-Coordinated universal time is a reference time scale that is maintained by the Bureau International de l'Heure (BIR). UTC is commonly used as a basis of a coordinated dissemination of standard frequencies and time signals used in communication systems.

COP3-Code of Practice #3

COPS-Common Open Policy Service

COPW-Customer Owned Premises Wire

Copy Protection-Copy protection is the use of technologies and/or processes that are developed to prevent or reduce the ability to copy data or media.

Copycat Competitor-Copycat competitors are companies and/or people whom copy systems or services from other companies to sell or provide competing products or services.

Copyright-A copyright is a monopoly which may be claimed for a limited period of time by the author of an original work of literature, art, music, drama, or any other form of expression-published or unpublished. Copyrights are Intellectual Property Rights, which give the owner, or assignee, the right to prevent others from reproducing the work or derivatives, including reproducing, copying, performing, or otherwise distributing the work. Copyrights can, in many countries be claimed without registration. Most countries, or regions, however, have a copyright office where copyrights can be officially registered.

Copyright Infringement-Copyright infringement is the unauthorized use of intellectual property (copyrighted content) by a person or company.

Copyright Protection-Copyright protection is the legal rights and/or physical protection mechanisms (security system) that provide control of the use of an original work of literature, art, music, drama, or any other form of expression

Copyright Registration-Copyright registration is the process of recording the creation of a new work with the copyright office or intellectual property rights centre in the country of origin. Registering a copyright is not required and a work may be considered copyright protected when it is initially created (such as in the United States). Copyright registration does provide additional benefits to the creator or rights holder especially when attempting to enforce copyright protection of the content. There may be a time limit between the creation of the work and registration (such as 5 years in the United States).

Copywriting-Copywriting is the creation of text content that explains concepts or promotes people, companies, products, or services.

CORBA-Common Object Request Broker Architecture

Cordless-Cordless systems are short range wireless telephone systems that are primarily used in residential applications. Cordless telephones typically use radio transmitters that have a maximum power level below 10 milliwatts (0.01 Watts). This limits their usable range of a 100 meters or less.

Core-(1-general) The central part of a device or system. (2-magnetic) A magnetic core is a memory system that uses magnetic field levels to store information. (3-optical) Fiber optic cores are the portion of an optical cable that conducts (transfers) light from the entry to exit points.

Core Applications-Core applications are the set of programs that is provided with or used by a device or system.

Core Dump-A core dump is the transfer of all of the information contained within a memory device or system.

Core Requirements-Core requirements are the components of a system or service that are essential to operate and provide the necessary features and services.

Core Specification-A core specification is a document or set of documents that are intended primarily to define the operation and essential technical requirements for items, materials, or services.

Core Switch-A backbone switch that interconnects to other core switches and edge switches. Core switches maintain information about virtual paths that are connected through the network.

Core System-The core system is the functional parts and processes to enable basic operation or services of the system.

Corporate LAN-A corporate LAN is a private data communication network that uses high-speed digital communications channels for the interconnection of computers and related equipment in geographic areas that are managed by a corporation.

Corporation-A corporation is a business entity that is owned by shareholders.

Corrective Actions-Corrective actions are the steps or processes that have been performed to resolve a conflict or problem between two or more parties.

Correlation-Correlation is an analysis function that defines the association or relationship between information or data sets.

COS-Class Of Service

COSMOS-Computer System For Mainframe Operations

Cost-Cost are the fees our resources that are required to produce or provide a product or service. Costs can include product development, marketing, license or franchise fees, transaction and service support costs.

Cost and Freight (CFR)-Cost and freight is a shipping agreement that a seller will obtain delivery services which deliver the products to an agreed location (such as to the buyers warehouse). For CFR, the shipper does not have to obtain insurance for the products.

Cost Based Pricing-Cost based pricing is the setting or assigning of prices for products or services that are based on the costs for providing those products or services.

Cost Code-A cost code is a unique identifier that is assigned to fees or costs that have a defined set of characteristics such as the business unit, product or process they are associated with.

Cost Component-A cost component is the value of each element of a product or service that comprises a unit or configuration.

Cost Efficiency-Cost efficiency is a measure of the effectiveness of a program or project to meet its financial performance objectives.

Cost Metric-A cost metric is a measurement value that is used to identify the fees or charges associated with a service or product.

Cost of Ownership-Cost of ownership is the acquisition cost (purchase price) of hardware and software along with the costs for operation and maintenance.

Cost Of Removal-Cost of removal is the combined fees or charges that are associated with removing a system or equipment. Cost of removal may include the disassembly labor and transportation costs.

Cost of Sale-The cost of sale is the combination of costs that are incurred by the acceptance of an order for a product or service. Some of these costs include wholesale product costs, packaging cost, transaction cost and sales commissions.

Cost Of Service-The total cost of providing utility service and includes operating expenses, depreciation, taxes, and a rate of return adequate to service investment capital.

Cost of Service Pricing-Cost of service pricing is process that determines the pricing of a product or service through formulas that use of the cost basis of the products or services offered.

Cost per Action (CPA)-Cost per action is the combination of costs that can be attributed to the efforts that lead customers or people to perform a desired action (such as list registration).

Cost Per Acquisition (CPA)-The cost of each customer or person who has completed a desired action (such as list registration) that is paid to an advertiser.

Cost Per Click (CPC)-Cost per click is the fee paid to a web host (such as a search engine web site) for each visitor that selects a specific link.

Cost per Gross Ad (CPGA)-Cost per gross ad are the total resources that are used to acquire (add) a new customer.

Cost Per Inquiry (CPI)-Cost per inquiry is the total cost of a promotion divided by the number of inquiries received as a result of the promotion.

Cost Per Lead (CPL)-Cost per lead is the fee paid to a web host (such as a search engine web site) for each visitor that provides contact information as a potential sales lead.

Cost Per Order (CPO)-Cost per order is the total cost of a promotion divided by the number of orders generated as a result of the promotion.

Cost per Phone Hour-Cost per phone hour is a call center resource unit efficiency measurement. Cost per phone hour is determined by dividing the total costs for the call center divided by the number of workstation call hours that are achieved.

Cost Per Piece (CPP)-Cost per piece is the cost of each item (such as a mailing piece) that is sent in a promotional campaign. The cost per piece is calculated using the total cost of the mailing promotion (list rental, production and mailing) divided by the number of pieces sent.

Cost Per Plant Unit-Cost per plant unit is an average cost for performing a task or work process in a plant.

Cost Per Rating Point-Cost per rating point is the combined media costs that are required to change the performance of sales or other measurable characteristics to customers or a segment of customers per one percentage point.

Cost Per Sale (CPS)-Cost per sale is the fee paid to a web host (such as a search engine web site) for each visitor that orders a product or service.

Cost Per Targeted Thousand (CPTM)-The cost for each targeted thousand is the cost of a promotional campaign that is divided by the number of targeted recipients of the campaign.

Cost Per Thousand (CPM)-Cost per thousand is a price that is used to calculate the cost of media such as TV time, radio spots, Internet ad impressions, print ads, etc. For example, a television ad might sell for $200 CPM. An ad for television show with 2,000,000 viewers will therefore cost $200 x 2,000 or $400,000 dollars.

Cost Per User (CPU)-Cost per user is the combination of fees or charges that can be attributed to each user of a system or service.

Cost per Visitor (CPV)-Cost per visitor is the combination of fees and charges that can be attributed to the efforts that lead customers or people to go to (visit) a location or function (such as logging onto a web page).

Cost Per Work Unit-Cost per work unit is the average cost for performing a task or work unit.

Cost Recovery-Cost recovery is the allocation or assignment of costs to accounts, services or projects to reimburse or reclaim the initial or continued outlays or investment into products or services.

Cost, Insurance and Freight (CIF)-Cost, insurance and freight is a shipping agreement that a seller will obtain delivery services and shipping insurance which deliver the products to an agreed location (such as to the buyers warehouse).

Costa Model -The Costa Model is a representation used to help define the way that telephone companies (telcos) use different departments and functions to optimally attract and keep customers. Key layers of the Costa Model include awareness, familiarity, attraction, preference and selection (sales).

Costing-Costing is the measuring of the production or creation value of goods or services.

COT-Customer Originated Trace

COTS-Customer Off The Shelf

Country Code-A 1-, 2-, or 3-digit number that identifies a country or numbering plan to which international calls are routed. The first digit is always a world zone number. Additional digits define a specific geographic area, usually a specific country.

Coupon-A coupon is a code or form of media that offers a buyer or a potential buyer an added incentive to purchase, use or try a product or service. Coupons may be provided as a code or certificate that enables a person, company or device can use to obtain a change in value or offer for a product or service.

Coupon Manager-A coupon manager is a program or system that coordinates the setup, fulfillment and tracking of codes or other items that offer buyers added incentive to purchase,

Courseware-Courseware is a combination of software programs that are used to provide an education presentation or learning services. Courseware may be a combination of online and offline materials.

Coverage-(1-general) The geographical area over which the signal strength of a given radio frequency is available for service. (2-broadcasting) The "footprint" of a broadcast station's signal, usually measured in terms of percentage of homes covered in a designated geographic area or in terms of the signal strength. (See also: penetration.)

Coverage Area-A coverage area is a geographical area that has a sufficient level of radio signal strength from a transmitting tower to provide an acceptable level of signal reception. An acceptable level of signal reception may be determined by signal to noise ratio (for analog systems) or bit error rate (for digital systems).

Coverage Holes-Coverage holes are portions of radio coverage in a wireless system where the customer cannot receive radio signals due to low radio signal levels or distorted radio signals. Coverage holes are commonly caused by multipath fading where multiple signals are combined in such a way that the signal levels cancel causing reduced signal levels.

CP-Content Protection

CPA-Cost per Action

CPA-Cost Per Acquisition

CPC-Cost Per Click

CPE-Customer Premises Equipment

CPFR-Collaborative Planning, Forecasting and Replenishment

CPGA-Cost per Gross Ad

CPI-Cost Per Inquiry

CPL-Call Processing Language

CPL-Cost Per Lead

CPM-Cost Per Thousand

CPN-Calling Party Number

CPO-Cost Per Order

CPP-Calling Party Pays

CPP-Cost Per Piece

CPR-Continuing Property Record

CPS-Cost Per Sale

CPTM-Cost Per Targeted Thousand

CPU-Central Processing Unit

CPU-Cost Per User

CPV-Cost per Visitor

CR-Call Rate

CR-Conversion Rate

Crack-To crack is the process that enables the access to a computer or information system without authorization of the owner or operator of the system.

Cracker-A cracker is a person who obtains access to a computer or information system without a plan to damage or disable the system.

Cramming-Cramming is the fraudulent addition of charges for services that were not agreed to by the end customer.

Crapware-Crapware is a software program that performs operations or functions that are not essential to the specific needs or problems of a user. Crapware may use computer processing and resources that present the user with unwanted features or services.

Crash-A complete or partial failure of a hardware device or a software program or operation.

Cream Skimming-Cream skimming is the targeting of the easiest or most profitable types of customers to sell products or services.

Credentials-Credentials are information elements that are used to identify and authorize the use of products, services or intellectual property (content). Credentials may include identification codes, service access codes and secret keys.

Credibility-Credibility is the presentation of information to a person that increases the believability of subjective or unconfirmed components of a message or media presentation.

Credit-A credit is reduction in an obligation owed or an increase in the value of an asset.

Credit Adjustments-Credit adjustments are amounts or values that are entered into a billing system that changes the balance of an account. Credit adjustments usually have explanations of the reason for the issuance of the adjustment.

Credit App-Credit Application

Credit Application (Credit App)-A credit application is a form that contains information that can be used to determine the credit worthiness of a person or company.

Credit Approval-Credit approval is an authorization to provide payment terms to a person or company so they can purchase goods or services.

Credit Authorization-Credit authorization is the establishment of a value that can be used to purchase products or services where the payment for the items will come at a later time or the approval to give a reduction in an obligation owed.

Credit Bureau-A credit bureau is a company or agency that provides information about the credit worthiness of people or businesses.

Credit Card-A credit card is a financial instrument can be processed (charged) by companies for the payment for products or services. The charges accumulated on the card are paid for by the owner or user of the card.

Credit Card Network-A credit card network is a communication network that is used to securely transfer credit card transaction information between banks.

Credit Card Phone-A credit card phone is a telephone device that accepts credit cards as a form of payment.

This diagram shows the basic steps in online store credit card processing. This diagram shows that when the customer places an order, it goes through the web store server. The web store server then sends the credit card order amount to a merchant processor. The merchant processor identifies the bank associated with the credit card and sends the information to that bank for processing. If the bank authorizes the transaction, the merchant processor will begin the steps to transfer the money to the online store's bank. The merchant process will keep approximately 2% to 4% of the transaction for their processing service.

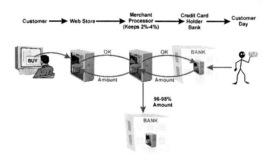

Online Store Credit Card Processing

Credit Check-A credit check is the process of obtaining credit history information for a specific person or company and reviewing the information to determine the ability for the person or company to pay products, services or debt repayment.

Credit Line-A credit line is the amount of authorization to obtain products or services where the payment for the items will come at a later time.

Credit Memos-A credit memo is a document issued by a seller of goods to a customer that indicates a value that can be used to pay for products or services (such as outstanding invoices). Credit memos are commonly issued for the return of goods or surrender of service rights.

Credit Rating-Credit rating is the assignment or use of a score or attributes to a customer or company to determine their ability or likeliness to pay bills.

Credit Score-A measure of a customer's credit worthiness that is based on information obtained from an independent credit bureau. Sometimes also termed "External Score".

Creditor Agency-A creditor agency is the company or person who is responsible for collecting a debt that is owed.

CRIS-Customer Record Information System

Critical Report Date-(1-testing) The date on which point-to-point testing is completed. (2-service order) The date on which implementation groups must report that all documents and material have been received to carry out a service order.

CRL-Certificate Revocation List

CRM-Customer Relationship Management

Cross Certification-Cross certification is a process of using other validation certificates to determine or extend the trust of an existing certificate or its' certifying authority.

Cross Database Access-Cross database access is the linking of data to multiple database storage locations.

Cross Docking-Cross docking is the process of receiving products from one source and redirecting some or all of the products for shipment to another location (it only moves across the shipping dock).

Cross Platform-Cross platform is the ability of a device, software application or service to operate in two or more different types of systems.

Cross Selling-Cross selling is a marketing activity that is designed to encourage customers to buy different products from the same company. In telecommunications, cross sell could include selling wireless or ISP service to wireline customers.

Cross Subsidization-Cross subsidization is the use of revenues from one product or service to apply to the cost of another product or service. An example of cross subsidization is the reduction of the retail cost of a mobile telephone through the application of a subsidy from the service revenue.

Cross Subsidization (Cross Subsidy)-Cross subsidy is the use of revenues or funds from one business unit or project to financially assist another business unit or project.

Cross Subsidy-Cross Subsidization

Crossover Cable-Another name for a null modem cable. A cross over cable reverses the transmit and receive communication lines to allow the direct connection of computers without the need for hubs, switches, or routers.

CS-Commercial Service

CS-Convergence Sublayer

CSC-Card Security Code

CSC-Customer Service Center

CSI or CSID-Called Subscriber Identification

CSL-Content Segment License

CSP-Commerce Service Provider

CSP-Communication Service Provider

CSP-Competitive Service Provider

CSR-Certificate Signing Request

CSR-Connection Success Rate

CSR-Customer Service Record

CSR-Customer Service Representative

CSS-Cascading Style Sheets

CSS-Cellular Subscriber Station

CST-Computer Supported Telephony

CSV-Comma Separated Value

CT-Computer Telephony

CTC-Click to Call

CTI-Computer Telephony Integration

CTN-Call Tracking Number

CTO-Chief Technology Officer

CTR-Click Through Rate

Cube Farm-An cube farm is a office that adds cubicles as new people are hired.

CUG-Closed User Group

Cumulative Discounts-Cumulative discounts are price reductions that are based on the accumulation of purchases or usage amounts over a period of time or an event.

Curing Threshold-The amount that a customer is required to pay in order for the collections process to be suspended. This amount is not necessarily the full amount due to-date and is dependent on: the type of customer, collectible amount, and behavior score.

Currency-Currency is a financial unit measurement that is defined by a country or other authorized entity.

Currency Codes-Currency codes are symbols or character sequences that identify monetary units of countries or other issuers of currency.

Currency Conversion-The process of converting the value of one currency (money) to the value of another currency.

Currency Conversion Table-A currency conversion table is a set of currency conversion rates that are used to convert prices or sales amounts from multiple currencies into a form that uses a single currency.

Current Billing-Current billing is the amount of new charges that are billed since the last billing invoice was sent.

Current Ratio-The current ratio is a comparison of a company's short-term financial obligations to the short revenue bookings. The current ratio indicates the financial health leverage a business has. The lower the current ratio, the more capability a company has to meet its current financial obligations and invest in new projects.

Cursor-A cursor is a small icon or spot that appears on a display. The cursor has a particular image display location associated with it (horizontal and vertical position). This location may be used to select items on a screen or to be used as a reference point (mark a spot) to be used for future reference (such as length calculations between cursor marks).

Custom Call Routing-Custom call routing is the processing and redirecting of calls based on the specific needs or desires of a user or application (such as a call center).

Custom Calling Features-Custom calling features are specific operations or characteristics of call processing features that is selected or customized for specific users.

Custom Calling Services-Custom local area signaling services (CLASS) are telephone service features available in a local access and transport area (LATA) that are primarily based on information that can be processed inside the telephone network. CLASS features include call forwarding, caller identification, and three-way calling.

Custom Local Access Signaling Services (CLASS)-A set of telephone services and enhanced features available in a local access customers that may include calling number delivery or calling name delivery (CND), message waiting, and other features.

Customer-An entity that has the ultimate responsibility for: signing contracts, making payments, allowing use of the services. Sometimes, this term also refers to prospects, or potential prospects.

Customer Account Record Exchange (CARE)-Procedure used for the exchange of customer records between the local exchange carrier (LEC) and the long distance (LD) carrier primary interexchange carrier selected (PIC'ed) by the customer.

Customer Account Record Exchange/Industry Standard Interface (CARE/ISI)-Customer account record exchange/industry standard interface is a set of guidelines and formats that are used in exchanges of equal access-related information between inter exchange carriers and telephone companies.

Customer Acquisition Cost (CAC)-Customer acquisition cost is the average cost to a carrier of signing up an individual subscriber. Some of the factors included in the cost of acquiring customers include handset subsidies, activation commissions, sales, marketing, advertising, and promotional campaigns.

Customer Advance-A customer advance is the providing of assets that will be used for the purchase of products or services in advance of when the products or services will be provided.

Customer Base-Customer base is the core or primary market for a product or service.

Customer Care-Customer care is the processes and communication that occurs between customers and companies to enable customers to resolve problems and successfully obtain products and services from the company.

Customer Care and Billing (CCB)-Customer care and billing is a system or a process that manages the communication with customers and their associated billing records.

Customer Care And Billing System (CCBS)-A system that provides customer account tracking, service feature selection, billing rates, invoicing and details.

Customer Care Center-A customer care center is a facility that is used to receive or originate communication with customers to process orders or to support requests for information.

Customer Care System-A customer care is a combination of hardware (such as telephones and computers) and software (call distribution) that enable, record and organize communication between customers and companies.

Customer Centric-Customer centric is the primary focus of a company or system is to provide features and services to customers.

www.AlthosBooks.Com

Customer Data Rate-Customer data rate is the information transmission speed that a customer provisioned to receive.

Customer Database-A customer database is a set of records (customer list) of people or companies that have purchased products or services.

Customer Demarcation Point (CDP)-Customer demarcation point is the connection point that separates the customer's communication system from the network that it is connected to.

Customer Desired Due Date (CDDD)-The date the customer desires delivery or operation of a product or service.

Customer Dialed Operator Serviced-Calls that are partially handled by an operator. Services performed include recording of credit card numbers, collect calls, coin collection, and person-to-person calls. This function is also called "0+" dialing.

Customer Experience Indicator (CEI)-Customer experience indicators define the perception of customer for a service provided by a communication services provider. CEIs are useful for determining the experience of the customer and the impact of the experience on the revenues and profits.

Customer Feedback-Customer feedback is information sent by customers regarding their purchase or other related experiences with a company or service provider.

Customer Group-A common term for a group of customers with similar characteristics (e.g. business customers.)

Customer Identification-Customer identification is the process of identifying a customer or user of a system or service through the inspection of information gathered about a service or usage characteristic.

Customer Incentive-Customer incentives are value offerings to customers to help motive them to take desired actions. Customer incentives can range from free product offers to customized product options.

Customer Initiated Contact-Customer initiated contact is an incoming communication (e.g. service information request) that is received from a customer.

Customer Inquiry-The process or record of a contact that is made by a customer to a company or person regarding the status of their order(s).

Customer Interaction Software-Customer interaction software is an application that manages communication and relationships with customers.

Customer Interconnection Record-A customer interconnection record is a setup of event information that is recorded when interconnection services occur.

Customer Interface Management (CIM)-Customer interface management is the process or system that coordinates information that is sent and received between companies and customers. CIM systems are used to manage inbound (e.g. customer inquiries) and outbound communication (e.g. sales promotion calls) with customers.

Customer Lifetime Value-Customer lifetime value is expected contributions (revenue) that will be received from a customer during the customer's total association (lifetime) with a company or its services.

Customer List-A customer list is a group of contacts that have already purchased products or services from a company or person.

Customer Loop-A dedicated communications channel, usually a pair of wires and/or a digital loop carrier channel, between a customer's telephone and a serving central office.

Customer Loyalty-Customer loyalty is the willingness or desire of a person or company to purchase products or services again from the same person or company.

Customer Loyalty Program-A customer loyalty program is a marketing process that identifies and rewards customers for repeated visits or purchases.

Customer Management-Customer management is the process that coordinates information that is sent and received between companies and customers. Customer management systems are used to schedule activities, allocate resources, and help control the sales activities within a company.

Customer Network Management-All activities associated with the planning, operation, administration, and maintenance of the communications network of a corporate customer. This term also refers to an arrangement that enables customers to manage their own networks, for example, to move Integrated Services Digital Network (ISDN) access channels from circuit-switched to packet-switched service.

Customer Off The Shelf (COTS)-Customer off the shelf is a program or device that is ready to use or to install directly after it is purchased.

Customer Originated Trace (COT)-A Custom Local Area Signaling Services (CLASS) feature that allows a subscriber to initiate a call trace request message. This call trace feature temporarily stores the dialed digits and alerts the telephone service operator to "tag" the originator's number to allow authorities to investigate the originator of the unwanted or unauthorized call. Some of the call trace activation codes include *57 on a touchtone phone or the dialing of 1157 on a rotary (pulse) phone. If the call trace of the last call was completed successfully, an announcement should be heard. The service operator will usually release the call trace information to law enforcement agencies and a signed authorization from the subscriber may be required.

Customer Owned Premises Wire (COPW)-Customer owned premise wires are the communication lines that are owned by a person or company.

Customer Portal-Customer portals are Internet web sites that act as an interface between a user and a company that provides them products or services. Customer portals typically allow users to log onto their accounts and may allow them to change features or services.

Customer Premise-The location owned or controlled by a customer.

Customer Premises Equipment (CPE)-All telecommunications terminal equipment located on the customer's premises, including telephone sets, private branch exchanges (PBXs), data terminals, and customer-owned coin-operated telephones.

Customer Problem Handling-Customer problem handling is the organizational unit responsible for taking trouble reports from customers and making sure that the problems are resolved.

Customer Profile-A customer profile is a set of characteristics that are associated with a user or account. These characteristics may include services, feature sets, rate plans, payment patterns, service limits along with many other characteristics.

Customer Provisioning-Customer provisioning is the process of delivering products and/or services to the customer. It is the operational aspects of delivering the service, from setting up accounts, user authentication, billing/return policies, channel deployment, packaging and customer support.

Customer Rating-Customer rating is the process of evaluating the credit worthiness of a customer to determine what services they are authorized to use and the amount of deposit the customer may be required to pay.

Customer Ratings-Customer ratings are the postings of feedback on products or services that customers have purchased.

Customer Record Information System-An electronic data processing system for keeping customer records and billing information, generally using magnetic tape.

Customer Record Information System (CRIS)-A customer record information system is a data format that is used by some telephone companies for billing end user customers.

Customer Relationship Management (CRM)-Customer relationship management is the process or system that coordinates information that is sent and received between companies and customers. CRM systems are used to schedule activities, allocate resources, and help control the sales activities within a company.

Customer Retention-Customer retention is the processes and actions to keep an existing customer and/or to achieve repeat business from an existing customer.

Customer Retrial-A customer's subsequent attempt, within a measurement period, to complete a call or a request for a service.

Customer Segmentation-Customer segmentation is the grouping of customers into groups based upon shared characteristics such as buying patterns, product types purchased and frequency of purchases.

Customer Self Care-Customer self care is the process of a allowing the customer to review and/or activate and disable services without the direct assistance of a customer service representative (CSR). Customer self care can be as simple as providing account billing information to the customer by telephone through the use of an interactive voice response (IVR) system to providing interactive service activation menus on an Internet web site.

This figure shows that a customer self-care system can be used to review product or service options, check billing records, and change customer feature options (called "provisioning"). This diagram shows that the customer can contact the billing and customer care sys-

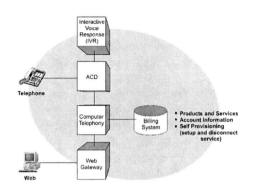

Customer Self Care Operation

tem via a gateway. The gateway may contain Internet web access (for graphic displays) or interactive voice response (IVR) systems to allow the customer to select their account, receive billing and customer care information, and possibly change feature options.

Customer Service-Customer service is the processes and systems that are used to communicate with customers for inquiries, orders, service support and dispute resolution.

Customer Service Center (CSC)-A customer service center is a facility that is responsible for installing, verifying, and maintaining customers and support of customer requests.

Customer Service Class-The customer category of services, including business, residence and public.

Customer Service Record (CSR)-A customer service record is a stored history of a customer's contact with a company. A customer service record entry will usually include the contact date and time, identification of the customer service representative who communicated with the customer, and notes about the contents of the call.

Customer Service Representative (CSR)-A company representative who manages customer communication for account inquiries, complaints, follow-up support, or other service related issues.

Customer Support-Customer support is the processes and communication that occurs between customers and companies to enable customers to resolve problems and successfully obtain products and services from the company.

Customer Tier-Customer tiers are the grouping (partitioning) of customers into specific category levels that have similar product or service buying characteristics.

Customer Tracking-Customer tracking is the process of gathering and organizing information related to customer activities. An example of customer tracking is the gathering of click through information for customers as they navigate through a web page or a series of web sites.

Customer Trouble Report Analysis Plan-An administrative procedure for recording trouble reports, furnishing a database from which analyses can be performed, and summarizing closed trouble data for exchange carrier administrative reports.

Customer Type-A customer type set of characteristics associated with customers who share a common interest in buying or using products or services.

Customer Value Assessment-Customer value assessment is the analysis of how much value a customer provides to a company.

Customizability-Customizability is the ability of a system or program to be modified by a user or system operator to have unique (customized) attributes.

Customizable Field-A customizable field is a value label and/or its characteristics that can be changed by the user or operator of the program or service. An example of a customizable field is the ability to ad a field value such as eye color to a database table.

Customizable Interface-A customizable interface is a display and/or menu system that can be modified by the operator or owner of the system.

Customs Duty-Customs duty is a fee or assessment that may be required for the import or export of goods that are exported or imported into a country. Customs duties can vary dramatically based on the type of products and companies that are sending or receiving them.

Cut Off Date-(1-development) A date after which major redesign or reprogramming of a project cannot be achieved without serious effects on costs or completion date. (2-service) The day a service authorization is terminated or scheduled to be terminated.

Cut-Off Call-A call that has been disconnected by any means other than user intended.

Cutoff Date-The last date that transactions will be included in the current billing cycle.

Cutover-Cutover is the process or date where new systems or equipment are used to provide services to customers (old systems are cut off).

CUTV-Catch Up Television

CVV2 Code-Card Verification Value 2 Code

CW-Call Waiting

CWA-Communication Workers of America

CWID-Call Waiting Identification

cXML-Commerce Extensible Markup Language

Cyber-Cyber is a term that is commonly used to identify data networks, computer related technologies, or high-technology processes.

Cyber Squatting-Cyber Squatting is the registering of an Internet domain name with the expectation of selling the domain name at a later time for a profit.

Cyberbusiness-A cyberbusiness is a company that sells and/or provides its products and/or services through the Internet.

Cyberbusiness Plan-A cyberbusiness plan is a document that defines the objectives of an online business (or segment of a business that uses online services to perform business), what resources it needs and how it expects to use these resources to achieve its objectives.

Cybercash-Cybercash is a value unit that can be used to purchase products or services. Cybercash may be used in electronic payment systems.

Cybermall-A cybermall is a group of online storefronts.

Cyberspace-A term that is commonly used to describe an interconnected public data network (such as the Web) that has tools available to allow users to find and retrieve data from the network.

Cycle-A cycle is the time interval a periodic signal (repetitive) to complete one period.

Cycle Pools-A cycle pool is a gathering of billing records that are identified for a billing cycle.

D

DA-Destination Address

DA-Discontinued Availability

DAC-Digital To Analog Converter

Daemon-Disk and Execution Monitor

Dailup-Dialup is a data connection that requires the user to initiate (dial up) a connection.

DAK-Deny Any Knowledge

DAL-Dedicated Access Line

DAM-Digital Asset Management

DAML-Digital Added Main Line

DAO-Data Access Objects

Dark Fiber-Dark fiber is a fiber strand in a cable that is unused by other light sources. Dark fiber may be used by customers to send any form or optical transmission (analog or digital.).

DAS Tape-Data Acquisition System

Dashboard-A dashboard is a screen or an online location where you can get all the information you need for an account or service. A dashboard usually offers the user the ability to manage account information (such as changing features or preferences for an account).

DAT-Digital Audio Tape

Data-Data is any representation of information that has an assigned meaning and is suitable for processing, transmission, communication, or interpretation by humans, systems, or computers.

Data Access Objects (DAO)-Data access objects are an application program interface between software program objects.

Data Acquisition System (DAS) Tape-The magnetic tape that is used to record information, communication, time change, tape management, and billing records for a cellular carrier system.

Data Aggregation-Data aggregation is the process of combining multiple data sources for distribution through other higher-capacity communication channels.

Data Analysis-Data analysis is the process of reviewing and analyzing information (data) for the purpose of identifying common characteristics that may be useful for other purposes.

Data Application Engine-A data application engine is a software program that can receive, process and respond to an end user's (client's) requests for selecting, processing and retrieving information from databases.

Data Availability-Data availability is the ability of a user or device to identify, retrieve or store data.

Data Backup-Data backup is the copying of data or information to an alternate storage device. The data backup information is usually stored on a device that reliably stores and operates relatively long periods of time.

Data Base Administration System-An operations support system that maintains a database of billed-number screening data. The system also handles auditing and fraud reporting as well as other duties, including calling card, collect, and bill-to-third number information.

Data Card-A data card is an adapter card that connects electronic communications equipment (such as a fax machine or computer) to a mobile telephone, typically via a standard PCMCIA type II connector.

Data Center-A data center is a place where data records are collected, stored, processed, and distributed.

Data Centric-Data centric is the central part or core focus of a process or system, which is based on data (e.g. data transmission) services.

Data Channel-A data channel is a transmission path for data from a transmitter to a receiver. A data channel may be composed of one or more logical channels.

Data Circuit Terminating Equipment (DCE)-Data circuit terminating equipment is the communication channel terminating equipment that allows data terminal equipment (DTE) to be connected to a network.

Data Collaboration-Data collaboration is the process of allowing multiple users to simultaneously access and edit the same data file.

This diagram shows how data collaboration allows multiple users to edit, append or modify the same information or data file. This example shows that three people; Betty, John and Susan are all editing the same word processor file. Betty is working on the top section, John is working on the middle section and Susan is working on the end section. As each of these people make changes, they are immediately displayed to other users on their computer screen.

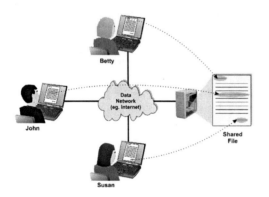

Data Collaboration

Data Collection-Data collection is the process of gathering data from one or more points to a central point.

Data Collection Site-A data collection site is a location where information is gathered or stored until it can be processed or transferred to another location (such as a data processing center).

Data Communications-Data communication is the transmission and reception of binary data and other discrete level signals that can be represented by a carrier signal that can represent the discrete (usually on-off levels) for signal transmission. There are two basic types of data communications: circuit-switched data and packet-switched data. Circuit-switched data provides for continuous data signals while packet-switched data allows for rapid delivery of very short data messages.

Data Communications Equipment (DCE)-The equipment that establishes, maintains, and terminates a data connection, as well as performs the signal conversion and coding required for communication between data terminal equipment (DTE) and a data circuit.

Data Communications Exchange-The hardware that enables data to be transmitted, received, switched, in real time to several different destinations over several different routes and channels, simultaneously.

Data Compression-Data compression is a technique for encoding information so that fewer data bits of information are required to represent a given amount of data. Compression allows the transmission of more data over a given amount of time and circuit capacity. It also reduces the amount of memory required for data storage.

Data Consolidation-Data consolidation is the merging of multiple data information sources into a shared information resource.

Data Contamination-Data contamination is the undesired changing or insertion of data into a file or media segment.

Data Control Block-A data block is a section of information that is used to define or control the interaction with other data or systems.

Data Conversion-Data conversion is the process of selecting information or media from one format or system, processing the data and storing data in the new system or format.

Data Corruption-Data corruption is the modification or changing of data as it passes from a sender to a receiver. Data corruption may be caused by transmission problems (transmission noise) or it may be a result of an attempt to modify data by an unauthorized person.

Data Coupler-An interface (interconnection) device that connects a data device (such as a computer) via a telephone device (such as a telephone or mobile radio) with a telecommunications network. A coupler typically includes circuitry that protects the telecommunications network from damage that may result from failure of the coupler or data device that is connected to the data coupler.

Data Definition Language (DDL)-Data definition language is a set of commands and processes that are used to setup, manage and delete databases.

Data Descriptive Language (DDL)-Data descriptive language is a set of commands and procedures that are used to identify and define data elements.

Data Dictionary-A data dictionary provides the identification, descriptions and characteristics of data elements in a database.

Data Discrepancy Detection System-A data discrepancy detection system is the combination of hardware (such as a computer server) and software programs that can be used to identify data that does not conform to defined rules or formats. A data discrepancy detection system may be used by billing systems to identify discrepancies before billing records are entered into the guiding and rating systems.

Data Element-A data element is the lowest or smallest common denominator in a database management system or media file. A data element is considered indivisible, representing a specific element such as customer names, phone numbers, and addresses.

Data Element Directory-A data element directory is a list of data elements and their locations.

Data Encryption Standard (DES)-The data encryption standard is an encryption algorithm that is available in the public domain and was accepted as a federal standard in 1976. It encrypts information in 16 stages of substitutions, transpositions and nonlinear mathematical operations.

Data Enrichment-Data enrichment is the processing of data to improve its value or ability to be used.

Data Entry-Data entry is the process of entering information into a computer or database.

Data Entry Field-A data entry field is an area on a form that a person uses to enter information.

Data Feed-A data feed are supply of data that is in a format that is suitable to be processed by the recipient. An example of a data feed is a news service that regularly provides articles and stories that can be received, processed and displayed by the recipient.

Data File-A data file is a group of information elements (such as digital bits) that are formatted (structured) in such a way to represent information. While the term data file is commonly associated with databases (tables of data), a data file can contain any type of information such as text files or electronic books.

Data Flow Chart-A data flow chart is a graphic portrayal that shows the sequence in which data functions are performed, from the beginning of a job to the end.

Data Formats-Data formats are the positional and semantic structures that are used to separate information items (records) and the elements (fields).

Data Integrity-Data integrity is the accuracy of data information as compared to its original information source. Data integrity can be verified through the use of error detection codes that are sent along with the original data information. Error detection codes relate to the original data information through the use of mathematical processes. To verify data integrity, the received data information is processed using the same error detection coding mathematical process and comparing the output to the received error detection bits. If they match, the data integrity has been verified.

Data Line Interface-An interface, assembly, or connection point where a data line is connected to a telephone network.

Data Link (DL)-A transmission path between communications devices. The data link includes all equipment and signals in the connection.

Data Link Connection Identifier (DLCI)-A temporary channel identifier used in a communication system to identify a specific circuit along with its required communication parameters (such as peak data rates). The DLCI in a frame relay system is 10 bits. It is pronounced ("dill-see").

This diagram shows how a data link connection identifier (DLCI) is used in a data communications network to route information packets through a network and local data system to reach their destination. This example shows that the DLCI uses access devices to determine which virtual circuit is being used. This diagram shows how a computer that is connected to access device 3 is sending a print file to the printer connected to access device 1 using permanent virtual circuit (PVC) 1. Access device 2 is coordinating a digital voice

connection to access device 1 using switched virtual circuit (SVC) 2. This example shows that the DLCI is only used by the access device. The data network uses its own routing tables to provide a virtual path connection through the network.

Data Link Layer (DLL)-The data link layer is the second layer (layer 2) of the seven-layer Open Systems Interconnection (OSI) protocol model that facilitates the detection of and recovery from transmission errors and a single data link. A data link connection is built on one or more physical connections.

Data Local Exchange Carrier (DLEC)-A service provider that specializes in transferring data. A DLEC is a competitive local exchange carrier (CLEC) that competes with other carriers in the local area such as the local exchange carrier (LEC). DLECs commonly use DSL for data transmission.

Data Loss-Data loss is the inability to access or use data that has been previously accessible.

Data Management-Data management is the process of identifying, describing and applying rules to the information instances (records) and their attributes (fields).

Data Manipulation Language (DML)-Data manipulation language is a set of commands and procedures that are used to retrieve, process and store data elements.

Data Mapping-Data mapping is the process of defining the relationships between a set of fields (elements) in a file, stream or data table to the set of fields (elements) in another file, stream or data table. An example of data mapping is the assigning of the field labeled First_Name in data table 1 to the field labeled F_Name in data table 2.

Data Mart-A data mart is a specialized form of data warehouse. Data marts are typically smaller and more focused in purpose than the larger, more generalized data warehouses.

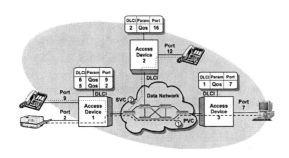

Data Link Connection Identifier (DLCI)
Operation

Data Miner-A data miner is a person or a computer program that sorts through data to detect patterns and establish relationships.

Data Mining-Data mining is the process of reviewing and analyzing information (data) for the purpose of identifying common characteristics that may be useful for other purposes. Data mining is commonly used in marketing programs to identify people or customers that have specific types of needs or buying patterns.

Data Mirroring-Data mirroring is a storage system that contains an exact copy of the data that is stored in another storage device or system.

Data Model-Data models are the set of elements and relationships between the data to be stored in a database. Data models are typically used to define how a database will be constructed in order to guarantee builders the best possible organization of data, indexes and interfaces.

Data Network-Data networks is a system that transfers data between network access points (nodes) through data switching, system control and interconnection transmission lines. Data networks are primarily designed to transfer data from one point to one or more points (multipoint). Data networks may be composed of a variety of communication systems including circuit switches, leased lines and packet switching networks. There are predominately two types of data networks, broadcast and point-to-point.

This figure shows the basic types of data networks. This diagram shows that the types of data networks range from very short range personal area networks (PANs) to wide area data networks (WANs). This example shows that a user is transferring data between a PDA and a computer within their personal area network (PAN) at their office in Chicago. This data can be transferred through their company's local area network (LAN), through a city wide metropolitan area network (MAN) and through a wide area network (WAN) so it

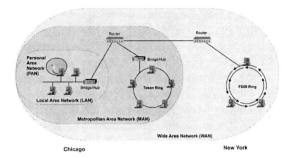

Data Networks

can reach other locations such as New York. When the data reaches New York, it is transferred back to a fiber distributed data interface (FDDI) local area network (LAN) so it can reach its destination computer.

Data Over Cable Service Interface Specification (DOCSIS)®-Data over cable service interface specification is an industry standard used by cable television systems for providing Internet data services to users. The DOCSIS standard was developed primarily by equipment manufacturers and CATV operators. It details most aspects of data over cable networks including physical layer (modulation types and data rates), medium access control (MAC), services, and security.

Data Packet-In Internet Protocol (IP) networks, the smallest amount of data that can packaged and transmitted from source to destination. In more general communications terms, any group of data which has been organized and labeled for transmission on a network or serial line. Each networking technology has minimum and maximum packet sizes which it supports.

Data Packet Switch-A data packet switch is a device in a data transmission network that receives and forwards packets of data. The data packet switch receives the packet of data, reads its address, searches in its database for its forwarding address, and sends the packet toward its next destination.

Data Partitioning-Data partitioning is the dividing of data or media information into segments that may be located on different areas of memory and/or on different storage devices.

Data Port-A data port is a connection point (usually a logical channel) that allows a computer or device to connect to other devices.

Data Processing Equipment-Data processing equipment is the equipment, assemblies, or peripherals (e.g. magnetic or other data storage media) that is used to receive, process, and output data information.

Data Propagation-Data propagation is the transfer of information or media from one system or format to another system or format.

Data Rate-Data rate is the amount of information that is transferred over a transmission medium over a specific period of time.

Data Rate-Data Transmission Rate

Data Rate Mismatch-Data rate mismatch is a condition where the data transmission rate of a received signal does not match the data rate (the receiver clock or local oscillator) of the receiver.

Data Security-Data security is the protection of data against unauthorized disclosure, transfer, modification, or destruction, whether accidental or intentional.

Data Service Unit (DSU)-A data service unit is a device that interfaces between data terminal equipment (DTE) and a data communication network. A DSU is the digital equivalent of the analog modem. DSUs are commonly used or combined with channel service units (CSUs) to allow specific channels of a communication line to communicate with specific DTE devices.

Data Service Unit/Channel Service Unit (DSU/CSU)-Devices that combine the functionality of data service units (DSU) and channel service units (CSU) to adapt data from user communication systems to communication lines with multiple channels. The DSU portion as an interface between a customer's data terminal equipment and a data communication network. DSU are the digital equivalent of the analog modem and are translation codecs (COde and DECode) coupled with a network termination interface (NTI). The CSU portion is used to coordinate communication from one or more data terminal equipment (DTE) devices to logical channels on a multichannel communication circuit.

Data Services-Data Services are communication services that transfer information between two or more devices. Data services may be provided in or outside the audio frequency band through a communication network. Data service involves the establishment of physical and logical communication sessions between two (or more) users that allows for the non-real time transfer of data (binary) type signals between users.

Data Silo-A data silo is a storage of information that is not accessible by one or other information systems.

Data Speed-The data transfer rate of a signal. Data speed is also called baud rate or bit rate.

Data Storage-Data storage is the retaining of information over a period of time (typically in digital form).

Data Storage Fees-Data storage fees are charges for the allocation and/or usage of storage media. Storage fees may be charged for raw data storage or for the storage of particular types of media (such as movies, pictures and voice mailboxes). Data storage fee rates may vary based on the reliability (such as backup) and data transfer performance available from the data storage device.

Data Stream-A data stream is a continuous flow of digital information (data).

Data Telephone-A data telephones is a telecommunication device that integrates analog telephone functions with a data communication interface. Because many data telephones use voice over Internet (VoIP) protocols, they are often referred to as an Internet telephones.

Data Terminal (DT)-Data terminals are data input and output devices that are used to communicate with a remotely located computer or other data communication device. Data terminals frequently consist of a keyboard, video display monitor, and communication circuitry that can connect the data terminal with the remotely located computer.

Data Terminal Equipment (DTE)-In a data communications network, the data source, such as a computer, and the data sink, such as an optical storage device. (See also: data sink, data source, network channel terminating equipment, channel service unit, data service unit.)

This diagram shows data terminals that are connected through data communication equipment (DCE) to allow a user to receive and send communication through a network to a remote computer. In this diagram, the data terminal allows the user to view information on a monitor and enter information through the keyboard. This data terminal is connected through data communication equipment (a modem) that converts the data terminal's digital signals into an analog form (audio signal) that can be sent through the telephone line to a remote computer.

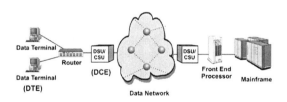

Data Terminal Equipment (DTE)

Data Throughput-Data throughput is the amount of data information that can be transferred through a communication channel or transfer through a point on a communication system.

Data Transfer-Data transfer is the process of selecting, transferring and storing information or media from one device or system to another. Data transfer from one location to another may require several data transfers such as from a database to a storage disk to another database. Data transfer may also require data processing functions such as transferring billing records from one system to another system that uses different data storage format.

Data Transfer Adapter (DTA)-A data transfer adapter (DTA) converts the data bits from a computing device into a format that is suitable for transmission on a communication channel that has a different data transmission format.

Data Transfer Cost-Data transfer cost is the fee paid for the transferring of data into or out of a data network.

Data Transfer Rate-Data transfer rate is the number of bits, characters, or blocks of time passing between corresponding equipment in a data transmission network.

Data Transfer Time-Data transfer time is the duration between the initial request, or transmission of data from a device, to the end of the data transfer session.

Data Transmission-Data transmission is the transfer of data (information) from one point to another point. The data may be transferred using one or more types of communication channels. The physical transfer of the information can be sent in analog or digital form.

Data Transmission Rate (Data Rate)-Data transmission rate is the amount of digital information that is transferred over a transmission medium over a specific period of time. Data transmission rate is commonly measured in the amount of bits that are transferred per second (e.g. bps, Mbps).

Data Type-Data types are the characteristics of data elements such as length, numerical value, logical value (yes or no) or the method of representation (such as text characters).

Data Universal Numbering System (DUNS)-The data universal numbering system assigns unique numbers to identify organizations. The DUNS system is managed by the company Dun and Bradstreet and the numbers are used as the electronic data interchange (EDI) address for companies.

Data Validation-Data validation is the process of reviewing data elements (such as a form entry field) to determine if it conforms to specific format or structure requirements (such as date format).

Data Warehouse-(1-information management) An information management service that stores, analyzes, and processes information that is derived from transaction systems. (2-system) A specialized database system, dedicated to the storage and retrieval of information for purposes of analysis and business intelligence investigation.

Database-A database is a collection of information items (records) and their associated elements (fields).

Database Adapter (DBA)-A database adapter is an application program that can select, process and change the format of data to allow it to be used or interact with other programs or databases.

Database Administrator (DBA)-A database administrator is a person who is responsible for the organization, design, and implementation of a company's databases.

Database Connections-Database connections are logical links between applications or services and information that is stored within databases.

Database Connectivity-Database connectivity is the ability to link applications or services to databases through the use of application programming interfaces (APIs). Database connectivity allows software application develop-ers to write applications that can dynamically access information that is stored in a database.

Database Dictionary-A database dictionary is a system table that stores information that defines the structure of a database. A database dictionary may be used in relational databases to define the names and data types of the data files (tables) and their fields (columns). A database dictionary may also be called a lookup table.

Database Dip-A process of an information search within a database. An example of a database dip used in a telecommunication system is the searching in a signaling control point (SCP) database to find the actual telephone associated with a toll free (800) or freephone (0800) telephone number.

Database Directory-A database directory is a collection of data that describes the location of and other possible characteristics (media types) of the files that are stored in memory or on media storage device.

Database Driven-Database driven is an application (such as a web site) that adjusts its interface information based on information from a database.

Database Formats-Database format is a structure of items (data records) and their associated elements (fields).

Database Grooming-Database grooming is the modification or updating of an existing database to correct or sort information and/or to remove redundant or unwanted information.

Database Integration-Database integration is the process of adapting and linking databases with systems and/or other databases.

Database Lookup-A database lookup is a software process that enables a requesting device or service to search and find information in a database.

Database Management System (DBMS)-A database management system is a collection of hardware and software that controls access to, organization of, security, and application interfaces to information or data.

Databases-Databases are collections of data that is interrelated and stored in memory (disk, computer, or other data storage medium). Database systems are typically accessed and controlled by computer terminals that are connected to the same data network as the database system.

Datacard-A datacard is a fee schedule and the terms of use for the rental of mailing lists. A datacard main contain a description of the list(s), source of the list, minimum order quantities, pricing and fees for additional selection, or list processing services.

Datagram-A datagram is a self-contained packet of data that represents a message and/or contains user information. A datagram is sent independently through a network to its destination where it is reassembled with other packets into the original message.

Datagram Service-Datagram service is service at the network layer in which successive packets may be routed independently from end to end. There is no call setup phase. Datagrams may arrive out of order. In Internet Protocol parlance, datagram service generally implies the use of UDP or non-assured delivery.

Date and Time Stamps-Date and time stamps are time references that indicate when an event occurred. Date and time stamps may be stored with, appended to or entered into a separate data file. An example of data and time stamps is storing the date and time information when voice mail systems have received a voice message.

Date of Issuance-The date of issuance is the date when a proposal (such as an RFP) has been released or when a contract has been awarded.

Date of Submission-A date of submission is the recorded date that a document or proposal (such as an RFP) has been received from an RFP responder.

Day-A day is a 24-hour period. For many systems, a day is defined to begin at 0100 Greenwich mean time (GMT) and end at 0100 GMT.

Day of Week (DOW)-The day of week is the date an event, service or action occurs within a specific week.

Day Rate Period-The period of time for which day rates apply. It may vary for different services, including interstate, intrastate, and overseas services. (See also peak time).

DB-25-A DB-25 is a D-type connector that has 25 pins that can be used for parallel or serial connections.

DBA-Database Adapter

DBA-Database Administrator

DBA-Doing Business As

DBMS-Database Management System

DCAS-Downloadable Conditional Access System

DCE-Data Circuit Terminating Equipment

DCE-Data Communications Equipment

DCPS-Digital Copy Protection System

DCTI-Desktop Computer Telephone Integration

DDC-Direct Department Calling

DDD-Direct Distance Dialing

DDL-Data Definition Language

DDL-Data Descriptive Language

DDoS-Distributed Denial of Service

DDR-Digital Disk Recorder

DDS-Digital Data Storage

Dead Freight-Dead freight is the shipping fee that is charged for freight services when the shipping container or facility is not used or is empty.

Dead Link-A dead link is a hyperlink that no longer is able to redirect a connection to another resource.

Dead On Arrival (DOA)-A term used to refer to equipment that is not working when it is first used or received by a customer.

Dead Spots-Dead spots are portions of radio coverage in a wireless system where the customer cannot initiate or receive calls or radio signals due to low radio signal level.

DeAuthentication-DeAuthentication is a process that is used to remove the association between authenticated devices.

Debit Card-A debit card is a financial instrument that can be processed (charged) by companies for the payment for products or services and the charges accumulated on the card are paid for by the owner or user of the card from funds they have available.

Debit Telecard-A debit telecard is a financial instrument that can be used to pay for telecommunication services where the payment for the services and the charges accumulated on the card are paid for by funds the owner has already paid or transferred into the card's account.

Debt Financing-Debt financing is the process of selling loans in a company as a way to obtain funds that will be used to run or develop the business.

Debugger-A debugger is a program that assists software programmers in discovering errors or undesired operations in software programs.

Debugging-Debugging is the process of finding, analyzing, and fixing software errors (bugs) in a software program.

Decapsulation-The process of removing protocol headers and trailers to extract higher-layer protocol information carried in the data payload. See also Encapsulation.

Decay Constant-Decay constant is an estimate of the rate of reduction in a process (such as the amount of sales after a promotional campaign) after the original source has stopped.

Deceased Identification-Deceased identification is the process of identifying people on lists or within databases that have died.

Decentralized Architecture-Decentralized architecture is a system that is designed to use a distributed intelligence to coordinate the operation of the system.

Decimal-Decimal is a numbering system that references digits with a base of 10.

Decipher-Deciphering is the process of using information (such as keys) and processes (such as a decryption algorithm) to recover information that has been previously enciphered.

Decision Criteria-Decision criteria are a set of evaluation criteria elements (such as installation time) along with the rating values or processes that are used to assign a value to each criteria element.

Decision Makers-Decision makers are the individuals or group of individuals who can authorize the purchase of your products or services.

Decision Making Unit (DMU)-A decision making unit is a group of people or a process that accesses information and decides on actions to take.

Decision Matrix-A decision matrix is a table or related set of data that contains qualifying rules and quantifying processes (rating) of criteria that is used to make a decision (such as selecting a vendor to supply a product or service).

Decision Support Systems (DSS)-A decision support system is a computerized system that can receive and process information (such as usage statistics) to assist in the making of business decisions.

Decision Tree-A decision tree is a hierarchical set of rules or conditional processes that are used to assist in decisions for a company or person.

Declarative Content-Declarative content is information that describes an item, file or media program. Declarative content may be contained within the same data file (media item) or it may be stored in a separate file or storage area.

Declarative Programming-Declarative programming is the describing of the desired outcomes or required results without specifying the operations that are required to achieve the results.

Decoding-Decoding is the process of converting encoded words into the original signal. For example, in pulse code modulation (PCM), decoding is the conversion of 7-bit (Dl type) or 8-bit (including D2, D3, D4, and D5 type) pulse code modulation (PCM) words to analog signals. Decoding is the inverse of encoding.

Decompression-Decompression is the processing of compressed digital information to convert it to its original uncompressed format.

Decryption-Decryption is the conversion of a coded signal into clear or readable form.

Dedicated Access-Dedicated access is the permanently assigned (non-switched) channel or circuit for the exclusive use of a particular subscriber.

Dedicated Access Line (DAL)-A communication line that has its data transmission capability reserved (dedicated) for a specific user.

Dedicated Bandwidth-A configuration in which the communications channel attached to a network interface is dedicated for use by a single transmitter or receiver and does not have to be shared.

Dedicated Line-A dedicated line is communications circuit that is setup to provide continuous information transfer between points. The resources for the communication circuit are dedicated exclusively to that connection.

Dedicated Plant Assignment Card-An outside plant record that gives the particulars of a distribution plant served by a dedicated outside plant control point, or serving area interface. The card lists the address of every housing unit served from the control point or interface, as well as the telephone number, central office equipment, type of service, feeder and distribution pair numbers, and the address of the distribution terminal. The dedicated plant assignment card generally has been replaced by mechanical loop assignment systems, such as the Facilities Assignment and Control System (FACS).

Dedicated Server-A dedicated server is a computer or data processing device that performs specialized network tasks, such as storing files for a specific user or application. A web hosting company may offer the option to assign a dedicated server to a company or person to allow them to ensure no other companies will share their processing resources.

Dedicated Trunk-A communication trunk that is connected directly through to a particular phone or hunt group. Dedicated trunks bypass attendant consoles.

Defacto Standards-Defacto standards are specifications that are widely used or accepted which are not officially recognized or controlled by standards organizations. An example of a defacto standard is Microsoft Windows operating system.

Default-Default is the initial value(s) or process(es) for a particular service or application.

Default Parameters-Default parameters are the specific values of the variables that are used for a particular service or application. An example of a default parameter is an "Include on Mailing List" option on a web page that is default set to "include" on the mailing list.

Default Port-A physical or logical port that is pre-configured or defined to be the communication port for traffic when a port assignment is not provided or when a process fails. Applications and protocols in communication systems such as the Internet use default ports so they can start specific applications or protocol communication when a packet is first received. For example, when an IP packet is received with the port number 21, it specifies file transfer protocol (FTP) should be used.

Defects Per Million (DPM)-Defects per million (DPM) is the number of calls lost per million calls processed. DPM is particularly useful for measuring the network availability of switched virtual circuits (SVC) services in a multiservice switch, where connections are constantly made, sustained, and torn down.

Defense Communications System-A communication system that is used in a defense systems network

Defensive Programming-Defensive programming is the process of developing an assembly of instructions for a computer to enable it to carry out a particular job where the program tasks or processes are able to detect and/or adjust to changing external influences (such as the loss of memory resources or the alteration of instructions due to virus programs).

Deferment-Deferment is the delaying or postponing of something, usually referring to payment.

Deferred compensation-A sum of money to be paid that is delayed in transaction due to non-occurrence of events required to make payment possible.

Defragmenter-A defragmenter is a software program or process that reads data files from a storage system and rewrites (stores) the data files into contiguous areas of the storage system. Defragmenting can reduce the amount of time necessary to read and write data files.

Deinstallation-Deinstallation is the process of removing software programs or equipment (such as a PBX system) so that original programs are systems are returned to their original state.

Delay-The amount of time it takes for a signal to transfer or for the time that is required to establish a communication path or circuit.

Delay Announcement-A delay announcement is a message that is played to a caller to a call center when no call center representatives are immediately available to answer the call.

Delay Time-Delay time is the duration of time that occurs from the requesting of a service to when the service is provided.

Deliberate Churn -Deliberate churn is a type of voluntary churn. Deliberate churn occurs when the customer decides to terminate service because they have found a more attractive service elsewhere.

Delimiter-A delimiter is a character or identifier that is used to separate information or media segments. Delimiters are used to separate fields in data records.

Delinquency Threshold-The amount owed by the customer that will trigger the collection process to be initiated.

Delinquent-Delinquent is the status of a claim or debt which has not been paid by the date specified in the agency's written notification or applicable contractual agreement, unless other satisfactory payment arrangements have been made by that date, or, at any time. Thereafter, the debtor has failed to satisfy an obligation under a payment agreement with the agency.

Deliverable-Deliverable is a document, product or service that will be provided as part of a contract or agreement.

Delivered Price-Delivered price is the offer value that is assigned to a product or service, which includes all costs to provide the product to a defined location. Delivered price may include the cost of packaging, insurance, and freight.

Delivery Attempts-Delivery attempts are the total number of messages that were transmitted as part of a marketing campaign (such as an email campaign).

Delivery Context-Delivery context is the set of parameters a device or recipient currently has which can influence the reception and processing of content or services.

Delivery Methods-(1-shipping) Delivery methods are the shipping options that are available for delivering products or services. Examples of delivery methods include ground, air, next day and downloadable. (2-network) The delivery method is the systems and processes that are used to transfer information or media from a sender to a receiver.

Delivery Platform-A delivery platform is the combination of system hardware and software that is used to process and distribute signals or content.

Delivery Point Bar Code (DPBC)-A delivery point bar code is a sequence of codes (typically black vertical bars separated by white spaces) that may be assisting in the delivery of items.

Delivery Point Validation (DPV)-Delivery point validation is a system that verifies the accuracy or ability of an item to be delivered to its address. The United States postal services has a DPV system that can confirm an address already exists which can be used by mailing house companies to reduce the number of nondeliverable items that are sent.

Delivery Report-A delivery report is a listing of the emails that were sent and the status of their delivery. Delivery reports may include the number of emails sent, how many were successfully delivered, the number of returned (bounced) emails, the number of emails opened, and the status of actions taken by the recipients (e.g. link clicks).

Delivery Schedule-Delivery schedule is the sequence of time events that will occur for the delivery of products or services.

Delphi Forecasting-Delphi forecasting is a method of forecasting the future by obtaining the opinions of a group of experts and averaging their opinions.

Delta Certificated Revocation List (Delta CRL)-A delta certificated revocation list (Delta CRL) is list of certificates that have been revoked since a previous list or updated has been issued.

Delta CRL-Delta Certificated Revocation List

Demand Analytics-Collection of data mining models that can be utilized to help define how changes in operational expense (OPEX) spending (advertising, marketing, customer service etc.) will impact the customers demand for telecommunications services.

Demand Based Build Out-Demand based build out is the process of installing and adding equipment in areas as customers (subscribers) are added to the system. This allows service providers to invest in their systems as their revenue grows rather than installing large and complex networks with the anticipation that customers will eventually subscribe to their services.

Demand Chain Management-Demand chain management is the identifying and managing of the companies and processes that are involved in the consumption or use of products and services by customers.

Demand Forecast-Demand forecasting is the process of estimating the future consumption or purchases of products or services.

Demarc-Demarcation Point

Demarcation Point (Demarc)-The physical and electrical boundary between an end user's telecommunication equipment and the telecommunications network. The demarcation point establishes point of ownership and accountability.

DeMilitarized Zone (DMZ)-A demilitarized zone (DMZ) is trusted part of a communications network, typically behind a firewall, that allows unrestricted access and transfer of information between devices. Information that passes in or out of a DMZ may be delayed and filtered. This may cause challenges with real-time communication applications such as IP telephony and media streaming.

Demographics-Demographics are the statistical characteristics of groups of people or companies.

Demurrage-Demurrage is an amount of time that exceeds the allowable shipping time. Demurrage can occur due to the detainment of cargo ships.

DEN-Directory Enabled Network

Denial Of Service (DoS)-A denial of service attack is a process that inhibits or reduces the ability of authorized users from gaining access to communications systems through the continual transmission of service requests or messages that disable communication sessions.

Deny Any Knowledge (DAK)-A claim by a customer that a call was not made (as in "I deny any knowledge of this call").

Department Of Communications (DOC)-A government agency of countries throughout the world that sets policies and rules regarding telecommunications within their country. In the United States, the FCC is the equivalent of the DOC in some other countries.

Dependent Charge Rate-A dependent charge rate is the service fee or charge unit for a specific service or service category, which is determined by a category or condition of other services or events.

Deployment-Deployment is the installation of products and services or the application of a process.

Deployment Options-Deployment options are the system equipment selections and their configurations that can be combined to provide desired services and features.

Deposit Processing-Deposit processing is the tasks and functions that are used to transfer payments from the buyer of products and services into bank accounts or other asset accounts.

Depreciated-Depreciated is the removal or plan to remove the support for a software program or application in the future.

Depreciation-Depreciation is the assignment of cost of an asset or acquisition that is consumed or devalued over the time period that the asset or acquisition is owned or used.

Deproductize-Deproductization is the removal of a product from production and/or distribution systems.

DES-Data Encryption Standard

Descrambling-Descrambling is a process of converting an encoded or encrypted signal back into its original form through the use of keys and/or descrambling algorithms.

Design Life-The design life is the operational time a device or system that is expected to operate within a performance level or set of characteristics. For satellite systems, the design life may be limited by the amount of propulsion fuel that is used to keep the satellite in its desired orbital plane.

Designated Authority-Any person, panel, or board which has been authorized by rule or order to exercise authority under the Communications Act.

Designated Market Area (DMA)-Designated market area is a defined geographic area that is used to test or define market characteristics. DMA is used by the Nielsen Media Research for the purpose of rating the viewership of commercial television stations.

Desktop Compatibility-Desktop compatibility is the ability of a device (such as a mobile telephone) to exchange information with desktop computer applications. Desktop compatibility commonly refers to the ability to synchronize data between portable devices and desktop computers.

Desktop Computer Telephone Integration (DCTI)-Desktop computer telephony integration is the merging of telephony services with computer technologies.

Destination Address (DA)-The address for whom a message is intended.

Destination Based Discount-Destination based discounts are price reductions that are based on the destination of a service.

Detailed Billing-A type of billing in which the details of each message, such as the date of call, number called, and charge, are listed as separate items on a toll statement that is included with a customer's bill.

Detailed Regulatory Monthly Allocation System-A software system that uses data from the Plug-In Inventory Control System (PICS) to develop central-office investments in the categories required by the FCC's Jurisdictional Separations Procedures. Each month, the investments in PICS are matched to an accounting cost record to provide cumulative investment in the required separations categories. These investments are put into the various separations allocations used by operating telephone companies.

Detariffed-A billing and regulatory term designated by the FCC and state public utility commissions that often refers to deregulated rate structures on subscriber-owned inside premise wiring and related CPE.

Developed Portion Of A Wire Center-In long-range outside plant planning, land that is either completely built up, built up with scattered vacant lots, vacant but surrounded by developed land, or a cluster of more than 200 living units in an otherwise undeveloped part of a wire center. Such areas are also called developed areas.

Development Fee-A development fee is an amount of money paid to one in order to compensate them for talent or services rendered in connection with the development stage of a project.

Development Plan-A development plan is a document that defines the key steps and processes that are likely to be used during the development of a product or service.

Development Stage-Development stage is the period of time it takes to complete a project from beginning to end.

Development Timeline-A development timeline is a sequence of events that occur over a time period.

Deviation-(1-general) A departure from a standard or specified value. (2-modulation) The peak difference between the instantaneous frequency of an AM signal and the carrier frequency. (3-phase) The peak difference between the instantaneous angle of a phase-modulated wave and the angle of the carrier.

Device-A device is a product, assembly, network unit that performs one or more processes. Examples of devices are modems, printers, serial ports, disk drives, routers, bridges, and concentrators. Some devices require special software or device drivers to control or manage them while others have built-in intelligence.

Device Authentication-Device authentication is a security process that verifies the identity of a device that is requesting access to a network, system and/or computer.

Device Capability-Device capability is the set of features, functions and options that are available within a device. Device capabilities can include display, processing, control, and other options.

Device Certificate-A device certificate is information contained within a device (usually in digital form) that can uniquely identify its identity along with other characteristics of the device. Device certificates are usually issued by an authority (a trusted party) that guarantees the information in the certificate is correct.

Device Class-Device class is a parameter that indicates the capabilities of a device. Device class capabilities may include maximum and minimum transmitter power levels, available modulation and coding types, and which services the device supports.

Device Driver-A device driver is a software program that can translate and link information between another software program (such as an operating system) and the components within a system (such as a hard disk drive).

Device Key-A device key is a unique code that is assigned or used by a device to decode media or information.

Device Management-Device management is the process of identifying, adding, and configuration devices that are part of a system. Device management may be a manual process or it may be automatically performed through the use of protocols that can discover and configure equipment that is added to a network and remove registrations from devices that are removed from a network.

Device Manager-A device manager is a controller in an assembly (such as a microprocessor) that coordinates the overall operation of the assembly.

Device Registration-Device registration is a process where a device transmits information to a system that informs it that it is available and operating in the system. Device registration allows the system to determine what services are authorized and what capabilities the device currently has available to process those services.

DFS-Distributed File System

DHCP-Dynamic Host Configuration Protocol

DHN-Digital Home Network

DHS-Digital Home Standard

DHTML-Dynamic Hypertext Markup Language

Diagnostic Mode-A diagnostic mode is a stateful condition of a device, service or system that is used to help test and evaluate the condition of devices, services or systems.

Diagnostic Program-A diagnostic program is a software application that has the capability to test and evaluate the condition of devices, services or systems.

Diagnostic Records-Records relating to a call or communication session that was processed by a call manager or a call server.

Diagnostics-Diagnostics are tests or programs (often built into a device or system) that test the functionality of the system and report the results. Diagnostic systems that are separate and simply monitor the operation of the subject system are considered to be non-intrusive.

Dial Around-The process of dialing a phone number via a pay phone via a toll free for freephone access number. Dial around has the effect of reducing the toll charges that may be collected by a pay telephone.

This figure shows how dial-around access occurs for pay telephones. This diagram shows that a caller is using a calling card with a toll free number on a pay telephone. The use of toll-free (or freephone) numbers on a pay telephone allows a caller to place the call without using any coins or payment cards on the pay telephone. The call is routed to the calling card center where the calling card account is

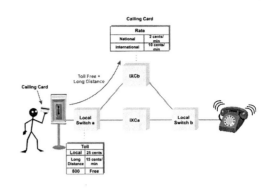

PSTN Dial Around

decreased based on the amount of usage. This diagram shows that the calling card charges may be substantially less than the charges available to the caller when directly dialing through the local telephone company. This ability to dial-around reduces the pay telephone revenue to the owner of the public telephone. The owner of the public telephone may be the local telephone company or an independent company.

Dial By Name-The ability to dial a person by spelling their name out on the telephone or communication device keypad.

Dial Level-Dial level is the selection of which stations or services can be associated with a calling plan that is based on the initial digit(s) dialed.

Dial Line Connection-Dial lines, plain old telephone service (POTS), are 2-wire, basic line-side connections from an end office with limited signaling capability. Because dial lines are line-side connections, call setup time may be longer than those connections that employ trunk-side supervision.

Dial Map-A dial map is the systematic use of certain prefix digits to dial a destination via user selected routing. An example is the use of the dialed prefix "9" from within a PBX to first select an outside local telephone line so that the originator can then dial a (typically 7 digit) local city telephone number. Similarly, a PBX may use the dialed prefix "8" to select a tie line to another PBX.

This figure shows how a basic dial map operates. This diagram shows that there are several dial plan rules that are used each time a number is dialed. The first step in the dial map is to determine if the first digit is a 0, 3, 8, or 9 are dialed. This first rule allows the system to determine if the caller desires to reach the attendant (0), is calling an internal number (3+), long distance (8), or outside line or emergency services (9). This rule changes how the next digit is processed. If the first digit is a 3, it is an internal call (4 digits for this system) and the system will wait for 3 more digits before attempting to connect the call to another unit in the system. If the first digit is 8, the system will capture multiple digits and analyze the call as a long distance public telephone number (country code, city code, exchange code, and extension). If the first digit is a 9, the system will analyze the following digits to determine if it is an emergency call (for example 911) or a local telephone call. If it is a local telephone call, the system will wait until it has sufficient digits and connect the call to a local gateway. If the next 3 digits were 911, it would connect the call to a local gateway and route to the emergency services number.

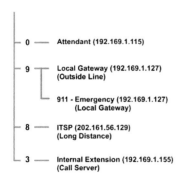

Dial Map Operation

Dial Plan-A dial plan (also called a dialing scheme) is the numbering system that is used by a company to identify devices within their network by unique numbers. After a system has been setup, a dialing plan is developed for each communication unit (or groups of communication units).

Dial Plan-Dialing Plan

Dial Tone-A signal tone provided from a local telephone service provider that indicates that the telephone network is ready to send a call (to receive dialed digits). The dial tone signal is usually a combination of 350 Hz and 440 Hz signals.

Dial Up Line-(1-general) A communication line that is established by sending the dialed digits to the network that allows the connection to be made. The term dial up is commonly used with the process of manually connecting a computer to the Internet through a standard telephone line. (2-humor) An ancient method of connecting to the Internet using MODEMS that signal over a telecom system known as the plain old telephone service or POTS. This method of communication was outlawed in the mid 2000s as part of the Computers With Disabilities Act.

Dialed Number Identification Service (DNIS)-A call identification service typically provided by a toll free (800 number) network. The DNIS information can be used by the PBX or automatic call delivery (ACD) system to select the menu choices, call routing, and customer service representative information display based on the incoming telephone number.

Dialed Number Recorder (DNR)-A dialed number recorder is a device that records telephone dial digital digits. Early DNRs stored the information as dashes on paper tape.

Dialing Plan (Dial Plan)-A dial plan (also called a dialing scheme) is a systematic use of certain prefix digits to dial a destination via user selected routing. An example is the use of the dialed prefix "9" from within a PBX to first select an outside local telephone line so that the originator can then dial a (typically 7 digit) local city telephone number. Similarly, a PBX may use the dialed prefix "8" to select a tie line to another PBX. A dialing plan differs from a numbering plan by being used inside a particular private telephone system, and also the specifics of different dialing plans are different in different PBXs or different private networks in the same country, while a numbering plan is uniform throughout an entire country.

Dialing Report-A dialing report is a listing of data that summarizes the results of telephone numbers that are dialed from a company or call center. Dialing reports may be used to help confirm or evaluate the performance of a call center for specific telemarketing campaigns.

Dial-Up Connection-An Internet connection that is created by a modem that dials the telephone number of an Internet Service Provider (ISP), usually via a local telephone number.

Dial-Up Network (DUN)-A software portion of Microsoft Windows 95, 98, NT, and 2000 that allows the user to connect the computer to a data network (such as the Internet). Because DUN is actually a process of establishing and maintaining a communications session, DUN is sometimes used for establishing connections on "always-on" circuits (such as DSL).

Dialup Networking Profile-Dialup networking profile is the protocols and procedures used by devices such as modems and cellular phones for implementing the usage model called "Internet Bridge." Among the possible scenarios for this model is the use of a cellular phone as a wireless modem for connecting a computer to a dialup Internet access server, or the use of a cellular phone or modem by a computer to receive data.

Dialup Switch-A dialup switch is a category of switching equipment that can manage the dialup connections between the PSTN and other systems such as the Internet or LANs.

Diameter-Diameter protocol is an evolution of Radius protocol. Diameter protocol was orig-inally a provisioning protocol, which was expanded to include authentication, authorization and accounting capability when it was determined that evolving the Radius protocol would be impractical. While diameter protocol is not backward compatible to Radius protocol, it can provide Radius capabilities along with many new features.

Diameter protocol features include the ability to use reliable transport protocols (such as TCP or SCTP), transport security (e.g. IPSEC or TLS), a large number of attribute value pairs (AVPs), both client side and service side service request ability, dynamic recovery, service negotiation capability, error notification, extensibility, and application layer acknowledgements.

Diameter Applications-Diameter applications are software programs that use Diameter protocol to perform authentication, authorization and accounting (AAA) functions for a specific type of usage or application.

Dictionary Attack-A dictionary attack is a process of attempting to gain access to a communications resource or to obtain valid email addresses by sequentially cycling through a list of words (such as a dictionary) that are likely to be used for passwords or account information.

DID-Direct Inward Dialing

DID Assignments-Direct inward dialing (DID) assignments are the mapping (connecting) of extensions in a private telephone system to the incoming calls on a common telephone line that identifies the incoming direct dial telephone numbers.

Differential Charging-Differential charging is the applying or assessing of costs to usage that is based on the difference between rated values.

Differentiated Services Code Point (DSCP)-Differentiated services code point is a code that is used to identify differentiated services.

Digest-A digest is a collection of information elements for a data file or message. A digest can also be used to validate that the contents of the file have not been corrupted or altered.

Digest Authentication-An authentication process that processes a user identification code and password identification without sending the raw data through the communication network to validate the true identity of the user. The digest authentication process begins by the challenger (usually the service provider) sending a random number (nonce value) that is used by the user (usually the client) along with other previously known information (secret key) to calculate a response using an encryption algorithm. The end result is sent back to the challenger who also has access to the previously known information and random number that is used to calculate the same result. If the result matches, the authentication passes.

Digest Method-A digest method is the process that is used to produce the digest value (e.g. hash value).

Digicash-Digicash is a name that describes electronic money that is electronically transmitted (such as payments that are sent through the Internet).

Digit Collection-Digit collection is the process of storing digits as they are entered or dialed and initiating a service (such as connecting a telephone call) after all the digits have been collected.

Digital-Digital electronic devices use electric currents or voltages that are intentionally restricted to take on a limited set of values for their intended use, rather than allowing continuous variation of the current or voltage. Typically, only two voltage values are used, for example, having values of approximately zero or five volts. Because undesired signals (noise, interference) are much smaller (typically less than 0.1 volt, for example) the digital signals often can be transmitted or recorded without errors because the presence of small deviations of the signal do not confuse the device from correctly interpreting the voltage as definitely representing one of the two intentional

voltage levels. When a digital coded representation of an analog signal is used, and the digital part of the system does not introduce any errors, the only degradation of the signal is due to the inherent inaccuracy of the initial encoding device (codec) that converts the signal from analog to digital representation. This inaccuracy can be controlled by the design of the codec.

This figure shows a digital signal that is in the form of a series of bits and these bits are combined into groups of 8 bits to form Bytes (B). In this example, the bits 01011010 are transferred in 1 second. This results in a bit (transmission) rate of 8 bps.

Digital Added Main Line (DAML)-A local loop access line system that uses ISDN digital transmission to provide two communication circuits on a standard copper wire pair. DAML differs from the standard ISDN basic rate interface (BRI) as the D signaling channel is not included. The D channel is simply not used. The line control (e.g. off-hook) is sensed by DAML modem. The modem creates the line control signaling messages. DAML allows a telephone service provider to add more lines on existing copper access lines without the need to add ISDN software upgrades to their switching offices.

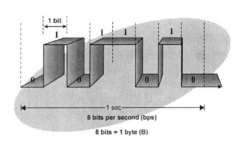

Digital Signal

Digital Asset-A digital asset is a digital file or data that represents a valuable form of media. Digital assets may be in the form of media files, software (e.g. applications) or information content (e.g. media programs).

Digital Asset Management (DAM)-Digital asset management is the process of acquiring, maintaining, distributing and the deletion of information (electronic) assets.

Digital Audio-Digital audio is the representation of audio information in digital (discrete level) formats. The use of digital audio allows for more simple storage, processing, and transmission of audio signals.

Digital Audio Tape (DAT)-A digital audio tape (DAT) is a magnetic tape storage format that uses a cartridge that has a 4 mm wide metal coated tape that stores audio information in digital form. The standard sample rate for DAT is 44.1 kHz and a single DAT can provide storage over 3 GB of storage capacity.

Digital Broadcasting-The process of transmitting the same digital data signal to all users that are connected to the digital broadcast network. Digital broadcast signals may be encoded in a way that only some of the users may be capable of decoding digitally broadcast messages (e.g. a specific pay-per-view movie channel).

Digital Cable Television System-A digital cable system distributes television (and other information services) via a cable television distribution system in digital modulated form.

Digital Cash-Digital cash is a form of value that is stored in electronic form, which can be used on systems that accept the digital cash form of payment.

Digital Certificate-A digital certificate is information in binary (digital) form that is used to identify the identity and possibly the capabilities of a device, system or software application. Digital certificates are usually issued by an authority (a trusted party), which ensures that the information in the certificate is correct.

This figure shows how digital certificates can be used to validate the identity of a requestor of content. This diagram shows that users of digital certificates have a common trusted bond with a certificate authority (CA). This diagram shows that because the content owner and content user both exchange identification information with the CA, they have an implied trusted relationship with. The content user registers with the CA and receives certificate information and the owner registers with the CA and receives information that can validate certificates. When the user requests information from a content owner, the user sends their certificate along with their request for information. The user's certificate can be validated by the content owner using information that is provided by the CA.

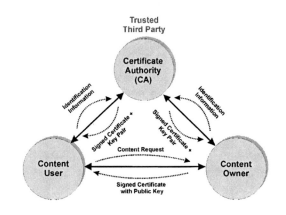

Digital Certificate Operation

Digital Cinema-A digital cinema is a place or building (such as a movie theatre) where digital moving pictures are displayed.

Digital Compression-Digital compression is a process that uses a computing device (such as a digital signal processor) to analyze a digital signal and create a new data signal that represents of the original signal using a lesser number of digital bits. Digital compression allows more information to be transmitted on a communication channel.

Digital compression devices use mathematical formulas and code book tables to compress the data. Mathematical formulae transform the original signal into it characteristic parts such as frequency and amplitude. Code book tables contain blocks of high occurrence information (such as particular tones used in fax machines). When transmitting digital information that has been compressed, only the parameters (such as the frequency, amplitude and code book word) are sent on the transmission channel. When the digital information is received, the compression process is reversed by a decoder to produce the same (or similar) initial signal.

Digital Content-Digital content is information or media that is stored in digital form.

Digital Copy Protection-Digital Copy protection is the mechanism and/or information that is sent or embedded within content to help ensure digital content is used in conformance with its usage rules and it can provide an auditable trail of how an asset is used within a consumer device. Copy protection defines the following rules for use of the content.

1.Copy Freely. User is free to make as many copies as they want.

2. Copy Never. The user is not able to make any copies.

3. Copy Once. Any device that makes a copy of a piece of content, will set the copy bits on the content to copy never on the copy and where possible on the original as well.

For video signals copy protection bits may be embedded in content in a number of ways including: 1. Within the video blanking interval (VBI) lines of the analogue output (CGMSA) 2. Within the media stream format (such as MPEG)

Digital Copy Protection System (DCPS)-Digital copy protection system is a defined a set of connections, hardware and processes that enable the connection of digital media equipment while controlling the copying of digital media. There are several systems that can be used for DCPS including DTCP, HDCP, HDMI and XCA.

Digital Data Storage (DDS)-Digital data storage is a digital information storage format that is used to store information on magnetic tape. The DDS format for data storage using 4 mm DAT tape cartridges were created in 1989 by Hewlett Packard and Sony corporations. There are several DDS storage formats. DDS-1 can store up to 2 GB of data, DDS-2 can store up to 8 GB of data, DDS-3 can store up to 24 GB of data and DDS-4 can store up to 40 GB of data. Over time, the information stored on DDS tapes begins to degrade and the tapes will need to be recreated. The number of uses of DDS tapes is also limited to approximately 2000 short recording sessions or 100 full recording sessions.

Digital Disk Recorder (DDR)-A digital disk recorder is a memory device that can store (record) digital video directly into memory storage (typically a hard disk).

Digital Dispatch-Digital dispatch is the combining of a dispatch communication system with information services. Digital dispatch systems can range from automated messaging devices to complex computer systems that display maps and vehicle positions on a computer monitor.

Digital Distributor-A company or person who distributes (sells and transfers) digital products. An example of a digital distributor is an online eBook retailer.

Digital Fingerprint-A digital fingerprint is a unique set of characteristics and data that is associated with a particular data file, transmission system or storage medium. Digital fingerprints may be codes that are uniquely embedded in a media file or they may be unique characteristics that can be identified in the storage or transmission medium such as the particular variance of digital bits that are stored on a DVD.

Digital Goods-Products or services that are delivered in digital form.

Digital Home Network (DHN)-A digital home network is a system that can transferring digital media in a customer's home or personal area. Digital home networking is used to connect computers, media players and other media sources and players to and each other and to send and receive media from wide area network connections.

Digital Home Standard (DHS)-Digital home standard is a system specification that defines the signals and operation for data and entertainment services that can be provided through electric power lines that are installed in homes and businesses. Development of the DHS specification is overseen by the UPA. More information about DHS can be found at www.upaplc.org.

Digital Ingest-Digital ingest is the process of transferring digital media into a computer storage system or network.

Digital Living Network Alliance (DLNA)-Digital living network alliance is a group of companies that work to create a set of interoperability guidelines for digital media devices that operate in the home. More information about DLNA can be found at www.DLNA.org.

Digital Loop Carrier (DLC)-A highly efficient digital transmission system that uses existing distribution cabling systems to transfer digital information between the telephone system (central office) and a users telephone and/or computer equipment. A DLC system usually includes a high-speed digital line (e.g. T1) from a central office and a remote digital terminal (RDT). The RDT converts the high-speed digital line to low speed lines (analog or digital) for routing to the end customers.

Digital Media-Digital media is the format of information that is used to express information (media) which is represented in a form that can have levels or signal composition of specific discrete levels (digital).

Digital Millennium Copyright Act (DMCA)-The digital millennium copyright act is a regulation (statute) that covers the technological methods to protect copyrights. A key aspect of the DMCA is restrictions on the creation or selling of products or systems that are designed to allow users to get around copy protection. The DMCA became law in October 1998.

Digital Object-A digital object is a group of data (digital) bits that represent data, information or images.

Digital Object Identifier (DOI)-A digital object identifier is a unique number that can be used to identify any type or portion of content. DOI numbers perform for long term (persistent) and locatable (actionable) identification information for specific content or elements of content. This content can be in the form of bar codes (price codes), book or magazine identification numbers or software programs. The DOI system is managed by the International DOI foundation (IDF) that was established in 1998. More information about DOI numbering can be found at www.DOI.org. DOI numbers point to a DOI directory, which is linked to specific information about a particular object or information element. The use of a DOI directory as a locating mechanism allows for the redirecting of information about identification information as changes occur in its identifying characteristics. For example, a book identification number may belong to the original publisher until the copyright of the work is sold to another publisher. At this time, the owner of the item content changes. The item number on the book can remain the same while the publisher information can change.

This figure shows digital object identifier (DOI) structure. This example shows that a DOI number is composed of a prefix that is assigned by the registration agency (RA) of the international DOI foundation and suffix that is assigned by the publisher of the content. This example shows that the first part of the prefix identifies the DOI directory that will be used and the second part identifies the publisher of the content.

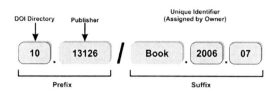

Digital Object Identifier Structure

Digital PBX (DPBX)-A PBX system that uses digital transmission for control and audio signals. Most PBX systems available in the late 1980s were digital PBX systems.

Digital Power Line-Digital power line is a term that refers to the sending of digital information through electric power lines.

Digital Pricing Extortion-Digital pricing extortion is a practice that is used by television service providers to artificially inflate the price of analog programming tiers to create small incremental pricing differences between analog and digital tiers, thereby encouraging digital upgrades by subscribers

Digital Property Rights Language (DPRL)-Digital property rights language is a set of instructions and procedures that are used to define rights of digital media. It was invented by Mark Stefik of Xerox's Palo Alto research center in the mid 1990s and has transformed into extensible rights markup language (XrML). The initial version of DPRL was LISP based and 2nd version (2.0) was based on XML to allow the flexibility of describing new forms of data and processes.

Digital Rights Management (DRM)-Digital rights management is a system of access control and copy protection used to control the distribution of digital media. DRM involves the control of physical access to information, identity validation (authentication), service authorization, and media protection (encryption). DRM systems are typically incorporated or integrated with other systems such as content management system, billing systems, and royalty management. Some of the key parts of DRM systems include key management, product packaging, user rights management (URM), data encryption, product fulfillment and product monitoring.

Digital Service Unit (DSU)-A device that interconnects the customer's digital telephone equipment to a telephone network.

Digital Service, Level 0 (DS-0)-A 64,000 b/s channel, the worldwide standard public telephone industry bit rate for digitizing one voice conversation. There are 24 DS-0 channels in a 1.544 Mb/s DS-1 digital multiplex bit stream, and 30 DS-0 traffic channels plus two additional DS-0 channels used for synchronization and signaling (a total of 32 DS-0 channels) in a 2.048 Mb/s E-1 digital multiplex bit stream.

Digital Signage-Digital signage is the use of digital systems and electronic displays to provide promotional messages to users or potential customers.

Digital Signal-Digital signals consist of a series of ones and zeros, most often represented in telecommunications signals by two different voltages. For example a +5 Volt level could represent a logical 1 (one) and 0 Volt level could represent a logical 0 (zero). The ones and zeros are called bits. Several bits (usually eight) are grouped into a byte and each byte is defined to have a specific meaning, such as a specific letter on a keyboard. Digital signals are used to represent specific levels on an analog signal. While a digital signal cannot represent every point on an analog wave, they can come close enough to be almost indistinguishable. Digital signals are much easier to process by computer systems and they are able to resist the effects of noise better than analog signals.

Digital Signal 1 (DS-1)-The primary rate telephone industry digital multiplexing system used in North America and Japan. It combines 24 DS-0 (64 kb/s) channels and a single 8 kbit/s synchronizing bit stream for a total of 1.544 Mb/s. A different primary rate multiplexing system that combines 30 DS-0 channels with a 64 kb/s channel for signaling and another 64 kb/s channel for synchronization and control, for a total of 2.048 Mb/s, is used elsewhere. The 2.048 Mb/s system is sometimes named E-1, or MIC, or CEPT Primary Rate Multiplexing. Both the 1.544 Mb/s DS-1 system and the 2.048 Mb/s system are recognized ITU standards. The 2.048 Mb/s system was designed to have some improvements and a slightly larger channel capacity than the 1.544 Mb/s DS-1 system, and was intentionally incompatible, a result attributed by some industry observers to a motive of protecting the European market from imported product competition. (Later, manufacturers in different countries changed their objectives from the former strategy of setting intentionally incompatible standards in different countries to the present strategy of setting internationally compatible standards in all countries.) DS-1 is not a trade name of any manufacturer. T-1 is effectively a synonym of DS-1 in North America, and was originally a trade name of just one manufacturer, but today it is widely used for all compatible products regardless of manufacturer.

Digital Signal 3 (DS-3)-A standard digital transmission line that is divided into twenty eight DS1 (T1) channels. The gross transmission rate for a DS3 channel is 44.736 Mbps. A single DS3 provides for 672 standard (64 kbps) voice channels.

Digital Signal 4 (DS-4)-Digital signal level 4 is a standard digital transmission line that has a gross data transmission rate of 274.176 Mbps.

Digital Signal Level (DSx)-Digital signal (DS) transmission is a hierarchy of digital communication channels and lines that range from 64 kbps to 565 Mbps. Lower level DS structures are combined to produce higher-speed communication lines. There are different structures of DS levels used throughout the world with significant variations between North American and European systems. DSx has been used to represent the digital transmission standards where the "x" denotes which service is under discussion.

Digital Signal Processing-Digital signal processing is the manipulation of digital signals into other forms using computing circuits or systems. Digital signal processors use software programs to allow them to perform complex signal processing operations such as filtering, modulation, data compression, and shaping of the information (such as digital audio signal) that are represented by digital signals.

Digital Signal Processor (DSP)-A digital signal process is an integrated circuit (chip) that is designed specifically for high-speed manipulation of digital information. DSP chips operate using software programs to allow them to perform complex signal processing operations such as filtering, modulation, data compression, and information processing. The use of DSPs in communication circuits allows manufacturers to quickly and reliably develop advanced communications systems through the use of software programs. The software programs (often called modules) perform advanced signal processing functions that previously complex dedicated electronics circuits. Although manufacturers may develop their own software modules, DSP software modules are often developed by other companies that specialize in specific types of communication technologies. For example, a manufacturer may purchase a software module for echo canceling from one DSP software module developer and a modulator software module from a different DSP software module developer.

This figure shows typical digital signal processor that is used in a digital communication system. This diagram shows that a DSP contains a signal input and output lines, a microprocessor assembly, interrupt lines from assemblies that may require processing, and software program instructions. This diagram shows that this DSP has 3 software programs, digital signal compression, channel coding, and modulation coding. The digital signal compression software analyzes the digital audio signal and compresses the information to a lower data transmission rate. The channel coding adds control signals and error protection bits. The modulation coding formats (shapes) the output signal so it can be directly applied to an RF modulator assembly. This diagram also shows that an optional interface is included to allow updating of the software programs that are stored in the DSP.

Digital Signal Regeneration-Digital signal regeneration is the process of reception and restoration of a digital pulse or lightwave signal to its original form after its amplitude, waveform, or timing have been degraded during transmission. The resultant signal is virtually free of noise or distortion.

Digital Signal, level 1 Combined (DS-1C)-A digital multiplexing system in North America and Japan. The total bit rate is 3.152 Mb/s. It comprises two T-1s (two J-1s in Japan), at 1.544 Mbps each, which are interleaved to support 48 DS-0 channels. The additional 64 kb/s is overhead used to support additional signaling and control requirements. DS-1C is seldom used, outside of limited telco applications. It has mostly been supplanted by DS-2 multiplexing, which has 96 channels. There is no European equivalent and DS-1C is not included in ITU standards.

Digital Signaling Tone (DST)-A tone that is sent on the analog radio channel to indicate a change in status (e.g. end call).

Digital Signature-A digital signature is number calculated from the contents of a file or message using a private key and appended or embedded within the file or message. The inclusion of a digital signature allows a recipient to check the validity of the file or data by decoding the signature to verify the identity of the sender.

Digital Subscriber Line (DSL)-Digital subscriber line is the transmission of digital information, usually on a copper wire pair. Although the transmitted information is in digital form, the transmission medium is usually an analog carrier signal (or the combination of many analog carrier signals) that is modulated by the digital information signal.

This figure shows a simplified ADSL communication system that consists of a digital subscriber line access multiplexer (DSLAM), local distribution lines that start from a main distribution frame (MDF) wire cabinet that brings the connection to the digital subscriber line (DSL) modem at the customer's location. Modems in the DSLAM convert the digital signals from the Internet to high frequency signals that travel down the telephone line to the DSL modem. The DSL modem converts the RF signals back to its original digital form so it can be provided to the customer's computer. Most DSL technologies (such as ADSL shown in this example) transmit the data information on frequencies above the audio channel. This

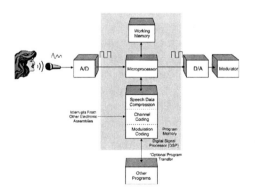

Digital Signal Processor (DSP) Operation

allows for the simultaneous transmission of analog and data signals on the same telephone line. The highest frequencies are used transmission from the DSLAM to the DSL modem and frequencies just above the audio band are used to transmit from the data from the customer to the DSLAM. Typical DSL technology allows up to 6 Mbps to be transmitted to the customer and up to 640 kbps can be received from the customer.

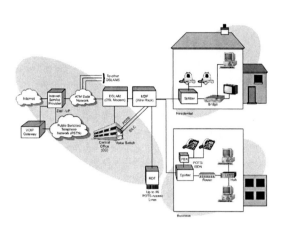

Digital Subscriber Line (DSL) Network

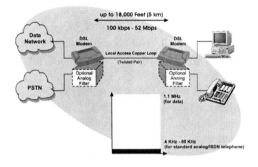

Digital Subscriber Line (DSL) System

This diagram shows that end user equipment of a DSL network adapts, or converts analog and digital signals to a high-speed DSL transmission signal via a DSL modem (an ATU-R for an ADSL system). The copper wire carries this complex DSL signal to a DSL modem at that connects to the central office (an ATU-C for an ADSL system) where it is converted back to its analog and digital components. The analog telephone portion of the signal (if any) is routed to the central office switching system. The high-speed digital portion is routed to a digital subscriber line access multiplexer (DSLAM). The DSLAM combines (concentrates) the signals from several ATU-Cs and converts and routes the signals to the appropriate service provider network.

This figure shows a simplified ADSL communication system that consists of a digital subscriber line access mulitplexer (DSLAM), local distribution lines that start from a main distribution frame (MDF) wire cabinet that brings the connection to the digital subscriber line (DSL) modem at the customer's location. Modems in the DSLAM convert the digital signals from the internet to high frequency signals that travel down the telephone line to the DSL modem. The DSL modem converts the RF signals back to its original digital form so it can be provided to the customer's computer. Most DSL technologies (such as ADSL shown in this example) transmit the data information on frequencies about the audio channel. This allows for the simultaneous transmission of analog and data signals on the same telephone line. The highest frequencies are used transmission from the DSLAM to the DSL modem and frequencies just above the audio band are used to transmit from the data from the customer to the DSLAM. Typical DSL technology allows up to 6 Mbps to be transmitted to the customer and up to 640 kbps can be received from the customer.

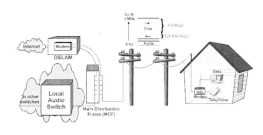

DSL Overview

Digital Subscriber Line Access Multiplexer (DSLAM)-A digital subscriber line access multiplexer is an electronic device that usually holds several digital subscriber line (DSL) modems that communicate between a telephone network and an end customer's DSL modem via a copper wire access line. The DSLAM concentrates multiple digital access lines onto a backbone network for distribution to other data networks (e.g. Internet).

This figure shows that a digital subscriber line access multiplexer (DSLAM) concentrates multiple digital subscriber lines onto a high-speed backbone network (e.g. ATM or

Ethernet) for distribution to other data networks (e.g. Internet). In this diagram, the DSLAM contains a backbone assembly (multiple sockets) that allow for the insertion of DSL modem line cards and that each DSL modem line card can provide service to more than one customer. This allows the service provider to add DSL modem line cards as the number of DSL lines increase. This DSLAM also contains the simple network management protocol (SNMP) communication capability that can be used to control the DSLAM and the DSLAM configuration information is stored in a management information base (MIB).

Digital Subscriber Line Modem (DSL Modem)-A DSL modem is an electronics assembly device that modulates and demodulates (MoDem) digital subscriber line (DSL) signals. DSL signals are usually transmitted on a twisted pair of copper wires. A DSL modem may be in the form of an internal computer card (e.g. PCI card) or an external device (Ethernet adapter). Most DSL modems have the ability to change their data transfer rates based on the settings that are programmed by the DSL service provider and as a result of the quality of the communication line (e.g. amount of distortion).

This figure shows a block diagram of a DSL modem. This diagram shows that a typical DSL modem connects to a two-wire telephone line and provides Ethernet and/or USB data connections. The modem has a hybrid assembly that separates the two-wire telephone line into a pair of transmit and receiver lines. The receiver demodulates the incoming DSL signals in the receiver frequency bands into data packets. The transmitter modulates the data packets from the Ethernet or USB connections into DSL signals on the transmitter frequency bands. A microprocessor (uP) coordinates the overall operation of the DSL modem using embedded software instructions that store in a memory section. The uP also can receive and send commands to the DSL system using simple network management protocol (SNMP).

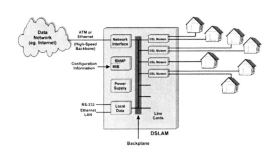

Digital Subscriber Line Access Multiplexer (DSLAM)

Using SNMP, the system operator can send configuration information to the DSL modem and this information is stored in the management information base (MIB) memory. The uP also controls status indicator lights such as power one, DSL sync (carrier detect), Ethernet or USB connection and an activity light.

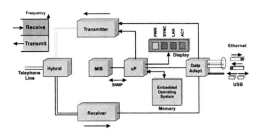

DSL Modem

Digital Switch-A digital switch is a device in a communication network that receives and forwards digital signals a data storage system. The digital switch receives a block of data from an input line, stores the data in a digital storage matrix or system, retrieves the block of data when the output line is ready, and transfers the block of data to the output line. This process occurs so frequently that the connection from the input line to the output line appears to be a continuous digital connection.

Digital Tape-Digital tape is a magnetic tape storage format that changes magnetic information on the tape to represent digital (discrete level) signals.

Digital Telephony-Digital telephony is a communication system that uses digital data to represent and transfer analog signals. These analog signals can be audio signals (acoustic sounds) or complex modem signals that represent other forms of information.

Digital Television (DTV)-Digital television is a process or system that transmits video images through the use of digital transmission. The digital transmission is divided into channels for digital video and audio. These digital channels are usually compressed. Digital television systems commonly use one of the motion picture experts group (MPEG) standards to reduce the data transmission rate by a factor of 200:1.

Digital Termination Service (DTS)-Digital termination service is the providing of communication service to customers using microwave radio transmission on a frequency band authorized by the regulatory authority (e.g. FCC).

Digital Terrestrial Television (DTT)-Digital terrestrial television is the broadcasting of digital television signals using surface based (terrestrial) antennas. DTT is also called digital video broadcasting terrestrial (DVB-T).

This figure shows a digital terrestrial television system. This diagram shows that a DTT system uses a single wide digital radio channel that is divided into multiple digital television channels. Television broadcasters are linked to the DTT system by a digital channel and they are assigned (mapped) to a specific

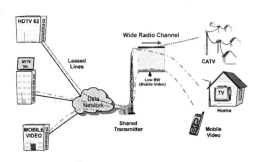

Digital Terrestrial Television System

portion (logical channel) on the digital radio channel. The DTT system combines (multiplexes) the incoming channels to form one digital transmission channel. DTT receivers receive and decode a DTT frequency, separate out (demultiplexes) the specific digital (logical) channel and converts the digital channel back into its original television (video and audio) form.

Digital To Analog Converter (DAC)-A digital to analog converter is a device or assembly that converts coded digital information into an analog signal.

Digital Transmission-Digital transmission is the process of sending information in digital (discrete level) form.

Digital Video-Digital video is a sequence of picture signals (frames) that are represented by binary data (bits) that describe a finite set of color and luminance levels. Sending a digital video picture involves the conversion of a scanned image to digital information that is transferred to a digital video receiver. The digital information contains characteristics of the video signal and the position of the image (bit location) that will be displayed.

This figure shows the basic process used by digital video to compress the video signal. This example shows that the first frame in a video sequence is a key frame. The next sequence of image data sent is the changes from the key

frame. This diagram shows a person who is waving. Because they are sitting still, only the hand changes are sent in frames after the key frame. The information (data) sent for the changed images are much smaller than the full image information. This allows digital video to be compressed by substantial amount for video that does not have rapid changes.

This figure demonstrates the operation of the basic digital video compression system. Each video frame is digitized and then sent for digital compression. The digital compression process creates a sequence frames (images) that start with a key frame. The key frame is digitized and used as reference points for the compression process. Between the key frames, only the differences in images are transmitted. This dramatically reduces the data transmission rate to represent a digital video signal as an uncompressed digital video signal requires over 270 Mbps compared to less than 4 Mbps

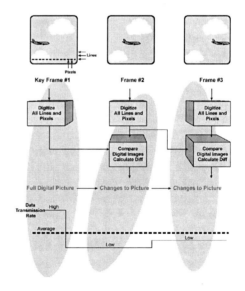

Digital Video

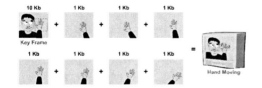

Digital Video Operation

for a typical digital video disk (DVD) digital video signal.

Digital Video Broadcasting Multimedia Home Platform (DVB-MHP)-Digital video broadcasting multimedia home platform is an industry standard for a digital television system that allows interactive applications.

Digital Video Recorder (DVR)-A digital video recorder is a device that stores video images in digital format.

Digital Wallet-A digital wallet is a storage of identification information, accounts and value(s) stored in the accounts in digital form that can be used to purchase items or enable a person or the device they control to perform actions or access services.

Digital Watermark-A digital watermark is a signal or code that is hidden (typically is imperceptible to the user) in a digital signal (such as in the digital audio or a digital image portion) that contains identifying information. Ideally a digital watermark would not be destroyed (that is, the signal altered so that the hidden information could no longer be determined) by any imperceptible processing of the overall signal. For example, a digital watermark should not be distorted or lost when the signal is passed through a conversion or compression process.

This figure shows how watermarks can be added to a variety of media types to provide identification information. This example shows that digital watermarks can be added to digital audio or video media by making minor changes to the media content. The digital watermark is added as a code that is typically not perceivable to the listener or viewer of the media. This example shows that digital watermarks can be added to audio signals in the form of audio components (e.g. high frequency sound) or video components (color shift) that cannot be perceived by the listener or viewer.

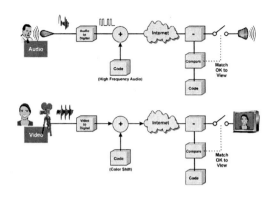

Digital Watermarking

Digital Wrapper-A digital wrapper is data or information that is added to media (such as a video program). Digital wrappers may provide descriptive and content protection information.

Digitization-Digitization is the conversion of analog into digital form. To convert analog signals to digital form, the analog signal is digitized by using an analog-to-digital (pronounced A to D) converter. The A/D converter periodically senses (samples) the level of the analog signal and creates a binary number or series of digital pulses that represent the level of the signal.

The common conversion process is Pulse Code Modulation (PCM). For most PCM systems, the typical analog sampling rate occurs at 8000 times a second. Each sample produces 8 bits digital that results in a digital data rate (bit stream) of 64 thousand bits per second (kbps).

Digital bytes of information are converted to specific voltage levels based on the value (weighting) of the binary bit position. In the binary system, the value of the next sequential bit is 2 times larger. For PCM systems that are used for telephone audio signals, the weighting of bits within a byte of information (8 bits) is different than the binary system. The companding process increases the dynamic range of a digital signal that represents an analog signal; smaller bits are given larger values that than their binary equivalent. This skewing of weighing value give better dynamic range. This companding process increases the dynamic range of a binary signal by assigning different weighted values to each bit of information than is defined by the binary system.

Two common encoding laws are Mu-Law and A-Law encoding. Mu-Law encoding is primarily used in the Americas and A-Law encoding is used in the rest of the world. When different types of encoding systems are used, a converter is used to translate the different coding levels.

This figure shows the basic audio digitization process. This diagram shows that a person creates sound pressure waves when they talk. These sound pressure waves are converted to electrical signals by a microphone. The bigger the sound pressure wave (louder the audio), the larger the analog signal. To convert the analog signal to digital form, the analog signal is periodically sampled and converted to a number of pulses. The higher the analog signal, the larger the number of pulses. The number of pulses can be counted and sent as digital numbers. This example also shows that

when the digital information is sent, the effects of distortion can be eliminated by only looking for high or low levels. This conversion process is called regeneration or repeating. This regeneration progress allows digital signals to be sent at great distances without losing the quality of the audio sound.

This figure shows the basic audio digitization process. This diagram shows that a person creates sound pressure waves when they talk. These sound pressure waves are converted to electrical signals by a microphone. The bigger the sound pressure wave (louder the audio), the larger the analog signal. To convert the analog signal to digital form, the analog signal is periodically sampled and converted to a number of pulses. The higher the analog signal, the larger the number of pulses. The number of pulses can be counted and sent as digital numbers. This example also shows that when the digital information is sent, the effects of distortion can be eliminated by only looking for high or low levels. This conversion process is called regeneration or repeating. This regeneration progress allows digital signals to be sent at great distances without losing the quality of the audio sound.

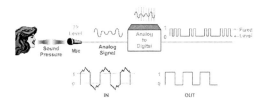

Audio Digitization

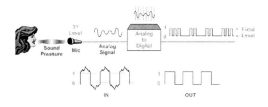

Digitization Process

Digtial Asset-A digital asset is a digital file or data that represents a valuable form of media. Digital assets may be in the form of media files, software (e.g. applications) or information content (e.g. media programs).

Direct Costs-Direct costs are the fees or bills that are directly associated with the production of a product or providing of a service.

Direct Department Calling (DDC)-Direct department calling is a telephone call processing service that directs (routes) incoming calls from specific communication lines (trunks) to specific phones or groups of phones that belong to a department.

Direct Distance Dialing (DDD)-The automatic completion of customer dialed long distance or toll calls in response to signals from a customer's telephone where no operator assistance is required.

Direct Distribution-Direct distribution is the process of transferring products or service between the manufacturer or producer and end users without the user of intermediary distributors. An example of direct distribution is the use of a fixed-satellite service to relay programs from one or more points of origin directly to terrestrial broadcast stations without any intermediate distribution steps.

Direct Expense-Expenses that are attributable to a particular category or categories of tangible investment described in and includes plant specific operations expenses in and plant nonspecific operations expenses.

Direct Inward Dialing (DID)-Direct Inward Dialing (DID) connections are trunk-side (network side) end office connections. The network signaling on these 2-wire circuits is primarily limited to 1-way, incoming service. DID connections employ different supervision and address pulsing signals than dial lines. Typically, DID connections use a form of loop supervision called reverse battery, which is common for 1-way, trunk-side connections. Until recently, most DID trunks were equipped with either Dial Pulse (DP) or Dual Tone Multifrequency (DTMF) address pulsing. While many wireless carriers would have preferred to use Multifrequency (MF) address pulsing, a number of LEC's prohibited the use of MF on DID trunks.

This figure shows the basic operation of direct inward dialing in a PBX system. This diagram shows how a caller has dialed a person in a company through the public telephone number. When this call is received by the end office (EO) switch in the public telephone company, the public telephone operator connects the call to one of the available incoming trunk lines between the telephone company switch and the PBX switch. This example shows that in addition to connecting the call and sending an alert (ringing) signal, the called number is also sent to the PBX. This allows the PBX system to lookup the called number to determine which extension the call should be connected to. The PBX system then connects the incoming trunk line to the correct telephone extension.

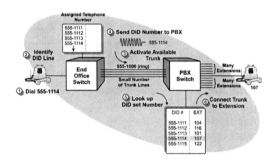

Direct Inward Dialing (DID) Operation

Direct Mail-Direct mail is a marketing program that sends media brochures direct to consumers through the postal (mail) system.

Direct Mail Cost-Direct mail cost is the combined costs of producing and sending materials via a postal system. Direct mail costs can include list rental, data processing, printing, envelope stuffing, mailing and service bureau fees.

Direct Marketing-Direct marketing is the process of promoting and selling products or services direct to customers. Direct marketing typically includes telemarketing, direct mail, email promotion, and fax broadcast.

Direct Order Entry (DOE)-Direct order entry is the acquisition of order data (contact information and product details) that comes from the customer or their agent.

Direct Outward Dialing (DOD)-A feature that allows private telephone systems users (PBX or Centrex) to directly call public telephone numbers without the need to use an attendant or operator. The use of "dial 9" to get an outside line is a direct outward dialing feature.

Direct Procurement-Direct procurement is the processes that are involved in the acquisition of products or services directly from the manufacturer or provider of the services.

Direct Sales-Direct sales are orders that are created as a result of a direct marketing effort.

Direct Sales Cost-Direct sales cost is the combined costs of contacting potential customers by direct employees or contractors and communicating advertising messages. Direct costs can include literature samples, list rental, telecommunication fees, travel costs, base pay and sales commissions.

Direct Selling-Direct selling is the use of people who are directly employed by a company to sell to customers.

Direct Store Delivery (DSD)-Direct store delivery is the shipment is the order fulfillment of a product from a manufacturer or a company other that the one that sold the product directly to the stores that sells the product.

Directories-(1-computer storage) File directories are portions of a computer storage area that are dedicated to the listing of file names and location pointers (starting addresses and file size) for the files contained on the computer storage system. (2-Internet) A web site directory is a listing of web pages that have specific characteristics (e.g. subjects). Web site search visitors enter key words (category search words) to find web pages that contain the characteristics associated with the search words. Directories are typically different than

search engines because URL submission and categorization process involves a human review process.

Directory-(1-computer) A list of all the files on a floppy diskette, hard disk, or web site. A directory may also contain other information such as the size of the files and the amount of free space remaining. (2-telephone listing) A telephone directory. (3-Internet listings) A web site portal (search engine or listing) that identifies companies, products, or listings of specific category types.

Directory Enabled Network (DEN)-A directory enabled network is the coordination of communications within a network through the use of a central information database that contains information about users, applications and network resources.

Disaster Recovery-Disaster recovery is the processes that are used to restore services after a significant interruption (disaster) in communications systems. Disaster recovery processes usually occur after events such as fires, floods, or earthquakes. However, disaster recover may also occur after critical equipment failures or information corruption that occurs from software viruses.

Disbursements-Disbursement is the transfer of value or assets to a person or company.

Disconnect Fee-A fee that is charged for the disconnection of service. If a disconnect fee is charged (in many cases there are no fees charged for the normal disconnection of service), it is usually a result of the customer's disconnection of service prior to the their committed service usage period.

Disconnect for Non-Payment (DNP)-Disconnect for non-payment is a process of terminating the authorization to use services for a user due to non-payment.

Disconnection Charges-Disconnection charges are fees that are assessed for the disconnection of a service. Disconnection fees may be charged if the user fails to pay their account balance and the service is disconnected for non-payment.

Disconnection Rate-Disconnection rate is the percentages of customers who unsubscribe or disconnect from services.

Discontinued Availability (DA)-Discontinued availability is a product or service (such as a communication circuit) that was once available but is no longer available for sale or use.

Discount-Discounts are reductions of pre-established fees or tariffs that are given for specific reasons. Discounts may be in the form of a specific amount or they may be based on a percentage of an item price or invoice amount. Discount types may be coded using specific identification codes and discount rates may be applied based on the specific type of sale or customer category using a discount schedule.

Discount Fee-A discount fee is a service charge that is assessed by a merchant processor or bank for the processing of transactions (such as a percentage of credit card transaction amounts).

Discount Qualification Rules-Discount qualification rules are requirements or conditions that must be satisfied for a price reduction to be applied.

Discount Rate-Discount rate is an adjusting value or percentage that is used to modify a price that is applied when certain conditions are met. Examples of discount rates are quantity discounts and wholesale discounts. Discount rates and their associated condition criteria may be listed in a discount schedule.

Discount Schedule-A discount schedule is an itemized list or table that provides pricing discount information for products or services. The price discount schedule will usually include the amount of discount based on quantity of product purchased and the types of customers that qualify to received the discounts (e.g. wholesale, education, retail).

Discount Schemes-Discount schemes are a structure that is used for pricing discount information for products or services. The price discount scheme may include discounts based on quantity, type of product, type of customer, product grouping, or other characteristics.

Discounted Payback Period-Discounted payback period is the time that a purchaser repays an initial investment where the payment values are discounted based on the amount of time that lapses between the purchase and each payment.

Discovery Metadata-Discovery metadata is information (data) that describes the content and attributes of the content that is contained in a collection of media (such as a media program).

Discrete Call Observing-The ability to monitor a call while it is in progress without indication to any of the call participants. Discrete call observing may be performed by managers who want to monitor the communication (typically sales or customer service) performance of employees.

Disk and Execution Monitor (Daemon)-A daemon is a program that can run in the background of computing operations to perform monitoring and processing functions. Common daemons include email monitoring, instant messaging and scheduling. Some daemons can be triggered by events to perform their processes and others can operate at timed intervals.

Disk Mirroring-A data protection strategy that uses redundancy of information on two (or more) storage devices (disks) to allow for real time backup and recovery of information. The process of disk mirroring is the storage of the same information on two disks. One disk is used as the primary source of information and the other disk is used when a failure is detected on the primary disk.

This figure shows the process of mirroring information on two disk storage devices to increase the reliability of information storage. Disk mirroring is also known as redundant array of inexpensive disks (RAID). This diagram shows that disk Mirroring is performed by a RAID controller card that controls both

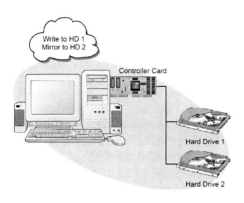

Disk Mirroring Operation

primary and secondary disk storage devices. Information is stored to both hard disks simultaneously. If the primary hard drive fails, the controller will automatically begin to use the secondary hard disk and alert the user that one of the hard disks has failed.

Disk Storage-Disk storage is memory device where the information can be stored on revolving materials that can retain data information. Disk storage may be fixed (disk sealed inside) or it may be removable.

Dispatch-Dispatch radio service allows a central controller (dispatcher) to send dispatch assignments to one or more receivers (typically many mobile radios). Dispatch radio systems normally involve the coordination of a fleet of users via a dispatcher. All mobile units and the dispatcher can usually hear all the conversations between users in a dispatch group. Dispatch operation involves push-to-talk operation by a group of users on the dispatch system. Using trunking technology, there can be several different dispatch groups that operate on the same system. Dispatch systems usually charge a fixed monthly usage fee for each user (mobile radio) without any airtime usage charges.

This figure shows a typical two-way SMR dispatch radio system. In this diagram, a single radio channel is shared by several mobile radios that operate within the radio coverage limits of a high power radio base station. A dispatcher communicates to all the mobile radios by using a high-powered radio transmitter. Because the mobile radios mounted in vehicles transmit at a lower power than the dispatcher's transmitter, this SMR system has several receiving antennas. When a mobile radio is communicating with the dispatcher, the receiver selects which antenna has the best signal (called a voting receiver).

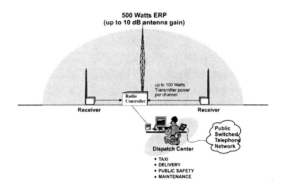

SMR Dispatch Operation

Dispatch Console-A dispatch console is a device or system that allows a person or group of people to access communication systems and services. The person who operates a dispatch console is a dispatcher.

Dispatch Point-A dispatch point is any place from which worker command messages can be originated under the supervision of a control point.

Dispatch Service-Dispatch service is a radiotelephone service comprising communications between a dispatcher and one or more mobile units. These communications normally do not typically exceed one minute in duration. For two-way radio systems, these signals are transmitted directly from a base station, without passing through mobile telephone switching facilities. For more advanced SMR systems, dispatch services may be established and coordinated by radio networks and switching systems.

Dispatch System-A communication system that allows a central controller (dispatcher) to coordinate communication between individual or groups of users in the system.

Dispatchers-Dispatchers are dispatch system users that have a high level of privilege to coordinate (interrupt and terminate calls) communication with other dispatch users.

Display Capability-Display capability is the ability of a device to render images into a display area. Display capabilities for IPTV include size and resolution (SD or HD), the type of video (interlaced or progressive) and display positioning (scaling and displaying multiple sources). Display capabilities for television systems are characterized in the MPEG industry standards and the sets of capabilities (size and resolution) are defined as MPEG profiles.

Display Device-A display device converts (renders) media or data into a format that can be viewed by a person or visual sensor.

Display Formatting-Display formatting is the positioning and timing of graphic elements on a display area (such as on a television or computer display).

Display Phone-A display phone is a telephone that has graphics or other display capabilities.

Disposable Content-Disposable content is the information contained within a message, call, media program or web site display that is of little or no value after it has been viewed or consumed.

Disposable Pay-Disposable pay is a portion of a paycheck after authorized deductions of amounts required by law to be withheld.

Dispute Management-Dispute management is a process of recording, tracking and reviewing differences between actions or services

that were supposed to be provided by a person or company and what a user or purchaser of products or services perceives that they have used or received.

Dispute Resolution-Dispute resolution is the actions that are agreed upon between parties that had a dispute between them.

Dispute Resolution Method-A dispute resolution method is the process that is used to identify, manage and resolve disputes between parties. Dispute resolution methods can range from self disputes that are automatically processed to complex legal proceedings that involve expert evaluation and testimony.

Disruptive Advertising-Disruptive advertising is the presentation of advertising messages that interrupt or alter the processes currently being performed by the recipient of the advertising message. An example of disruptive advertising is the creation of a pop-up advertising message on a computer display when the user is working with other processes or applications.

Disruptive Technology-A new technology that significantly reduces or eliminates the value of existing technology implementations by performing or providing the service faster and at lower cost.

Distance Based-Distance based is a type of service that is determined by a distance between service points.

Distance Learning-Distance learning is the process of providing educational training to students at locations other than official learning centers (schools). Distance learning has been available for many years and is now used in elementary education (grades K-12), higher education (college), professional (industry), government training and military training. In the early years, distance learning was provided through the use of books and other printed materials and was commonly referred to as "correspondence courses".

Distance learning has evolved through the use

of broadcast media (e.g. televisions) and moved onto individual or small group training through the availability of video based training (VBT) or computer based training (CBT). These systems have developed to interactive distance learning (IDL) as the computer allowed changes in the training.

Distance Sensitive Pricing-Distance sensitive pricing is the setting or assigning of prices that are based on distance that is involved or related to the providing of the product or service. An example of distance sensitive pricing is the changing of an airline ticket price that is based on the distance between the originating and terminating locations.

Distinctive Ringing-A service feature that alerts a customer via a special ring (usually short, long or rapid ring) that an incoming call is received that has a different purpose or priority from others that are received on that same telephone line. Distinctive ringing is used for sharing multiple phone numbers on a single line or for priority ringing.

This diagram shows the operation of a telephone system that has distinctive ringing feature. This diagram shows a single telephone that is assigned two different telephone numbers even though the telephone operates on one telephone number (one switch port). In

this example, when an incoming call is received for the registered number 555-6234, it is re-directed (forwarded) towards the actual destination number 555-1234 along with information that allows the system to uniquely identify the call with a dual ring (2 rings in the 2 second ring period). When calls are received to 555-1234, the ring is a single 2 second/4 second cadence. This allows the receiver of the call to determine which telephone number was dialed by the distinctive ring sound.

Distributed Architecture-Distributed architecture is a system that is designed to use a distributed intelligence to coordinate the operation of the system.

Distributed Billing-Distributed billing is a network that is designed to receive and process call detail and service usage information where the reporting and reconciling of billing records may be performed at various billing concentration (distribution) points within the billing system. Distributed billing transfers some of the intelligence (bill processing functions) to various points (distributed intelligence) to reduce the number of billing records that reach the central billing system.

Distributed Call Center-A call center that has several telephone workers (customer service representatives) that are located at different places. Distributed call centers usually allows the worker to have access to information systems even though they may be independently located at remote places (such as home workers).

Distributed Denial of Service (DDoS)-A distributed denial of service attack is a process that inhibits or reduces the ability of authorized users from gaining access to communications systems through the continual transmission of service requests from multiple locations (distributed locations) or messages that disable communication sessions.

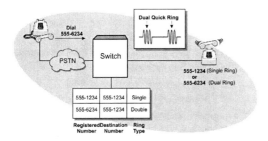

Distinctive Ringing Operation

Distributed File System (DFS)-A distributed file system is a naming and organizational structure that is used to group and store files at multiple locations.

Distributed PBX-A PBX system that has multiple switching systems that are interconnected with each other.

Distributed Scheduling (DSCH)-Distributed scheduling is the process allowing nodes within a system to independently select data or content that will be transmitted, transferred, made available (accessible) or streamed from and/or through nodes without requiring the direct involvement or approval from another network management system.

Distributed Switching-Distributed switching is a network that is designed to interconnect (switch) lines at various points within a network where there can be various interconnection concentration (distribution) points.

Distributed Verification-Distributed verification is the providing of credentials through a successive chain of distribution points so that direct access the root authority is not needed to authenticate or validate data, devices or users.

Distribution-(1-switching network) The capability in a switching system of connecting an input to any of several outputs. (2-traffic network) The separation of calls on incoming trunk groups at a toll or tandem office and their recombination on outgoing trunk groups. (3-data network) The process of distributing information from one station to one or more stations.

Distribution Channel-A distribution channel is the route that a product service uses to get from the original manufacturer or supplier to the customer or end user.

Distribution Control-Distribution control is the processes that are used to route products or service uses to get from the manufacturer or supplier to the customer or end user.

Distribution Fee-Distribution fee is an amount of money charged by most film and television distributors in order to be compensated for selling or licensing of programming on behalf of producers.

Distribution Hub-A distribution hub is a data interchange point where data is received and redistributed to other locations.

Distribution Network-(1-general) A system of cables and terminals that interconnect communication devices. (2-cable television) The portion of a cable television system that links the head end to the end customer's televisions. (3-telephone) Cables from a main telephone switching or distribution junction that usually contain from 25 to 200 pairs of wires.

Distribution Plant-Distribution plant is the physical property and facilities of a company (e.g. a telephone or cable TV company) that is used to distribute signals or services.

Distribution Rights-Distribution rights are the authorized allowable methods that can be used by a content distributor (such as a broadcaster) for to store, process and transfer content.

Distribution Service (DS)-Distribution is the transfer of information throughout a geographic area or through a network. Distribution services include broadcast, multicast, and point-to-point communication.

Distribution System (DS)-(1-communication) A system or network that is composed of transmission lines and switching or hub equipment that allows signals are messages to be transferred from one point to one or many other points in the network. (2-802.11 WLAN) The process or system that is used by access points (APs) to communicate with each other.

Distributor-(1-signal) The module within a link aggregator responsible for assigning frames submitted by higher-layer clients to the individual underlying physical links. (2-Product) A company or individual that sells products they receive from other companies to retailers or customers.

Distributor Accounts-Distributor accounts are companies or organizations that supply products or services to retailers. Distributor account sales support needs may include order management, order tracking and the supply of product media and literature.

DIUC-Downlink Interval Usage Code

DL-Data Link

DLC-Digital Loop Carrier

DLCI-Data Link Connection Identifier

DLEC-Data Local Exchange Carrier

DLL-Data Link Layer

DLL-Dynamic Link Library

DLNA-Digital Living Network Alliance

DMA-Designated Market Area

DMCA-Digital Millennium Copyright Act

DML-Data Manipulation Language

DMU-Decision Making Unit

DMZ-DeMilitarized Zone

DNIS-Dialed Number Identification Service

DNP-Disconnect for Non-Payment

DNR-Dialed Number Recorder

DNS-Domain Name Server

DNS Lookup-Domain Name Server Lookup

Do Not Call Registry-The do not call registry is a list of people who register that they do not want to be solicited by telephone offers by unauthorized companies. The list does not restrict calling from some types of organizations including ones that have received written permission to call, non-commercial calls that do not include advertisements, or calls by or on behalf of tax-exempt non-profit organizations. More information on the do not call registry can be found at www.FCC.gov.

DOA-Dead On Arrival

DOC-Department Of Communications

DOCSIS®-Data Over Cable Service Interface Specification

Document Commenting-Document commenting is software application program feature (such as a word processor program) that allows a user to add comments to a document or a media file.

Document Management-Document management is the processes and systems that are used to index, store and retrieve documents. To enable document management, each document is uniquely identified and the contents may be in the form of text, scanned pages, spreadsheets, web pages or another type of media file.

Document Object Model (DOM)-Document object model is a standard developed by W3C to allow software programs and scripts to interface, interact and format the presentations of documents.

Document Type Definition (DTD)-Document type definition is a industry specification that describes the content of a standard generalized markup language (SGML). The DTD defines the structural component of a document as distinct from the actual data or content of the document.

Documentation Server-A documentation server is a computer that stores reference materials, brochures or the locations of documents. A documentation server is typically attached to a network so users, managers or order processing systems can select, access and distribute documentation.

DOD-Direct Outward Dialing

DOE-Direct Order Entry

DOI-Digital Object Identifier

Doing Business As (DBA)-Doing business as is a name that is used by a company other than its original registered name.

DOM-Document Object Model

Domain-(1-general) A domain is a region, area or portion of a system which can be defined or controlled. (2-magnetic) The region in a magnetic material where the direction of magnetization is uniform. (3-parameters) The range of values assumed in a given document.

Domain Holder-A domain holder is a company or person who has the right to use a domain name for a specified period of time.

Domain Name-A domain name is the unique text or sequence of characters that is used to identify an address where information can be accessed on a data communication network. A domain name is associated with one or more IP addresses through the use of Domain Name Service (DNS).

Domain Name Pointing-Domain name pointing is the process of assigning a destination address (such as an IP address) to a domain name.

Domain Name Registration-Domain name registration is the process of submitting a domain name registration to a domain registrar. A domain registrar is a company or organization that is authorized by the Internet corporation for assigned names and numbers (ICANN) to issue and manage domain names.

Domain Name Server (DNS)-A domain name server is a data processing device (e.g. a computer) that translates text and numeric names for an Internet addresses. A DNS uses a distributed database containing addresses of other DNS servers that may contain the Internet address.

Domain Name Server Lookup (DNS Lookup)-A domain name server lookup is a process that obtains an IP address or a domain name from a DNS server.

Domain Registrar-A domain registrar is a company or organization that is authorized by the Internet corporation for assigned names and numbers (ICANN) to issue and manage domain names.

Dongle-A dongle is a small assembly or device that attaches to another object or system. Dongles can contain active electronic circuits that allow them to be used as part of an encryption or digital rights management (DRM) system to enable access or decoding to information.

Doorway Page-A doorway page is a web site that is developed to attract users or visitors of a specific type. Doorway pages are typically optimized to have high ranking in search engines and act as a portal for visitors to be transferred to other web sites that sell related products or services.

DoS-Denial Of Service

Dot Com Company (.Com Company)-A dot com (.Com) company primarily operates its business on or through the Internet.

Double Booking-A double booking is the assignment of programs or services from two projects that use the same resources at the same time.

Double Play-Double play refers to providing of two main services such as voice and data, data and video, or video and data on one network. For cable MSOs, this usually means building out the next generation network to DOCSIS 2.0 specifications, for Carriers this often means building out fibre or VDSL (very fast DSL) networks. Usually it is the larger MSOs and Telecom Carriers that roll out triple play services, and the advantage is that they can sign customers to a bundle of two services, thereby increasing revenue and customer loyalty.

DOW-Day of Week

Down Sizing-Down sizing is the reduction of the number of jobs at a company to better match the business or financial needs of the company.

Down Time-Down time is the duration between the end of the operation of a system to when the system becomes available for use again.

Downlink Interval Usage Code (DIUC)-The downlink interval usage code is information that is contained in the downlink MAP (DL-MAP) message that defines when information will be transmitted on the downlink and what formats it is supposed to use (burst profile).

Download Agent-A download agent is a software program that assists in the transferring (and possibly the installing) of programs or media items from a source (e.g. server) to a target device (e.g. a personal computer).

Download and Play-Download and play is a process of downloading a media program (an audio or video file) and then playing it after the file has completely downloaded.

Download Sites-Download sites are web sites that are setup to allow people to find and download data or media files. Mobile download sites may provide ring tones, games, screen backgrounds or other types of application programs.

Download Time-Download time is the time it takes to transfer a file or web page from a server (such as a web site) to a client (such as a user's computer). Download time depends on the modem/data transfer speed, file size, and protocols that are used to coordinate the transfer of data. Because a web page can contain links with several large files (such as images or video clips), download time can be several minutes with slow speed data connections.

Download to Own-Download to own is the process of downloading data or media with the intent to keep the content for an indefinite time period.

Download to Rent-Download to rent is the process of downloading data or media with the intent to keep the content for limited period of time or use.

Download Version-A download version is an application program or service that is provided on a system where the program is operated and/or managed by the company that provides the applications or services.

Downloadable-Downloadable is the capability to transfer data or a program from a source (such as a web server) to another device (such as a computer that is connected to the Internet).

Downloadable Advertising-Downloadable advertising is the communication of a message or media content to one or more potential customers by downloading media or programs that can be used to present one or more messages to the recipient. Downloadable advertising messages may be accessed at a later time during certain trigger events (such as when starting a particular service or software application).

Downloadable Client-A downloadable client is a software program that can be transferred from a source (e.g. a server) to a communication device (e.g. a computer or mobile telephone) which can be used to request services from a network.

Downloadable Conditional Access System (DCAS)-A downloadable conditional access system is a security process that is used in a communication system (such as a broadcast television system) to limit the access of media to authorized users that can be modified or updated.

Downloading-Downloading is the transferring of a program or of data from a computer server to another computer. Download commonly refers to retrieving files from a web site server to another computer.

Downstream-Downstream is the direction of transmission from the source to the destination (such as from a network to an end customer).

Downtime-An amount of time that a communication network or computer system is not available to users. Downtime usually occurs from hardware failure, software crashes, or operator errors.

DPBC-Delivery Point Bar Code

DPBX-Digital PBX

DPM-Defects Per Million

DPRL-Digital Property Rights Language

DPV-Delivery Point Validation

Draft Proposal-A draft proposal is an unfinished or unapproved document that contains some of the elements and content of a proposal document.

Drag And Drop-Drag and drop is a graphical interface that allows users to select and move object into an area or onto another object to initiate a process or change in data. For example, a user can drag a media element into a program during a time period schedule to allow the item to begin playing when that time period is reached.

Draw-A draw is a value that is provided to a person or company that is paid in advance of scheduled payments.

Drayage-Drayage is a fee that is charged for transferring packages from a receiving area to a specific location within a building or industry event.

Drill Down-Drill down is the process of selecting information (data) for the purpose of getting additional information or identifying related data or characteristics about it that may be useful for other purposes.

Drive Letter-A drive letter is a label that is used to identify a storage device in a computer or on a network.

Drive Mapping-Drive mapping is the assignment of a drive label (e.g. drive letter) to a storage device in a computer or on a network.

Driver-(1-general) An electronic circuit that supplies an isolated output to drive the input of another circuit. (2-fiber optic) The electric circuit that drives the light-emitting source, modulating it in accordance with an intelligence bearing signal. (3-software) A software module that controls an input/output port or external device, such as a keyboard or a monitor.

DRM-Digital Rights Management

Drop-(1-cable) A drop is a wire or cables that is the final connection point of a system (e.g. a telephone pole) to a building or user connection location. (2-wireless) The loss of a connection with a wireless system.

Drop Point-A drop point is the physical point where microwave service is provided to a customer.

Drop Rate-Drop rates are the percentages of customers or potential customers who unsubscribe or disconnect from services.

Drop Shipment-A drop shipment is the order fulfillment of a product from a manufacturer or a supplier directly to the customer who purchased the product.

Drop Shipping-Drop shipping is an order fulfillment process where the retailer who sells a product provides the customer purchase information to the manufacturer or supplier of goods so they can directly ship the products to the buyer. While drop shipping can reduce the total shipping cost, manufacturers or suppliers often charge a processing fee or reduce the purchase discount for companies that use drop shipping services.

Drop Shop-A drop shop is a store that accepts products that can be sold by other stores (such as online stores).

Drop Wire-A drop wire is the wire or pairs of wires that are connected between a customer's premises and a nearby network line. Although the first drop wires were connected from a telephone pole to a building, drop wires can be buried or aerial.

Dropped Calls-Cellular telephone calls that are inadvertently disconnected from the system because of interference, inadequate coverage or lack of capacity.

Dry Circuit-A communication line that does not provide electrical power with the information signal. The use of a dry line requires the terminating equipment to supply its own power.

DS-Distribution Service
DS-Distribution System
DS-0-Digital Service, Level 0
DS-1-Digital Signal 1
DS-1C-Digital Signal, level 1 Combined
DS-3-Digital Signal 3
DS-4-Digital Signal 4
DSCH-Distributed Scheduling
DSCP-Differentiated Services Code Point
DSD-Direct Store Delivery
DSL-Digital Subscriber Line
DSL Bonding-DSL bonding is the combining of multiple DSL communication lines to provide for higher data rate. For example, if eight 1.5 Mbps DSL lines are combined, the data transmission rate is 12 Mbps. This is also called inverse multiplexing.

DSL Concentrator-An interface that allows more local loop telephone lines to share a digital subscriber line access multiplexer

(DSLAM) that is allowed by the number of DSL modems that are installed in the DSLAM. The DSL concentrator acts as a mini-switch connecting the local loop to the DSL modem when data service is requested.

DSL Forum-A forum that was started in 1994 to assist manufacturers and service providers with the marketing and development of DSL products and services. The DSL forum was previously called the ADSL forum.

DSL Microfilter-A DSL microfilter is a blocking filter device that attaches to a telephone jack that blocks unwanted high-speed data signals from entering into the telephone.

DSL Modem-Digital Subscriber Line Modem

DSLAM-Digital Subscriber Line Access Multiplexer

DSP-Digital Signal Processor

DSS-Decision Support Systems

DST-Digital Signaling Tone

DSU-Data Service Unit

DSU-Digital Service Unit

DSU/CSU-Data Service Unit/Channel Service Unit

DSx-Digital Signal Level

DT-Data Terminal

DTA-Data Transfer Adapter

DTD-Document Type Definition

DTE-Data Terminal Equipment

DTMF-Dual Tone Multi-Frequency

DTMF Cut Through-A process that allows the interruption of a service (such as an announcement message) when the system detects a DTMF tone that a caller has pressed. This feature allows for better control of interactive voice response (IVR) and auto attendant systems.

DTMF Decoder-A device or process that converts dual tone multi-frequency (DTMF) signals into another form (such as data digits).

DTS-Digital Termination Service

DTT-Digital Terrestrial Television

DTV-Digital Television

Dual Mode-Dual mode is the ability of a device or system to operate in two different

modes (not necessarily at the same time). For wireless systems, it refers to mobile devices that can operate on two different system types.

This figure shows how the IS-95 CDMA system can provide both digital and analog "dual mode" operation. This diagram shows that a dual mode CDMA mobile radio can originate and receive calls in two different system types; CDMA and AMPS analog. The mobile device typically searches for CDMA channels first. If it cannot find the CDMA channels, it will then scan for analog channels.

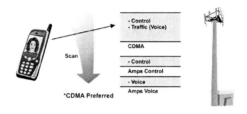

IS-95 CDMA Dual Mode Operation

Dual Tone Multi-Frequency (DTMF)-DTMF signaling is a means of transferring information from a user to the telephone network through the use of in-band audio tones. Each digit of information is assigned a simultaneous combination of one of a lower group of frequencies and one of a higher group of frequencies to represent each digit or character. There are 8 tones that are capable of producing 16 combinations; 0-9, *, #, A-D. The letters A-D are normally used for non-traditional systems (such as the military telephone systems). This diagram shows how DTMF tones are created. This example shows that DTMF tones consist of two frequencies that are combined. These frequencies are selected from a matrix of 4 (low group) by 4 (high group) frequencies.

Each button on a telephone is assigned one frequency from a high group and one from a low group. This diagram shows that DTMF signals are used to represent numbers, characters and extended digits on a telephone keypad.

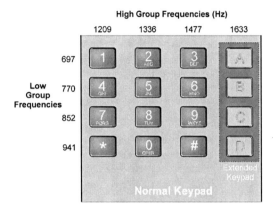

High Group Frequencies (Hz)

DTMF System

This diagram shows how dual tone multi-frequency (DTMF) tones can be used to send dialing information from a telephone to a telephone system. There are 8 different frequen-

cies that can be combined to represent 16 keys. The keys A-D are not usually included on standard telephone sets. To represent each button, two tones are combined. In this example, the button 3 is pressed, followed by a pause, then button 2 is depressed. Button 3 is represented by the combined tones 1477 Hz and 697 Hz. Button 2 is represented by the combined tones 1336 Hz and 697 Hz. To determine if the user is finished dialing, a timer is used. When the user has stopped dialing, the digits can be sent to the call processing section of the telephone system to initiate the call.

Duct-(1-general) A pipe or conduit, installed underground or in a building, whose purpose is to protect the cables installed therein. (2-height) The height above the Earth of the lower boundary of an elevated propagation duct. (3-metal floor wiring) One of various proprietary schemes of cable ducting to provide flexibility in equipment installation in office areas. (4-nest) A number of cable ducts provided for and laid in one trench. (5-propagation) A layer of cold air under warm air, experienced in some areas, that causes microwave signals to propagate further than normally possible. (6-surface) A radio duct whose lower boundary is the surface of the earth. (7-thickness) The difference in height between the upper and lower boundaries of a tropospheric radio duct.

Duct Rental-Duct rentals are fees that are charged for the authorization to use a duct (e.g. wiring ducts).

Due Diligence-Due diligence is a process of examining information about a company, product or service to validate or assign a value. Due diligence is commonly performed by potential investors or their agents to gather information about companies that they plan to invest in. During the due diligence process, the managers and key staff at a company are interviewed along with financial statements and other reference information.

Dumb Number-A dumb number is a reference identifier that is composed of characters

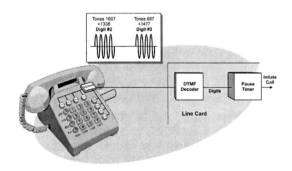

DTMF Dial Operation

that have a random structure and there are no required inferences or relationships between characters.

Dumb Terminal-A computer display terminal that serves as a slave to a host computer. A dumb terminal has a keyboard for data entry and a video display, but no computing power of its own.

Dump-Dumping is the process of copying or transferring the entire contents of a file or information.

DUN-Dial-Up Network

Dunning-Unique treatment of customers for the purpose of collection of service charges or account balances.

DUNS-Data Universal Numbering System

Duopoly-A duopoly is an industry where two companies supply or control all the customers. An example of a duopoly is the licensing of two companies to provide public services such as mobile telephone systems.

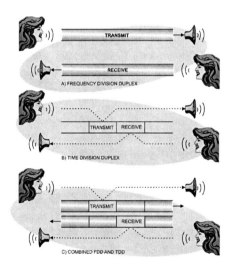

Duplex Systems

Dup-Elim-Duplication Elimination

Dupes-Duplicates

Duplex-Duplex communication is the transmission of voice and/or data signals that allows simultaneous 2-way communication.

This figure shows the different types of duplex systems. Frequency division duplex is a communications channel that allows the transmission of information in both directions (not necessarily at the same time) via separate bands (frequency division). Time division duplex (TDD) is a process of allowing two-way communications between two devices by time-sharing. When using TDD, one device transmits (device 1), the other device listens (device 2) for a short period of time. After the transmission is complete, the devices reverse their role so device 1 becomes a receiver and device 2 becomes a transmitter. The process continually repeats itself so data appears to flow in both directions simultaneously. It is also possible to combine FDD and TDD.

Duplicate Record Elimination-Duplicate record elimination is the process of deleting or segregating records that contain the same data or are created to represent the same event information.

Duplicate Records-Duplicate records are data entries in a database or list that are duplicates of other entries.

Duplicated Audience-A duplicated audience is a group of people or companies where a communication message or marketing promotion campaign reaches recipients by more than one communication channel.

Duplicates (Dupes)-Duplicates are multiple records or information elements within a file or database, which identify the same item or have duplicated information. Dupes can occur when the key element (key field) in a record uses a name or identifier that can be expressed in multiple ways (such as David or Dave).

Duplication Elimination (Dup-Elim)- Duplication elimination is the process of removing duplicate records in a list or database.

Duplication Rights-Duplication rights is the permission given to duplicate content or portions of content in specific formats (such as duplicating an article for the employees of a company).

Duration-Duration is the time that elapses between the start of an event (such as answering a call) and the termination of the event (such as the ending of a call).

Duration Based Charging-Duration based charging is the rating of billing cost that is determined by the duration (start to end) time of the service regardless of the amount of data or service used.

Dutch Auction-A Dutch auction is a reverse auction where the price is lowered until the product sells.

Duties (Duty)-Duty is a fee or assessment that may be required for the import or export of goods that are exported or imported into a country. Duties can vary dramatically based on the type of products and companies that are sending or receiving them.

Duty-Duties

Duty Drawback-Duty drawback is the refund of customs duties that were paid on goods that were imported and subsequently exported.

DVB-MHP-Digital Video Broadcasting Multimedia Home Platform

DVR-Digital Video Recorder

Dynamic-A process, item, or information element that has parameters that can change at unplanned times.

Dynamic Address-An address that is assigned to a device or service, usually at the beginning of a communication session.

Dynamic Allocation-Dynamic allocation is the assignment of a resource on an unscheduled basis. Dynamic allocation is a process of sensing the need for a communication resource (such as a channel) and the assignment (allocation) of the resources that are required by that communication need.

Dynamic Call Routing-Dynamic call routing is the process of automatically re-routing a call based on specific criteria (such as time of day or as the network traffic levels of congestion) change.

Dynamic Capacity Allocation-Dynamic capacity allocation is the evaluation and provisioning of capacity resources. An example of dynamic capacity allocation is the increasing of the number of mobile telephone radio channels in a geographic area (such as a car accident area) and the reduction of radio channels in adjacent areas.

Dynamic Data Rates-Dynamic data rates are the processes used or ability to change data transmission rates at unscheduled times. Data transmission rates are dynamically changed to optimize the data transmission rate for a communication channel or to share the available data transmission channels with other users.

Dynamic Host Configuration Protocol (DHCP)-Dynamic host configuration protocol is a process that dynamically assigns an Internet Protocol (IP) address from a server to clients on an as needed basis. The IP addresses are owned or controlled by the server and are stored in a pool of available addresses. When the DHCP server senses a client needs an IP address (e.g. when a computer boots up in a network), it assigned one of the IP addresses available in the pool.

This figure shows how a computer uses DHCP to obtain a temporary IP address when it requires an Internet communication session. In this example, the computer requests a connection with an Internet service provider (ISP) via a modem that is connected to a universal serial bus (USB) line. When the Internet service provider receives the request for connection, it assigns an IP address from its list of available IP addresses. The computer will then use this IP address for all of its commu-

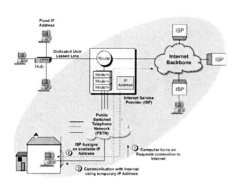

Dynamic Host Configuration Protocol (DHCP)
Operation

nications with the Internet until it disconnects
the connection to the ISP.

**Dynamic Hypertext Markup Language
(DHTML)**-Dynamic hypertext markup language is an evolved version of HTML that ads
scripts and features that allows for the creation of web pages that have content that is
processed after the user has selected some
options.

Dynamic IP Addressing-Dynamic IP
addressing is a process of assigning an
Internet protocol address to a client (usually
and end user's computer) on an as needed
basis. Dynamic addressing is used to conserve
on the number of IP addresses required by a
server and to provide an enhanced level of
security (no predefined address to use for
hackers).

Dynamic Licensing-Dynamic licensing is
the process of rapidly (e.g. automatically)
defining, authorizing and compensating for
the rights to develop, use or sell products and
services.

Dynamic Link Library (DLL)-Dynamic link
libraries are executable software code modules
that can be loaded into memory on demand
and linked when the applications begin to
operate (run time). This enables the software
code to access the latest parameters in its
related software applications. DLL files are
unloaded when they are no longer needed
(when the application is closed). Each DLL
applications that is initiated is copied into the
working memory of the computer. A key benefit of using dynamic linked libraries is that the
executable program files are not as large
because the frequently used routines can be
put into DLL files.

Dynamic Loading-Dynamic loading is the
process of obtaining files or media as the need
for the data is determined. An example of
dynamic loading is the obtaining of a digital
video clip file and loading it into a media player after a person has selected the link to that
file.

Dynamic Pricing-Dynamic pricing is ability
to increase or decrease in reaction to certain
conditions such as changing competitive
prices.

Dynamic Traffic Routing-Dynamic routing
is the automatic selection of alternative communication routes or systems. The choice of
alternative routes may be dependent on cost of
service, traffic congestion, line failure, or other
criteria that may change the choice of routing
path.

E

E&O-Errors And Omissions

E.164 International Public Telecommunications Numbering Plan- The International Telecommunications Union (ITU), a division of the United Nations, has defined a world numbering plan recommendation, "E.164." The E.164 numbering plan defines the use of a country code (CC), national destination code (NDC), and subscriber number (SN) for telephone numbering. The CC consists of one, two or three digits. The first digit identifies the world zone. The number of digits used for telephone numbers throughout the world varies. However, no portion of a telephone number can exceed 15 digits. There are several "E" series of ITU numbering recommendations that assist in providing unique identifying numbers for telephone devices around the world.

E.164 International Public Telecommunications Numbering Plan 1- This diagram shows the world (telephone) numbering plan recommendation, "E.164" developed by the International Telecommunications Union (ITU).

This diagram shows the numbering plan divides a telephone number into a country code (CC), national destination code (NDC), and subscriber number (SN) for telephone numbering. The CC consists of one, two or three digits and the first digit identifies the world zone. This diagram shows that the local number can be divided into an exchange code (end office switch identifier) and a port (or extension) code.

E411-Enhanced 411

E911-Enhanced 911

Early Adopter-An early adopter is a person or company who is more willing to buy or use new systems or services than the average person or company.

Early Termination Option-An early termination option is a term in an agreement that allows for the termination of some or all of an agreement provided that some condition exists.

Earned Company-An earned company is an organization or entity that has performed a service for another company or person and is entitled to revenue or assets from that company. An example of an earned company is a local telephone service provider that has completed a call to a customer from a long distance company. The local telephone search provider has earned a portion of the revenues that are collected for using their network to complete the call to the customer.

Earned Rate-Earned rate is the rate charged for advertising that is based on the frequency and number of ads that are run.

Earnings Before Interest, Taxes, Depreciation, and Amortization (EBITDA)-EBITDA is used by finance and investment analysts to gain a more accurate appraisal of the true net worth and/or cash flow of a company. EBITDA reflects the ability of a business to achieve profitability without the influence of financial positions that may vary between different companies.

Earnings Per Share (EPS)-Earnings per share are the amount of earnings money that is paid to shareholders based on the number of

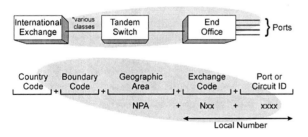

E.164 Telephone Numbering System

shares. It is calculated by dividing a company's post-tax profits by the company's number of common shares outstanding.

EAS-Emergency Alert System

EAS-Extended Area Service

Easement-Easement is the right of use of another person's land that is granted by the owner of the land or by an authorized government agency.

Easyware-Easyware is equipment and/or software that is simple to setup and operate.

EB-Exabyte

EBB-Electronic Bulletin Board

EBCDIC-Extended Binary Coded Decimal Interexchange Code

EBITDA-Earnings Before Interest, Taxes, Depreciation, and Amortization

EBPP-Electronic Bill Presentation and Payment

EBS-Emergency Broadcast System

e-Business-Electronic Business

ebXML-Electronic Business Extensible Markup Language

EC-Exchange Carrier

eCash-Electronic Cash

ECML-Electronic Commerce Modeling Language

E-commerce or ECommerce-Electronic Commerce

E-Commerce Solution-A software program and or hosted services that provide the tools and processes necessary to setup an online storefront.

Economic Bypass-A form of bypassing communication services where the cost of the bypass service is lower than that of an equivalent telephone company service.

Economic Conditions Churn -Economic conditions churn is a type of voluntary, incidental churn that occurs when customers loose their jobs or face other economic crises, and must terminate service.

Economic Value Added (EVA)-Economic value added is a measure of how much financial value has been added from the investment in equipment, plants, factories or other assets.

Economy of Scale-Economy of scale is the process of decreasing item or service cost as the volume of items produced or services provided increases.

ECPM-Effective Cost Per Thousand

ECSA-Exchange Carriers Standards Association

EDC-Electronic Draft Capture

EDI-Electronic Data Interchange

EDI File-Electronic Data Interchange File

EDI Translation-EDI translation is the conversion of electronic data interchange (EDI) messages into a format other systems can use.

EDM-Electronic Document Management

EDSL-Extended Digital Subscriber Line

Educational Access Channel-An educational access channel is a media source that is dedicated to education. Educational programming may come from public or private schools. Examples of educational programming include student programming, school sporting events, distance learning classes, student artistic performances and the viewpoints and teachings of instructors.

EFF-Electronic Frontier Foundation

Effective Cost Per Thousand (ECPM)-The effective cost for each thousand units associated with media such as TV time, radio spots, Internet ad impressions and print ads to determine the actual cost for each thousand events. For example, a company may pay $1 per click and 1% of viewers may click so this provides an effective cost per thousand impressions of $10 (100 viewers per $1).

Effective Date-An effective date is the day that a service or process begins operation or is authorized to start.

Effective Speed Of Transmission-The rate at which information is processed by a transmission facility, expressed as the average rate over some significant time interval. The quantity usually is stated as the average number of characters or bits per unit time.

Efficient Selling-Efficient selling is the process of creating more sales using less time and resources. Efficient selling uses effective process for identifying, qualifying, assessing, fact finding and order processing by salespeople to help customers to obtain the products or services that they want or need.

EFS-Error Free Seconds

EFT-Electronic Funds Transfer

EIS-Executive Information System

Elapsed Call Timer-A time display that indicates that amount of time that has elapsed since a call was initiated or received.

eLearning-Electronic Learning

Elective Calling-Elective calling is a process of calling another device where signals are transmitted using prearranged code to gain the attention of the device that is being called.

Electronic Bill Presentation and Payment (EBPP)-Electronic bill presentation and payment is the systems, applications and processes that are used to group, process and display billing information along with the ability for the user to interact and pay invoices and balances that are due on accounts.

Electronic Bulletin Board (EBB)-A electronic bulletin board is a computer server that users can connect to and access files, post messages and share resources.

Electronic Business (e-Business)-Electronic business is the operation of business functions through electronic systems (usually through the Internet).

Electronic Business Extensible Markup Language (ebXML)-Electronic business extensible markup language is an e-business platform that can be used by companies to exchange electronic business documents or data over electronic communication networks. eBXML systems may be used instead of or in addition to the electronic data capture (EDI) system. More information about eBXML can be found at www.eBXML.org.

Electronic Cash (eCash)-Electronic cash is a form of value that is stored in electronic form, which can be used on systems that accept the eCash form of payment.

Electronic Commerce (E-commerce or ECommerce)-A shopping medium that uses electronic networks (such as the Internet or telecommunications) to present products and process orders.

Electronic Commerce Modeling Language (ECML)-Electronic commerce modeling language is a set of commands and processes that are used to format and process e-commerce web sites.

Electronic Data Interchange (EDI)-Electronic data interchange is industry standards or the processes used by companies to exchange electronic business documents or data over electronic communication networks. Accepted EDI standards include ANSI X.12, ISO 9735, and CCITT X.435.

Electronic Data Interchange File (EDI File)-An electronic data interchange file is a file that holds business data records (such as invoices and shipping records).

Electronic Document Management (EDM)-Electronic document management is the process of acquiring, maintaining, distributing, and the elimination of documents. Documents may be in the form of scanned documents, word processor documents and other electronic forms.

Electronic Draft Capture (EDC)-Electronic draft capture is a financial transaction process that uses electronic devices or systems to capture account (such as a credit card number) or other financial payment information.

Electronic Fingerprint-An electronic fingerprint is a unique identifier in electronic format (typically digital bits) that is associated with a specific device, user, media or a combination of these items. Electronic fingerprints can be a representation of the characteristics of a physical fingertip or it can be group of characteristics that uniquely identifies a specific device or equipment. Examples of electronic fingerprint characteristics include the combination of software and hardware serial numbers located on a specific computer.

Electronic Frontier Foundation (EFF)- The electronic frontier foundation is a foundation that was established in 1990 with a focus to help ensure that the principles within the Constitution and Bill and Rights are protected as new communications technologies emerge.

Electronic Funds Transfer (EFT)- Electronic funds transfer is the process of transferring value through the transfer of electronic messages or data between the parties (or their agents).

Electronic Invoice-An electronic invoice is a grouping of billing charges for a time period or event that are associated with a specific user or account that can be presented on a display device (such as a computer or television).

Electronic Invoice Data (EID)-Electronic invoice data is a billing exchange format developed by the GSM Association that enables GSM network operators to electronically exchange invoice data about roaming customer usage via a pre-defined format according to a standardized business process. EID may supplement or replace the customary, and in some countries, the legally binding process of manual paper invoicing.

Electronic Learning (eLearning)- eLearning is the providing of lectures or presentations of learning materials to people who receive materials and/or participate in the training sessions through the computerized systems (such as via a CDROM or through an Internet portal page).

Electronic Mail (Email or e-Mail)-E-mail is a process of sending messages in electronic form. These messages are usually in text form. However, they can also include images and video clips.

Electronic Mail Payments (Email Payments)-Electronic mail payments are authorizations to transfer financial amounts from a sender's account (such as a bank account) to an account associated with an email (such as an online commerce account).

Electronic Outsourcing (E-Outsourcing)- Electronic outsourcing is the use of an outside firm to produce products, assemblies or to perform specific business functions that would or could be conducted internally where the outside firm is found, communicated with or produces products that are delivered through the Internet.

Electronic Payment-An electronic payment is a transaction that can be performed using information transactions. Some electronic payment systems can have instant or near instant transfers and others such as wire transfers may take hours or days to process.

Electronic Procurement (E-Procurement)-Electronic procurement is the processes that are involved in the acquisition of products or services via the Internet.

Electronic Rate (E-Rate)-Electronic rate is a service classification rate plan that is setup for schools and libraries. The e-rate is a discounted service rate that was established as part of the Telecommunications Act of 1996.

Electronic Retailers (E-Tailers)-Electronic retailers are companies that sell products on the Internet.

Electronic Service Guide (ESG)-Electronic service guides are an interface (portal) that allows a customer to preview and select from possible lists of available services. ESGs can vary from simple program selection guides to interactive filters that dynamically allow the user to search through services by theme, time period, or other criteria.

Electronic Signature-An electronic signature is a number that is calculated from the contents of a file or message using a private key appended or embedded within the file or message. The electronic signature can be used to validate that a specific person or device has created the file and that the file data has been unaltered.

Electronic Storefront-An electronic storefront is a web site that enables visitors to find, order and pay for products and services using data networks (such as the Internet).

Electronic Switching System (ESS)-A system that can connect incoming and outgoing digital lines together through the use of temporary memory locations. For an ESS system, a computer controls the assignment, storage, and retrieval of memory locations so that a portion of an incoming line (time slot) can be stored in temporary memory and retrieved for insertion to an outgoing line.

Electronic Telephone Directory Service-Electronic telephone directory service is a telephone system feature that stores and provides a directory of all phone numbers when requested.

Electronic Tokens-Electronic tokens are information elements (e.g. electronic messages) that are used to indicate a value or authorization that can be used to access or consume data or media items.

Electronic Wallet (E-Wallet)-An e-wallet is a storage of identification information, accounts and value(s) stored in the accounts that can be used to purchase items or enable a person or the device they control to perform actions or access services.

Element-An element is the smallest identifiable component of an object, system or program. (1-atom) A substance that consists of atoms of the same atomic number. Elements are the basic units in all chemical changes other than those in which atomic changes, such as fusion and fission, are involved. (2-information) A unit of structure in media document such as a metadata item. (3-marketing) An influencing component of a marketing campaign.

Eligible Bidder-An eligible bidder is a person or company that has been recognized by the bid issuing company that meets the qualifications of the bidding process.

Eligible Vendor-An eligible vendor is a provider of products or services that has been recognized by the RFP issuing company that meets the qualifications of the RFP process.

Email Broadcasting-Email broadcasting is a process of sending messages to relatively large number or email recipients.

Email Campaign-An email campaign is the marketing activities that are designed to send messages to customers about products, services, and options via email messages.

Email Forwarding-Email forwarding is the process of redirecting emails that are received at one address to another address.

Email Marketing-Email marketing is the process of sending marketing and sales information using email messaging systems. Email marketing systems generally combine advanced message broadcast systems along with tracking systems that can monitor the reception, opening, and response to email messages that have been sent. The email marketing process typically involves a list aquisition phase, offer testing phase and promotion and retention phase.

This figure shows an example email marketing process with some sample numbers. This example shows that email marketing is the process of managing lists, developing ad campaigns, creating promotional offers, broadcasting email messages, and tracking the results. This diagram shows an example of lists that cost 1 to 10 cents per name. Names in these lists are grouped to match specific marketing campaigns. These marketing campaigns have several promotional offers. The emails are broadcast and some of the emails are returned

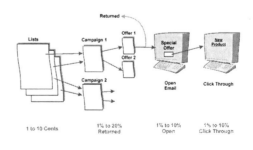

Internet Marketing Email Broadcast

and 1% to 20% returns is typical. Of the emails that make it to the recipient, some of the emails are never opened. Of the 25% to 80% of the emails that are opened, approximately 2% to 10% click through to the specific product or service order page.

Email Marketing Cost-Email marketing cost is the combined costs of producing and distributing advertising messages via email. Email marketing costs can include email creation, list rental, transmission fees, and banner network exchange fees.

E-Mail Notification-E-mail notification is the process of sending a subscriber an email message that notifies them that information has changed (such as a new voicemail).

Email Opt-in-Email Option In

Email Option In (Email Opt-in)-Email option in is a process that requires that a person to agree (select an option) to be added to an email list.

Email or e-Mail-Electronic Mail

Email Payments-Electronic Mail Payments

Email Service Bureau-An email service bureau is a company that provides email broadcasting services to for other companies. The services provided can range from email transmission, list filtering, list suppression and other list management and transmission services.

Email Template-An email template is a message and/or associated content that can be used as an example for a message that will be sent by email. Email templates may be used to assist and standardize how customer service representatives (CSRs) communicate with customers.

Embedded Customer Premises Equipment-Embedded customer premises equipment are devices or systems that are provided by a company. For United States telephone companies, embedded customer premises equipment includes devices and systems that were owned by the telephone company as of December 31, 1982 and installed and used by the customer at their premises.

Embedded Linux-Embedded Linux is a version of the Linux operating system that is used in devices that have limited processing capability (such as mobile telephones).

Embedded Operating System (EOS)-An embedded operating system is a group of software programs and routines that directs the operation of a device in its tasks and assists programs in performing their functions that is not generally accessible by the user. The embedded operating system software is responsible for coordinating and allocating system resources. This includes transferring data to and from memory, processor, and peripheral devices. Software applications use the embedded operating system to gain access to these resources as required.

Emergency Alert System (EAS)-Emergency alert system is a system that coordinates the sending of messages to broadcast networks of cable networks, AM, FM, and TV broadcast stations; Low Power TV (LPTV) stations and other communications providers during public emergencies. When emergency alert signals are received, the transmission of broadcasting equipment is temporarily shifted to emergency alert messages.

Emergency Broadcast System (EBS)-An abbreviation for emergency broadcast system, an alerting scheme developed by the FCC in which radio and TV stations cooperate to deliver timely emergency information to their audiences.

Emergency Dialing-Emergency dialing is the use of abbreviated calling numbers to enable people to more rapidly call numbers for police, fire department or ambulance services.

Emergency Response Systems-Emergency response systems are devices, software and processes that are used to identify, analyze and assist in the recovery of failed systems or services.

Emergency Response Team (ERT)-An emergency response team is a defined group of people who work together during emergency situations to identify, analyze and assist in the recovery of failed systems or services.

Emergency Ringback-A call return feature enables a public service answering point (PSAP) or emergency service attendant to call back to a caller who has terminated the call or left the phone off-hook. The emergency ringback service may produce a loud "howling" on the phone to get the attention of the caller.

Emergency Service-Emergency service is the providing of communication services for a need that is unforeseen by the service provider. Emergency service users may be given higher access and call control priority than other users on the communication systems.

This figure shows that when an Internet telephone is connected to the Internet, it does not know its exact location. This can lead to problems when the user dials for emergency services (such as dialing "911"). If a call is received to a public telephone emergency number, it is not possible to determine which emergency call center should receive the call.

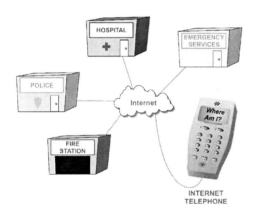

Internet Telephone Emergency Location

Emergency Service Number (ESN)-An emergency service number is a three to five digit number that used to initiate or route calls to public safety answering points.

Emergency Services Surcharge-Emergency services surcharge is the fee that is collected from a user of communication services to help operate emergency services (such as public safety access points).

Emergency Session Detection (ESD)-Emergency session detection is the identification of a communication session request that is a request for urgent assistance.

Empirical-A conclusion that is based not on pure theory, but on practical and experimental work.

Employee Advances-Employee advances are the providing of funds or assets for work from an employee that is pledged to be provided in the future.

EMR-Exchange Message Record

EMS-Enhanced Messaging Service

Encapsulation-(1-connection) Encasing an electrical connection in a container with protective material to seal it to the environment (e.g. keep out water and air). (2-software program) The grouping of software code into a single entity, object or program. The encapsulation of a software program isolates the inner workings of the program from the programmers and other programs that use it. (3-layered protocols) The adding of additional protocol layers to a protocol data unit (PDU) to add new control or routing capabilities to the PDU. (4-data packet) A process of inserting the entire contents of a data packet into the payload (data portion) of another packet.

This figure shows how an IP packet is encapsulated into the data portion of another packet that is sent through a GPRS network. As the IP packet is sent from the data device (e.g. portable computer) into the wireless device, packets are individually addressed with their destination address (the web site host). When the packet is received at the SGSN, the packet is embedded into another packet and this packet that has the address of the GGSN. This packet is forwarded to the destination GGSN on the virtual path (tunnel) through the GPRS data network. Both the SGSN and GGSN have their own IP address. When the data packets arrive at the GGSN, the mobile device IP pack-

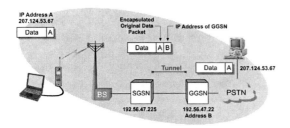

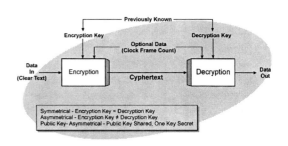

Packet Encapsulation

Encryption Operation

et is extracted from the data payload so it can be sent through the internet to its final destination (the web host).

Encipherment-The process of encrypting data with a cipher key.

Encryption-Encryption is a process of a protecting voice or data information from being obtained by unauthorized users. Encryption involves the use of a data processing algorithm (formula program) that uses one or more secret keys that both the sender and receiver of the information use to encrypt and decrypt the information. Without the encryption algorithm and key(s), unauthorized listeners cannot decode the message. When the encryption and decryption keys are the same, the encryption process is known as symmetrical encryption. When different encryption and decryption keys are used (such as in a public encryption system), the process is known as asymmetrical encryption.

This diagram shows how encryption can convert non-secure information (clear text) into a format (cyphertext) that is difficult or impossible for a recipient to understand without the proper decoding keys. In this example, data is provided to an encryption processing assembly that modifies the data signal using an encryption key. This diagram also shows that additional (optional) information such as a frame count or random number may be used along with the encryption key to provide better information encryption protection.

Encryption Algorithm-An encryption algorithm is a mathematical process that modifies data or information using keys or private information to prevent unauthorized companies or people from being able to make use of the information, media or data.

End Of Message (EOM)-End of message is a code or frame of data that indicates the end of a message data segment.

End User-A customer or device that uses communications services.

End User Licensing Agreement (EULA)-An agreement between an end-user of a product (usually a software product) and the owner of the product (e.g. software developer) that defines the terms of how the end user must abide by when they use a product. The end user is often prompted to enter into an EULA prior to installing and using a software application.

End-To-End-(1-general) End-to-end is the complete process or responsibility for a system or service. (2-communication) The overall process from the origination of the request or service to the connection or completion of the service or command. End-to-end commonly is used to refer to commands or communication over multiple communication systems and/or processes.

Energized-The condition when a circuit is switched on or powered up.

Enhanced 411 (E411)-Enhanced 411 (information) is a National Directory Assistance service that finds any listing in the U.S., Canada, or Puerto Rico and Entertainment Guides get you Movie Listings, Stock Quotes, Flight Times, Sports, Horoscopes, and More.

Enhanced 800 Services-A call processing service that allows for the routing of 800 (or other toll free or freephone numbers) to be routed to different locations based on other criteria such as day, time of day, or caller location.

This diagram shows an example how an Enhanced 800 number translation service operation allows a company to route calls to different service centers depending on the time of day and location of the caller. This diagram shows that a company has 3 offices; Boston, San Francisco, and Dallas and the company has a single toll free line 800-227-9681. Callers from east coast exchanges are routed to the Boston office from 9:00am - 5:00pm EST. Callers from west coast exchanges are routed to the San Francisco office from 9:00am - 5:00pm PST. All other calls are routed to the Dallas office. This allows the corporate office to handle calls after the regional offices are closed.

Enhanced 911 (E911)-E911 is an emergency telephone calling system that provides an emergency dispatcher with the address and number of the telephone when a user initiates a call for help. The E911 system has the capability of indicating the contact information for the local police, fire, and ambulance agencies that are within a customers calling area.

Enhanced Messaging Service (EMS)-Enhanced messaging service is an evolution of short messaging service (SMS) that adds the capabilities of text formatting, animation, pictures, and sound to be transferred in short messages. The 3GPP standard for EMS services is TS.23.040.

Enhanced Permissions-Enhanced permissions are authorizations to access, use or process data or objects that go beyond the typical assigned authorizations.

Enhanced Record-An enhanced record is an assemblage of data elements that have been in some way modified (enhanced) by another process (such as a call detail record being modified by a rating engine).

Enhanced Service Provider (ESP)-A provider of communication services that enhance the services provided by existing carriers. ESPs often provide information processing services such reformatting data or information management services. Because ESPs use existing communication services, ESPs do not typically file tariffs.

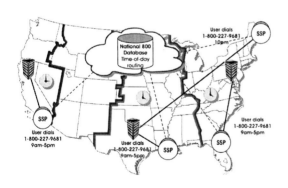

Enhanced 800 Service

Enhanced Telecom Operations Map (eTOM)-Enhanced telecom operations map (eTOM) is a business system structure that allows communication service providers to define, develop, and procure business and operational support systems. The eTOM framework was created by the telemanagement forum.

Enhancement Layer-An enhancement layer is a stream or source of media information that is used to improve (enhance) the resolution or appearance of underlying (e.g. base layers).

Enhancer-An enhancer is a device or processing function that modifies or enhances information (such as adding billing rates to a service usage record).

Enterprise Computing-Enterprise computing is the information processing that is performed for or within business. Examples of enterprise computing applications include order entry, accounting and production management.

Enterprise Network-The set of Local, Metropolitan, and/or Wide Area Networks and internetworking devices comprising the communications infrastructure for a geographically-distributed organization.

Enterprise Resource Planning (ERP)-Enterprise resource planning is a set of processes that are used to identify, forecast and schedule the acquisition or development of resources that are used to produce products or services in the future for a business or company.

Enterprise Service Bus (ESB)-An enterprise service bus is an abstraction layer that can use enterprise communication systems (such as messaging system) to integrate multiple applications or services with a reduction in the amount of modifications or software coding that is required to link the systems.

Enterprise Switch-A switch used within an enterprise backbone. Enterprise switches are generally high-performance devices operating at the Network layer that aggregate traffic streams from sites within an enterprise.

Enterprise Video-Enterprise video is the transmission of video signal to audiences within a company or organization.

Entertainment Application-Entertainment applications are software programs and/or services that provide a diversion or an amusement experience.

Entertainment Expense-An entertainment expense is a fee or assessment that occurs as a result of activities that are performed to help motivate or assist in the development of business or project related activities.

Entitlements-Entitlements are rights to use a product, service or media.

Entity-(1-general) An entity is something or someone that exists as a unit. (2-OSI) An active element within an OSI (open systems interconnection) layer or sublayer. (3-ISDN) In the Integrated Services Digital Network (ISDN), the logical representation of a function in a piece of equipment.

Entrance Facilities-The structure or conduit assembly that permits communication cabling to enter a building or facility.

Entrepreneur-An entrepreneur is a person who identifies business opportunities and takes actions (such as establishing and running a company) to develop business transactions for that business.

ENUM-tElephony NUmber Mapping

Enumeration-Enumeration is the process of discovering the communication parameter and configuration state of a data device that has been connected to a computer. When the computer (the host) discovers a new device, it determines what type of data transfers it anticipates transferring. Enumeration is a process used in the many computing systems including the Universal Serial Bus (USB) system.

Envelope-(1-data) An envelope is block of data that contains addressing information and can hold data. (2-signal) The limits of the variations of a complex wave form. (3-mathematics) The outer boundary limits of a function.

Enveloping-Enveloping is the process of sealing a message within a file or message in a way that recipients require a key or other information to open and view the enveloped information.

EOM-End Of Message

EOS-Embedded Operating System

E-Outsourcing-Electronic Outsourcing

E-Procurement-Electronic Procurement

EPS-Earnings Per Share

Equal Access-A telephone service that provides a communication user with an equal choice of long distance carriers.

Equal Access Code-An equal access code is a number that identifies a long distance communication service provider that is connected to a telephone line. Equal access code numbers can be entered before the dialing of a long distance telephone call to allow a customer to have equal access to all the long distance companies that could be connected to their telephone line. In the United States, to discover which long distance service provider is connected to a telephone line, dial 1-700-555-4141.

Equipment Cabinet-An enclosure assembly that allows the installation of electronic assemblies or components. Common equipment racks are 7 feet tall by 24 to 26 inches wide. The inside mounting area is often 19 or 22 inches wide (the rack). Some equipment cabinets come with power supplies and cooling fans.

Equipment Configuration-Equipment configuration is the process of sending information to a device that is used to adapt the equipment or software program to its environment (configuration).

Equipment Provisioning-Equipment provisioning is the identification of devices or functional capabilities within devices and the sending of information or configuring of these devices to enable them to perform services and/or run applications for customers.

Equipment Rack-A container that holds and interconnects electronic or electrical assemblies. A common size for an equipment rack is 19 inches (48.26 cm) wide. Electronic assemblies (called "modules") typically slide or are guided into the equipment rack where connectors located in the back of the equipment rack make contact with the modules.

This diagram shows an equipment rack that is used for mounting and interconnecting network communication equipment. This diagram shows that several modules (line cards) can be mounted in each equipment rack. This diagram shows that the back (call the "backplane") of the equipment rack contains connectors that can interconnect the modules to each other and to cables from other equipment racks or devices.

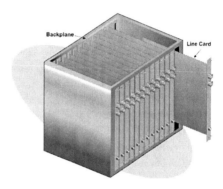

Equipment Rack Diagram

Equipment Requirements-Equipment requirements are a set of resources and their associated characteristics that are necessary to meet the needs of equipment to operate or provide types of services.

Equipment Revenue-Equipment revenue is income that is generated from selling of products.

Equipment Room (ER)-A room or space that is dedicated for communication equipment and cable connection points. Equipment rooms may be centralized points that allow one or many companies to access telephone equipment, data communication equipment, and communication line connection points.

Equity-Equity is the ownership interest in a company or business venture.

Equity Financing-Equity financing is the process of selling equity interest in a company as a way to obtain funds that will be used to run or develop the business.

Equity Loan-An equity loan is the temporary providing of cash or assets where the loan is secured by value in real property.

ER-Equipment Room

Erase-The process of discarding, obliterating, or marking information as deleted from a storage medium.

E-Rate-Electronic Rate

Ergonomics-Ergonomics is the science of understanding how humans interact with machines.

Erlang-A measurement unit of the average traffic usage of a telecommunications facility during a period of time (normally a busy hour) with reference to one hour of continuous use. The capacity of Erlangs is the ratio of time during which a facility is occupied to the time the facility is available for occupancy with reference to one hour. For example, a 12 minute call is 0.2 Erlangs. 1 Erlang equals 36 centum call seconds (CCS).

ERP-Enterprise Resource Planning

Error Blocks-Error blocks are groups of image bits (a block of pixels) in a digital video signal that do not represent error signals rather than the original image bits that were supposed to be in that image block.

Error Checking-Error checking is a process or processes that are used to determine if information that is transferred or stored contains errors. Error checking can be performed by sending or storing additional data (error check bits) that is related is the original data by a formula or algorithm where the processing of the relationship can determine if the same error check bits are recreated.

Error Code-An error code is a calculated value or information element that can be used to detect differences between a signal that is transmitted and the same signal at a receiver.

Error Detection-The process of detecting for bits that are received in error during data transmission. Error detection is made possible by sending additional bits that have a relationship to the original data that can be verified. See also: error correction.

This figure shows the basic error detection and correction process. This diagram shows that a sequence of digital bits is supplied to a computing device that produces a check bit sequence. The check bit sequence is sent in addition to the original digital bits. When the check bits are received, the same formula is

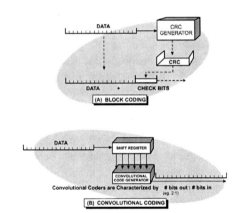

Error Detection Operation

used to check to see if any of the bits received were in error.

Error Free Seconds (EFS)-Error free seconds is the amount of time (in seconds) that a signal or media is received without any errors.

Error Handling-Error handling is the processes that are performed when a program encounters unexpected data or data that is received in error.

Error Listing-A data set or printout of errors and adjustments. The listing can include totals, comparisons with other data, and a coded description of an error condition. An error listing is sometimes called an exception report.

Error Rate-Error rate is a ratio between the amount of information received in error as compared to the total amount of information that is received over a period of time. Error rate may be expressed in the number of bits that are received in error of the number of blocks of data (packets) that are lost over a period of time.

Error Reporting Utility-An error reporting utility is a software program that is available to act on behalf of a system (an agent) that can detect and potentially add information (e.g. the operating conditions) to define when and how an error occurs. The error reporting utility software program sends the error reporting message to a recipient (such as the owner of a software application that caused the error).

Error Resilience-Error resilience is the ability of a processing system to detect, correct or adjust for errors.

Error Suspense-Error suspense is the process of holding transactions or data in a separate file (an error suspense file) to await how the errors will e resolved.

Errors-(1-general) A collective term that includes all types of edit rejects, inconsistencies, transmission deviations, and control failures. (2-difference) The difference between the measurement of a particular quantity and the true value of the quantity. (3-digital system) Data that has been lost or damaged through either system problems or media defects. The most common general types of errors are hard errors where the errors occur over a long time period and are repeatable and soft errors that are non-reoccurring and may be hard to identify and duplicate for verification.

Errors And Omissions (E&O)-A type of insurance that is provided to production companies and studios to offer security in case the copyright of a film might be jeopardized.

ERT-Emergency Response Team

ESB-Enterprise Service Bus

Escalation-The process of taking a trouble call or ticket up through increasing levels of management until the problem gets resolved.

Escalation Clause-An escalation clause is a term in a contract that defines a new requirement or process that occurs (such as the changing of charging rates) when a condition occurs (such as payment default).

Escrow-Escrow is an account or process of using a third party as an intermediary between a seller and a buyer to help ensure that the terms of a transaction or agreement are fulfilled. The escrow agent will receive assets (transfer money, property, or other valuable items) from a buyer and informs the seller they have received the asset. After the terms or conditions of the acquisition are completed (the item is received in the condition as promised), the escrow agent releases the funds to the seller.

Escrow Services-Escrow services is the providing of third party intermediary support between a seller and a buyer to help ensure that the terms of a transaction or agreement are fulfilled. Escrow services can receive assets (transfer money, property, or other valuable items) from a buyer and will release the funds to the seller when the terms or conditions of the acquisition are completed.

ESD-Emergency Session Detection

ESG-Electronic Service Guide

E-Sign Act-E-Sign Act is a regulatory act that defines how electronic signatures are authorized to be used.

ESN-Electronic Serial Number

ESN-Emergency Service Number

ESP-Enhanced Service Provider

ESS-Electronic Switching System

Essential Service Protection-Essential service protection is an access method that provides priority access (e.g. dialtone) to essential users (such as public safety and emergency personnel).

E

Estimated Load-The usage activity of a system during a measurement period, as determined by available data and traffic theory.

E-Tailers-Electronic Retailers

Ethernet-Ethernet is a packet based transmission protocol that is primarily used in LANs. Ethernet is the common name for the IEEE 802.3 industry specification and it is often characterized by its data transmission rate and type of transmission medium (e.g., twisted pair is T and fiber is F).

Ethernet systems in 1972 operated at 1 Mbps. In 1992, Ethernet progressed to 10 Mbps data transfer speed (called 10 Base T). In 2001, Ethernet data transfer rates included 100 Mbps (100 BaseT) and 1 Gbps (1000 Base T). In the year 2000, 10 Gigabit fiber Ethernet prototypes had been demonstrated.

Ethernet can be provided on twisted pair, coaxial cable, wireless, or fiber cable. In 2001, the common wired connections for Ethernet was 10 Mbps or 100 Mbps. 100 Mbps Ethernet (100 BaseT) systems are also called "Fast Ethernet." Ethernet systems that can transmit at 1 Gbps (1 Gbps = 1 thousand Mbps) or more, are called "Gigabit Ethernet (GE)." Wireless Ethernet have data transmission rates that are usually limited from 2 Mbps to 11 Mbps.

Originally created by an alliance between Digital Equipment Corporation, Intel and Xerox, Ethernet DIX, is slightly different than IEEE 802.3. In Ethernet the packet header includes a type field and the length of the packet is determined by detection. In IEEE 802.3, the packet header includes a length field and the packet type is encapsulated in an IEEE 802.2 header. Most modern day "Ethernet" devices are capable of using both protocol variation, however, older equipment was not able to do this.

This figure shows several types of Ethernet LAN systems and the approximate distance devices can be connected together in these networks. Thicknet Ethernet uses a low loss coax-

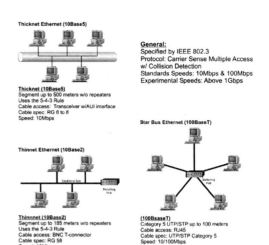

Ethernet System

ial cable to provide up to 500 meters of interconnection without the need for repeaters. Thinnet systems use a relatively thin coaxial cable system and the typical signal loss in this cable restricts the maximum distance to approximately 185 meters. 100 BaseT systems use category 5 UTP cable and the maximum distance is approximately 100 meters.

Ethernet Switch-A device which relies on enhancements to the original Ethernet specification that transfer data directly between ports without the need to re-broadcast the information to all ports of the device. Switched Ethernet incorporates modified layer 2 (Data Link Layer) electronics in order to allow an individual 10 Mbps (for Ethernet) or 100 Mbps (for fast Ethernet) user to transfer data directly to an end segment.

eTOM-Enhanced Telecom Operations Map

ETSI-European Telecommunications Standards Institute

EULA-End User Licensing Agreement

European Telecommunications Standards Institute (ETSI)-An organization that assists with the standards-making process in Europe. They work with other international standards bodies, including the

International Standards Organization (ISO), in coordinating like activities.

EVA-Economic Value Added

Event-An event is an occurrence that happens at a specific time or time period. An example of an event is a process or data transfer that has initiated, changed, or completed in a communication system.

Event Based Charging-Event based charging is the measurement of a service that is provided as a result of an event (such as purchasing the 24 hour right to view a pay per view movie).

Event Category-An event category is a classification or category of an occurrence that happens at a specific time or time period. An example of an event category is the connection of a communication link for a voice communication service.

Event Code-An event code is a category or label that defines the causes of an event.

Event Correlation-Event correlation is the determination of the relationship between definable or measurable occurrences that occur at specific times to other items, users or accounts.

Event Driven-Event driven is an action or software program operation that starts when a specific event or trigger has occurred.

Event History-Event history is the records and related information of the activities that have been performed or related to a customer or prospect (a customer record). Example of items in an event history include phone calls to customers, items mailed to customers and responses received from questionnaires sent to customers.

Event List-An event list is group of time related functions or occurrences (such as trade shows) that are related (such as to a specific industry, technology or project).

Event Logging-Event logging is the recording of event information. Event logging is commonly used to record usage information to be used by billing systems. An example of event logging is the recording of television media that is accessed or transferred.

Event Management-The recording, organizing, and displaying events that occur in communication systems. These events may include the type of event such as a program activated by a software application or information that is generated by event triggers (some preset level). Event management is used to determine the status and changes (such as increased bandwidth or CPU usage) that are occurring in communication systems to help find potential problem areas before they cause system failures.

Event Origination-Event origination is the location or device for which an event of information originated (e.g. circuit connection).

Event Processing-Event processing is the evaluation and potential modification that may occur to event information.

Event Record-Event record is a data record that holds information related to a specific event. Event records for communication systems contain information about the specific usage of a network element or service. Examples of event records include the path and connection time used by a specific switch in a communication network. A communication session (such as a voice call) typically involves multiple events (such as multiple switch connections). Call detail records (CDRs) are typically created by combining multiple event records.

Event Record Segregation-Event record segregation is the process of separating or redirecting the storage or processing of data that holds event information. An example of event record segregation is the filtering of billing data records that have no usage fees (free service).

Event Recorder-An event recorder is a device that can store information about activities, transfers or processes (events).

Event Time-Event time is when an event (such as the beginning of a service billing period) occurred.

Event Track-An event track is a portion of a storage medium (such as a section of a video-tape or digital media file) that provides information on the occurrence of events that are related specific media essence (such as voice, data or video).

Event Trigger-An event trigger is a condition that initiates a process (such as sending an event message).

Events-(1-general) External information that is received that indicates a change in status within a communication system. (2-Bluetooth) Incoming messages to the L2CA layer along with any time-outs. Events are categorized as Indications and Confirms from lower layers, Request and Responses from higher layers, data from peers, signal Requests and Responses from peers, and events caused by timer expirations.

Evolutionary Standard-An evolutionary standard is a set of operational descriptions, procedures or tests that are part of an industry specification document or series of documents that adds or is designed to allow the addition of improved capabilities or features to an existing standard or planned future standards.

E-Wallet-Electronic Wallet

Ex Gratia (Gratis)-Ex gratia is providing a product or service as a courtesy without a fee.

Ex Works (EXW)-Ex works is a shipping option where the seller makes the goods available at the their own location (sellers works).

Exabyte (EB)-An exabyte is one billion billion (quintillion) bytes of data. When exabyte is used to identify the amount of data storage space (such as computer memory or a hard disk), an exabyte commonly refers to 1,152,921,504,606,846,976 bytes.

Exact Match-An exact match is a comparison between a reference (such as a keyword) and another item (such as a search word) where the two items exactly match character for character. Exact matches for keyword searches usually do not require an exact matter of case (capital letters).

Exception Reports-Exception reports are tables, graphs or images that represent data or information that was not able to be processed using existing programs (e.g. could not find or match data records) or had results that were unexpected or which fall outside allowable ranges.

Exchange-(1-switch) The term used in some countries to refer to a network switch (See "Switch"). (2-company) A communication company or companies for the administration of communication service in a specified area, which usually embraces a city, town, or village and its environs, and consisting of one or more central offices, together with the associated plant, used in furnishing communication service in that area. The first exchange was installed in 1878.

Exchange Area-An exchange area (telephone number switch code) is a geographic area defining the territory in which a local operating company may offer its exchange telecommunications and exchange access services at a single, uniform set of charges. An exchange area may be served by one or more central offices. A call between any two points within an exchange area is a local call.

Exchange Carrier (EC)-A company that provides local telecommunications services. Exchange carriers are generally regulated by a national regulatory agency for interstate or inter-regional services and are regulated by state or local agencies for internal local access and transport area (InterLATA) services.

Exchange Carriers Standards Association (ECSA)-Formed in 1980 to issue standards for the telephone industry, it was later supplanted by Alliance for Telecommunications Industry Solutions (ATIS).

Exchange Message Record (EMR)-An exchange message record is a standard data format for the exchange of messages between telecommunications systems that is often used for billing records. The records may be exchanged by magnetic tape or by other medium such as electronic transfer or CD ROM.

Exchange Network-An exchange network is an organization or company that assists or manages the exchange of services or items through its system or facilities.

Exchange Point-An exchange point is a location, device or facility where several networks may interconnect with each other (a common switch or node).

Exchange Service-Exchange service is the providing of communication services (such as local telephone service) within an exchange (local system) area.

Exclusion Rules-Exclusion rules or conditions or requirements that inhibit, restrict or alter an offer, process, or operation. An example of an exclusion rule is the limitation that promotional offers cannot be combined.

Exclusive Distributor-An exclusive distributor is a company or individual that has the exclusive right to sell products they receive from other companies to retailers or customers. Exclusive rights may be limited to specific types of products and/or geographic areas.

Exclusivity-Exclusivity is the assignment (licensing) of rights to a single person or company.

Executable Short Message-A message that is received by a Subscriber Identity Module card in a wireless system (such as a mobile phone system) that contains a program that instructs the SIM card to perform processing instructions.

This diagram shows how an executable short message can be sent by a system operator to add a new feature into a mobile phone. The executable short message is a program that is stored in the SIM card and interacts with the operation of the mobile phone to allow the new feature to operate. The system simply sends the file as an executable message directly to the mobile phone identification. When the complete executable short message has been received in the SIM card, it is stored in memo-

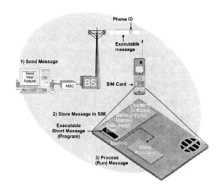

Executable Short Message Operation

ry and this program can complete (run) instructs that allows the new feature to operate.

Executive Busy Override-A feature that allows an executive or other user with a high priority to interrupt lower priority calls so the executive can initiate a call.

Executive Information System (EIS)-An executive information system is an information application can identify, analyze and processes information (such as sales statistics) to assist corporate leaders in the making of business strategies.

Executive Override-Executive override is a feature that permits certain users (such as executives) to connect to or listen to conversations on other extensions.

Exit Strategy-Exit strategy is the planned set of actions that a company or person plans to take to sell or end a business or project.

Expandability-The ability of a system to supply processing or services without significant changes to its fundamental assemblies. Measures of expandability in communication include the maximum number of customers that can receive service, number of radio transmitter towers that can be connected to a switching system, and maximum data transfer rates on a communication line.

Expandable Ads-Expandable ads are images or video clips that can be enlarged or extended in time and/or content. Expandable ads may first appear as images, links or banner ads.

Expanding-Expanding is a process that increases the amount of amplification (gain) of an audio signal for smaller input signals (e.g., softer talker). The use of expanding allows the level of audio signal that leaves the modulator to have a larger overall range (lower minimum and higher maximum) regardless if some people talk softly or boldly. As a result of expanding, high-level signals and low-level signals output from a modulator may have a different conversion level (ratio of modulation compared to input signal level). This could create distortion so expanding allows the modulator to convert the information signal (audio signal) with less distortion. Of course, the process of expanding must be initiated at the transmitting end, called companding, to recreate the original audio signal.

Expansion Loop-An expansion loop is a short length of cable that is part of a communication line (such as a pole mounted cable TV or telephone lines) that provides additional cable length that may be necessary to compensate for changes in the length of the cable due to temperature changes or cable loading (wind or ice).

This figure shows how an aerial expansion loop provides protection to a cable for the expansion and contraction of the messenger wire. This diagram shows that the expansion loop is placed at the telephone phone. As the messenger cable moves, the cable will flex taking from or returning cable to the expansion loop.

Expense-An expense is a fee or assessment that occurs as a result of the creation of revenue or performance of an action.

Expense Allowances-Expense allowances are values that are pre-assigned for expense categories that can be charged without budgeting or authorization.

Expense Approval-Expense approval is the process of authorizing or enabling expense costs to be recorded and paid. The expense approval process may be a combination of expense filing requirements (expense reports) along with manager approvals.

Expense Policies-Expense policies are a set of rules or guidelines that define acceptable types of expenses and how the expenses should be recorded and processed.

Expense Report-An expense report is a form or document that identifies, organizes and totals expenses for an employee, contractor or company.

Experimental Standards-Experimental standards are documents that describe proposed methods of performing actions or processes to allow people, groups or companies to design and build hardware, firmware, software, or combinations there of and have them inter work with similar products designed and built by others.

Expert Witness-An expert witness is a person who is qualified to give an opinion or other information that can be used as part of legal proceedings.

Expiration Date-An expiration date is the day that an authorization, validity or offer period or process will end.

Exploration Phase-Exploration phase is a portion of a development process that is used to determine the scope and needs of a project.

Export Restrictions-Export restrictions are a set of rules or lists items or must receive spe-

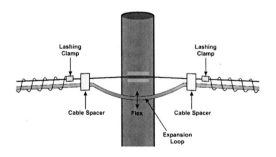

Aerial Cable Expansion Loop

cial authorization or testing to be allowed to be shipped out of a country. Export restrictions may include items such as technology, chemicals or weapons.

Exporting-Exporting is the process of identifying and processing data or media into a format that can be transferred to another file or system.

Exposure-(1-business) Exposure is the amount of risk associated with a transaction, project or business agreement. (2-marketing) Exposure is the amount of media publicity or promotion.

Express Consent-Express consent is an indication that a person or company has specifically and deliberately given permission to perform specific actions (such as sending an email message) that use their contact information.

Extended Allocation-Extended allocation is the assignment of a resource that exceeds the basic assignment allocation. Extended allocation is a process of determining an increased need or the availability of a resource and the assignment (allocation) of the resources that is in excess of a typical resource assignment.

Extended Area Service (EAS)-Extended area service is a billing arrangement that extends a rate plan into other service areas. An example of an extended area service is the providing of an unlimited number of calls to areas outside of a local service area.

Extended Binary Coded Decimal Interexchange Code (EBCDIC)-Extended binary coded decimal interexchange code is a digital encoding scheme used for digital transmission and storage of letters, numerals, signs, and punctuation marks. EBCDIC accommodates twice the symbols and functions of American Standard Code for Information Interchange (ASCII).

Extended Digital Subscriber Line (EDSL)-Extended digital subscriber line combines 23 B-channels and one 64-Kbps D-channel on a single line.

Extended Header-An extended header is a part of a data packet that typically contains address, routing, and origination information, which has been extended with additional information or control fields.

Extensibility-The ability to upgrade existing systems or services without significant changes to the existing systems.

eXtensible Business Reporting Language (XBRL)-Extensible business reporting language is a software standard that is used to define how business and financial data can be reported. More information about xBRL can be found at www.xbrl.org.

Extensible Hypertext Markup Language (xHTML)-Extensible Hypertext Markup Language (XHTML) Basic is a text based software communication standard that is used to allow the web software developer to define new (extensible) elements of a Internet web pages. xHTML was created by the World Wide Web Consortium (WC3) in 1996 to provide a common markup language for wireless devices and other small devices with limited memory. It is a widely supported open technology (i.e. non-proprietary technology) that is used for data exchange between any type of application that can understand XML. The combination of XML with Hypertext Markup Language (HTML) produces a web presentation language that is flexible (extensible) based on the needs of the application it is being used for.

This figure shows an example of how XHTML Basic operation can be used to process information requests from mobile devices. This example shows that a user has sent movie time request from a Wireless device to their preferred entertainment information provider. This request is routed through the cellular tower and to the cellular system that forwards the request to the Internet using the selected Internet address of the information provider. The Internet routes the request to a WAP server. The WAP server determines that this request is for a document or information that

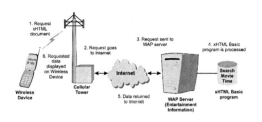

xHTML Basic Operation

is written in XHTML Basic and stored on the WAP server. The XHTML Basic program is accessed, and the requested data is sent via the Internet through the cellular tower to the wireless device for display. The requested data could be any web site that, for example, has been converted to XHTML Basic from HTML so that it can be displayed correctly on any wireless device.

Extensible Markup Language (XML)- Extensible markup language is a software standard that is used to define exchangeable elements of a web (HTML) page. Extensible Markup Language was developed in 1996 by the World Wide Web Consortium (W3C). It is a widely supported open technology (i.e. non-proprietary technology) for data exchange. XML documents contain only data, and applications display that data is various ways. XML permits document authors to create their own markup for virtually any type of information. Therefore, authors can use XML to create entirely new markup languages to describe specific types of data, including mathematical formulas, chemical formulas, music and recipes.

Extensible Query Language (XQL)- Extensible query language is a set of commands and processes that can organize and process data and item information contained in an XML document.

Extensible Rights Markup Language (XrML)-Extensible rights management language is a XML that is used to define rights elements of digital media and services. Extensible Rights Markup Language was initially developed by Content Guard and its use has been endorsed by several companies including Microsoft. The XrML language provides a universal language and process for defining and controlling the rights associated with many types of content and services.

Because XrML is based on extensible markup language (XML), XrML files can be customized for specific applications such as to describe books (ONIX) or web based media (RDF). For more information on XrML see www.XrML.org.

This figure shows XrML system. This example shows how a media file can contain XrML meta tags can be used to identify, describe and control access to media. This diagram shows that the XrML data can be appended as part of the media file that is sent to the user.

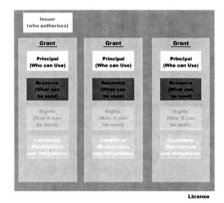

Extensible Rights Markup Language (XRML)

Extensible Solution-An extensible solution a system, product or service that is capable of being changed or updated to allow for added services and benefits.

Extension-(1-software) A set of commands or protocols that are used to extend the capabilities of another application or protocol. Software extensions are commonly used to rapidly extend the capabilities of an existing application or protocol without changing the underlying application or protocol. (2-telephone) An additional telephone connected to a line. Allows two or more locations to be served by the same telephone line or line group. May also refer to an intercom phone number in an office. (3-file name) The optional second part of a PC computer filename. Extensions begin with a period and contain from one to three characters. Most application programs supply extensions for files they create.

Extension Dialing-Extension dialing is a process that allows callers to enter the extension number of the person they are calling. The extension number may be entered by DTMF tones or by audio (voice) commands.

External Alliances-External alliances are relationships that are formed with other companies or entities. An example of an external alliance is the providing of frequent flyer miles from airline companies purchases of products or services.

External Data Representation (XDR)-A standard information format that is used to define data that is transferred between networks or network parts. XDR was developed by Sun.

External Interface-A device that adapts internal networks to external networks. The use of an external interface allows devices and applications that are not directly connected to a network to request and receive some or all services from the network.

Extranet-An extranet is a network (typically an Internet network) that is only accessible to customers, affiliates, or members rather than the general public. Extranets may use the public Internet as its transmission system. However, extranet data connections through the Internet are usually encrypted and accounts are password protected to prevent access by unauthorized users.

EXW-Ex Works

Eye Candy-Eye candy is images or objects that are pleasing or attention getting to viewers.

E

F

FA-Functional Acknowledgement

FAB-Fulfillment, Assurance, and Billing

FAC-Feature Access Code

Facilitators-Facilitators are individuals or a group of individuals who evaluate or provide information within the company that is relative the purchase of your products or services.

Facilities-Facilities are the parts (elements) of a network or system that are owned or leased by a company or person. The term "facilities" is sometimes generalized to describe buildings, utilities, or communication equipment.

Facilities Contract-A facilities contract is an agreement that defines the acquisition, modification or management of buildings or structures.

Facilities Control Plan-A facilities control plan is a document that describes the method for identifying and managing facilities (such as cable pairs in a telephone system).

Facility-(1-telephone) Any one of the elements of the physical telephone plant needed to provide service, such as switching systems, cables, and microwave radio transmission systems. (2-line) Transmission plant between offices. (3-outside plant) The outside plant from a central office mainframe to a point of termination on a customer's premises, but not including customer equipment. (4-packet switching) A specific X.25 service, such as Closed User Group and/or a protocol element that conveys information about that service.

Facsimile (Fax)-Facsimile is the representation of optical images by electrical signals for transmission over communication systems. When the electrical signal is received, it is converted back to an optical format for display or printing of the original image. This process is commonly called FAXing.

Fax usually involves the transmission of still images (typically black and white with no intermediate shades of gray). The most widely used type of FAX service today is described by ITU standard T.30, Group 3. Group 3 facsimile uses a digital modem over a voice grade public telephone channel to transmit the FAX signal. Group 4 FAX also permits user-selectable pixel size and typically operates via a 64 kb/s ISDN channel. Group 1 was an early analog fax system, typified by the Xerox Telecopier. Group 2 was an early digital FAX system. Groups 1 and 2 are obsolescent, although some Group 3 FAX machines are backward compatible with Group 2 machines.

Facsimile Service-Facsimile service is the transmission of still images from one place to another by means of a communication channel.

Fact Finding-Fact finding is the gathering of information that is necessary for the prospect to purchase your product or service.

Factoring-Factoring is the selling of receivable assets (sales that have not been collected) in exchange for a discount on the amount that is to be received (you get less money now than what is owed to you).

FACTS-Fully Automated Collect And Third Number Service

Fail Safe Operation-A type of system control that allows for the automatic operation of additional equipment or reconfiguration of existing equipment to prevent the improper functioning of communications or systems in the event of circuit loss or impairment.

Failure-A detected cessation of capability to perform a specified function or functions within previously established limits. A failure is beyond adjustment by the operator through means of controls normally accessible during routine operation of the system. (This requires that measurable limits be established to define "satisfactory performance.")

Failure Rate-Failure rate is a comparison of the number of failures to the number of times each item has been subjected to a set of specified conditions.

Fair Dealing-Fair dealing is a principle of copyright law that permits the use of copyrighted material under certain circumstances. An example of fair use is the quoting of a book review in a magazine. The term fair dealing is known as "fair-use" in the United States.

Fair Market Value (FMV)-Fair market value is an estimation of the worth of an asset that is based on market conditions.

Fair Reasonable and Non-Discriminatory (FRAND)-Fair, reasonable and non-discriminatory are licensing terms (fees and restrictions) that are offered by owners of intellectual property that allows other companies to build products that incorporate their technology while providing the potential for sustainable profits.

Fair Use-Fair use is a principle of copyright law that permits the use of copyrighted material under certain circumstances. An example of fair use is the quoting of a book review in a magazine.

Fairness Scheduling-Fairness scheduling algorithm that coordinates the sequences of processes or information so that the data transmission rate or application processing time is fairly distributed to users of the system or services. Fairness scheduling can be used to overcome the differences in capabilities of the access or systems for each user.

False Advertising-False advertising is the use of statements or media that are not accurate or are misleading with the intention of gaining a commercial advantage.

False Rejection Rate (FRR)-False rejection rate is a measure of how many valid users or potential customers requested services to the total number of people that requested services.

FAQ-Frequently Asked Questions

FAR-Federal Acquisition Regulation

Far End-Far end is a reference point that is at the end of a transmission line or circuit (near the exit or final termination point).

Fast Busy-An alert tone that indicates communication resources (such as switch trunks) are not available. A fast busy signal operates at twice the normal busy signal rate (120 tones/minute).

Fast Reconnect-Fast reconnect is a process of re-establishing a communication session within a relatively short amount of time. Fast reconnection typically involves the ability to skip over some of the initial steps that occur in the setup of a communication connection because some of the parameters of the session are already known (such as an IP address) and can be used in the re-establishment of the connection.

Fast Recovery-Fast recovery is a process of restarting or continuing a service within a relatively short amount of time. Fast recovery typically involves the ability to skip over some of the initial steps in the setup of a service or software application as some of the parameters of the service are already known (such as the type of media and file pointers) and can be used in to restart the service or application.

Fatal Exception-A fatal exception is a process that may occur in the operation of a software program where the program can no longer operate.

Fault-A condition that causes a device, a component, or an element to fall to perform in a required manner. Examples include a short circuit, a broken wine, or an intermittent connection.

Fault Management-Fault management identifies the network problems, failures, events and corrects them. Fault management is the reactive form of network management. SNMP traps, syslog, and RMON typically are used in fault management. Fault management is one of the five functions defined in the FCAPS model for network management.

Fault to Resolution Process-Fault to resolution process is the capturing of the flow of events as well as the interaction points between the point where a fault is either proactively logged onto a ticketing application and the point where the fault is resolved and the contracted service restored.

Fault Tolerance-Fault tolerance is the ability of a network or sub-system to continue to operate in the event of a hardware or software failure. Fault tolerant systems are typically able to identify the fault and replace the failed component or sub-system with another piece of equipment.

Fault-management, Configuration, Accounting, Performance, and Security (FCAPS)-Fault-management, Configuration, Accounting, Performance, and Security is a categorical model of the working objectives of network management and is a standard adopted by the International Telecommunications Union (ITU). There are five levels, called the fault-management level (F), the configuration level (C), the accounting level (A), the performance level (P), and the security level (S).

Fax-Facsimile

Fax Back-A service that allows callers to request information that they will be delivered to them by fax. Fax back service usually involves the use of an interactive voice response (IVR) system to provide information to the caller of the fax back service. The user selects appropriate documents they desire to have sent, usually making selections with the keypad. The system then requests the caller to enter the destination fax number so the documents can be sent to the caller.

Fax Broadcasting-Fax broadcasting is a process of sending messages to a group of recipients via fax transmission.

Fax Mail (FxM)-A communications service that delivers documents in fax format by converting a document (such as a word processor document) to a fax image format delivering it to a fax machine.

Fax Mailbox-A fax mailbox is a portion of memory, usually located on a computer hard disk, that receives and sends fax images.

Fax Modem-An adapter that is capable of both modem and facsimile transmission. A fax modem may be an internal card (such as a PCI card) or an eternal assembly. The fax modem can receive and convert digital fax images from their analog transmission form. The fax modem is capable communicating with both modem and fax protocols. There are several versions of modem and fax protocols and a fax modem may be capable of communicating with some (e.g. slow speed) or all of them.

Fax On Demand (FOD)-A telecommunications transmission service that sends previously stored faxes to a user that has requested a specific fax message. For FOD service, the caller listens to the options available and requests that additional information is delivered by fax. FOD service is often used to deliver previously stored instructions or marketing materials.

This diagram shows how a fax on demand (FOD) system can store and automatically deliver faxes. The manager of this system has previously stored operating manuals in fax mailboxes at the company. Each fax document that is stored is assigned to a fax mailbox. The verbal (audio) name that is associated with this fax mailbox is programmed into the interactive voice response (IVR) system. This

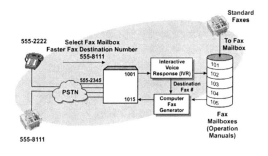

Fax on Demand (FOD) Operation

allows the user to hear their options and to make a fax mailbox selection. This example shows how a caller dials a telephone number that this company uses for automatic fax back service (555-2345 in this example). When the phone system receives a call on this line, it automatically routes the call to the IVR system at extension 1001. The IVR system prompts the caller to select the fax mailbox and to enter the destination fax number (555-8111). The fax mailbox (usually a storage area on the computer hard disk) retrieves the data and sends it to a computer fax generator. The destination digits (destination fax number) are also sent to the fax generator. This allows the computer to create the fax from the fax mailbox data and send the fax to the destination fax machine.

FC-Feedback Control

FCAPS-Fault-management, Configuration, Accounting, Performance, and Security

FCC-Federal Communications Commission

FCC Surcharge-Federal communication commission surcharges are additional fees (surcharges) that are imposed on communication services to assist in the costs of management and development of public communication services.

FCC Type Approval-A approval from the FCC that identifies the radio equipment manufactured has passed tests certifying it meets the minimum FCC requirements for that type of radio equipment. Most radio devices must meet several FCC specification requirements to receive FCC type approval. Companies typically use an independent testing lab to certify that equipment meets FCC requirements.

FDDI-Fiber Distributed Data Interface

Feasibility Analysis-A feasibility analysis is a review of information to determine if the potential success of a product, project or business. Some of the information used in a feasibility analysis includes definitions and scope of anticipated products or services and target markets and their size that have a need or want for the products or services. This may also include the resources such as skills, equipment and capital that will be required to develop and provide these products or services.

Feasibility Study-A feasibility study is a document that defines the requirements for the development of a product or service, the potential revenue or benefits of developing the product or service, and if the resources required to develop the product match the company capabilities and objectives.

Feature-(1-general) Features are a specific operations or characteristics of a piece of equipment or a service provided to a user. (2-end office) Features are specific call processing features that are provided by the switching system that is located at the end office (also called the end office).

Feature Access Code (FAC)-The codes assigned (usually numeric codes) that users may use to access the features of communication system.

Feature Benefits-Feature benefits are the relationship between the features a product or service has to the benefits that a user of the product or service can receive.

Feature Extensibility-Feature extensibility is the ability to upgrade or extend the capabilities of existing features without significant changes to the existing systems.

Feature Group-A switched access service offered by a local exchange carrier to an interexchange carrier, in North America.

Feature Group A (FGA)-A line-side (end user side) access connection to a local exchange carrier end office with an associated local telephone number. The objective is line access to and from any interexchange carrier. This type of access is not used today but was the only type of access first available to competitive interexchange carriers. It requires the originator to dial an access number before dialing the destination number (and often an identification number as well), and does not have positive supervision signals for call timing and billing. The FGA line is the first point of switching in a telephone network.

Feature Group B (FGB)-A trunk-side connection for interexchange carriers access when the originator dials 950-XXXX, where XXXX represents a Carrier-specific Access Code. It also offers terminating access with multifrequency signaling. Although still used to some extent, this has largely been supplanted by dialing the 1010xx...x access prefix.

Feature Group D (FGD)-A trunk-side access to exchange carrier end-office switching systems and tandems. It provides the equal-access service that the former Bell operating companies must offer, as required by the Modification of Final Judgment. For an interexchange carrier, Feature Group D offers positive call connect and disconnect supervision for accurate call time billing, a uniform access code option (by dialing 1010XX...X), optional calling party identification, recording of access charge billing details, and pre-subscription to a customer-specified interexchange carrier.

Feature Matrix-A feature matrix is a listing of features and the key areas the features appear or apply. A feature matrix may be used to compare a list of features to how they are included in competing companies' products or services.

Feature Provisioning-Feature provisioning is the adding, modifying or deletion capabilities for services and applications.

Feature Race-A feature race is a competitive environment where manufacturers or vendors continually add new features to gain perceived advantages.

Feature Service Access Code-A code, in the form -X or -XXX, used by customers for controlling access to custom calling services.

Features-Features are characteristics of a product or service that perform specific functions or processes for the customer or user.

Federal Acquisition Regulation (FAR)-Federal acquisition regulation is a set of rules that define the requirements that apply when supplying products or services to the United States government and its agencies.

Federal Communications Commission (FCC)-The federal communications commission (FCC) is a government agency of the United States that establishes and enforces laws and regulations regarding interstate radio and wired communications services. The agency was established by the Communications Act of 1934. The FCC must certify (FCC type approval) radio and computer equipment before it can be sold in the United States.

Federal Universal Service Fee-A federal universal service fee is a tax that is assessed on telecommunications services.

Federated Architecture-A federated architecture distributes the data and the application logic across many systems, using an array of manageable databases rather than one massive store

Fee-A fee is a cost or assessment of value for goods or services.

Feedback-(1-general) The return of a portion of the output of a device to the input. (2-acoustic) The howling of a public address system caused by feedback into a microphone of some of the output from a loudspeaker in the same audio system. (3-marketing) The gathering of information from customers or people related to experiences they have had to products, services or other contacts they had with companies or people.

Feedback Channel-A feedback channel is a return channel that is part of a system or service that can return information from users or devices that are related to experiences or services they are or have received.

Feedback Control (FC)-Feedback controls are actions or processes taken by a device, system or network regulate or adjust it operation.

Feedback Loop-A feedback loop is a path that connects a signal or information back to a controlling component, device or system.

Feng Shui-Feng Shui is the practice of arranging objects in an environment to achieve better success, wealth and happiness.

FF-Form Feed

FG-Functional Group

FGA-Feature Group A

FGB-Feature Group B

FGD-Feature Group D

FGI-Finished Goods Inventory

Fiber-An optical fiber is thin filament of glass or plastic (usually smaller than a human hair) that is used to transmit voice, data, or video signals in the form of light energy (typically in pulses).

Fiber Distributed Data Interface (FDDI)-Fiber distributed data interface (FDDI) is a computer network protocol that utilizes fiber optic or copper cable as the transmission medium to provide a token-passing, logical ring topology network operating at 100 Mbps. FDDI also provides for a mode of operation whereby two counter rotating rings are used to provide immediate fail-over and ring recover should a fiber cut occur. FDDI was commonly used as a backbone network to interconnect lower speed Ethernet and Token Ring networks within an enterprise. The American National Standards Institute standard X3T12 defines the protocol.

This figure shows FDDI system that uses dual rings that transmit data in opposite directions. This diagram shows one dual attached station (DAS) and a dual attached concentrator (DAC). The DAS receives and forwards the token to the mainframe computer. The DAC receives and token and coordinates its distribution to multiple data devices that are connected to it.

Fiber Distribution Panel-A fiber distribution panel is an array of switches or connectors (optical patch panels) that can be used to reroute or reconfigure an optical transmission system. Optical patch panels are commonly installed to allow reconfiguration of connections such as in an office building where people are moved or reassigned without the need to install new wiring or network distribution equipment.

Fiber Optic Transmission-Fiber optic transmission is the transmission of information using lightwaves as a carrier signal that travel through an optical waveguide (optical fiber).

This figure shows a fiberoptic communication system that is composed of two end-nodes and a fiber optic cable transmission medium. This diagram shows two optical network units (ONUs) that connect data networks together using fiber cable. This diagram shows that two fiber strands are needed: one for transmitting and one for receiving.

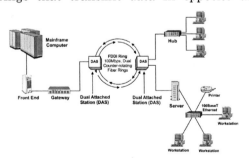

Fiber Distributed Data Interface (FDDI)

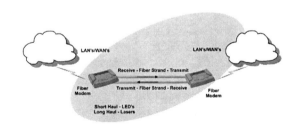

Fiber Optic Transmission System

Field-(1-data) A specified number of bits in a data record that is designated for a particular type information. (2-energy) Electric and/or magnetic lines of force in a specific area or region. (3-video) Half of an NTSC picture frame which has half of the lines required to produce a picture image. This is to interlace video signals where adjacent lines are contained in alternate fields for each picture.

Field Label-The label (name) of a field within a database record.

Field Sales-Field sales are orders that are created as a result of marketing efforts in places outside company facilities (such as at the customer's location).

Field Testing-Field testing is a process of testing a device, assembly, or system at a location that typically involves its normal operation.

Field Trials-Field trials are sets of tests that are conducted by users of a product or service in their native (field) environment to determine if the services or systems are operating or performing to specification levels that have been set by the buyer of the system or services.

Field Upgrade-A field upgrade is a physical and/or software change that may occur when a product change takes place at a location other than the manufacturer's facility.

Field Upgradeable-Field upgradeable is a product or system that can be updated without the need to return the product to the manufacturer or to a repair facility.

Field Validation-Field validation is the process of reviewing the data that is stored in a data field (such as the date in a billing record) to determine if conforms to rules, formats or correctly relates to other information.

FIFO-First-In-First-Out

File-A file is a collection of information or data (such as digital bits). A file can contain software instructions (a program), information or media (such as text or digital video) or a combination of software instructions with information or media (a file container).

File Container-A file container is a collection of data or media segments in one data file. A file container may hold the raw data files (e.g. digital audio and digital video) along with descriptive information (meta tags).

File Directory-File directories are portions of a computer storage area that are dedicated to the listing of file names and location pointers (starting addresses and file size) for the files contained on the computer storage system.

File Downloading-File downloading is the transfer of a program or of data from computer server to another computer. File download commonly refers to retrieving files from a media server or a web site server to another computer.

File Format-File formats are the sequencing and grouping of information elements (e.g. digital bits) within a block of data (file) or as organized on a sequence (stream) of information. File formats can range from simple linear (time progressive) sequences to indexed multiple file formatted blocks of data (containers).

File Integrity-File integrity is the amount of data or information that remains unchanged as compared to its original form.

File Permissions-File permissions are a set of parameters that are associated with a file (or a directory) that determine who or what devices can access the file and how they may interact with the file. File permissions may include the ability to read, write or run (execute) the file.

File Server-A file server is a computer that stores data centrally for network users and manages access to that data. File servers may be dedicated so that no processes other than network management can be executed while the network is available or non-dedicated so that standard user applications can run while the network is available.

F

File Sharing-File sharing is the providing of access rights and the ability to identify, access and potentially modify files between computing equipment which are usually connected to a network.

File System-A file system is a naming and organizational structure that is used to group and store files.

File Transfer Access, And Management (FTAM)-The protocol of open systems interconnection (0SI) that enable users to transfer files between computers, regardless of manufacturer. The FTAM standards were developed by the International Standards Organization (ISO).

File Transfer Protocol (FTP)-A protocol that is used to manage the transfer of data files between computers and networks. Because FTP is a standard protocol, it permits the transfer of any type of data file between different types of computers or networks.

File Type-File types are the formats of information contained within a block of data (a file). File types may be determined by the file extension (end of the file name), through identifying information within the file (typically in the beginning header area of the file) or through a combination of both (such as a jpg image file that uses specific types of data encoding and compression).

File Upload-File upload is the transfer of a program or of data from one computer to a computer server. File upload commonly refers to sending files from a computer to a web site server.

Filing Date-The date upon which a document must be filed after all computations of time authorized by this section have been made. In the United States, a filing date may be obtained by using a so-called express mail practice. The filing date of the first application for patent is considered as the "priority date" for subsequent applications.

Filing Period-Filing period is the time that is allowed or prescribed by statute, rule, order, notice or other regulatory authority (e.g. FCC) to meet the document filing requirements.

Fill Rate-Fill rate is the percentage of how many orders are filled as compared to the total number of orders received over a time period.

Filter-(1-general) A network that passes desired frequencies but greatly attenuates other frequencies. (2-active) An RC filter that uses solid-state amplifiers to produce a desired frequency shaping characteristic. (3-band elimination) A band-stop or band-rejection filter that passes, with negligible loss, all signals except those in a specified band. (4-band-stop) A band elimination filter. (5-bandpass) A filter that greatly attenuates signals of all frequencies above and below those in a specified band. (6-capacitor-input) A common type of power-supply smoothing filter. The output from a rectifier is shunted by a large capacitor as the first element in the smoothing circuit. (7-cavity) A filter with precise characteristics for separating microwave frequencies using cavity resonance. (8-choke input) A low-pass power-supply smoothing filter with an inductance as its first element. (9-comb) A filter with several sharp band-stop sections for different frequencies. (10-composite) An m-derived filter made up of several filter sections, calculated to give the required impedance and sharp frequency changeover characteristics. (11-constant-k) A filter in which the product of the impedance of shunt components and of series components is a constant, independent of frequency. (12-crystal) A filter with sharp cutoff or changeover characteristics, obtained through the use of quartz crystal components in resonant circuits. (13-high-pass) A filter that attenuates signals below a specified frequency but passes with minimal attenuation all signals above that frequency. (14-LC) A filter with inductance (L) and capacitance (C) circuit elements. (15-longitudinal suppression) A filter designed to suppress unwanted noise signals flowing in the same direction on the two wires of a pair. (16-low-pass) A filter that greatly attenuates signals higher than a specified frequency, but passes with minimal attenuation all signals lower in frequency. (17-notch) A bandpass filter in which the upper cutoff frequency is twice the lower cutoff frequency. (18-power interference) A filter in series with the utility power input to a rectifier that passes the fundamental frequency of the power supply but greatly attenuates higher interfering frequencies. (19-software) A software routine that separates computer data according to specified criteria.

This figure shows typical signal processing for an audio filter. In this example, the audio signal is processed through a filter to remove very high and very low frequency parts (audio band-pass filter). These unwanted frequency parts are possibly noise and other out of audio frequency signals that could distort the desired signal. The high frequencies can be seen as rapid changes in the audio signal. After an audio signal is processed by the audio band-pass filter, the sharp edges of the audio signal (high frequency components) are removed.

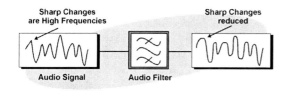

Audio Signal Filtering Operation

Filtering Agent-A filtering agent is a program or process that is used to restrict the reception or completion of delivery of items that have specific characteristics.

Filtering Database-A data structure within a bridge that provides the mapping from Destination Address to bridge port (in a D-compliant bridge), or from the combination of Destination Address and VLAN to bridge port (in a Q-compliant bridge).

Filtering Rules-Filtering rules is a set of requirements or conditions for restricting or allowing data or signals to pass through a point or function.

Final Billing-Final billing is gathering of all charges, payments, adjustments, and other account postings and combining them to produce a final invoice or account statement.

Final Signing-Final signing is the day that the contract to provide products and services defined by the RFP becomes binding on the RFP responder.

Final Value Fee (FVF)-A final value fee is a value that is paid to a selling company (such as an online auction company) for the completion or fulfillment of the auction process. A FVF may be a fixed fee or a percentage of the selling price.

Finance Charges-Finance charges are the fees or interest charged on an account or invoice balance for the option to delay payment for products or services beyond their payment due date.

Finance Department-A finance department is a group within a company or organization that is responsible for the financial operations. Financial operations may include accounting systems, billing systems, financial reports, tax filing, and financial planning.

Finance System-A finance system is a combination of software and hardware that receives and processes accounting information, grouping this information for specific assets, liabilities, revenue, expenses or other types of accounts.

Financial Information-Financial information is data that represents the financial performance of a company or person. Financial information may include income statements, balance sheets, cash flow statements and estimates on how these financial indicators may change over the next few years (financial projections).

Financial Information Exchange (FIX)- Financial information exchange is an industry standard protocol that is used to perform real time security transactions.

Financial Management Information System (FMIS)- A financial management system gathers, organizes, and processes financial information for a department or company. FMIS systems are developed and used by companies to manage its financial and accounting needs.

Financial Posture - The strategic positioning established by upper management that defines the way that it wants to be perceived by investors, regulators and the public in financial terms. The indication of a strong financial posture would communicate to outsiders that the company was highly successful and cash rich and a weak posture would communicate the opposite. Financial posture objectives can be defined and controlled by the company through major marketing and public relations activities and they may be different than the actual financial performance of the company.

Financial Products Markup Language (FPML)- Financial products markup language is an XML vocabulary for representing financial transaction information. FPML is a standard information format that can be used to transfer over the counter transactions which can be sent via communication systems (such as through the Internet). More information about FPML can be found at www.FPML.org.

Financial Projections- Financial projections are estimates of the future performance of a project or business that is based on existing data and analysis.

Financial Qualifications- Financial qualifications are materials or information that can validate the ability of a person or company to perform projects or services.

Financial Strength- Financial strength is a measure of the ability of a company to withstand and potentially recover from changes in its ability to produce revenue.

Financing Options- Financing options are a set of delayed payment options along with the terms (such as payment intervals and interest rates) associated with each payment option.

Financing Requirements- Financing requirements are the terms that are required (or desired) that a vendor will provide if the buyer selects their products or services.

Find Me Service- A process of forwarding calls to a sequence of numbers so that the call will find the recipient. Follow me service could be password privileged to screen unwanted callers from being forward to private numbers such as home telephones or mobile phones.

Fingerprint- A fingerprint is a unique identifier that is associated with a specific device, user, media or a combination of these items. Fingerprints can be used to identify or assist in the identification of devices or users.

Finished Goods Inventory (FGI)- Finished goods inventory is the amount of products or goods that have been produced and are available for shipping or to fill orders.

Firefighting- A process of rapidly addressing many problems that are occurring in a system. Solving a specific problem is referred to as putting out a fire.

Firefighting Plan- A plan that addresses how to rapidly resolve critical problems within a network when normal procedures cannot be followed. A firefighting plan usually prioritizes which systems receive priority for corrective action (such as voice communication) and who will be responsible for the non-standard procedures to correct the problems.

Firewall- A firewall is a data filtering device that is installed between a computer server or data communication device and a public network (e.g. the Internet). A firewall continuously looks for data patterns that indicate unauthorized use or unwanted communications to the server. Firewalls vary in the amount of buffering and filtering they are capable of providing. An ideal (perfect) firewall is called a "brick wall firewall."

This figure shows how a firewall works. This diagram shows that a user with address 201 is communicating through a firewall with address 301 to an external computer that is connected to the Internet with address 401. When user 201 sends a packet to the Internet requesting a communications session with computer 401, the packet first passes through the firewall and the firewall notes that computer 201 has requested a communication session, what the port number is, and sequence number of the packet. When packets are received back from computer 401, they are actually addressed to the firewall 301. Firewall 301 analyzes the address and other information in the data packet and determines that it is an expected response to the session computer 201 has initiated. Other packets that are received by the firewall that do not contain the correct session and sequence number will be rejected.

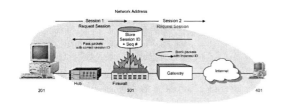

Firewall Operation

Firewall Compatibility-Firewall compatibility is the ability of a service or media connection to operate through existing or specific types of firewalls.

Firm Offer-A firm offer is an item and/or characteristic of a product or service that is offered to a customer for a defined period of time.

Firmphone-A firmphone is a softphone in a multimedia computer relies on its own circuitry for the telephone signal processing so that it does not burden the computer's operating system when processing the call. More importantly, if the operating system is busy servicing other software applications (such as a word processor), it does not degrade the quality of the telephone call.

Firmware (FW)-Firmware is software program instructions that are stored in a hardware device that performs data manipulation (e.g. device operation) and signal processing (e.g. signal modulation and filtering) functions. Firmware is stored in memory chips that may or may not be changeable after the product is manufactured. In some cases, firmware may be upgraded after the product is produced to allow performance improvements or to fix operational difficulties.

First Choice Route-The first choice trunk group, or series of groups, between two switching systems.

First Generation (1G)-A term commonly used to describe the first technology used in a new application. In cellular telecommunications, the first generation used analog (usually FM) radio technology. For first generation cordless telephones, the first generation of products used single channel (using AM) radios.

First In Still Here (FISH)-First in still hear is an accounting process that maintains inventory value based on the value of the items that are first to arrive into the inventory system.

First Mile-The signal path between a program origination site and its entry point to the communication network or a private satellite uplink. The first mile is usually a terrestrial RF link or a local telco loop.

First Mover-A first mover is a person or company that performs the first new action in an industry to gain a competitive advantage.

First-Dollar Gross-In reference to all income shares starting at the first dollar, usually acquired by the distributor or by the studio.

First-In-First-Out (FIFO)-First-in-first-out is the process of tracking or accounting for products that are produced where the first products entering into a process or area (such as into an inventory) are counted if the first products exiting the process or area.

FISH-First In Still Here

Fishing-Fishing is the process of pushing a stiff wire or tape through the hollows of a wall or conduit. Once the fish wire has reached its destination (e.g. an outlet hole in the wall), the cable or wires are attached to one end and they are pulled through the wall or conduit until the cable or wire is pulled through the hole.

Five Nines (99.999%)-A measurement of reliability expected from a public telephone network. It is the equivalent of having a total down time of less than 5 minutes per year.

FIX-Financial Information Exchange

Fixed Allocation-Fixed allocation is the assignment of a resource for a predetermined or scheduled amount of time.

Fixed Charges-Fixed charges are recurring and non-recurring charges that have a cost that is fixed and independent of usage quantities.

Fixed Compensation-An unconditional set rate sum of money insured to be issued to one in trade for ones services.

Fixed Costs-Fixed costs are costs that occur regardless of the production of products or services.

Fixed Length Records-Fixed length records are sets of data that have the same field structure and data length. Fixed length records are simpler to process than variable length records.

Fixed Mobile Convergence (FMC)-Fixed mobile convergence is the process of using or providing the same services to fixed and mobile users, regardless of location, access or terminals using existing core infrastructure and back office systems.

Fixed Public Service-Fixed public service is a radio communication service that is provided between fixed radio transceivers that are available for public use or subscription.

Fixed Service-Fixed service is the providing of radio communication services between fixed geographic points.

Fixed Wireless Access (FWA)-Fixed wireless access is the process of using a radio link to provide communication services to fixed locations such as homes or businesses.

Flags-(1-computer) An indicator used to signal a condition or to mark information for further attention. The flag may be "raised" or lowered" through the results of computation, or specifically controlled through software. (2-ISDN, PPSN) In the Integrated Services Digital Network (ISDN) and the Public Packet-Switched Network (PPSN), a unique digital pattern that is used by a link layer to delimit frames. Each link layer frame starts with an opening flag and ends with a closing flag. (See also: Call Reference FIag). (3- SS7) In the Signaling System 7 protocol, a unique pattern on the signaling data link used to delimit a signal unit. The binary flag sequence used in ISDN, PPSN, SS7 and also X.25 and Frame Relay packet systems is the HDLC binary flag 01111110. To prevent false premature end of packet indication, the contents of every packet are processed before transmission to insert a binary 0 following any natural occurrence of 5 consecutive binary 1 bits. At the receiving end, a binary zero inside a packet is deleted if it follows 5 consecutive binary 1 bits, thus accurately restoring the original packet data.

Flaming-The process of sending one or many flame messages to the sender of unwanted emails. Flaming typically has the intent of intimidating or disabling the sender of unwanted emails.

Flash-A system special service request feature that is used to indicate that a subscriber has a desire to recall a service function or to activate a custom calling feature (such as a call transfer request).

A flash feature service request can be created when the user initiates a short on-hook interval or through the sending of a special service request message. The short on-hook interval is created by a momentary operation of the telephone switch hook, in the midst of a prolonged off-hook period. This momentarily turns off the loop current in the subscriber loop. The duration of the current-off state must be typically more than 1 second but less than 2 seconds in most systems. The special service request message can be sent by a button on a telephone (such as a PBX telephone) sometimes labeled with such words as FLASH or BREAK or SERVICE, or by pressing the SEND or TALK key on a mobile telephone. In a mobile telephone or an ISDN telephone, a digital message is sent when the subscriber presses the button, instead of momentarily turning off the loop current.

This diagram shows how the flash signal (special service request) can be sent to the telephone system to activate additional call processing features. This example shows a flash feature is sent on an analog line by momentarily opening the current loop connection. When the loop current sensing circuit senses a brief open (no current flow) period, it creates a flash message that is sent to the call process-

ing section of the telephone system. For digital telephones, the flash message is sent via a signaling message on the digital channel. This diagram shows that on an ISDN line, the flash message is sent on the D (signaling) channel.

Flash Hook-Flash hook is a signaling process that is created by the rapid changing of the on-hook condition of a telephone set.

Flash Memory-Flash memory is a type of memory storage that has the ability to be reprogrammed multiple times and can retain information without power.

Flash Messaging-Flash messaging is the automatic displays of a message as soon as it is received. An example of a flash message is an important news alert or weather bulletin that is immediately displayed on a mobile telephone display.

Flash Request-A request to initiate a special processing function. Dual Mode cellular allows flash requests in both directions while analog systems only allow flash requests from the mobile phone to the cell site.

Flash With Info-A flash message (special call processing message) that is sent over a communication channel that provide an indication that the originator needs special processing along with some additional information (parameters) that are associated with the call processing request.

Flat Fee Referral-A flat fee referral is a fixed amount that is charged for each person or company that is referred to another person or company.

Flat File-A flat file is a collection of information or data (such as digital bits) where all of the information is located within the file boundaries (not referenced to other locations).

Flat Rate-Flat rate is a service fee that charges customers the same fee for services (such as telephone calls) regardless of the number of service uses or the duration of each service use. Usually flat rates are single monthly or periodic charges.

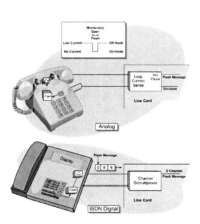

Flash Operation

Fleet Management-Fleet management is the dispatching, routing and tracking of vehicles to provide shipment or other services that involve the transport of materials and people.

Flexible Pricing-Flexible pricing is the ability of a system or service to charge different prices for the sale or use of products or services that are provided to different types of users or systems.

Flexible Pricing Tariffs-Flexible pricing tariffs are rate structures that can be changed quickly so they can meet market conditions (such as increased competition).

Flexible Reporting-Flexible reporting is the process that provides a user or system with the options on select data, how the data is processed and how the information will be presented.

Flocking-Assembling adHoc teams of expert resources in response to high priority issues.

Floor-A floor is a value or payment minimum.

Flow Chart (Flowchart)-A graphic portrayal that shows the sequence in which functions are performed, from the beginning of a job to the end.

Flow Control-A hardware or software mechanism or protocol that manages data transmissions when the receiving device cannot accept data at the same rate the sender is transmitting. Flow control is used when one of the devices communication cannot receive the information at the same rate as it is being sent; this usually occurs when extensive processing is required by the receiver and the receive buffers are running low. Examples are flow control algorithms are IEEE 802.3x used in Ethernet networks, Forward Explicit Congestion Notification and Backward Explicit Congestion Notification used in ATM networks and XON/XOFF used is RS-232 serial communication.

Flowchart-Flow Chart

Flyer-A flyer is a single sheet media item that includes promotional information about a company, product or service.

FM-Forensic Marking
FMC-Fixed Mobile Convergence
FMIS-Financial Management Information System
FMV-Fair Market Value
FOB-Free On Board
FOB-Freight on Board
FOC-Full Operational Capability

Focus Group-A focus group is a selection of people that share some market characteristics (e.g. business users) who review or evaluate a concept, product or service.

FOD-Fax On Demand
FOD-Free on Demand

Folder-A folder is an area of memory that holds files or media.

Folder Rights-Folder rights are the authorizations to create, view or rename folders (directories).

Folksonomy-Folksonomy is the practice and method of collaboratively creating and managing tags to annotate and categorize content. Folksonomy is also known as collaborative tagging, social classification, social indexing, social tagging, and other names. In contrast to traditional subject indexing, metadata is not only generated by experts, but also by creators and consumers of the content. Usually, freely chosen keywords are used instead of a controlled vocabulary.

Follow Me Call Forwarding-Follow me call forwarding is a call processing feature that changes the forwarding number as a user moves to different telephone numbers in a call forwarding list.

Follow Me Television-Follow me television is a video delivery feature that changes the destination of a video program as a user moves to different television display devices.

Follow-Me Phone Service-A service that allows calls to be routed to a customer's choice of forwarding phone numbers. Follow-me service may be automatic (e.g. when a cellular telephone automatically registers with a visited systems) or manually set (e.g. when a customer calls in with a hotel phone number where they will be temporarily located).

This diagram shows how follow-me service can allow calls to be automatically routed to one of three telephone numbers that a customer has provided to the follow-me system. In this example, the follow-me service is automatic. When the incoming call is received by the system, the system first tries the customer at the office number. Because there is no answer at the office, the system will then call the home number. When the user answers the call at home, the system will automatically move the home telephone number to the top of the follow-me calling list.

Follow-Me Roaming-Follow-me calls are automatically forwarded to the area outside their home area where the mobile subscriber

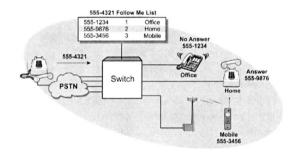

Follow-Me Phone Service Operation

has registered. A cellular call generated to a user when the mobile subscriber has informed the local system that the user is roaming in another area.

Font-A font is a set of information (images or formulas) that is assigned to alphanumeric and other characters that can be used to present (render) the characters in different formats. Examples of fonts include Helvetica, Times Roman or Century Schoolbook.

Footprint-(1-satellite) The minimum radio coverage signal strength boundaries from a satellite over in a geographic area. A single satellite may have several antennas that have different footprints. (2-equipment) The floor area occupied by a given piece of equipment. (3-software) The amount of memory required to run an operating system or software application.

Force Majeure-Unforseen or unanticipated abrupt occurance putting productions to a stop.

Forced Account Code Billing-Forced account code billing is a call processing feature that requires a user to enter a billing code before a call connection can be completed. If the user does not select or enter a billing code, the call will not be processed.

Forced Call-In reference to contact taken by ones form of employment, requiring one to return to work prior to time initially arranged by party's.

Forced Rerouting-In the Signaling System 7 protocol, a procedure for transferring signaling traffic from one signaling route to another when the route in use fails or must be cleared of traffic.

Forecasted Load-Forecasted load is the predicted resource utilization for a specified future period.

Forecasting-Forecasting is the process of estimating the future performance for a product or service. Forecasting may be performed using historical and/or analysis information.

Foreign Exchange Office (FXO)-Foreign Exchange Office (FXO) interface or channel unit that allows an analog connection (foreign exchange circuit) to be directed at the PSTN's central office or to a station interface on a PBX. The FXO sits on the switch end of the connection. It plugs directly into the line side of the switch so the switch thinks the FXO interface is a telephone. (See also: foreign exchange station.)

F

Foreign Exchange Station (FXS)-A type of channel unit used at the subscriber station end of a foreign exchange circuit. A foreign exchange station (FXS) interface connects directly to a standard telephone, fax machine, or similar device and supplies ring, voltage, and dial tone. (See also: foreign exchange office.)

Foreign Key-A foreign key is a label or code that is used to uniquely identify objects in lists or sets of data in a different table or data file.

Forensic Accounting-Forensic accounting is the process of using recorded information to track the source or reasons for actions or financial performance.

Forensic Marking (FM)-Forensic marking is the addition of information to a media component, product or service that assists in the identification and/or measure information related to the use, operation or actions on that item by a company, system or person.

Forensics-Forensics is the ability to identify and/or measure information related to the use, operation or actions on a product or service by a company, system or person.

Forklift Upgrades-An upgrade that requires the removal of old equipment (possibly by a forklift) and installation of new equipment.

Form-A form is a set of information areas that a user (such as a web site visitor) can enter data that is stored or transferred when the user selects the Submit button on the form.

Form Data-Form data is information that is formatted to a specific type or format (such as text or numeric).

Form Factor-(1-product) The physical shape and size of a product. (2-mathematic) The ratio of the root-mean-squire value of a periodic function to the average absolute value, averaged over a full period of the function.

Form Feed (FF)-A form feed is a printer command that instructs the printer to advance paper in the printer to the top of the next form (next page). The ASCII form feed character has a decimal value of 12.

Form Posting-Form posting is the transferring of the information that has been entered into a form screen to another location (such as sending as a text message or adding the data to a new record in a database).

Format-(1-computer) To prepare or preprogram a storage medium, such as a floppy disk, so that it can receive and store data. (2-document) The shape, size, and general makeup of a document. (3-interconnection) A specific grouping of bits that facilitates transmission of data through a digital system. (4-video) The specific form of the elements that make up a video signal (such as a component vs. composite format).

Forms-Forms are structured displays of screen information that usually allows a user to enter and edit information in predefined fields on specific areas of the form.

Formula Translator (FORTRAN)-A computer language that was developed by IBM to allow programmers to code instructions in a manner similar to ordinary mathematical equations.

FORTRAN-Formula Translator

Forum-A forum is a group of people or companies that help to develop, promote, or assist with various aspects of a technology, product, or service.

Forward Auction-A forward auction pricing system is a process of allowing buyers to increase their buying offers until the end of the auction period or until there are no more bidders product at which time the product is sold.

Forward DNS Lookup-A forward DNS lookup is a process that obtains an IP address from a DNS server by providing a domain name (a URL).

Forward Messages (Forwards)-Forward messages are received items (such as email) that are forwarded by the recipient to one or more people.

Forward Prediction-Forward prediction is the process of estimating the likely changes or occurrences that may occur within future media, images or media components in a sequence of media (such as audio packets or video frames).

Forward Secrecy-Forward secrecy is the ability of a security system to have current or future security keys or processes that are not related to past security keys or processes.

Forwarding-The process of taking a frame received on an input port of a switch and transmitting it on one or more output ports.

Forwards-Forward Messages

FOSS-Free Open Source Software

Fourth Generation (4G)-Fourth Generation wireless networks with bandwidth reaching 100 Mbps that allow for voice and data applications that will run 50 times faster than 3G. This capacity will enable three dimensional (3D) renderings and other virtual experiences on the mobile device.

FPA-Front Panel Assembly

FPML-Financial Products Markup Language

FQDN-Fully Qualified Domain Name

Fractal-Fractal is the process of dividing or breaking an item or data into components or shapes that have some similarity.

Fractional T-1 (FT-1)-A digital transmission service that provides a customer with multiple 64 kbps channels but less than the full 24 channels offered by a T-1 channel.

FRAD-Frame Relay Access Device

Fragmented Marketplace-A fragmented marketplace is a group of customers that purchase or use specific types of products and services that are available from different vendors. These products and services are proprietary or incompatible with each other.

Frame-(1-general) A frame is a basic repeated bit pattern in time division multiplexing systems, and/or the time duration of this pattern. (2-frame relay) A packet of data in Frame Relay systems. (3-video) In video processing, a frame is a single still image within the sequence of images that comprise the video. Note that, in an interlaced scanning video system, a frame comprises two fields. Each field contains half of the video scan lines that make up the picture, the first field typically containing the odd numbered scan lines and the second field typically containing the even numbered scan lines. (4-equipment) An electronic rack that is used to interconnect and hold electronics assemblies. (5-audio) In audio processing, a frame is a group of audio samples that are processed together (to determine an instantaneous frequency spectrum for a codec like MP3, for example), though two adjacent frames might contain common samples in an overlap region. (6-HTML) An HTML frame is a division of a web page into separate framed areas.

Frame Relay-Frame relay is a packet-switching technology that provides dynamic bandwidth assignments. Frame relay systems are a simple bearer (transport only) technology and do not offer advanced error protection or retransmission. Frame relay were developed in the 1980s as a result of improved digital network transmission quality that reduced the need for error protection. Frame relay systems offer dynamic data transmission rates through the use of varying frame sizes.

This figure shows a frame relay system. This diagram shows a local area network (LAN) in San Francisco is connected to a LAN in New York. A virtual path is created through the frame relay network so data can rapidly pass through each frame relay switch as its path is previously established. When data is to be

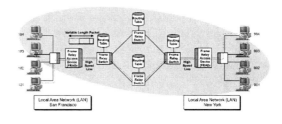

Frame Relay System

transferred through the LAN (e.g. a large image file), the data file passes through a FRAD that is the gateway to the frame relay network. The FRAD divides the data file from the LAN into variable length data frames. The FRAD sends and receives control commands to the frame relay network that allows the FRAD to know when and if additional data frames can be sent.

Frame Relay Access Device (FRAD)-A frame relay access device is a communications access device that converts data from a user's network into the format that is required by a frame relay network.

Frame Relay Network Device (FRND)- The FRND is a packet switch that also operates as a gateway to the frame relay network. The FRND passes frames it receives from the frame relay access device (FRAD) to other frame relay switch that forward packets toward their destination network. Frame relay switches have buffer memory that allows them to hold, prioritize packets before they are retransmitted. Packet switches can selectively discard packets if network congestion occurs. The FRAD and FRND provide information about the priority of the frames (e.g. non-essential discard eligibility) and status of the system (e.g. network congestion notification).

Frame Structure-(1-data transmission) Frame structure is the division of defined length of digital information into different field (information) parts. Frame structure fields typically include a preamble for synchronization, control header (e.g. address information), user data, and error detection. A frame may be divided into multiple time slots. (2-web page) Frame structure is the process or arrangement of frames that hold text, images or other media on a web page.

Frame Switch-A device that forwards data frames based on layer 2 addresses contained within the received data frame. Frame switches may directly switch (cut-through) or operate as store and forward switches.

Frames-Frames are divisions of viewable areas of web pages whose contents can be separately displayed and controlled. An example of the use of frames is the use of a toolbar frame for overall site navigation, a title bar frame to holds a company logo and contact information, and a main frame to display the selected items.

Framework-Framework is a theoretical structure that defines objects and relationships between them to allow the development or modeling of systems or processes.

Franchise-A franchise is an authorization issued by a franchising authority that defines how a franchisee can use or provide products or services and what value is exchanged for the franchise rights.

Franchise Fee-Franchise fees are an amount charged or assigned to an account for the authorization to use a product, service, or asset. Franchise fees can be a combination of fixed fees or a percentage of sales.

Franchising Authority-A franchising authority is an entity (such as an agency or company) that is authorized (e.g. by a government or company) to grant a franchise.

FRAND-Fair Reasonable and Non-Discriminatory

Fraud-Fraud is the process of obtaining or using products or services without authorization.

Fraud Based Churn-A common form of involuntary churn. Fraudsters are customers who steal services without payment and who are terminated by the telephone company (Telco).

Fraud Detection-Fraud detection is the processes that are used to determine if a product or service has been acquired or used by an unauthorized person.

Fraud Management-Fraud management is the processes and steps taken to identify, minimize and correct the unauthorized use of products or services.

Free On Board (FOB)-Free on board is a shipping agreement that the seller will provide the products to an agreed location (such as to a ship) at no additional cost to the buyer.

Free on Demand (FOD)-Free on demand is a service that provides end users to interactively request and receive video services without service usage fees.

Free Open Source Software (FOSS)-Free open source software is one or more computer programs that is provided without a fee that includes the original source code. FOSS programs allow other developers to make changes to the software to meet their specific application needs.

Free Software Foundation (FSF)-The free software foundation is an organization founded by Richard Stallman that develops GNU software, which is freely available.

Free Trade Zone (FTA)-A free trade zone is an area that can be used to exchange goods without the need to perform or obtain customs clearance. A FTA may be used to allow products to travel through a country towards its destination without the need to complete complex customs documentation or paying customs duty.

Free Units-Free units are a specified quantity in terms of which other quantities can be measured that are offered or provided at no cost.

Freephone-A service that allows callers to dial and telephone number without being charged for the call. The toll free call is billed to the receiver of the call. In Europe and other parts of the world, toll free calls are called Freephone and typically begin with 0-800. In the United States, freephone calls are called toll free calls and they are preceded by a 1-800, 1-888 or 1-877 exchange.

FreeWare-Freeware is software that can be transferred (e.g. downloaded from the Internet) for use without paying a fee for its acquisition. Freeware may be copyrighted and there may be restrictions on the use of the program as part of the freeware distribution process. Freeware was trademarked by Andrew Fluegelman who won an inventor of freeware and shareware award.

Freight Bill-A freight bill is the charges that are assessed for shipping products or goods.

Freight Note-A freight note is an invoice that contains the shipping costs for a product or goods.

Freight on Board (FOB)-Freight on board is a shipping agreement that the seller will provide the products to an agreed location (such as to a ship) at no additional cost to the buyer.

Freight Terms-Freight terms are the rules or conditions that are applied to the shipping of products or services.

Frequency-(1-radio) The frequency of an electrical or optical wave is the number of complete cycles or wavelengths that the wave has in a given unit of time (second). The standard measurement for this is number of cycles per second, also known as Hertz (the scientist, not the car company), abbreviated Hz. (2-marketing) The number of times a message or advertisement appears in communication media (such as a magazine or web page) over a period of time.

This diagram displays how frequency is measured. In this example, there three cycles of a wave that are transmitted over a 1 second period. This equals a frequency of 3 Hertz.

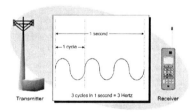

Frequency

Frequently Asked Questions (FAQ)- Frequently asked questions are a set of commonly asked questions and answers. FAQs may be located on a web site to help reduce the number of customer service inquiries or to assist customers in selecting products and completing sales orders.

Fringe Area-The area outside the designated boundary of the communications system, such as a zone for radio, television, or mobile communication service may be provided.

FRND-Frame Relay Network Device

Front End-Front end is the part of a system that is closest to the user or client device. Front end systems may control user interfaces and pre-process data for back end systems.

Front Panel Assembly (FPA)-The front panel assembly is the front portion of a device (such as a television set top box) that is accessible to users.

Frontload-Frontload, referring to a modified cost agenda, is an amount of money paid in percentages prior to aggreement or at the time services are rendered.

Frozen Standard-A frozen standard is an industry specification that has been accepted by a group (typically the standards body) that members or users agree that no additional changes will occur for a period of time or indefinitely.

FRR-False Rejection Rate

FSF-Free Software Foundation

FSN-Full Service Network

FT-1-Fractional T-1

FTA-Free Trade Zone

FTAM-File Transfer Access, And Management

FTP-File Transfer Protocol

Fulfillment-Fulfillment is the process of gathering the products and materials to complete an order and shipping the products or initiating the services that were ordered.

Fulfillment House-A fulfillment house is a company that processes and tracks orders for another company.

Fulfillment, Assurance, and Billing (FAB)-The three major functions performed by communications companies. Fulfillment is the process of preparing the network for customer access, taking orders for service, and initiating that service. Assurance is the process of making sure that customers receive continuous and high quality service. Billing is the process of receiving call detail and service usage information, grouping this information for specific accounts, producing invoices, and recording payments made for those invoices.

Full Access-An arrangement in which all traffic offered to a group of servers (trunks) has access to all the servers in the group.

Full Operational Capability (FOC)-Full operational capability is ability of a system to provide all of its defined services within the defined operational limitations of the system.

Full Period Service-In reference to the tariff of services, a service, circuit, facility, or piece of equipment that is continuously available for use by a customer.

Full Service Network (FSN)-A full service network is the infrastructure and support systems that are capable of providing all of the communication services required by the end user.

Full Services Access Network Forum-A forum that was established in 1995 to help identify technologies and network architectures that can cost effectively provide narrowband and broadband telecommunications services.

Full Trunk Group-A normally high-usage trunk group on which no overflow to an alternate route is permitted. Enough trunks must be provided for acceptable service.

Fully Automated Collect And Third Number Service (FACTS)-Fully automated collect and third number service system is a set of equipment and software that can process, bill and collect third-number-billed calls with no or minimal operator intervention.

Fully Qualified Domain Name (FQDN)-A fully qualified domain name is a name that includes all the sub directory paths.

Functional Acknowledgement (FA)-Functional acknowledgement is a process or control code that is used to confirm a message has been received and has been successfully interpreted or understood. A FA message is used in the electronic data interchange (EDI) system to acknowledge that the contents of an EDI message could be translated.

Functional Block-In defined process or service that contains one or more processes that interact with other defined functional blocks.

Functional Group (FG)-A functional group is a set of data that share common characteristics.

Functional Requirements-Functional requirements are a set of processes and their associated characteristics that are necessary to meet the needs of a business or system.

Functional Specification-A functional specification is a document that provides an operational description and requirements of a device, equipment or system.

Functional Tests-Functional tests are observations and/or measurements that are performed during normal operating conditions of a device, service or system to determine if it can perform its designed functions.

Functions-Functions are processes that accept one or more forms of information inputs or arguments and produces a single output or value that is determined by the combination of the inputs and the formal specification of the function.

Funding Request-A funding request is a statement or document that defines how much money and/or resources are needed to perform a project or develop a business. A funding request usually includes how the money or resources are going to be used and how the investor of the resources will be repaid or rewarded.

Future Proofing-Future proofing is the process of designing a system to allow for future improvements in technology or capabilities. An example of a future proof design is a video compression device, which can download new compression decoding or encoding programs.

Fuzzy Logic-A branch of mathematics, used in some artificial intelligence computers, in which decisions are based on ideas and approximations rather than on mathematically rigid calculations.

FVF-Final Value Fee

FW-Firmware

FWA-Fixed Wireless Access

FxM-Fax Mail

FXO-Foreign Exchange Office

FXS-Foreign Exchange Station

G

GAAP-Generally Accepted Accounting Principles

Game Merchandising-Game merchandising is the selling of items that are branded or associated with a game. Examples of game merchandise t-shirts, hats, back packs and figurines.

Game Server-A game server is computing device that can process requests for and deliver game programs and services.

Games On Demand (GOD)-Games on demand is a service that provides end users to interactively request, obtain and use gaming programs or services. These gaming services are from previously stored media (entertainment movies or education videos) or they can have a live connection (with other gaming players).

Gaming-Gaming is an experience or actions of a person that are taken on a skill testing or entertainment application with the objective of winning or achieving a measurable level of success.

Gantt Chart-A gantt chart is a diagram that shows project tasks as timeline horizontal bars with starting and ending dates.

Gap Analysis-Gap analysis is the review and assessment of differences between events (such as the time different between when a service is started and ended).

Garbage-Garbage is data or programs that are unused or poorly written.

Garbage Collection-Garbage collection is the process of searching and identifying programs or data that are no longer necessary in order to reclaim that storage space.

GARP-Growth At a Reasonable Price

Gatekeeper (GK)-A gatekeeper is a server that coordinates access to other servers. The gatekeeper receives requests from clients, determines the destination server that it needs to communicate with, and coordinates access with that server. For packet voice systems, the gatekeeper translates user names or telephone numbers into physical address for H.323 conferencing.

This figure shows how a gatekeeper sets up connections between Internet telephones and telephone gateways. The gatekeeper receives registration messages from an Internet telephone when it is first connected to the Internet. This registration message indicates the current Internet address (IP address) of the Internet telephone. When the Internet telephone desires to make a call, it sends a message to the ITSP that includes the destination telephone number it wants to talk to. The ITSP reviews the destination telephone number with a list of authorized gateways. This list identifies to the ITSP one or more gateways that are located near the destination number and that can deliver the call. The ITSP sends a setup message to the gateway that includes the destination telephone number, the parameters of the call (bandwidth and

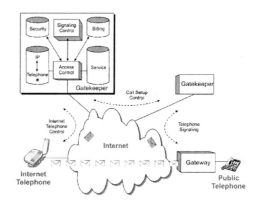

Gatekeeper Basic Operation

type of speech compression), along with the current Internet address of the calling Internet telephone. The gatekeeper then sends the address of the destination gateway to the calling Internet telephone. The Internet telephone then can send packets directly to the gateway and the gateway initiates a local call to the destination telephone. If the destination telephone answers, two audio paths between the gateway and the Internet telephone are created. One for each direction and the call operates as a telephone call.

-Gatekeepers are individuals or group of individuals who control your access to individuals or groups of individuals within a company or organization.

Gateway (GW)-A gateway is a communications device or assembly that transforms data that is received from one network into a format that can be used by a different network. A gateway usually has more intelligence (processing function) than a bridge as it can adjust the protocols and timing between two dissimilar computer systems or data networks. A gateway can also be a router when its key function is to switch data between network points.

Gateway Page-A gateway page is a web page that is specifically designed for to interact with another web page or system (such as interfacing to a search engine).

Gateway Server-A gateway server is a computer with communication processing capability and associated software that can be used to adapt communication information that is sent between networks (such as signaling protocols and/or media).

GB-Gigabyte

Gbps-Gigabits Per Second

GCC-Gulf Cooperation Council

GCR-Group Call Register

GE-Gigabit Ethernet

Geek-A geek is a person who is focused on technology, typically computers who does not tend to conform to mainstream habits such as dressing for success and/or regular bathing.

General Ledger Interface-A general ledger interface is a connection between the general ledger in an accounting system and another system (such as a television system). The general ledger interface allows for the recording of accounting entries that result from the operation of other systems.

General Parameters-In specification description language, the basic operating parameters of a unit or system.

General Policies-General policies are the rules processes that are applied to all or groups of device, service or users

Generalized Mark-up Language (GML)-Generalized mark-up language is a set of text based communications messages that describe the item selection and formatting features to be transferred independent of the type of computer system used. GML is a precursor to SGML and HTML.

Generally Accepted Accounting Principles (GAAP)-Generally accepted accounting principles are industry standard procedures are rules and processes that are used to record, process and report on the creation and use of services or resources.

Generic Mapping-Generic mapping is the process of designing or setting relationships (maps) between items on a group or general basis.

Generic Requirement (GR)-A specification document controlled by Telcordia (formerly Bellcore) that defines requirements of systems that connect within a telephone network.

Geographic Information System (GIS)-Geographical information system is a data processing system that is capable of gathering, analyzing and providing geographic information. GIS systems are used to create maps or display the results of queries for the location of objects or systems.

Geolocation-Geolocation refers the geographic location of person (e.g. web site visitor), company or device.

GHz-Gigahertz

GIF-Graphics Interchange Format

Gift Certificate-A gift certificate is document or code that entitles the recipient or holder to obtain a value of merchandise or services.

Gift Policy-A gift policy is a set of rules or guidelines on how people may give or receive items or services.

Gift Registry-A gift registry is an account that is maintained by a vendor (such as a retail store) for the purposes of allowing people to see what gifts they desire. Gift registries are commonly associated with events such as weddings.

Gifting-Gifting is the process of one person or company purchasing or committing to purchase products or services that are provided to another person for a company to use.

Gifts-Gifts are the providing of an item or service to a person or company without requiring direct compensation or reward from the recipient.

Giga-A prefix that represents one billion units.

Gigabit-One billion bits of data.

Gigabit Ethernet (GE)-A data communications system primarily used for computer networks based on the Ethernet IEEE standard 802.3 that transmits at 1000 Mbps (1 Gbps).

Gigabit Passive Optical Network (GPON)-A gigabit passive optical network (GPON) is an ATM and Ethernet based optical network that combines, routes, and separates optical signals through the use of passive optical filters that separate and combine channels of different optical wavelengths (different colors).

Gigabits Per Second (Gbps)-Gigabits per second is the transfer of is one thousand million (10^9) bits per second.

Gigabyte (GB)-A gigabyte is one billion bytes of data. When gigabyte is used to identify the amount of data storage space (such as computer memory or a hard disk), a gigabyte commonly refers to 1,073,741,824 bytes of information.

Gigahertz (GHz)-Gigahertz is one billion cycles per second.

GIS-Geographic Information System

GK-Gatekeeper

Glass Ceiling-A glass ceiling is an invisible boundary that limits specific types of people or workers for progressing to jobs or roles up in a business organization. An example of a glass ceiling is the inability of foreign nationals or minorities to be assigned to jobs in high-level executive positions in a company.

Gleaning Resource Descriptions from Dialects Language (GRDDL)-Gleaning resource descriptions from dialects is a set of commands and processes that can be used to identify, define and extract information from XML documents.

Glitch-A glitch is an unexpected condition or a temporary interruption of a signal or period of distortion.

Glitching-Glitching is a process or event that temporarily interrupts the normal operation or reduces the capabilities of a system or service.

Global-Applies to all users or devices connected to a network or system.

Global Catalog-A global catalog is a group of products, services or data file that are accessible by all users who desire to obtain information about the items.

Global Positioning System (GPS)-The global positioning system is a location determination network that uses satellites to act as reference points for the calculation of position information. GPS is used extensively by the military and aircraft GPS chipsets are now being incorporated into wireless devices, including phones, personal digital assistants (PDAs), as well as automotive applications. Each GPS satellite transmits two frequencies; 1575.42 MHz (the L1 carrier) and 1227.6 MHz (the L2 carrier). A GPS receiver compares the

G

signals from multiple GPS satellites (4 satellite signals are usually used) to calculate the geographic position.

This figure shows a global positioning satellite (GPS) system. This diagram shows how a GPS receiver receives and compares the signals from orbiting GPS satellites to determine its geographic position. Using the precise timing signal based on a very accurate clock, the GPS receiver compares these signals from 3 or 4 satellites. Each satellite transmits its exact location along with a timed reference signal. The GPS receiver can use these signals to determine its distance from each of the satellites. Once the position and distance of each satellite is known, the GPS receiver can calculate the position where all these distances cross at the same point. This is the location. This information can be displayed in latitude and longitude form or a computer device can use this information to display the position on a map on a computer display.

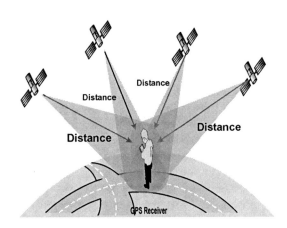

Global Positioning System (GPS) Operation

Global Roaming-Global roaming is the capability to move from one carrier's system area to another carrier's service area and obtain service in different countries. While it is desirable to roam without loosing functionality of the phone or device, some communication systems offer advanced features (such as high-speed data) that other systems may not have installed. This may limit the operation of advanced features when global roaming.

GML-Generalized Mark-up Language

GMT-Greenwich Mean Time

GNU-GNU's Not Unix

GNU's Not Unix (GNU)-GNU is operating system software that is freely available.

GOD-Games On Demand

Golden Handcuffs-Golden handcuffs are incentives that are offered to employees (such as senior executive staff) to remain in a job or assignment. Golden handcuffs may be used during a merger or company acquisition to keep key employees at the company.

Golden Parachute-A golden parachute is a clause in an employment contract or an incentive that is available to an employee (such as a senior executive) to compensate them if their job is lost for certain reasons (such as a hostile takeover).

Goodwill-Goodwill is a value assigned to intangible assets of a business such as brand value.

Googolplex-A googolplex is a number that is 10 to the power of googol (10 to the power of 100).

Governing Laws-Governing laws are governmental rules and regulations that are used to enforce the terms of an agreement. Agreements commonly specify which government entity or jurisdiction shall apply to the agreement and other laws are requirements that may apply.

Government Access Channel-A government access channel is a media source that is dedicated to informing citizens of public related information. Examples of government programming include legal announcements, property zoning, public worker training programs, election coverage, health related disease controls and other public information that is related to citizens.

GPON-Gigabit Passive Optical Network

GPS-Global Positioning System

GR-Generic Requirement

GR-303-A set of technical specifications that help define the next generation of digital loop carrier (DLC) interconnection.

Grant-A grant is the providing or awarding an amount or value when specific terms and conditions have been satisfied.

Grant Management-Grant management is the allocation (granting) of resources (such as transmission time or bandwidth) to a device or system.

Granularity-(1-network operation) The steps of network resource allocation. This is typically bandwidth allocation that is adjustable per communication session. (2-digital conversion) The increments (step size) of conversion from an analog signal to a digital signal.

Graphic User Interface (GUI)-A graphic user interface is the presentation of graphics on a monitor or display that allows a system to convert the output or requested input of a software application to a format the can be understood by a user. The use of buttons, icons and dynamically changing windows are typical examples of a GUI. Sometimes pronounced "gooey interface."

Graphics-Graphics is the portrayal of information through the use of graphs, letters, lines, drawings, or pictures.

Graphics Artist-A graphics artist is a person who has developed skills at creating or assisting in the creation of graphics images that express or provide interpreted meanings.

Graphics Illustrator-A graphics illustrator is a person who has developed skills at converting sketches or other forms of media into a graphic form (such as a cartoon).

Graphics Interchange Format (GIF)-A graphics file data compression format that produces relatively small graphic files. A GIF image may contain up to 256 colors and uses a lossless data compression method. A version of the GIF format GIF89a permits the addition of animation features, image interleaving, and the use of transparent backgrounds.

Gratis-Ex Gratia

Gray Market-A gray market is a trading environment or place that is used to obtain goods that are hard to obtain or have a limited amount of supply.

GRDDL-Gleaning Resource Descriptions from Dialects Language

Great Understanding Relatively Useless (Guru)-A consultant or expert with a reputation for being helpful to other less knowledgeable users.

Green Application-A service application or software program that is designed to efficiently use system resources.

Green Light-(1-business) Green light is an agreement or contract allowing a studio to commence with project development. (2-financial) An indication that financial criteria or research results indicate that a financial transaction is valid or likely to be valid.

Greenfield-Greenfield is the introduction of new systems or equipment without the requirement of keeping, updating or interfacing to existing systems or equipment.

Greenwich Mean Time (GMT)-Greenwich mean time is the mean solar time at Greenwich England. This time is in accordance with 0' latitude that passes through Greenwich and it is commonly referred to as coordinated universal time (UCT).

Grooming-Grooming is the process of selecting or managing the resources (such as bandwidth) within a network to allow the facilities or communication circuits to be used as effectively as possible.

Gross Add-Gross add is the total addition of new customers or subscribers to a service (such as adding new users to a communication system) or list.

Gross Participation-Gross participation is the sharing of gross revenue that is generated by the sale of a program (e.g. movie), product or service.

G

Gross Revenues-Gross revenues are monies (or equivalent values) that are earned through the sale or providing of products and services without and reductions for costs associated with developing, producing or selling the products or services.

Gross Sales-Gross sales are the total recorded revenue for products or services. Gross sales may be defined in agreements as the total amount of invoices before applying discounts, allowances or returns.

Group Address-A group address is a code that is used to identify a set of stations or a number of devices as the destination for transmitted data.

Group Call-A call from one mobile radio, telephone or dispatcher to a predefined group of receivers in a group that are capable of receiving and decoding the messages.

This figure shows how voice group call service may operate in a GSM system. In this diagram, a single voice message is transmitted on GSM radio channels in a pre-defined geographic area. Several mobile radios are operating within the radio coverage limits (group 5 in this example) of the cells broadcasting the group message. In this example, a user is communicating to a group. Each user in this group (including the dispatcher) listens and decodes the message for group 5. Other handsets in the area are not able to receive and decode the group 5 message.

Group Call Register (GCR)-A group call register is a network database that holds a list of group members and the attributes that allow the set-up and processing of calls to and from group members. The GCR holds the membership lists, account features, priority authorization and the current location of group members.

Group Delay-(1-general) A condition in which the various frequency elements of a given signal suffer differing propagation delays through a circuit or a system. The delay at a lower frequency is different than the delay at a higher frequency, resulting in a time related distortion of the signal at the receiving point. (2-optical) The transit time required for optical power, traveling at the group velocity of a given mode, to traverse a specified distance. The measured group delay of a signal through an optical fiber has a dependence on specific wavelengths because of the each wavelength reacts differently to dispersion mechanisms in the fiber.

This figure shows how group delay can cause pulsed signals, such as in digital transmission systems, can cause signal distortion. This diagram shows that a digital pulse signal is actually composed of many low, medium, and high frequency components. As the pulse is transmitted through the transmission line, some of the frequency components are delayed more

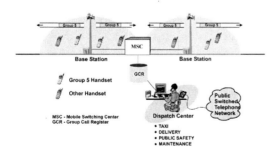

GSM Dispatch Operation

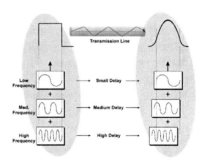

Group Delay Operation

than others. This results in a distorted pulse at the receiving end of the transmission line.

Groupware Applications-Groupware applications are software programs that are designed to perform operations using commands or information from other sources where information is gathered and or distributed to multiple users. Popular groupware applications include whiteboard sharing, video conferencing and online seminars.

Growth At a Reasonable Price (GARP)- Growth at a reasonable prices is an investment concept that prefers picking stocks or investments that provide growth that have a risk that is relatively lower than other investment options.

Guaranteed Service-Guaranteed service provides a specific bandwidth with a set maximum end-to-end transmission delay time.

Guarantor-A guarantor is a person or company who agrees to be financially responsible for a bill for another person or company in the event the person or company fails to meet their obligations.

Guarded Object-Guarded objects are software components or hardware devices that have their information or access channels protected or controlled by a guarding program or device.

Guerrila Marketing-Guerrila marketing is the process of promoting and selling products or services through efficient communication processes that create excitement or interest that is spread by people.

Guest Book (Guestbook)-A guest book is a book or form that requests visitors to leave a record of their visit. These guest book records are commonly used for promotional mailings such as news or discount stories.

Guest Mailbox-A guest mailbox is a temporary message storage area (a mailbox) that is temporarily assigned to a person (such as a hotel guest). A guest mailbox may have a limited set of features along with added message prompts to help the guest understand how to use the mailbox features.

Guestbook-Guest Book

GUI-Graphic User Interface

Guiding Billing Records-Guiding is a process of matching call detail records to a specific customer account. Guiding uses the call detail record identification information such as the calling telephone number to match to a specific customer account.

Guiding Records-Guiding records is a process of matching call detail records to a specific customer account. Guiding uses the call detail record identification information such as the calling telephone number or user identification login code to match to a specific customer account.

Guild-A guild is an organization, society or association. A form of unity among members as to form a Union.

Gulf Cooperation Council (GCC)-The gulf cooperation council is an economic group of countries created in 1981 in the Middle East. It consists of Bahrain, Kuwait, Oman, Qatar, Saudi Arabia and the United Arab Emirates.

Guru-Great Understanding Relatively Useless

GW-Gateway

H

Hacker-A hacker is a person or a machine that attempts to gain access into networks and/or computing devices. Hackers may perform their actions for enjoyment (satisfaction), malicious reasons (revenge), or to obtain a profitable gain (theft).

Hacking-Hacking is the process that is used when attempt to gain unauthorized access into networks and/or computing devices. The term hacking has also been used by programmers to solve their programming problems. They would continually change or hack the program until it operated the way that they desire it to operate.

HAL-Hardware Abstraction Layer

Half Duplex-Half duplex communication is the process of transferring voice or data information in either direction between communications devices but not at the same time. The information may be transmitted on the same frequency or divided into different channels. When divided into different channels, one channel of frequency is used for transmitting and the other channel or frequency is used for receiving.

The use of different frequencies is common in half duplex radio transmission because the transmitter and receiver are commonly connected to the same antenna. If the same transmitter and receiver frequency were used, the high transmitter power would probably destroy the receiver circuitry.

Handheld Device Markup Language (HDML)-HDML is an internet web browsing language that evolved from hypertext markup language (HTML). HDML is optimized for low bandwidth operation using limited screen size and limited user input keys. HDML allows soft keys to be dynamically defined to allow a phone and return the key presses to the network application.

Handle-A handle is a label or a string of characters (alphabetic, numeric, or any combination including the underscore) that is used to define or associate a link or service to a particular person, device or communication channel.

Handset Churn-Type of voluntary churn unique to, and common in the wireless industry. Handset churn occurs when a customer changes carriers in order to get a newer and better (and often free) handset.

Hard Currency-Hard currency is a monetary exchange value that can be exchanged with other currencies throughout the world.

Hard Disk-A hard disk is a data storage system or device that uses a rigid disk as the storage medium.

Hard Disk Drive (HDD)-A hard disk drive is an information storage device that maintains information on a hard disk of material. Hard disk drives typically store and retrieve the information by rotating the disk (a platter) and using a set of magnetic conversion devices (heads) that can detect and write magnetic information to areas of the disk.

Hard Drive Surfing-Hard drive surfing is the connecting, viewing and transferring of files from hard disk drives of other computers that are attached to a network (such as computers logged onto Wi-Fi networks).

Hard Launch-Hard launch is the process of making a product available for purchase and distribution with significant promotional efforts. Companies may perform a hard launch strategy after a soft launch or it may use a hard launch campaign to maximize the effects of new product publicity.

Hardware-Hardware is physical equipment or devices.

Hardware Abstraction Layer (HAL)-Hardware abstraction layer is a collection of related processing functions that are used to isolate the upper layer functions (such as audio, data or video applications) from the underlying hardware system (such as a computer or set top box).

Hash Function-A hash function is a mathematical process that converts (transforms) a block of data or information into an output value. Hash functions are one-way processes which result in the output value (called a hash or digest) not being convertible back into its original data or information form.

Hash Table-A hash table is a list of codes that assist in the finding of information.

Hash Value-A hash value is a calculated number or value that is created from converting a password or information element into a fixed length (hashed) code. A hash value can be used to check that a password or information element is correct without the need to send the actual passwords or information elements.

Hashing-Hashing is a computational process that converting a data or a message into a fixed length data message. Hashing can be used to convert text based passwords into fixed length digital password codes.

HAVi-Home Audio Video Interoperability

HD-High Definition

HDD-Hard Disk Drive

HDLC-High-Level Data Link Control

HDMI-High Definition Multimedia Interface

HDML-Handheld Device Markup Language

HD-PLC-High Definition Power Line Communication

HDTV-High Definition Television

Head Hunter (Headhunter)-A headhunter is a job recruiter who looks for people that meet certain qualification criteria.

Head On Collision-A head on collision is the simultaneous occurrence of an attempt to obtain access to the same resources (such as a communication trunk line).

Headend-A headend is part of a television system that selects and processes video signals for distribution into a television distribution network. A variety of equipment is used at the headend, including antennas and satel-

lite dishes to receive signals, preamplifiers, frequency converters, demodulators and modulators, processors, and scrambling and descrambling equipment.

This figure shows a diagram of a simple head-end system. This diagram shows that the head-end gathers programming sources, decodes, selects and retransmits video programming to the distribution network. The video sources to the headend typically include satellite signals, off air receivers, microwave connections and other video feed signals. The video sources are scrambled to prevent unauthorized viewing before being sent to the cable distribution system. The headend receives, decodes and decrypts these channels. This example shows that the programs that will be broadcasted are supplied to encoders and modulators to produce television channels on multiple frequencies. These channels are combined onto a single transmission line by a channel combiner.

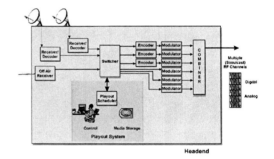

CATV Headend System

Headend Management System (HMS)-A headend management system is a combination of equipment and software that is used to setup, control, monitor and manage the operation of a headend (media center) of a communication system.

Header-(1-communications) The part of a data packet that contains address, routing, and origination information. (2- record) A record that contains administrative, physical and electrical data describing a cable count, carrier facility or type of equipment. (3-style element) A pre-defined style format that has a hierarchical structure that allows for the multi-level organization and presentation of documents through the use of header element types. Headers can have different font sizes, position structure (left, center, or right), and font types.

Headhunter-Head Hunter

Headroom-(1-signal level) The difference, in decibels, between the typical operating signal level and a peak overload level. (2-bandwidth) The amount of available data transmission bandwidth that exists for applications and services.

Headset-A headset a transducer device that can be worn or positioned on a persons head to allow users to extend the audio portions of their communication devices to portable and flexible external speakers. Headsets may be capable of producing audio and/or capturing (via a microphone).

Heartbeat-A heartbeat is a communication test function that uses a repeated transmitted signal that travels through a system or network which when received confirms that the network is still operating (it is alive).

Heartbeat LED-Heartbeat Light Emitting Diode

Heartbeat Light Emitting Diode (Heartbeat LED)-A heartbeat LED is flashing indicator that confirms that a repeated test signal (heartbeat) that travels through a system or network has been received which confirms that the device and the network is operating (it is alive).

Hecto-A prefix meaning 100.

Help Desk-A help desk is an accessible location where questions about the operation of a product or service can be answered. A help desk may be reached by a combination of voice, instant messaging or email.

Help Desk Support-Help desk support is the providing of interactive information services that assist in the operation or implementation of a product or service. Help desk support may be provided by a combination of voice, instant messaging or email.

Help Engine-A help engine is a software program that can receive, process and respond to an end user's (client's) requests for help on topics or key index words.

Help Facilities-Help facilities are the parts (elements) of a network or system that help to provider or present information that can help agents or users to answer questions, discover solutions or solve problems.

Help Screens-Help screens are displays that provide information to users to help them answer questions, discover solutions or solve problems.

Hertz (Hz)-A measurement unit for frequency that is equal to the number of cycles per second. This measurement unit was named after the German physicist Herrich R. Hertz (1857-1894).

This figure shows how to measure a signal wave in cycles per second (Hertz). This diagram shows a signal wave that has three cycles that moves past a point in 1 second. This equals a frequency of 3 Hertz. Radio waves typically have several million cycles per second that is called a Megahertz (MHz).

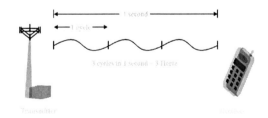

Measuring a Wave in Hertz

Heterogeneous-Heterogeneous is a system or network that is composed of different technologies or components.

Heuristic-Heuristic is a solution seeking process that involves trial and error to discover the solution to a problem.

Hexadecimal-Hexadecimal is a numbering system that is based on 16 values ranging from 0 to F.

HFC-Hybrid Fiber Coax

HFS-Hierarchical File System

Hiatus-A hiatus is a break given to staff or employees between working hours.

HID-Human Interface Devices

Hidden Cost-A hidden cost is unforeseen additional charges or reductions in revenue potential that occur resulting from the purchasing of products or services.

Hierarchical File System (HFS)-A file system that is structured like trees where files are associated with higher level files or directories (roots).

High Definition (HD)-High definition (HD) television is the resolutions of enhanced analog television and digital television. The resolutions of HD range from 480/60p-480 pixels (vertical) by 728 pixels (horizontal) with 60 progressive fields (60p) per second to 1080/60p-1080 pixels (vertical) by 1920 pixels (horizontal) with 60 progressive fields per second.

High Definition Multimedia Interface (HDMI)-HDMI is a specification for the transmission and control of uncompressed digital data and video for computers and high-speed data transmission systems. HDMI is a combination of a digital video interface (DVI) connection along with the security protocol HDCP. HDMI has the ability to determine if security processes are available on the DVI connection and if not, the HDMI interface will reduce the quality (lower resolution) of the digital video signal.

This figure shows an HDMI connector diagram. This diagram shows that the HDMI connector is has 19 pins that are used to carry digital signals and is 13.9 mm wide by 4.45 mm high. The HDMI connector has 3 high-speed TDMS® data connections (0, 1 and 2).

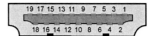

Pin	Function	Pin	Function
1	TMDS Data 2+	11	TMDS Clock Shield
2	TMDS Data 2 Shield	12	TMDS Clock-
3	TMDS Data 2-	13	CEC
4	TMDS Data 1+	14	Reserved
5	TMDS Data 1 Shield	15	SCL
6	TMDS Data 1-	16	SDA
7	TMDS Data 0+	17	DDC/CEC Ground
8	TMDS Data 0 Shield	18	+5V
9	TMDS Data 0-	19	Hot Plug Detect
10	TMDS Clock+		

HDMI Connector Diagram

High Definition Power Line Communication (HD-PLC)-High definition power line communication is communication system developed by Panasonic that uses high frequency signals over a power line to transmit data and digital media signals. The HD-PLC system transmits high-speed data signals using frequencies between 4 MHz to 28 MHz. The HD-PLC system supports the use of 128 bit encryption to ensure data privacy. It has a maximum transmission distance of approximately 150 meters.

High Definition Television (HDTV)-High definition television (HDTV) is a TV broadcast system that proves higher picture resolution (detail and fidelity) than is provided by conventional NTSC and PAL television signals. HDTV signals can be in analog or digital form.

High Level Language-A high level language is a set of commands and associated parameters that can be converted into specific sequences of computer processing instructions. High level languages are more intuitive to humans than the machine code instructions they represent.

High-Level Data Link Control (HDLC)-An ISO communication protocol that is located in the data link layer in that delineates the beginning and end of a data frames. HDLC is used in X.25 packet switching communication networks to transfer bit-oriented, synchronous protocol that provides error correction at the data-link layer. HDLC systems allow messages to be transmitted in variable-length frames.

Hijacking-A process of gaining security access by the capture of a communication link in mid-session after the session has already been unauthorized.

This figure shows how hijacking may be used to obtain access to an authorized media session to gain access to protected media. This example shows that an unauthorized user has obtained information about a media session request between a media provider (such as an online music store) and a user (music listener). After the media begins streaming to the validated user, the hijacker modifies a routing table distribution system to redirect the media streaming session to a different computer.

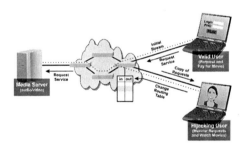

Hijacking Operation

History-History is data (such as sales history or billing charges) that is stored for future use (inquiry, adjustments, analysis, ect.).

Hit-(1-Internet) A request for a file transfer on a web site (web page access). (2-telephone signaling) A short off-hook or on-hook supervisory signal on a telephone line. Telephone applications may process or ignore hits by identifying signals that are shorter than a minimum specified time period, typically 10 ms to 400 ms. (3-data) A short disruption in the transmission of a data stream. (4-lightning) A lightning strike.

Hit Counter-A hit counter is device or software program that counts the number of web page access requests (hits) that occur over a period of time.

HKSW-Hook Switch

HLR-Home Location Register

HMI-Human Machine Interface

HMS-Headend Management System

HMS-Hosted Media Server

Hold-(1-general) A temporary mode of operation that is typically entered into by a device when there is no need to send voice or data information for a relatively long time. The hold mode allows the device audio to be muted or the transceiver to be turned off in order to save power. (2-Bluetooth)The Bluetooth hold mode is used to release devices from actively communicating with the master. This allows the devices to sleep for extended periods and allows the master control device to discover or be discovered by other Bluetooth devices that want to join other Piconets. With the hold mode, Piconet capacity can be freed up to do other things like scanning, paging, inquiring, or attending another Piconet sessions.

Holiday-Saturday, Sunday, officially recognized federal legal holidays and any other day on which the Commission's offices are closed and not reopened prior to 5:30 p.m. For example, a regularly scheduled Commission business day may become a holiday if its offices are closed prior to 5:30 p.m. due to adverse weather, emergency or other closing.

Home-The geographic location where a communication service is based.

Home Audio Video Interoperability (HAVi)-Home audio visual interoperability is a specification for home networks comprised of consumer electronics devices such as CD players, televisions, VCR's, digital cameras, and set-top boxes. The network configuration is automatically updated as devices are plugged in or removed. The IEEE 1394 protocol, also known as FireWire, is used to connect devices on the wired HAVi network at up to 400 Mbps.

Home Carrier-A home carrier is the communication service provider (such as a mobile telephone company) that a user has registered for service use.

Home Coverage-Home coverage is the available area (for wireless) or number of outlets (for wired systems) over which the signal strength of a premises distribution network is sufficient to transmit and receive information, media or data.

Home Location Register (HLR)-The home location register is a part of a communication network (typically cellular or PCS) that holds the subscription and other information about each subscriber authorized to use the wireless network.

Home Media Management-Home media management is the systems and processes that are used to identify, select, adapt and control media distribution in a home or premises.

Home Network-A home network is the equipment and software that is used to transfer data and other media in a customer's facility or home. A home network may be used to connect terminals (computers) and media devices (such as TV set top boxes) to each other and to wide area network connections. Home networks may use a mix of wired Ethernet, wireless LAN, powerline, coaxial and phone lines to transfer data or media.

Home Office-A home office is the location where the central decisions or coordinating functions are performed for a business or person.

Home Page (Homepage)-A home page is the first web page that is accessed on a web site. The default home page file name is typically index.htm.

Home Phoneline Networking Alliance (HomePNA)-The HomePNA is a non-profit association that works to help develop and promote unified information about a phoneline technologies, products and services.

Home Phoneline Networking Alliance Specification (HPNA)-The HPNA specification defines the signals and operation for data and entertainment services that can be provided through telephone lines that are installed in homes and businesses. The HPNA specification is designed to co-exist with other communication systems including POTS, ISDN and ADSL. More information about HomePNA can be found at www.HomePNA.org.

Home Subscriber Server (HSS)-A home subscriber server is a computing system that can store and process subscription data and other information about each subscriber authorized to use the network.

Home System-The home system is a communication network where a customer has registered for service.

Homepage-Home Page

HomePlug Audio Visual (HomePlug AV)-The HomePlug Audio Visual (HomePlug AV) specification was ratified in 2005 by the HomePlug Board of Directors to provide home media networking. HomePlug AV was designed to give priority to media that requires time sensitive delivery (such as IPTV) while allowing reliable data communication (such as web browsing) to simultaneously occur. HomePlug AV uses a mix of random access (unscheduled) and reserved access (scheduled) data transfer. The carrier sense multiple access with collision avoidance (CSMA/CA) protocol, which provides for efficient transfer of bursty data while the scheduled TDMA system ensures real time media (such as digital video and audio) will be delivered without delays and will take priority on the wire over CSMA/CA traffic.

HomePlug AV-HomePlug Audio Visual

HomePNA-Home Phoneline Networking Alliance

Honoraria-Honoraria is the providing of value to someone or to a company in recognition of their service or assistance in business or projects.

Hook-A hook is an option or design feature in a software program or system that simplifies or assists in the adding of new features or capabilities.

Hook Flash (Hookflash)-A special feature service request signaling process that allows the user to momentary disconnect circuit disconnection (temporary off-hook state) as a means if indicating a special service request (e.g. call forwarding request).

Hook Switch (HKSW)-A hook switch is a electrical connection device that is located in the handset cradle of a telephone device that is connected when the telephone handset is lifted from the cradle.

Hookflash-Hook Flash

Hop-(1-communication) The single path between a transmitter and a receiver. (2 - propagation) A single refraction of a radio signal from the ionosphere.

Horizontal Application-A horizontal application is a program or software that is used across multiple industries or by different types of users.

Horizontal Discount-A horizontal discount is a price reduction in the cost of media (such as broadcast time (or space in a print publication) that is defined in an agreement to advertise over a long period of time.

Horizontal Market-A horizontal market is the offering of products or services where the buyers have different sets of characteristics (multiple vertical market segments).

Hospitality IPTV-Hospitality IPTV systems are IP television systems that have been designed or setup to be used in private facilities such as in hotels, cruise ships or college campuses.

Host-A host is a computer or other type of data information processing device that is connected to a network that processes requests and provides information services to remote users.

Host Switching System-A switching system with centralized control over most of the functions of one or more remote switching units. The host usually provides trunk access to an exchange carrier's intraLATA network and access tandems.

Host Terminal Interface-The interface (communication link and commands) between a host and a unit that is using the services of a host.

Hosted Application-A hosted application is a software program that is designed to perform operations using commands or information from other sources (such as a computer connected to the Internet) that operates on a system that is not owned by the company or person who manages the application.

Hosted Data-Hosted data is information that is stored on devices that are owned or managed by another company. An example of hosted data is web pages that are stored on a computer server that is owned by a web hosting company.

Hosted E-Commerce-Hosted e-commerce is shopping medium that uses electronic networks (such as the Internet or telecommunications) to present products and process orders where the product information is stored and orders are processed on devices that are owned or managed by another company.

Hosted iPBX-Hosted iPBX systems are IP telephone systems that are operated at a remote location (hosted) to provide private telephone services that are similar to private branch exchange systems (PBX). Hosted iPBX systems can be operated by independent companies and systems or they can be operated by telephone companies by sharing central exchange (IP Centrex) telephone switch functions.

Hosted Media Server (HMS)-A hosted media server (remote server) may be used by companies who do not have the technical expertise to setup and manage computer servers. This solution is a bit more expensive than stand alone systems (the hosting company must cover its costs and make a fair profit). However, this solution allows companies to more rapidly setup mobile video services and it may be an excellent solution for companies to start gaining experience in providing mobile video services.

Hosted PBX-Hosted PBX systems are telephone systems that are operated at a remote location (hosted) to provide private telephone services that are similar to private branch exchange systems (PBX). Hosted PBX systems can be operated by independent companies and systems or they can be operated by telephone companies by sharing central exchange (Centrex) telephone switch functions.

Hosted Podcast-A hosted podcast is a shared Internet resource that allows media owners to upload their content to the host (owner or manager of the service) so the materials can be automatically transferred to recipients as soon as they become available (such as a weekly audio program).

Hosted Solution-A hosted solution is an application or service that is provided by a system or service that is operated by a different company (a host).

Hosted Telephony-Hosted telephony is a managed IP Telephony communication service (also known as IP Centrex). Hosted telephony is a communications service for users who prefer to use a communication system that is managed by another company. The term "hosted" or "managed" is used instead of "Centrex" because an IP-based feature set is generally broader than what existing Centrex offerings provide. Furthermore, hosted or managed solutions do not just cater to the traditional Centrex market-typically campus-based, such as universities or government. The flexibility of IP allows for very compelling offerings to other markets, such as SOHO and greenfields.

Hosted Version-A hosted version is an application program or service that is provided on a system where the information and/or software programs are stored on devices that are owned or managed by another company.

Hostile Program-A hostile program is a software application or module that performs undesired processes such as using system resources or disabling communication capabilities.

Hosting-The providing of application services (such as virtual telephone service or the providing of web service) for the benefit of a client or customer.

Hot Links (Hotlinks)-Hotlinks references (such as a web link) and can connect information from one document to another, regardless of the type of application used.

Hot Patch-A method of patching a failed digital line, such as a T1 line, onto a spare facility. This technique does not use bridging repeaters. Service is interrupted briefly when the patches are removed. This type of patching is used only when bridging repeaters are not available for full-service patching.

Hot Standby-Hot standby is a mode of operation of a device, equipment or system where it is fully configured and ready for operation but not in service. A hot standby device or equipment can rapidly replace other devices or systems in the event of a failure.

Hot Swaping-Hot swapping is the process of removing and replacing cards or assemblies while a system is on and operating.

Hotel System Integration-Hotel system integration is the process of selecting, installing, testing, and validating equipment and/or software programs that communicates or interacts with hotel business systems.

Hotline-A restricted calling class that forces a telephone (usually a wireless telephone) to be connected to an operator regardless of the digits actually dialed. Hotline is typically used when a telephone is first sold or activated to allow activation after the customer has provided the information to register for service or when the customer has not paid their bill.

Hotlinks-Hot Links

House Account-A house account is a customer account that is directly managed by the company that produces or provides the products or services.

House List-A house list is a group of contacts that has been created or is managed directly by the company (the house) that produces or provides products or services to customers.

House Phone-A call routing feature commonly used in private telephone systems (PBX) to allow the connection of a call directly to an attendant or announcement system when the phone goes off the hook (handset is lifted).

HPNA-Home Phoneline Networking Alliance Specification

HSS-Home Subscriber Server

HSTB-Hybrid Set Top Box

HTML-Hypertext Markup Language

HTML Format-HTML format is a text based document or message that contains formatting and item selection features that are used to modify the presentation (the format) of the document or message.

HTTP-Hypertext Transfer Protocol

HTTPS-Hypertext Transfer Protocol Secure

Hub-A hub is a communication device that distributes communication to several devices in a network through the re-broadcasting of data that it has received from one (or more) of the devices connected to it. A hub generally is a simple device that re-distributes data messages to multiple receivers. However, hubs can include switching functional and multi-point routing connection and other advanced system control functions. Hubs can be passive or active. Passive hubs simply re-direct (re-broadcast) data it receives. Active hubs both receive and regenerate the data it receives.

This figure shows how a hub distributes communication to several devices in a network through the re-broadcasting of data that it has received from one (or more) of the devices connected to it. A hub generally is a simple device that re-distributes data messages to multiple receivers. However, hubs can include switching functional and multi-point routing connection and other advanced system control functions. Hubs can be passive or active. Passive hubs simply re-directs (re-broadcasts) data it receives. Active hubs both receive and regenerate the data it receives.

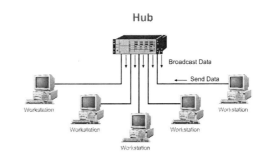

Hub Operation

Human Factors Engineering-Engineering that applies the study of ergonomics to the design of equipment and software to create safe, easy-to-use systems.

Human Interface Devices (HID)-Human interface devices (HID) convert signals and information into forms that are accessible to humans.

Human Machine Interface (HMI)-The human machine interface is the method of how a user will enter (input) and receive information from a device or system. For the GSM/GPRS system, the HMI defines the requirements for a WAP browser (micro browser) on a mobile device.

Human Testable-Human testable is the ability of a device or service to be tested by people (such as quality assessment by a group of people).

Humanware-Humanware is hardware or software that is can be controlled or used by humans.

Hunt Group-A hunt group is a list of telephone numbers that are candidates for use in the delivery of an incoming call. When any of the numbers of the hunt group are called, the telephone network sequentially searches through the hunt group list to find an inactive (idle) line. When the system finds an idle line, the line will be alerted (ringing) of the incoming call. Hunt lines are sometimes called rollover lines.

Hunting-A telephone call-handling feature that causes a transferred call to "hunt" through a predetermined group of telephones numbers until it finds an available ("non busy") line.

This diagram shows the process of hunting (also called "roll-over") for an available telephone. This diagram shows that an incoming call enters into the telephone switching system and is attempted to be delivered to the main telephone line extension (or dialed telephone number port). The switching system is programmed with a hunt list that allows the switching system to determine where to redirect a call if it is unable to deliver a call. This

example shows that the system first tries 1001 that is off-hook (unavailable). The hunt group table shows that the call should be routed to 1002. Because 1002 is also unavailable, the hunt group list instructs the switch to try extension 1003. This telephone (or port) is available and the call can be delivered (telephone rings).

Hybrid-(1-Device) A device that combines transmit and audio signals from two-pairs of lines to one pair of lines. (2-Network) The combination of two different types of network technologies (such as fiber and coax) to form a combined (Hybrid) network.

This figure shows how a typical analog telephone transmission line operates. In this diagram, audio from customer #1 is converted to electrical energy by microphone #1. This signal is applied to the telephone line via the hybrid adapter #1. A portion of this signal is applied to the handset speaker to produce sidetone (so the customer can faintly hear what they are saying). This audio signal travels down the telephone line to hybrid #2. Hybrid #2 applies this signal to speaker #2 so customer #2 can hear the audio from customer #1. When customer #2 begins to speak, microphone #2 converts the audio to an electrical signal. This signal travels down the line to hybrid #1. Hybrid #1 subtracts the energy from microphone #1 (the combination of both signals are actually on the line) and applies the different (audio from customer #2) to the speaker #1.

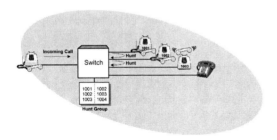

Hunting Operation

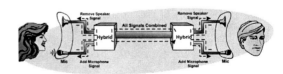

Hybrid Telephone Operation

Hybrid Fiber Coax (HFC)-The hybrid fiber coax (HFC) system is an advanced CATV transmission system that uses fiber optic cable for the head end and feeder distribution system and coax for the customers end connection. HFC are the 2nd generation of CATV systems. They offer high-speed backbone data interconnection lines (the fiber portion) to interconnect end user video and data equipment. Many cable system operators anticipating deregulation and in preparation for competition began to upgrade their systems to Hybrid Fiber Coax (HFC) systems in the early 1990's.

Hybrid ITSP-An Internet telephone service provider that combines packet voice with switched voice services.

Hybrid Key System-A key telephone system (KTS) that combine some of the advanced features of PBX systems with traditional KTS features. An example of a hybrid key system is the ability to assign different telephone lines to each electronic telephones without the need to rewire the switching system.

Hybrid Set Top Box (HSTB)-A hybrid set top box is an electronic device that adapts multiple types of communications mediums to a format that is accessible by the end user. Hybrid set top boxes are commonly located in a customer's home to allow the reception of video signals on a television or computer. The use of HSTBs allows a viewer to get direct access to broadcast content from terrestrial or satellite systems in addition to accessing other types of systems such as interactive IPTV via a broadband network.

Hybrid System-A hybrid system is the combining of two or more systems that can accommodate multiple types of signals, physical channels, and services. Examples of hybrid systems include hybrid fiber coax (HFC) and combined analog and digital signal transmission.

Hypercompetition-Hypercompetition is a condition of business where competitive advantages only offer short benefits as competitors rapidly discover and adapt to compete with the new advantages.

Hyperlink-Hypertext Link

Hypertext Link (Hyperlink)-Hyperlinks are tags, icons, or images that contain a crossed reference address that allows the link to redirect the source of information to another document or file. These documents or files may be located anywhere the link address can be connected to.

Hypertext Markup Language (HTML)-Hypertext markup language is a text based communications language that allows formatting and item selection features to be transferred independent of the type of computer system. HTML is primarily used for Internet communication.

Hypertext Preprocessor (PHP)-Hypertext preprocessor is a scripting language that can be embedded on a server (such as a web server) to modify or process HTML documents.

Hypertext Transfer Protocol (HTTP)-A protocol that is used to transmit hypertext documents through the Internet. It controls and manages communications between a Web browser and a Web server.

Hypertext Transfer Protocol Secure (HTTPS)-Hypertext transfer protocol secure is a set of commands and processes that can securely transmit hypertext documents through the Internet. It securely controls and manages communications between a Web browser and a Web server.

Hz-Hertz

I

I/O-Input/Output

IAB-Interactive Advertising Bureau

IAB-Internet Architecture Board

IAD-Integrated Access Device

iAD-Interactive Advertisements

IAD-Internet Addiction Disorder

IANA-Internet Assigned Numbering Authority

IAP-Intercept Access Point

IAP-Internet Access Provider

IBC-Initial Billing Company

ICANN-Internet Corporation for Assigned Names and Numbers

ICE-Information and Content Exchange

I-Commerce-Internet Commerce

Icon-Icons are a small graphic or symbol that is linked to a software program or function. The purpose of the icon is to allow the user to quickly identify (find) their programs. Icons usually contain logos or unique product identification characteristics.

ICP-Integrated Communications Provider

ICP-Intelligent Call Processing

ICS-Implementation Conformance Statement

ICS-Intercompany Settlements

ID-Identifier

Ideagora-An ideagora is an application that allows multiple people to easily and quickly develop and exchange new ideas. An ideagora may be used by companies to obtain new ideas for product innovations.

Identifier (ID)-An identifier is a code or label that uniquely identifies (can be associated with) an item, object or other definable thing.

Identity-An identity is a name, symbol or information element that uniquely identifies a person, device or service that can be recognized by a process or system.

Identity Verification-Identity verification is a process of exchanging information that allows the confirmation of the true identity of the user (or device). Identify verification may

be used to allow a service provider to decide to enable or deny service to users that cannot be identified.

Idle URL-A URL address that is used by a device (such as an IP telephone) should the device come into an idle state. Using an idle URL allows for the display of customized status messages on device displays ("Lunch Special").

IEC-Interexchange Carrier

IEEE 1394-IEEE 1394 is a personal area network (PAN) data specification that allows for high-speed data transmission (400 Mbps). The specification allows for up to 63 nodes per bus and up to 1023 busses. The system can be setup in a tree structure, daisy chain structure or combinations of the two. IEEE 1394 also is known as Firewire or I.Link.

IFF-Interchange File Format

IGFs-International Gateway Facilities

IIF-Import/Export Interchange File

IIF-IPTV Interoperability Forum

IIS-Internet Information Server

IKE-Internet Key Exchange

ILE-ISDN Line Emulator

ILEC-Incumbent Local Exchange Carrier

Illegal-(1-general) Not lawful (2-humor) A sick bird.

Illustrator-An illustrator is a person who has developed skills at converting sketches or other forms of media or imagery into a graphic form (such as a diagram or cartoon).

Image Map-An image map is one or more predefined display areas on a web page that can be selected ("clickable") by a mouse. Image maps are commonly graphic images that are used to help users navigate through a web site.

Imaging-A process that may include converting an image into a different form (such as digital format), transferring, processing, storing, displaying, and reproducing (printing).

IMAP-Internet Message Access Protocol

Immunity-In telecommunications, a characteristic that permits equipment to operate normally when the equipment or any external lead or circuit is subjected to electromagnetic voltages, currents, or fields.

Impacted User Minutes (IUM)-Used as a network management measurement in conjunction with availability. Impacted User Minutes or IUMs is the total amount of unproductive user time due to network or server downtime and performance issues.

Impairment-(1-circuit) The loss of quality of service provided by an individual circuit when its transmission units are exceeded or signaling functions (such as seizure, disconnect, and automatic number identification) are experiencing intermittent failures. (2-signal transmission)Any distortion that affects a transmitted waveform or is perceivable by a video or audio user.

Impairment Emulator-An impairment emulator is a system that creates or simulates impairments to the operation or communication with a software program or hardware device. The use of an impairment emulator is to allow developers to simulate the operation of programs or devices under conditions that may happen to their products or services and to determine the changes in performance or operation that results from these impairments.

Impairment Resolution-Impairment resolution are the steps or processes that are taken to correct or adjust for a loss of quality of service or other type of changes that disrupt the providing of services.

Imperial Units-Imperial units are a system of measurements based on the foot, pounds and seconds.

Implementation Conformance Statement (ICS)-An implementation conformance statement is a document that is provided by a company or testing facility that states that the product or system provides and/or supports a specific set of capabilities.

Implementation Plan-An implementation plan is a sequence of events or tasks that are identified (and usually scheduled) as necessary to complete a project or development program.

Implicit Rights-Implicit rights are actions or procedures that are authorized to be performed based on the medium, format or type of use of media or a product.

Import Error Report-An import error report is a set of information that identifies and potentially summarizes (e.g. total number of errors) records or data could not be transferred into a file or database.

Import Restrictions-Import restrictions are a set of rules or lists of items that must receive special authorization or testing to be allowed to be shipped into a country. Import restrictions may include items such as technology, chemicals or weapons.

Import/Export Interchange File (IIF)-Import/export interchange file is a standard accounting record structure that is used by Quick Books to import and export accounting records and transactions.

Importing-Importing is the process of gathering and processing data or media into a format that can be stored within a file or system.

Impression Fraud-Impression fraud is the process of increasing the number of ad or search result impressions by users or systems who do not have an interest in using the search results for its intended purpose. Impression fraud may be performed by companies or people to increase the popularity of web links or to disable the display of pay per click ads as a result of low click through rates.

Impressions-Impressions are the number of times a media message has been viewed. For web pages, impressions are the number of times a web page has been requested and displayed.

Impulse Buying-Impulse buying is the purchasing of an item as a result of the immediate presentation of an item or service.

Impulse Pay Per View (IPPV)-Impulse pay per view is a service that allows a viewer to select a movie for immediate viewing without making previous arrangements to order the movie or service. IPPV fees are usually charged to the viewers account or credit card at the time the service is ordered.

IMS-IP Multimedia Subsystem

IMUX-Inverse Multiplexer

IN-Intelligent Network

In Band Signaling-In-band signaling occurs when control messages share the same communication channel as the information signals (e.g., within the audio signal bandwidth). In-band signaling requires the user's voice or data information to be momentarily interrupted or altered while signaling messages are being transferred. In-band signaling is sometimes called blank and burst signaling.

Historically, the term "in-band" is related to the use of different frequency bands with the allocated frequency band. However, in band can be applied to signals in digital multiplexing that occur in the same time slot, better named "in-slot." Two examples of true in-band signaling are DTMF (touch tone) and Multifrequency Signaling (MF). During the period of in-band signaling, the voice or data communication should be temporarily inhibited (muted) to allow the transfer of control messages without interference.

This figure shows how the basic process of in band signaling is used in analog cellular radio to deliver control messages sharing the same communication channel for voice and control signals. In this diagram, a radio base station desires to send a message to the mobile radio. The base initially sends a dotting sequence that indicates a synchronization word and message will follow. The mobile radio detects the dotting sequence. As a result, the mobile radio mutes the audio and begins to look for a synchronization word. The synchronization word is used to determine the exact start of the message. The mobile radio receives the

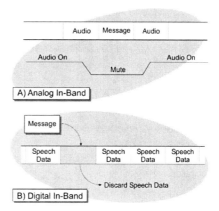

In Band Signaling

message and upon completion of the message, the mobile radio will then un-mute the audio and conversation continues. Because the sending of the message can be less than ¼ second, the user may not even notice a message has been received.

In Building Service-A classification of communication services that may be provided in buildings. For the 3rd generation wireless systems, in buildings, service has the capability of providing data transmission rates of up to 2 Mbps.

In Game Advertising-In game advertising is the inclusion of media within a game that is designed to be recognized and promote products or services.

In House-In house is the ability to produce or modify a product, service or system at or by the company that owns or operates it.

In Stock-In stock is the availability of products or goods that are available for shipment to new customers.

In the Black-In the black is a saying used to portray ones beneficial income, financial success, or profits .

In The Clear-In the clear is information or data that is not security protected.

In the Red-In the red is a saying used to portray an individuals financial disadvantage, debts, or losses.

Inactive Subscriber-An inactive subscriber is an end user of a service that is authorized to use the service but are not using the services.

Inbound Call Center-A call center (group of customer service agents) that receives telephone calls from customers. Inbound call centers (teleservice centers) are often used in response to advertisements and direct marketing campaigns. The call routing to inbound call centers can be fixed (established telephone lines) or dynamically controlled (based on activity or skills based routing.)

Inbound Telemarketing-The communication of product or service information (marketing) to customers who have initiated a call to a call center. The call may be initiated from an advertisement, direct mailing, or other outbound marketing program.

Incarnation Number-A unique name or number sent within a data unit to avoid duplicate data unit acceptance.

Incentive Program-Incentives programs are the defining of value components that will be included as part of a marketing campaign offers as added incentives to help motivate people or companies to take desired actions.

Incident Report-An incident report lists the current unresolved problems (incidents) that have occurred within a company or a communication system.

Incident Tracking-The process of recording and updating information about problems that occur in a communication system (incidents) and the steps taken to resolve the problem.

Incidental Churn-Incidental churn is a type of voluntary churn where the customer changes carriers because of changes in their situation or lifestyle, and where the change of carrier is only a secondary issue.

InCollects-Incollects are call detail records (CDRs) that are received by service provider A from service provider B for services provided by B to A's customers. An example of an Incollect is a roaming record.

Income Statement-An income statement is a summary list of revenues and expenses and the difference between these amounts (profit or loss) for a business over a period of time.

Incoming Call Restriction-In telephone call-processing feature that disables a telephone from receiving incoming calls.

Incoming Peg Count-A measurement of the number of attempts, counted at the incoming end of a trunk group, that have seized a trunk in the group. Incoming peg count measurements generally are used where no peg count measurement is available at the originating end of a trunk group.

Incompatibility-Incompatibility is the use of products or services that can only operate independently (cannot interact) or can only partially interact with other products or services.

Incoterms-International Commercial Terms

Incremental Request-An incremental request is a message that defines the additional amount of a resource (such as transmission bandwidth) is requested to provide for an application or service.

Incubator-An incubator is a company or facility that can be used to help people or companies to develop new products or services.

Incumbent-An existing system or service provider.

Incumbent Local Exchange Carrier (ILEC)-A telephone carrier (service provider) that was operating a local telephone system prior to the divestiture of the AT&T bell system.

InDate-Installation Date

Indemnification-Indemnification is the providing of protection against damage, loss or injury.

Independent Billed Message-An independent billed message is a toll message that is billed by an independent exchange carrier to a customer. The connecting exchange carrier company shares the toll revenue.

Independent Representative-An independent representative is a person or a company who performs sales functions for one or more companies. Independent representatives commonly use their own resources in the promotion and selling products or services in return for a fee and/or a percentage of the sales proceeds.

Index-An index is a value or a count of the relative number of iterations or processes that have occurred since the index value was established.

Index File-(1-general) An index file is a list of location pointers or reference values that are used to redirect software programs to the location of information for specific items such as media files or web pages. (2-database) A file that is used for indexing (reorganizing) the data in a database. Index files create a relationship between the record order of the database and the organization of values (sorted) desired by the user.

Indirect Costs-Indirect costs are the fees or bills that are associated with the operational support systems of a company (the overhead).

Indirect Procurement-Indirect procurement is the processes that are involved in the acquisition of products or services from distribution companies (not the original producer or manufacture).

Individual Booking-Individual bookings are scheduling of a program playout or reservations for resources that will be required to provide services (such as broadcasting television programs) for a specific company or project for one event.

Individual Policies-Individual policies are the rules are processes that are applied to a specific device, service or user.

Individualized Key-An individualized key is an access code or information that can be used to decode or process information that is unique to the specific user and potentially the configuration of their processing system (model and device numbers).

Inducement Agreement-An inducement agreement is a condition of agreement binding an individual to terms of his or her loan-out company.

Industrial Products-Industrial products are devices or materials that are purchased or used by companies that produce or manufacture products or services. Industrial products are designed to meet strong reliability and durability specifications.

Industry Association-An industry association is an organization that represents the interests of an industry or group of people that share common interests. Members of an industry association share common interests.

Industry Directory-An industry directory is a listing of companies or people who are involved in providing products or services for an industry.

Industry Dynamics-Industry dynamics are the characteristics of an industry or external factors that influence an industry that can be used to characterize or classify its performance or to estimate likely changes in its market potential. Examples of industry dynamics include political climate, new company starts, mergers and the adoption of industry standards.

Industry Standards-Industry standards are operational descriptions, procedures or tests that are part of an industry standard document or series of documents that is recognized by people or companies as having validity or acceptance in a particular industry. Industry standards are commonly created through the participation of multiple companies that are part of a professional association, government agency or private group.

I

Inertia Selling-Inertia selling is repeated and automatic creation of new sales orders for customers that are existing (and usually subscription) customers. An example of inertia selling is the providing of new books of a certain category to book subscription service customers.

Infomediary-An infomediary is a person, company or service (such as a web portal) that provides information that assists people or companies to buy or use products or services.

Information Access Service-A service offering that gives many telephone callers simultaneous access to selected prerecorded messages or databases furnished by private entrepreneurs known as information providers. This service also is called mass announcement network service.

Information and Content Exchange (ICE)-Information content exchange is an XML protocol that is used to describer and define content that is exchanged between networks and affiliated companies. ICE is used for automated delivery, content asset management and media backup. The ICE specification is governed by the international digital enterprise alliance (IDEAlliance). For more information about ICE, visit www.IDEAlliance.org.

Information Bartering-Information bartering is the exchanging of valuable information instead of using money or purchasing the information. An example of information bartering is the swapping of mailing lists.

Information Bulletin-A message directed only to amateur operations consisting solely of subject matter of direct interest to the amateur service.

Information Centric-Information centric is the central part or core focus of a process or system, which is based on information processing (such as software applications) services.

Information Highway-A term that refers to a common communication path (highway) for the transport of information. The world wide web (Web) is sometimes called the information highway.

Information Integrity-Information integrity is the accuracy of data information as compared to its original information source. Information integrity can be verified through the use of error detection codes that are sent along with the original data information.

Information Management-Information management is the process of gathering, organizing, storing and retrieving information.

Information Networking-The processing, delivery, and management of any type of information through public or private telecommunications networks.

Information Overload-Information overload is a condition where a person that is searching for information finds so much data it is difficult or even impossible to sort the information results into a useful form.

Information Rich Products (Information-Rich)-Information-rich products contain a relatively large quantity of information that is evaluated by the customer prior to purchase. Customers typically research and gather information about information-rich products prior to making a buying decision. These types of products typically increase the buying decision time.

Information Separator (IS)-An information separator is a character or code that is used to separate data elements.

Information Service-Information services involve the processing of information that is transferred through a communications system. Information services add value to information by generating, acquiring, storing, transforming, processing, retrieving, utilizing, or making available information via telecommunications. Examples of information services include fax store and forward, electronic publishing, text to voice conversion, and news services.

Information Systems-Information systems are combinations of data processing equipment and software that can store, transfer, and process information for specific purposes. Information systems consist of hardware (usually computers) and software (data and applications) that add value to information by generating, acquiring, storing, transforming, processing, retrieving, utilizing, or making available information via data and telecommunications connections. Examples of information systems include information transaction processing systems (TPS), management information system (MIS), decision support system (DSS), executive information system (EIS), online analytical processing (OLAP) and data mining.

Information Technology (IT)-Information technology is the processes or the study of processes used for information and data processing.

Information URL-A URL address that is used by a device (such as an IP telephone) should the user select an information button or icon. Using an information URL allows for the display of appropriate data should an information button be pressed or selected.

Informational Standards-Informational standards are documents that describe processing methods or provide supporting information that enable people, groups or companies to use other standards or technology to design and build hardware, firmware, software, or combinations thereof and have them inter work with similar products designed and built by others.

Information-Rich-Information Rich Products

Infotainment-Infotainment is a media program type that provides information in a way that is entertaining to the viewer.

Infrastructure-All parts of the communication systems, excluding the subscriber. This includes switches, radio carrier equipment, databases, and other network parts that enable telecommunications networks. In a traditional telephone network, the equipment includes the switches, multiplexers and other equipment used to manage the network and the copper cables that connect them together and to the users. In a cellular system, the equipment includes base stations and microwave links as well as wireline equipment. In an optical network, infrastructure includes optical analogues to the traditional switches and multiplexers, as well as the optical fiber among them and connections to the traditional and cellular networks.

Ingesting Content-Ingesting content is a process for which content is acquired, usually from a satellite downlink and loaded onto initial video servers (ingest servers). Once content is ingested it can be edited to add commercials, migrated to a playout server or played directly into the transmission chain.

Ingress-(1-signal) A process where a strong signal outside a communication system enters into a transmission line or system. (2-network) The process of data or signals entering into a communication network.

Inhibit-A process or control signal that prevents a device, circuit, or system from operating.

Initial Billing Company (IBC)-The initial billing company is a service provider that initiates a billing record that is exchanged with other companies that provide additional services to complete the billing record. An example of an IBC is a local telephone company that bills an interexchange carrier for the long distance portion of a telephone call.

Initial Operational Capability (IOC)-Initial operational capability is the amount of service ability a system has when it is initially activated.

Initial Operational Capability Date-The Initial operational capability date is the time when a system or service initially activated.

I

Initial Period-(1-billing) The unit of time for billing at the beginning of each message, as defined by a tariff. (2-default) The initial (default) time required before normal processing can occur.

Initial Rate-Initial rating is the process of providing an initial value for a usage of a product or service.

Initialization-A process of setting the initial values and settings of the software and electrical components within a circuit or equipment is when it is first activated.

Injunction-An injunction is a legal restriction that is imposed on a person or company that prohibits them from performing a specific action or set of actions.

InkML-InkML is an XML data format for representing digital ink data that is input with an electronic pen or stylus as part of a multimodal system.

In-Line Advertising Charge-An in-line advertising charge is a fee that is assessed for the actual viewing of an advertising message.

Inoperability-Inoperability is the condition of a device or program that makes it unusable to a user or system.

Input-(1-signal) The waveform fed into a circuit, or the terminals that receive the input waveform. (2-data) Any data being sent to a computer from a user, another computer, or other equipment.

Input Stream-An input stream is a flow of data or information into a device or system. A stream may be continuous (circuit based) or bursty (packetized).

Input/Output (I/O)-Input/output is the process of sending signals between systems or devices.

Insertion Fee-An insertion fee is a value that must be paid by a person or company for the insertion of an ad into a media item or program.

Insertion Order (IO)-An insertion order is a form or document that authorizes the insertion of an ad or other data into a media channel (such as inserting an ad into a magazine).

Insider Piracy-Insider piracy is the duplication and sale of program materials, particularly TV programs on DVD, videotape or other digital media format in violation of copyright laws. Insider privacy is performed or assisted by people who are employees, contractors or supporting personnel of companies that produce, distribute or manage the program materials.

Inspection Lot-A collection of units of product from which a sample is drawn and inspected to determine conformance with acceptability criteria.

Installation-Installation is the locating and configuration of equipment and/or wiring either inside or outside of buildings or facilities. When the equipment has passed operational and performance tests installation may be complete.

Installation Charges-Installation charges are fees or values that are assessed for the locating or configuration of equipment and/or wiring either inside or outside of buildings or facilities.

Installation Date (InDate)-The date on which a telecommunications link or equipment is installed, sometimes referred to as an indate.

Installation Income-Installation income is revenue that is generated from providing services for the installation of products or systems.

Installation Load-The installation load is the maximum amount of pulling force (tensile strength) a cable can withstand before it is deformed or damaged.

Installed First Cost-The installed cost of an outside plant item when it is placed in service.

Instant Activation-Instant activation is a process that allows users to obtain service immediately after applying for service. Instant activation may use the same data connection for requesting and activating communication services as for the transfer of the required services and features.

This figure shows how the instant activation process works. In this example, a customer selects an Internet telephone service provider (such as from the list www.ITSPdirectory.com) that offers the services they desire (such as voice mail and a specific area code). The customer directly enters the billing information along with payment information (a charge card). This information is added to the customer database and an account code is provided to the customer. When the customer receives the user identification code and password, they can immediately begin to use the service.

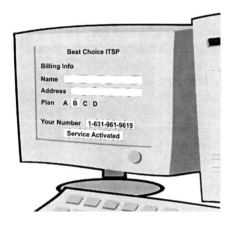

IP Telephony Instant Activation

Instant On-Instant on is the process of having a system or service starting the operation immediately (or perceived as immediate) after the system or service is requested. Examples of instant on is the ability of a computer to begin operating when it is turned on or the presentation of a media image or video as soon as it is requested.

Instantiation-Instantiation is the creation of an instance (unique attribute or set of related objects) from a service or program.

Insurance-Insurance is a promise or a commitment to provide something (such as cash) if an event occurs (such as the loss or damage to an item).

Insured Rate-An insured rate the bandwidth or service that a customer is guaranteed to receive.

Intangible Property-Assets that have no physical existence but instead have value because of the rights which ownership confers.

Integrated Access Device (IAD)-A device that converts multiple types of input signals into a common communications format. IADs are commonly used in PBX systems to integrate different types of telephone devices (e.g. analog phone, digital phone and fax) onto a common digital medium (e.g. T1 or E1 line).

This figure shows an integrated access device (IAD) combines multiple types of media (voice, data, and video) onto one common data communications system. This diagram shows that three types of communication devices (telephone, television, and computer) can share one data line (e.g. DSL or Cable Modem) through an IAD. The IAD coordinates the logical channel assignment for device and provides the necessary conversion (interface) between the data signal and the device. In this example, the telephone interface provides a dialtone signal and converts the dialed digits into messages that can be sent on the data channel. The video interface buffers and con-

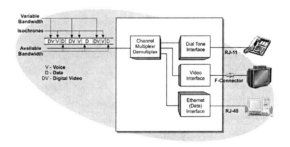

Integrated Access Device (IAD) Operation

verts digital video into the necessary video format for the television or set top box. The data interface converts the line data signal into Ethernet (or other format) that can be used to communicate with the computer. This diagram also shows that the IAD must coordinate the bandwidth allocation so real time signals (such as voice) are transmitted in a precise scheduled format (isochronous). The digital television signal uses a varying amount of bandwidth as rapidly changing images require additional bandwidth. The IAD also allocates data transmission to the computer as the data transmission bandwidth becomes available (what is left after the voice and video applications use their bandwidth).

Integrated Communications Provider (ICP)-An integrated communication provider is a company or business entity that provides multiple types of communication services where each of the systems or services can interact or use portions of the other systems or services.

Integrated Market Planning -The process of combining market research, advertising, direct marketing, sales, and public relations activities into one, cohesive united operational plan.

Integrated Marketing-Integrated marketing is the process of coordinating the promotion and selling of products or services through multiple communication channels.

Integrated Services Digital Network (ISDN)-A structured all digital telephone network system that was developed to replace (upgrade) existing analog telephone networks. The ISDN network supports for advanced telecommunications services and defined universal standard interfaces that are used in wireless and wired communications systems.

ISDN provides several communication channels to customers via local loop lines through a standardized digital transmission line. ISDN is provided in two interface formats: a basic rate (primarily for consumers) and high-speed rate (primarily for businesses). The basic rate interface (BRI) is 144 kbps and is divided into three digital channels called 2B + D. The primary rate interface (PRI) is 1.54 Mbps and is divided into 23B + D for North America and 2.048 Mbps and is divided into 30B + 2D for the rest of the world. The digital channels for the BRI are carried over a single, unshielded, twisted pair, copper wire and the PRI is normally carried on (2) twisted pairs of copper wire.

This diagram shows the different interfaces that are available in the integrated services digital network (ISDN). The two interfaces shown are BRI and PRI. These are all digital interfaces from the PSTN to the end customers network termination. Network termination 1 (NT1) equipment devices can directly connect to the NT1 connection. Devices that require other standards (such as POTS or data modems) require a terminal adapter (TA). This example shows that the NT2 interface works with the NT1 interface to allow the application layers (terminal intelligence) to communicate with the ISDN termination equipment.

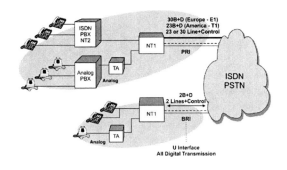

Integrated Services Digital Network (ISDN) System

Integrated Television (Integrated TV)-An integrated television is a viewing device that has the receiver, decoder and display assembly combined into one product.

Integrated TV-Integrated Television

Integrated Video Gateway (IVG)-An integrated video gateway is a device or assembly that can transform video that is received from a device or user into a format that can be used by another device or network. A video gateway can adjust the modulation, protocols and timing between two dissimilar communication devices or networks.

Integration-(1-component) The production of complete and complex circuits on a single chip, usually of silicon. (2-services) The combining of different services (such as voice, data, and video) onto a common communication system. (3-multiple networks) The process of interconnecting different types of networks through the use of portals or gateways.

Integration Framework-Integration framework is a structure that defines the functional parts of a system and how these parts will inter-operate with each other.

Integration Guide-An integration guide is a set of processes or procedures that can be used to assist a person or company to install and setup products or services with other systems.

Integration Tools-Integration tools are programs, systems or processes that can be used to help combine or link different or new applications or services.

Integrity Verification-Integrity verification is the process that confirms that a message or information has not been altered or changed from when it was originally created or transmitted.

Intellectual Property (IP)-Intellectual property is intellect that is produced that has some form of value. Intellectual property may be represented in a variety of forms and the copying, transfer and use of the intellect may be protected or restricted.

Intellectual Property Management and Protection (IPMP)-Intellectual property management and protection is a protocol that is used in the MPEG system to enable digital rights management (DRM).

Intellectual Property Rights (IPR)-Intellectual property rights are the privileges (such as exclusive use) for the owner or the assignee of the owner of patents, trademarks, copyrights, and trade secrets.

Intelligent Agent-An intelligent agent is a software program or service that performs actions or tasks that are based on rules or objectives that it has been provided. An example of an intelligent agent is a webbot that searches through web sites looking for specific types of projects and obtains the pricing and shipping details related to the product.

Intelligent Building-A building, often part of a commercial complex, in which local area networks, alarm circuits, and similar communications facilities are designed around a common communications strategy. Intelligent buildings are sometimes called smart buildings.

Intelligent Caching-Intelligent caching is a process by which information is moved to a temporary storage area to assist in the processing or future transfer of information to other parts of a processor or system and the decision to store the information is based on knowledge of the information type or expected future need for the information.

Intelligent Call Forwarding-Intelligent call forwarding changes the route of incoming calls to alternative destinations based on your preferred settings and the status of a telephone line or communication session when an incoming call is received.

Some of the advanced control features include transferring calls based on the time of day (home or work), amount of time an unanswered line is allow to ring before transfer (such as transfer to voice mail) or to transfer the call to another number where you last made a call (call following). Using intelligent call forwarding, you can setup your own hunt group (call rollover) anytime you want. The setup of intelligent call forwarding is usually accomplished via an Internet web page.

This figure shows an example of intelligent call forwarding that allows the destination of the call forwarding number to be changed based on time of day. In this example, these changes are made via web pages. This diagram shows that the user has setup intelligent call forwarding via a web page that changes the call forwarding number to an office number during normal working hours. When the call is received out of these hours, it is routed to a home office telephone.

-The dynamic processing of calls based on information obtained from or selected by the caller. This information can be used by automatic call distribution (ACD) systems to route calls to the next available agent, agents with specific qualifications, or other call centers.

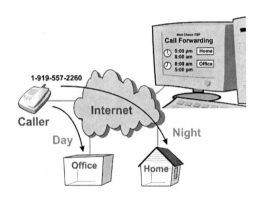

Intelligent Call Forwarding

Intelligent Content Distribution- Intelligent content distribution is the process of transferring content to one or more persons, companies or connection points where the distribution paths and/or storage within the network changes to better optimize the distribution of content reducing network resources.

Intelligent Modem-A modem that performs certain functions under computer control. As an example, an intelligent modem automatically dials a selected telephone number and hangs up when it detects a busy signal and may re-dial the same number. It is also called a smart modem.

Intelligent Network (IN)-A telecommunications network architecture that has the ability to process call control and related functions via distributed network transfer points and control centers as opposed to concentrating in switching system. (See. Advanced Intelligent Network.)

Interaction-The process by which a system accepts input, processes input requests, and (if necessary) returns appropriate response data to the originating terminal.

Interaction Channels-Interaction channels are communication links that are used to enable interactive controls to services.

Interactive Advertisements (iAD)- Interactive advertisements are media communication messages that allow a viewer to influence (control) or respond to the advertising message.

Interactive Advertising-Interactive advertising is the process of allowing a user to select or interact with an advertising message.

Interactive Advertising Bureau (IAB)- Interactive advertising bureau is a group that assists and tracks the progress of interactive advertising.

Interactive Applications-Interactive applications are software programs that are designed to perform operations using commands or information from users (such as a user selecting an option via a keyboard). Popular interactive applications include gaming, shopping and expandable advertising.

Interactive Content-Interactive content is the information contained within a message, call, or web site display that is created after the user or visitor has selected or interacted with the system or web site.

Interactive Entertainment-Interactive entertainment is a process that provides a diversion or an amusement experience where information is exchanged in both directions.

Interactive Programming Guide (IPG)- Interactive programming guides are an interface (portal) that allows a customer to preview, search and select from possible lists of available content media. IPGs can vary from simple program selections to interactive filters that dynamically allow the user to filter through program guides by theme, time period, or other criteria.

Interactive Services-Interactive services use two-way communication to successfully satisfy the information requirements of the end customer. Examples of interactive services include conversational service and online gaming (real time), messaging services and video on demand (near-real time) and retrieval services (non-real time).

Interactive Television (ITV)-Interactive television is the providing of video services that allows for the user to control part or all of the viewing experience. Interactive television has three basic types: "pay-per-view" involving programs that are independently billed, "near video-on-demand" (NVOD) with groupings of a single film starting at staggered times, and "video-on-demand" (VOD), enabling request for a particular film to start at the exact time of choice. Interactive television offers interactive advertising, home shopping, home banking, e-mail, Internet access, and games.

Interactive Video Services (IVS)- Interactive video services are processes that can automatically interact with a video viewer through the providing of audio and/or visual prompts to request information. The responses can be in the form of key presses, voice responses, or some other activity that can be sensed by the video delivery system. Interactive video responses are typically converted to digital form for processing the interactive video service platform.

Interactive Voice Response (IVR)-IVR is a process of automatically interacting with a caller through providing audio prompts to request information and store responses from the caller. The responses can be in the form of touch-tone(tm) key presses or voice responses. Voice responses are converted to digital information by voice recognition signal processing. IVR systems are commonly used for automatic call distribution or service activation or changes.

This figure shows a sample IVR system that is used to route an incoming call. When this call is received by the PBX system, an initial voice prompt informs the user of the system along with initial menu options. The user selects an option. This results in the playing of another prompt indicating new menu options. The user enters the data for the option and the IVR system retrieves data and creates a new verbal response.

Interactive Voice Response (IVR) Operation

Interactivity-Interactivity is the ability of a person to be able to identify, select and control some or all of the presentation of media.

Intercept Access Point (IAP)-A place in a communication network where information and/or content is intercepted for the purpose of passing it to a law enforcement agency.

Intercept Service-Intercept recording is the insertion of a recorded message into a call path that informs the caller that some action has occurred such as the "call cannot be completed as dialed" or the "system is busy."

Intercept Trunk Group-A trunk group that provides information about called numbers that are unassigned, changed, disconnected, or placed on trouble intercept.

Interception-Interception is the process of gaining access to information or a flow of data as it is transferred between a sender and a receiver.

Interchange-An interchange is an information transfer between trading partners in an electronic data interchange (EDI) system.

Interchange Control-Interchange control is the process of packaging business documents (such as an invoice or shipping document) into an electronic envelope that is sent through an electronic data interchange (EDI) system.

Interchange File Format (IFF)-The interchange file format is a common media format that was created by the Electronic Arts company that is used to create files that can be transferred between different types of programs and computers that read and decode files in the IFF format. Each IFF format begins with a 4 character ID code of FORM, LIST or CAT (the CAT code includes a space character at the end).

Interchangeable Numbering Plan Area (NPA) Codes-Codes in the format NNX previously used as central office codes but now available to supplement the supply of traditional numbering plan area (NPA) codes (NO/iX). (See also: Interchangeable Central Office Codes.)

Intercom Call-Intercom calls allows calls to directly communicate from one terminal toward another terminal. In a Bluetooth system, an intercom call between two terminals can be rapidly set up with gateway support if the two terminals are members of the same wireless user group (WUG).

Intercom Group-A group of telephone devices (the intercom group) that can be simultaneously accessed using an intercom feature. The use of the intercom group feature allows divisions or functional groups within a company to more effectively communicate with all the members of the group.

Intercompany Compensation-Intercompany compensation is the division of revenues between for services that are provided jointly.

Intercompany Settlements (ICS)-Financial settlements made between carriers for usage of each other networks (such as for collect calls or roaming charges).

Interconnect-Interconnection is the linking of communication devices or systems to each other.

Interconnection-Interconnection commonly refers to the connection of telephone equipment or communications systems to another network such as the public switched telephone network. Government agencies such as the FCC or department of communications usually regulate the interconnection of systems to the public switched telephone network.

This figure illustrates some of the different types of private to public telephone system interconnection. This diagram shows some groups of phone lines (e.g., dial line, Type 1) that provide limited signaling information (line-side) that primarily interconnect the PSTN with private telephone systems. Another group of lines (Type 2 series) are used to interconnect switching systems or to connect to advanced services (such as operator services or public safety services). The interconnection lines (trunk-side) provide more signaling information. Also shown is the type S connection that is used exclusively for sending

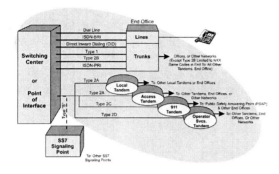

Private to Public Telephone System
Interconnection System

control signaling messages between switching system and the signaling system 7 (SS7) telephone control network.

Interconnection System-Any system of interconnection facilities used for distribution of media or programs to telecommunication systems and/or companies.

Interested Party-An interested party is a person or company who has made or responded to a request for proposal (RFP).

Interexchange-Services and functions relating to telecommunication originating in a LATA and terminating else-where. In common usage, it is synonymous with the term interLATA.

Interexchange Carrier (IEC)-A telephone service company that provides long distance (interLATA, interstate, and/or international telecommunications) service.

Interexchange Carrier (IXC)-Interexchange carriers (IXCs) interconnect local systems with each other. IXCs are also known as long distance carriers. In the US, from 1984 until 1997, IXC and LEC operating companies were legally required to refrain from engaging in directly competitive business operations with each other. Since 1997, one business entity can engage in both IXC and LEC business if it satisfies certain competitive legal rules. In Europe and throughout the rest of the world, the same PTT operators also usually provide inter-exchange service within their country. In any case, governments regulate how networks are allowed to interconnect to local and long distance networks.

For inter-exchange connection, networks as a rule connect to long distance networks through a separate toll center (tandem switch). In the United States, this toll center is called a point of presence (POP) connection.

Interexchange Circuit-A circuit or trunk between two exchanges that carry primarily message telecommunications service and WATS traffic, or a private line.

Interexchange Transmission Equipment-The combination of (a) interexchange cable and wire facilities, (b) interexchange circuit equipment and, (c) associated land and buildings.

Interexchange Trunks-Multiple channel communication lines (trunks) that interconnect switches between different exchanges.

Interface-(1-general) An interface is a device or circuit that is used to interconnect two pieces of electronic equipment (2-data) A shared boundary or point common to two or more similar or dissimilar command and control systems, subsystems, or other entities against which, at which, or across which useful information flow occurs. (3-network) A point of connection between a LAN through which information can be exchanged between systems.

Interface Equipment-The conversion equipment that enables circuits designed to one set of characteristics to communicate efficiently with circuits designed to meet different protocols and specifications.

Interim Bill-An interim bill is a list or charges or fees that are associated with a product or service that will be assigned to a customer or account. An interim bill may not contain all the charges that will be posted on the final bill.

Interim Standard (IS)-A designation of the American National Standards Institute usually followed by a number that refers to an accepted industry protocol; e.g., IS-95, IS-136, IS-54.

Interim Standard 124 Data Message Handler (IS-124)-A standard billing communication protocol that allows for the real time transmission of billing records between different systems. IS-124 messaging is independent of underlying technology and can be sent on X.25 or SS7 signaling links. The development of the standard is primarily led by CiberNet, a division of the cellular telecommunications industry association (CTIA).

Interim Standard 826 (IS-826)-An extension to the IS-41 intersystem signaling standard that allows for prepaid services.

interLATA-Telecommunication services that cross from a local access and transport area (LATA) into another LATA.

Interlata Service-telecommunications between a point located in a local access and transport area and point located outside such area.

Interlineation-An interlineation is a change in a line or sequence of words in a document.

Intermittently Aware Applications-Intermittently aware applications are software programs that are designed to perform operations using commands or information from users (such as a user at a keyboard), which also have the capability of communicating with other services or applications when a connection is available.

International Advertising-International advertising is the promoting of a product in a place other than its country of origin.

International Callback-International callback is a call processing service that reverses the connection of calls. International callback service is popular in countries that have high tariffs (fees) for outgoing (originating) international calls and have low tariffs for incoming (received) international calls. This process is divided into the call setup (dial-in) and callback stages. The international caller dials a number that provides access to the international callback service. This number may be local in the visited country or be an international number. The international callback gateway receives the call and prompts the caller to say or enter (e.g. by touch tone) the international number they desire to be connected to and the number they want the callback service to connect to. The international callback center then originates calls to both numbers and connects the two individuals to each other.

International Commercial Terms (Incoterms)-International commercial terms are a set of shipping and sales related terms that are used throughout the world.

International Denial-International denial is a call feature or restriction that prohibits or restricts a communication device from making international calls or connections.

International Gateway Facilities (IGFs)-Systems or equipment that provide access between telephone systems in different countries. International gateways may convert SS7 and other signaling formats between different signaling formats. These include ANSI standards, ITU standards, national variants of SS7 signaling standards, MF signaling, and R2 signaling. International gateways may also provide for transcoding services between mu-LAW PCM and A-LAW PCM speech coding.

This figure shows how two national SS7 systems interconnect using an international gateway between an ANSI based end office SSP in North America and an ITU based end office SSP in Asia. This example shows that an ANSI based SS7 system require address translation and circuit identifier code format changes as the messages are passed between the systems. The ANSI 24 bit destination point code (DPC) and origination point code (OPC) addressing must be translated to 14 bit DPC and OPC codes for the ITU system. It

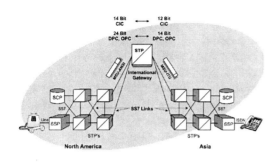

International Gateway Facilities

also shows that the 14 bit ANSI CIC code used in ISUP messages must be translated to 12 bit CIC codes used by the ITU system.

International Programming-International programming is the selection of shows and programs that are offered by a television network provider that provides media channels over multiple countries. An example of an international program is a television program that is created and network operator that is broadcasted by many local television systems which are located in multiple countries

International Record Carrier (IRC)-An international record carrier is a service provider of telecommunication services such as telex, data, voice, and video communications.

International Standards Organization (ISO)-An international body concerned with worldwide standardization for a broad range of industrial products, including telecommunications equipment. Members are represented by national standards organizations, such as ANSI in the United States. ISO was established in 1947 as a specialized agency of the United Nations.

International Telecommunication Union (ITU)-A specialized agency of the United Nations established to maintain and extend international cooperation for the maintenance, development, and efficient use of telecommunications. The union does this through standards and recommended regulations, and through technical and telecommunications studies. Based in Geneva, Switzerland, the ITLI is composed of two consultative committees: the International Radio Consultative Committee (CCIIR) and the Consultative Committee for International Telephony And Telegraphy (CCITT).

International Toll Free Service (ITFS)-International toll free is a service that allows callers to dial and telephone number from other countries without being charged for the call. The toll free call is billed to the receiver of the call.

Internet (Net)-The Internet is a public data network that interconnects private and government computers together. The Internet transfers data from point to point by packets that use Internet protocol (IP). Each transmitted packet in the Internet finds its way through the network switching through nodes (computers). Each node in the Internet forwards received packets to another location (another node) that is closer to its destination. Each node contains routing tables that provide packet forwarding information. The Internet evolved from ARPANET and was designed to allow continuous data communication in the event some parts of the network were disabled.

Internet Access-Internet access is the ability for a user or device to connect to the Internet. An example of Internet access is the requesting and connection to the Internet via a dial-up or broadband connection through an Internet service provider (ISP).

Internet Access Provider (IAP)-A company that provides an end user with data communication service that allows them to connect to the Internet. Internet access providers are also called Internet service providers (IAPs.)

Internet Addiction Disorder (IAD)-Internet addition disorder is a condition where an Internet user becomes addicted to interacting (surfing) the Internet.

Internet Address-An Internet address is a unique binary digital number that identifies a specific connection point within the Internet. An internet address is 32 bit for version 4 and 128 bits for version 6.

Internet Architecture Board (IAB)-A technical advisory group that is part of the Internet Society (ISOC) that manages Request for Comments (RFCs) publication standards and documents. The IAB also serves as an board to hear appeals and provides other services to the ISOC.

Internet Assigned Numbering Authority (IANA)-The Internet assigned numbering authority is a group that is responsible for the assignment and coordination of Internet addresses and key parameters such as protocol variables and domain names.

Internet Audio-Internet audio is the transmission of audio information in digital (discrete level) formats that are transferred using IP data packets (datagrams) which are sent through the Internet.

Internet Billing System-An Internet billing system is a combination of software and hardware that receives Internet usage information, grouping this information for specific accounts or customers, produces invoices, creating reports for management, and recording (posting) payments made to customer accounts.

Internet Broadcasting-Internet broadcasting is a process that sends digital voice, data, or video signals simultaneously to a group of people or companies in a specific geographic area or who are connected to the Internet. Internet broadcasting is typically associated with online radio channels that send the same digital audio signal to many receivers who connect to a common server or group of servers.

Internet Commerce (I-Commerce)-A shopping medium that uses a Internet to present products and process orders.

Internet Corporation for Assigned Names and Numbers (ICANN)-Internet corporation for assigned names and numbers is an international organization that is responsible for the issuing and managing of Internet addresses. The ICANN has assigned the responsibility of issuing and managing IP addresses and domain names to Internet registries.

Internet Information Server (IIS)-An Internet information server is a computer server that is used to provide data to users that request information through Internet protocols such as Hypertext Transfer Protocol (HTTP).

Internet IPTV-Internet IPTV is the delivery of digital television services over broadband Internet data connections.

Internet Key Exchange (IKE)-Internet key exchange is a set of commands and processes (a protocol) that used setup and exchange key information through data networks that use Internet protocol. IKE is defined in request for comments 2409 (RFC 2409).

Internet Marketing-Internet marketing is the process of providing or sending marketing and sales information using the Internet. Internet marketing systems generally combine email broadcast systems with tracking systems that can monitor the reception, opening, and response to email messages.

Internet Marketing Campaign-Any marketing activities that are designed to send messages to customers about products, services, and options via the Internet.

Internet Message Access Protocol (IMAP)-Internet message access protocol is a set of commands and processes that defines the access and storage procedures for electronic mail (e-mail) messages. IMAP is used with simple mail transfer protocol (SMTP) to move e-mail messages between email servers and mailboxes. IMAP is defined in RFC 2060.

The IMAP protocol defines message headers to allow the recipient to better search, select, and download specific messages or parts of messages. IMAP also includes security authentication procedures.

Internet Number-The dotted-quad address used to specify a certain system within the Internet.

Internet on Television (Internet on TV)- Internet on TV is the ability of a user to access the Internet through the use of television equipment.

Internet on TV-Internet on Television

Internet Phone (IPhone)-One of the first commercial Internet telephones that was introduced in February 1995. The IPhone product was developed by VocalTec Inc.

Internet Portal-An Internet portal is an Internet web site that acts as an interface between a user of a product or services and a system that provides the product or services.

Internet Protocol (IP)-A low-level network protocol that is used for the addressing and routing of packets through data networks. IP is the common language of the Internet. The IP protocol only has routing information and no data confirmation rules. To ensure reliable data transfer using IP protocols, higher level protocols such as TCP are used. IP protocol is specified in RFC-791.

This protocol defines the packet datagram that hold packet delivery addressing, type of service specification, dividing and re-assembly of long data files and data security. IP protocol structure is usually combined with high-level transmission control protocols such as transaction control protocol (TCP/IP) or user datagram protocol (UDP/IP).

Internet Protocol Address (IP Address)- The address portion of an Internet Protocol (IP) packet. For IP version 4, this is a 32-bit address and for IP version 6, this is a 128 bit address. To help simplify the presentation of IPv4 addresses, it is common to group each 8 bit part of the IP address is a decimal number separated from other parts by a dot(.), such as: 207.169.222.45. For IPv6 it is customary to represent the address as eight, four digit hexadecimal numbers separated by colons, such as 1234:5678:9000:0D0D:0000:5678:9ABC:8777.

This figure shows how different types of data network addressing systems can co-exist. This diagram shows a data connection that is com-posed of several parts. An end-user is connected to an application server through a company Ethernet network. The computers network interface card (NIC) has an address unique to the Ethernet hub. The Internet address is included as part of the data message after the Ethernet address. The company's network is connected to an ISP by a high-speed frame relay connection. The frame relay access device (FRAD) has a unique identifier to the ISP. The ISP connects the data connection via asynchronous transfer mode (ATM) to the ASP.

Internet Protocol Billing (IP Billing)-IP billing is the recording and processing of Internet protocol events for billing purposes.

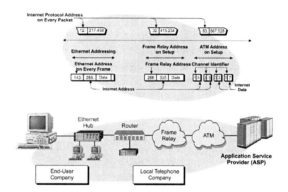

Internet and Network Numbering System

Internet Protocol Cable Television (IPCATV)-Internet protocol cable television (IP CATV) is the process of providing television (video and/or audio) services through the use Internet protocol (IP) delivered over cable television networks. These IP networks initiate, process, and receive voice or multimedia communications using IP protocol.

Internet Protocol Centrex (IP Centrex)- IP Centrex is the providing of Centrex services to customers via Internet protocol (IP) connections. IP Centrex allows customer to have and use features that are typically associated with a private branch exchange (PBX) without the purchase of PBX switching systems. These features include 3 or 4 digit dialing, intercom features, distinctive line ringing for inside and outside lines, voice mail waiting indication, and others.

Internet Protocol Connectivity (IP Connectivity)- IP connectivity is the ability of a device or system to setup communication sessions using Internet protocol (IP).

Internet Protocol Datacasting (IP Datacasting)- Internet protocol datacasting is the transmission of IP data (voice, data or video IP packets) to one or more people or companies who are connected to the IP addressable data network.

Internet Protocol Detail Record (IPDR)- A data record containing information related to an IP-based communication session. This information usually contains identification information of the users of the service, types of services used, quantity measurement unit type (e.g. kilobytes or time), quantities of services used, Quality of Service parameters, and the date/time (usually relative to GMT) the services were used.

Internet Protocol Device Control (IPDC)- Internet protocol device control is a set of commands and processes that are used to control devices that can communicate using Internet protocol.

Internet Protocol Phone (IP Phone)- An Internet protocol phone (IP phone) is a device (a telephone set) that converts audio signals and telephony control signals into Internet protocol packets. These stand alone devices plug into (connect to) data networks (such as the Ethernet) and operate like traditional telephone sets. Some IP Telephones create a dialtone that allows the user to know that IP telephone service is available.

Internet Protocol Private Branch Exchange (IPBX) or (IP PBX)- A private local telephone system that uses Internet protocol (IP) to provide telephone service within a building or group of buildings in a small geographic area. IPBX systems are often local area network (LAN) systems that interconnect IP telephones. IPBX systems use a IP telephone server to provide for call processing functions and to control gateways access that allows the IPBX to communicate with the public switched telephone network and other IPBX's that are part of its network. IPBX systems can provide advanced call processing features such as speed dialing, call transfer, and voice mail along with integrating computer telephony applications. Some of the IPBX standards include H.323, MGCP, MEGACO, and SIP.

IP PBX represents the evolution of enterprise telephony from circuit to packet. Traditional PBX systems are voice-based, whereas their successor is designed for converged applications. IP PBX supports both voice and data, and potentially a richer feature set. Current IP PBX offerings vary in their range of features and network configurations, but offer clear advantages over TDM-based PBX, mainly in terms of reduce Opex (operating expenses).

Internet Protocol Security (IPSec)- A part of the Internet Protocol that helps to ensure the privacy of user data. IPSec is part of the next generation internet, IPv6. IPSec is defined in RFC 1827.

Internet Protocol Set Top Box (IP STB)- An IP Set Top box is an electronic device that adapts IP television data into a format that is accessible by the end user. The output of an IP set top box can be a television RF channel (e.g. channel 3), video and audio signals or digital video signals. IP set top boxes are commonly located in a customer's home to allow the reception of IP video signals on a television or computer.

This figure shows a functional block diagram of an IP STB. This diagram shows that an IP STB typically receives IP packets that are encapsulated in Ethernet packets. The IP STB extracts the IP packets to obtain the transport stream (TS). The channel decoder detects and corrects errors and provides the transport stream to the descrambler assembly. The descrambler assembly receives key information from either a smart card or from an external conditional access system (e.g. via a return channel). Using the key(s), the STB can decode the transport stream and the program selector can extract the specific program stream that the user has selected. The IP STB then demultiplexes the transport stream to obtain the program information. The program table allows the IP STB to know which streams are for video, audio and other media for that program. The program stream is then divided into its elementary streams (voice, audio and control) which is supplied to a compositor that create the video signal that the television can display.

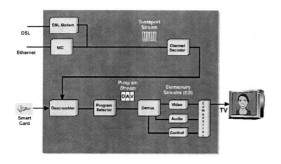

IP Set Top Box

Internet Protocol Suite (IP Suite)-IP suite is a group of network protocols that can interoperate with each other to provide packet data based communication applications and services. The layers of protocol suite include physical layer, network (or routing) layer, transport (or session) layer, and application layer.

This protocol suite is overseen by the Internet Engineering Task Force (IETF). Key protocols included in the Internet Protocol Suite include Internet Protocol (IP), Transaction Capabilities Protocol (TCP), and User Datagram Protocol (UDP). There are many other protocols that are part of the Internet Protocol suite.

Internet Protocol Telephony (IP Telephony)-IP telephone systems provide voice or multimedia communication services through the use Internet protocol (IP) networks. These IP networks initiate, process, and receive voice or multimedia communications using IP protocol. These IP systems may be public IP systems (e.g. the Internet), private data systems (e.g. LAN based), or a hybrid of public and private systems.

Internet Protocol Television (IPTV)-Internet protocol television (IPTV) is the process of providing television (video and/or audio) services through the use Internet protocol (IP) networks. These IP networks initiate, process, and receive voice or multimedia communications using IP protocol. These IP systems may be public IP systems (e.g. the Internet), private data systems (e.g. LAN based), or a hybrid of public and private systems.

Internet Protocol Version 4 (IPv4)-A revision of Internet protocol that uses 32 bit addressing.

Internet Protocol Version 6 (IPv6)-IP version 6 is a network packet routing protocol that uses 128 bit address. IPV6 is an enhanced version of Internet protocol version 4 (IPv4) that was developed primarily to correct shortcomings of IPv4 such as the 32 to bit address that limited the maximum number of devices that could be addresses and to extend the capabilities of IP to meet the demands of the future such as improved quality of service (QoS) capabilities. IPv6 addresses are denoted as 8 hexadecimal numbers, separated by colons. A typical address will look like this: 0800:5008:0000:0000:0000:1005:AABC:AD46 A short-hand notation that replaces one set of consecutive zeros with colons (::) may also be used. The above address can also be denoted by: 0800:5008::1005:AABC:AD46

IPv6 utilizes a hierarchical address, called an Aggregatable Global Unicast Address Format (AGUAF). In this format, each IP address is built by concatenating a Top-Level Aggregation (TLA) ID, a Next-Level Aggregation ID (NLA) and a Site-Level Aggregation ID (SLA) and an interface ID. This enables the IPv6 address space to be assigned in a logical manner by multiple address assignment authorities while still guaranteeing that all hosts have unique IP addresses and the addresses can be used to easily route packets without requiring switches to maintain enormous routing tables.

The IPv6 header has been simplified with the introduction of extension headers. The basic IPv6 header contains just seven fields such as hop count and destination IP address. Also included is a Next Header field, which points to the next header in the packet. This greatly simplifies the logic required to parse the packet size in hosts and routers. IPv6 natively supports functions to discover neighbors, assign IP addresses dynamically and to identify multicast participants.

The most useful application of IPv6 is in next generation wireless phone networks. Many companies are moving to IP based networks to replace existing cellular technology. Over the next few years the need for IPv4 and IPv6 to co-exist will be an important factor in the deployment of IPv6 networks.

Internet Protocol Video (IP Video)-IP video is the transfer of video information in IP packet data format.

This figure shows how a basic IP television system can be used to allow a viewer to have access to many different media sources. This diagram shows how a standard television is connected to a set top box (STB) that converts IP video into standard television signals. The STB is the gateway to a IP video switching system. This example shows that the switched video service (SVS) system allows the user to connect to various types of television media sources including broadcast network channels, subscription services, and movies on demand. When the user desires to access these media sources, the control commands (usually entered by the user by a television remote control) are sent to the SVS and the SVS determines which media source the user desires to connect to. This diagram shows that the user only needs one video channel to the SVS to have access to virtually an unlimited number of video sources.

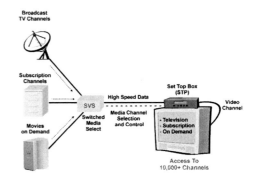

IP Video System

Internet Radio-Internet radio is the sending or broadcasting of digital audio signals through IP data networks (such as the Internet). Internet radios can be software programs that operate on multimedia computers or they can be dedicated devices (an audio player with a data plug instead of a radio antenna).

Internet Reference Model-The Internet reference model is a standard layer model was developed by the United States Department of Defense (DOD) to build a robust packet data network. The Internet reference model helps to standardize the inter-connection of computers and data terminals to their applications, regardless of their type or manufacturer. The protocols specify four layers: network access (physical), inter-networking (IP addressing), transport (flow control) and application (user presentation). Each layer performs specific functions for data exchange and is independent of the other layers.

Internet Relay Chat (IRC)-Internet relay chat is an Internet protocol that was developed to allow for direct instant messaging (IM) between members of chat group.

Internet Service Provider (ISP)-An Internet service provider (ISP) is a company that receives and converts (formats) information to and from Internet connections to Internet end users. An ISP purchases a high-speed link to the Internet and divides up the data transmission to allow many more users to connect to the Internet.

Internet Shopping Mall-An Internet shopping mall ("cybermall") is a group of online storefronts.

Internet Software-Software that is used to manage or communicate with or through the Internet.

Internet Streaming Server-An Internet streaming server is a computer or a device that efficiently and effectively performs continuous transmission (streaming) of digital media through the Internet.

Internet Tax Freedom Act (ITFA)-The Internet tax freedom act is a set of laws that was enacted in 1998 that placed a moratorium on taxes that would apply to Internet based services.

Internet Telephone (IP Telephone)-An IP telephone is a device that is specifically designed to communicate telephone signals through the Internet without the need for a voice gateway. Internet telephones contain embedded software that allows them to initiate and receive calls through the Internet using standard protocols such as H.323 or SIP.

Internet Telephony-Telephone systems and services that use the Internet to initiate, process and receive voice communications.

This figure shows how calls can be made between company telephones through the Internet to standard telephones anywhere in the world. In this example, an existing company PBX telephone system in Paris is connected to the Internet through a voice gateway. Each PBX telephone is registered with a public Internet telephone service provider (ITSP) that is located in New York. The ITSP is able to provide connections to gateways located throughout the world. In this example, when the caller in Paris dials a telephone number in Cairo, the dialed digits are first routed to the ITSP in New York. The ITSP server searches for the telephone number in its address list. If it finds that it has access to a voice gateway (the ITSP may not actually own the voice gateway) that is connected to the Internet near the destination telephone number in Cairo, it informs the destination voice gateway that an incoming call is to be received. The ITSP then provides the PBX gateway with the data net-

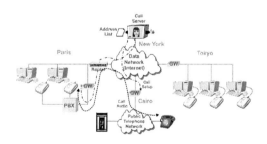

Calling Through the Internet

work address of the destination voice gateway in Cairo. The call can then proceed from the PBX telephone, through the PBX voice gateway, through the Internet, through the destination voice gateway (in Cairo), to the dialed telephone.

Internet Telephony Service Provider (ITSP)-Internet Telephony Service Providers (ITSPs) are companies that provide telephone service using the Internet. ITSPs setup and manage calls between Internet telephones and other telephone type devices.

An ITSP coordinates Internet telephone devices so they can use the Internet as a connection path between other telephones. ITSPs are commonly used to connect Internet telephones or PC telephones to telephones that are connected to the public telephone network. This is accomplished by using gateways. Gateways convert packets of audio data from the Internet into standard telephone signals.

This figure shows how an ITSP sets up connections between Internet telephones and telephone gateways. The ITSP usually receives registration messages from an Internet telephone when it is first connected to the Internet. This registration message indicates the current Internet address (IP

address) of the Internet telephone. When the Internet telephone desires to make a call, it sends a message to the ITSP that includes the destination telephone number it wants to talk to. The ITSP reviews the destination telephone number with a list of authorized gateways. This list identifies to the ITSP one or more gateways that are located near the destination number and that can deliver the call. The ITSP sends a setup message to the gateway that includes the destination telephone number, the parameters of the call (bandwidth and type of speech compression), along with the current Internet address of the calling Internet telephone. The ITSP then sends the address of the destination gateway to the calling Internet telephone. The Internet telephone then can send packets directly to the gateway and the gateway initiates a local call to the destination telephone. If the destination telephone answers, two audio paths between the gateway and the Internet telephone are created. One for each direction and the call operates as a telephone call.

Internet Television-Internet television (also called broadband television) is the delivery of digital television services over broadband Internet connections. They may be able to control and guarantee the quality of television services if the underlying broadband connections have enough bandwidth. Internet service providers or media management companies usually provide unmanaged IPTV systems through broadband Internet connections.

Internet Television Service Provider (ITVSP)-Internet Television Service Providers (ITVSPs) are companies that provide television or video services that connect through the Internet or other types of data networks. ITVSPs setup and manage television services between multimedia computers, televisions with adapters, or integrated IP television devices and media sources.

An ITVSP coordinates Internet television

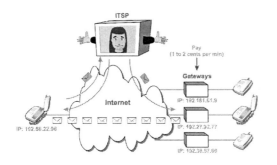

Internet Telephony Service Provider (ITSP)
Operation

devices so they can use the Internet as a connection path between television media sources. ITVSPs are commonly used to connect end users to television content providers that use media gateways. Media gateways convert packets of audio data from the television source into packets that can be routed through data networks to end users.

Internet Time Protocol (ITP)-Internet time protocol is a set of commands and processes that allow the requesting and delivery of time information through Internet protocol (IP) networks. ITP is defined by RFC 868.

Internet Video-Internet video is the transfer of video information through the Internet in IP packet data format.

Internet2-A second generation of the Internet that uses a high-speed backbone communications network. The Internet system is a result of the next generation Internet (NGI) initiative that is sponsored by the United States government. Internet2 is seen as the way to deliver multimedia content (e.g. video on demand) through the Internet.

Interoperability-Interoperability is the condition achieved among communications and electronics systems or equipment when information or services can be exchanged directly between them, between their users, or both.

Interoperability Testing-Interoperability testing is the performing of measurements or observations of a device, system or service to determine if the device will operate with other devices of a similar type or with devices that have been designed and tested to specifications (e.g. industry standards).

Interoperator Settlement-Interoperator settlement is the financial clearing process that is used to settle the usage charges for services they provide to visiting customers and the fees that other carriers have charged for services provided to their customer that have visited other systems.

Interoperator Tariff (IOT)-Interoperator tariffs are the fees charged by a system operator to another system operator for the services

used by a visiting subscriber within its system.

InterPBX-The direct connection of calls between PBX switching systems through the use of dedicated connection lines (tie lines).

Interpreter-An interpreter is a computer program that converts (translates) high-level language statements into language (e.g. machine code) that the computer can understand. Because the interpreter program translates instructions to computer instructions one line at a time, the execution of an interpreted program may take longer than executable programs, which are written, in the native (e.g. binary) computer processing language.

Interrupt-An interrupt is an event signal that is used to inform a processing system that suspension of the operation is required so another sequence of instructions can be processed. Interrupts are usually created by device under a processor's control, such as an accessory device, that interrupts normal processing to gain rapid processing response. Interrupts usually cause software processing routines to branch from their current processing steps to an interrupt service routine (ISR). Interrupts are sometimes classified to different types. These include internal hardware, external hardware, and software interrupts. Because there can be several types of interrupts in a system, interrupts can be prioritized. Interrupts used by Intel products allows up to 256 prioritized interrupts. Of these, the first 64 interrupts are reserved for use by the system or the operating system.

Interstate-The crossing of one or more state lines.

Interstate-The crossing of one or more state lines.

Interstate Access Surcharge-Interstate access surcharge are fees that are collected to help develop access systems that allow customers to select alternative communication access providers.

I

Interstate Communication-The providing of services, communication signals, or information services that cross one or more state lines.

Interstitial Ad-An interstitial ad is the temporary display of advertising messages "in between" the requesting and opening or closing of a web page. The interstitial ad is usually setup to display for a short period of time (e.g. 5 seconds). Some interstitial ads include a link that allows the viewer to bypass the interstitial ad message.

Interstitial Metadata-Interstitial metadata is information (data) that describes the attributes of other data that occurs between the beginning and end of a media or data file. Interstitial metadata or meta tags can be used to describe segments of program within a media file.

Intertoll Circuits-Intertoll circuits are communication links between toll centers in different toll center areas.

Interview Based Authentication-Interview based authentication is a process gathering information via questions or requests for information from a user by another person or company (e.g. such as a service provider) that verifies the identity of the user.

Interworking-Interworking is the process of adapting the communications between two different types of networks. This may include circuit switched, packet switched or messaging services.

Interworking Function (IWF)-Interworking functions are systems and/or processes that attach to communications network that are used to process and adapt information between dissimilar types of network systems.

Interworking Profile-An interworking profile is the processes, protocol adaptations and call processing requirements that are used to adapt the operation of communication systems between two different types of networks.

Intranet-A private network that is used within a company to provide company information to employees. Intranets may be connected to vendors and customers through private data connections or via public Internet connections. When Intranets are connected to the Internet, they are commonly connected through firewalls to protect the company's internal data.

Intrastate Access Services Tariff-Intrastate access service tariff is a rate plan that applies to services that are supplied across state boundaries.

Intrastate Service-A broad category that includes all telecommunications services and/or activities that are covered by an intrastate or state tariff. Such services normally include state toll, local exchange, extended area service, multiple message unit, most optional calling service plans, and access services under state jurisdiction.

Invalid Clicks-Invalid clicks are the selection of links ("clicks") by users or systems (such as automated scripts) for a purpose other than its intended purpose.

Invalid Record-An invalid record is a group of information (a record) that has some of its information mission or in a format that is not usable by a service or an application.

Inventory-Inventory is the amount of products or goods available for shipping or to fill an order. Inventory types include physical goods and electronic goods such as advertising space on a web page.

Inventory Available Date-Inventory available date is the day on which equipment and facilities will be available for sale or reuse.

Inventory Financing-Inventory financing is the process of using inventory value to obtain funds or resources that will be used to run or develop the business.

Inventory Inquiry-An inventory inquiry is a message that requests to know if a certain amount of inventory exists or is available.

Inventory Management-Inventory management is the process of identifying the quantity of inventory assets (stock) that are available, determining how many will be needed in the future and coordinating the ordering of the stock items. An Inventory management system may store information regarding the availability of products, services or content such as physical equipment (shelf, card, port, route), logical equipment (bandwidth, disk space available etc), and resources (ports/circuits of the equipment) that are in use/reserved/available. Inventory management can be used (queried) to process news to determine if or not the requested service can be supplied.

Inverse Multiplexer (IMUX)-A device that divides a single telephone or data communication channel into two or more channels to be transported over multiple communication links. Inverse multiplexing may be in the form of frequency division (e.g. multiple radio channels on a coax line), time division (e.g. slots on a T1 or E1 line) , or code division (coded channels that share the same frequency band) or combinations of these.

Inverse Multiplexing-The combining of information signals received on multiple communications channels to form a higher speed communication channel than is possible on a single independent communication channel. Inverse multiplexing has been used on wireless communication systems to allow high-speed digital video signals to be sent over cellular radio channels that have a limited maximum data transmission rate.

Inverter-An inverter is a device or electrical circuit that converts a direct current (DC) into an alternating current (AC).

Investor Relations (IR)-Investor relations are a part of a company that communicates or coordinates interaction with investors and potential investors.

Invite Message-A message that is used to invite a person or device to participate in a communication session. The invite message is defined in session initiation protocol (SIP).

Invoice-An invoice is a grouping of billing charges for a time period or event that are associated with a specific user or account.

Invoice Association-Invoice association is the linking or assigning of an invoice to an account or transaction.

Invoice Format-Invoice format is the organization and presentation of objects on an invoice (such as item name, product ID, unit cost and order details). Invoice formats can vary based on the type of product or service offered by the company. An example of an invoice format difference is the inclusion of a quantity field on product invoice format and the inclusion of an hours field on a service invoice.

Invoice Record-A telecommunications data record that contains control counts or totals that describe an accompanying data set. Excluded are indexes and header and trailer information.

Invoice Statement-An invoice statement identifies outstanding (unpaid) invoices that are associated with a user, device or account. This figure shows a sample invoice. This invoice statement provides the customer account information (name, address, and

Invoice Statement

account number), invoice charge totals, along with detailed billing information. This example shows that the customer pays recurring charges (monthly fees) plus additional charges such as taxes and communication costs that are not part of their rate plan. The detailed charges identify the category of the charge (rate), the amount of usage (time), and any additional charges (surcharges) that may apply.)

Invoicing-Invoicing is the process of gathering (aggregating) and adding up all of the billing records associated with a specific account during a billing cycle in a bill pool, applying recurring charges (e.g. monthly charges) and totaling all the charges.

Involuntary Churn-Involuntary churn is the termination of service or user accounts that occurs when a carrier decides to disconnect service from an existing customer regardless if the customer wants to disconnect service or not.

IO-Insertion Order
IOC-Initial Operational Capability
IOT-Interoperator Tariff
IP-Intellectual Property
IP-Internet Protocol

IP Access Network-Access network is a portion of a communication network (such as the public switched telephone network) that allows individual subscribers or devices to connect to the core network. Examples of networks that can be used to provide IP connections include a digital subscriber line (DSL), Cable TV, Wireless broadband (WBB) and optical fiber.

IP Address-Internet Protocol Address

IP Address Suppression-IP address suppression disables tracking, counting events or operations for visitors that have specific IP addresses. IP address suppression may be used to disable counting of visitors that come from a company's own IP address so that employee web site visits are not counted in marketing statistics.

IP Backbone-IP backbones are the core infrastructure of a network that connects several major IP network components together.
IP Billing-Internet Protocol Billing
IP Centrex-Internet Protocol Centrex
IP Centrex System-A system that provides Centrex services to customers using Internet protocol (IP) connections. IP Centrex allows customers to have and use features that are typically associated with a private branch exchange (PBX) without the purchase of PBX switching systems.

This figure shows a basic IP Centrex system that allows a local exchange company (LEC) in New York City to provide Centrex services to a company in Los Angeles. In this diagram, the LEC in New York City uses a class 5 switch to provide for plain old telephone (POTS) and Centrex services to their local customers. The Centrex software is installed in the switch and existing Centrex customers in the local area continue to connect their telephone stations directly to the Class 5 switch. To provide Centrex services to new customers located outside the geographic area, the LEC has installed a network gateway in New York that can communicate with the customer gateway in Los Angeles. Because the network gateway converts all the necessary signaling commands to control and communicate with the

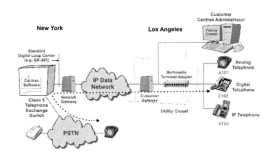

IP Centrex System

customer gateway, the class 5 switch does not care if the customer gateway is in Los Angeles or Tokyo. It simply provides the Centrex services as the users request.

IP Centric-IP centric systems use Internet Protocol (IP) for the central part or core focus of their system.

IP Connectivity-Internet Protocol Connectivity

IP Convergence-IP convergence is the process of adapting one or more transmission mediums (such as radio packet or circuit data transmission) into Internet protocol data transmission formats.

IP Core-An IP core network is a central network portion of a communication system that uses Internet protocol as the basis of its operation. The core network primarily provides interconnection and transfer between edge networks.

IP Datacasting-Internet Protocol Datacasting

IP Datagram-A packet of data that contains the ultimate network destination along with some control information and travels within (encapsulated) in the data portion of other network packets that are used to transport the IP datagram towards its ultimate destination.

IP Enabled Devices-IP enabled devices are electronic devices that have Internet protocol communication capability.

IP Enabled PBX-A non-IP PBX (such as a traditional TDM PBX) which allows for the support of IP phones or IP interconnect through the use of a special line or trunk card which converts TDM signals to IP signals. This is a common approach in migrating users of TDM PBXes to IP telephony technologies. This term often implies that IP to IP telephony audio may pass through a TDM switch.

IP Multicast-An Internet protocol that is used to broadcast the same message to multiple recipients. An IP multicast message is transferred to all the members within predefined group.

IP Multimedia Subsystem (IMS)-IP multimedia subsystem is service based architecture that uses Internet protocol (IP) based systems to provide enhanced multimedia services. IMS evolved from the evolution of the 3rd generation mobile telephone standards that enabled users to access multimedia services using any type of access network that could use Internet protocols.

IP Phone-Internet Protocol Phone

IP STB-Internet Protocol Set Top Box

IP Suite-Internet Protocol Suite

IP Telephone-Internet Telephone

IP Telephony-Internet Protocol Telephony

IP Television Service (IPTV)-IP television is the transmission of digital video and audio through data networks, usually through the Internet. IP television services may be on a subscription basis (paid for by the recipient) or may be funded by commercials or government agencies. IP television broadcasters transmit multimedia data signals to end users or to distribution points that redirect the digital television signals to end users.

IP Video-Internet Protocol Video

IPBX or IP PBX-Internet Protocol Private Branch Exchange

iPBX System-Internet protocol private branch exchange (IPBX) systems use Internet protocols to provide voice communications for companies. IPBX systems can be separate from data network or they may share the data network systems. When the iPBX system shares the local area network (LAN), it may be called LAN Telephony or TeLANophy.

IPC-ISDN To POTS Converter

IPCATV-Internet Protocol Cable Television

IPDC-Internet Protocol Device Control

IPDR-Internet Protocol Detail Record

IPG-Interactive Programming Guide

IPhone-Internet Phone

IPMP-Intellectual Property Management and Protection

IPPV-Impulse Pay Per View

IPR-Intellectual Property Rights

IPSec-Internet Protocol Security

I

IPTV-Internet Protocol Television

IPTV-IP Television Service

IPTV Accessories-IPTV accessories are devices or software programs that are used with IPTV systems or services. Examples of IPTV accessories include remote controls, gaming controllers and other human interface devices that are specifically designed to be used with IPTV systems and services.

IPTV Client-An IPTV client is a hardware device or software program that is configured to request and create (render) IP television services from a network.

IPTV Distribution Equipment-IPTV distribution equipment is the hardware (physical products) that route, switch or temporarily store and forward IPTV signals between the head end and the customer's IPTV equipment (such as a set top box). IPTV distribution includes IPTV DSLAM (with IGMP and IP Multistream capability), video distribution servers, IPTV edge routers and video switches.

IPTV Distributor-An IPTV distributor is a system or company that distributes information thorough a Internet protocol (IP) transmission system.

IPTV Ecosystem-The IPTV ecosystem is interrelationship of systems, services and business processes that influence the consumer needs and the ability to deliver voice, data and video services over IPTV networks. The IPTV industry is characterized by a high degree of flexibility to the operator in terms of such components of the end-to-end system from video servers and encoders, through network delivery, middleware and conditional access systems. In addition to the network distribution, the IPTV provider has multiple home distribution choices including coax, cat-5, power line, phone line or wireless. Each ecosystem can be tailored for the requirements of the region or operator to enable the level of functionality, performance and security that is appropriate.

IPTV Headend-An IP television headend is the part of an IP television system that selects and processes video signals for distribution into an IP distribution network. A variety of equipment may be included at the headend including satellite dishes to receive signals, content decoders and encoders, media servers and media gateways.

IPTV Interoperability Forum (IIF)-IPTV interoperability forum is a group that was created by the Alliance for Telecommunications Industry Solutions (ATIS) to review requirements for IPTV systems.

IPTV Network-IP television networks are television distribution systems that use IP networks to deliver multiple video and audio channels. IP Television networks are primarily constructed of computer servers, gateways, access connections, and end user display devices. Severs control the overall system access and processing of calls and gateways to convert the IP television network data to signals that can be used by television media viewers.

IPTV Roaming-IPTV roaming is the capability for an IPTV subscriber to access the content (e.g. television programs) offered by their IPTV service provider through other communication systems (such as broadband data connections in hotel rooms). While it is desirable for IPTV subscribers to roam without loosing access to some programming or functionality (e.g. remote control navigation), IPTV roaming may limit access to some programs due to content distribution restrictions and some navigation features may change.

IPv4-Internet Protocol Version 4

IPv6-Internet Protocol Version 6

IR-Investor Relations

IRC-International Record Carrier

IRC-Internet Relay Chat

IS-Information Separator

IS-Interim Standard

IS-124-Interim Standard 124 Data Message Handler

IS-826-Interim Standard 826

ISDN-Integrated Services Digital Network

ISDN Line Emulator (ILE)-An ISDN line emulator creates the signals necessary to simulate the digital signals that are created by an ISDN telephone device. An ILE senses and creates signaling messages on the ISDN D channel.

ISDN To POTS Converter (IPC)-A device that converts an ISDN basic rate interface (BRI) into a POTS analog telephone interface.

ISDN User Adaptation Layer (IUA)-ISDN User Adaptation layer is used to transport ISDN user signaling (Q.931) over IP between two signaling endpoints. The use of an IUA protocol eliminates the use of the MTP protocol portion in a signaling system.

ISDN User Part (ISUP)-The functional part of SS7 protocol that provides the call processing control signaling functions that are required to support basic communication services. Although is it based in the Integrated Services Digital Network (ISDN) signaling functions, ISUP is used for analog and digital call processing functions.

ISO-International Standards Organization

ISO 9000-Quality management standards that are defined by the international standards organization (ISO) to help companies define and improve their production and service processes. These standards require companies to define their own product and service processes and demonstrate to independent auditors that their processes provide the ability to track the quality of the products or services they provide.

ISO Date Format-ISO date format is an international standard that defines how date information should be structured. The ISO date format defines that the day should be referenced by day-month-year.

ISOC-Isochronous

Isochronous (ISOC)-Isochronous communication is the process of sending data between communication devices in continuous form (equal transmission time for all data). Isochronous signals are used in systems that require continuous data to be sent at specific time intervals (such as digital audio communication systems).

ISP-Internet Service Provider

Issuance-Issuance is the process of authorizing, selling and delivering.

Issue-An issue is a problem or identifiable attribute that can be analyzed and investigated as a cause of a condition that influences a performance area.

Issuing Authority-An issuing authority is an entity (such as an agency or company) that is authorized (e.g. by a government or company) to issue a regulation or document (such as a request for proposal).

Issuing Bank-An issuing bank is the financial institution that processes and provides the financial instrument (such as providing a credit card).

Issuing Carrier-A service provider (carrier) that files its own tariff with a regulatory agency for a given service offering.

Issuing Certificates-Issuing certificates is the process of gathering identification and qualifying information from a potential certificate holder, creating authentication keys and processes that are associated with a certificate holder and providing a unique identifying authenticated document (certificate) to the certificate holder.

ISUP-ISDN User Part

IT-Information Technology

Itemized Billing-Itemized billing is the process of including service usage details (such as listing all telephone numbers called and their duration) on an invoice.

Iterative Loop-An iterative loop is a sequences of instructions in that are repeated in a software routine.

ITFA-Internet Tax Freedom Act

ITFS-International Toll Free Service

ITP-Internet Time Protocol

ITSP-Internet Telephony Service Provider

ITU-International Telecommunication Union
ITV-Interactive Television
ITVSP-Internet Television Service Provider
IUA-ISDN User Adaptation Layer
IUM-Impacted User Minutes
IVG-Integrated Video Gateway
IVR-Interactive Voice Response
IVS-Interactive Video Services
IWF-Interworking Function
IXC-Interexchange Carrier

J

J2ME–Java 2 Micro Edition

J2ME–Java Version 2 Mobile Edition

Jack-A jack is a receptacle or connector that makes contact with a mating connector to allow the transfer of signals through the jack to cables or assemblies.

Jack Positions-A Jack position is a labeling system that identifies the location of a connection point (a jack) to a user or technician.

Jack Type–Jack type is the physical characteristics of a connection device. Examples or jack types include RJ-11, RJ-45 or BNC.

Java-An object-oriented programming language that works with a wide variety of computers. Created by Sun Microsystems, Java adds animation and interactivity to Web pages, granted you to have a Java-enabled browser. Java technology is a portable, object-oriented language that is well suited for web-based (Internet) and platform-independent applications. Java is a high level language and is architecturally neutral as it can operate on almost any underlying operating system.

Java 2 Micro Edition (J2ME)–Java 2 Micro Edition suns Java platform for developing applications for various consumer devices, such as set-top boxes, Web terminals, embedded systems, mobile phones and pagers.

Java Applet-A Java applet is a small software program written in the Java programming language. Java applets are designed to add specific capabilities or functions to a Web page. The applet is transferred into the user's computer and it executes (processes information) within the browser.

Java applets are inserted into a Web page using the <APPLET> tag, which indicates the initial window size for the applet, various parameters for the applet and the location from which the browser can download the bytecode file. Java applets are inserted into a Web page using the <APPLET> tag, which indicates the initial window size for the applet, various parameters for the applet, and the location from which the browser can download the bytecode file.

Java Database Connectivity (JDBC)–Java database connectivity is the linking of an application or service to databases through the use of a Java application-programming interface (API). JDBC allows software application developers to write applications that can dynamically access information in database using Java scripts.

Java Message Service (JMS)–Java message service is the transferring of small amounts of data or text (usually several hundred characters) between Java based applications or devices. JMS allows applications to directly share information with each other.

Java Script-Java Script is a scripting language that is embedded in the HTML code. It is used in web pages designed to be viewed by PC-based browsers.

Java Server Page (JSP)-A Java server page is a script interpreter (real time program operation) that allows the development and use of software program modules to operate on a web server that can run Java servlets.

Java Telephony Application Programming Interface (JTAPI)–Java telephony application programming interface is an industry standard that defines the application interface between computers and telecommunications devices based on the Java programming language. The JTAPI standard allows computers to control private telephone systems (such as PBX systems).

Java Version 2 Mobile Edition (J2ME)-A compact version of Sun's Java technology targeted for embedded consumer electronics.

Java Virtual Machine (JVM)-Java virtual machine (JVM) is a software program that operates on a computer that allows the computer to use java instructions to request, transfer and process information.

JavaBeans-Javabeans are software programming object modules that can be reused.

JavaScript (Jscript)-Javascript is a scripting language designed to allow the development of active online content on Web severs (web sites). Javascript was created by Netscape Communications and Sun Microsystems that allows developers to add a specific information processing capabilities Web pages.

JavaTV-JavaTV is an industry middleware standard used in television systems to allow the user to access additional interactive services such as Internet browsing and electronic programming guides. JavaTV is developed by Sun Microsystems and more information about JavaTV can be found at www.java.sun.com/products/javatv/.

JCL-Job Control Language

JD-Julian Day

JDBC-Java Database Connectivity

Jeopardy-A condition resulting from any schedule change that is likely to cause a service request to be completed later than a committed due date. Failure to update the status of an order on or before a critical report date can result in a jeopardy condition. The term applies to both installation and repair jobs.

Jini-A connection technology developed by Sun Microsystems that provides simple mechanisms which enable devices to plug together to form an impromptu community. The Jini system allows each device within a community system to provide services that other devices in the community may use without any required planning, installation, or human interaction. These devices provide their own interfaces, which ensures reliability and compatibility. Jini works at high (application and session layer) protocol layers.

JIT-Just In Time

JMS-Java Message Service

Job-A job is a process or task that is performed by a computer or person.

Job Control-Job control is the procedures or processes that are used to manage jobs. Job control, allowing them to be started, stopped, and moved between the background and the foreground.

Job Control Language (JCL)-A computer language that links an operating system and applications programs in order to define processing jobs and executable programs, and to provide for a job-to-job transition.

Job Trigger-A job trigger is the set of conditions or rules that when matched or exceeded initiate a process or action.

Join-To join items is a process that removes boundaries (such as field limiters) between two or more elements or items to create a single item or unit.

Joining-Joining is a process of attaching or linking two or more items.

Joint Access Costs-The costs associated with network access facilities that are used for services from two (or more) access systems (such as local and long-distance connections.) Included are costs for basic termination, installation labor, inside wiring, drop wire, a subscriber loop, and all non-traffic-sensitive equipment in a local central office. Joint access costs include those that are incurred regardless of whether a customer makes a call. Also included are those expenses that vary with the number of customers rather than the amount of use, such as commercial, directory monthly billing, and testing.

Joint Costs-Joint costs are the shared costs of operating systems and providing services.

Joint Technical Committee (JTC)-A joint technical committee is a group that is composed of members from multiple organizations or affiliations with the purpose of analyzing, recommending, solving technical issues or creating specifications.

Joint Trench-A joint trench is a narrow pit (a

trench) that is shared by multiple companies or utilities.

Joint Use-A mutual agreement among two or more telephone, power, cable television, or other utility companies to use common poles, trenches, and similar facilities.

Joint User-A joint user is a person or companies that are authorized or has shared access (joint access) to a system or network..

Joint Venture (JV)-A joint venture is a company or business unit that is setup by partners (companies) who work together to develop the business.

Journalization-The process of booking billing charges & credits to the appropriate financial accounts. A journal is the interface between the billing system and the general ledger.

Joy Clicker-A joy clicker is a person who constantly clicks with a mouse.

Jscript-JavaScript

JSP-Java Server Page

JTAPI-Java Telephony Application Programming Interface

JTC-Joint Technical Committee

Judgement-A judgement is court decision that defines a resolution of a petition or other matter of law. An example of a judgement is the legal decision for a petition to collect a debt that authorizes a person or a company to obtain access to assets that will satisfy the debt.

Julian Date-A julian date is a date format in which days of the year are numbered in sequence. The first day is 001, the second is 002, and the last is 365 (or 366 in a leap year).

Julian Day (JD)-Julian day is the number of days that have passed since January 1st, 4713 B.C.

Jump Page-A jump page is a display area that promotes a viewer to change (jump) to a new viewing area (such as a new web site).

Jurisdiction-A geographical area or identifying characteristic that is assigned to a regula-tory authority to determine the boundaries of their authority.

Just In Time (JIT)-Just in time is a process that provides materials, services or products to a person or process at or close to the time that the materials, services or products are needed for the next step in a process.

JV-Joint Venture

JVM-Java Virtual Machine

K

K-1024
k-kilo
kB-kilobyte
kbps-kilobits Per Second
kbyte-kilobyte
KDM-Key Delivery Message
KE-Knowledge Element
Keep It Simple Stupid (KISS)-A philosophy that states that simple things or processes are better than complex methods. KISS is commonly applied to business practices.
KEK-Key Encryption Key
Kermit-Kermit is a transmission and error-correction protocol that is used in asynchronous file transfer.
Kernel-A kernel is a software part of an operating system that is fundamental to its overall operation. The kernel software continually manages specific processes such as system memory, the file system, and disk operations. A kernel may also run processes such as the communication between applications, coordinating the input and output of information within an application. Once loaded, the kernel software usually remains stored and operates from computer memory and is typically not seen by the users of the system.
Key-(1-connector) A short pin or other projection that slides into a mating slot or groove to guide two parts being assembled. The key is used to prevent a connector interface from rotating. (2-database) An attribute that uniquely identifies the sets of ordered elements in a database relationship. (3-encryption) A word, algorithm, or program used to encrypt and decrypt a message. (4-video) A signal, also called key source or key cut, that can be used to electronically cut a hole in a video picture to allow for insertion of other elements, such as text or another video image. The key signal is a switching or gating signal for controlling a video mixer, which switches between or mixes the background video and the inserted element. Key also may refer to the composite effect created by cutting a hole in one image and inserting another image into the hole. (5-radio) To initiate radio transmission.
Key Assignment-Key assignment is the process of creating and storing a key so a user or device is able to decode content. A key assignment may be performed by sending a command from a system to a device or by providing information to a user for direct entry into a device or system that creates a new unique key.
Key Delivery Message (KDM)-Key delivery messages are commands and associated data that are used to deliver, update or delete security keys.
Key Encryption Key (KEK)-A key encryption key is a code or value that is used to encrypt another key.
Key Exchange-Key exchange is the process used to exchange the key value between to or more devices or users. Exchanging key information does not necessary involve the physical transfer of the key. It may only involve the transfer of information parameters that are used to create and/or validate the key.
Key Generation-Key generation is a process of creating a code or value that is used as part of another process to decode or validate information.
Key Hierarchy-A key hierarchy is a structure of key levels that assigns higher priority or security level to keys at higher levels. Upper layer keys may be used to encrypt lower level keys for distribution.
Key Length-Key length is the number of digits or information elements (such as digital bits) that are used in an encryption (data privacy protection) process. Generally, the longer the key length, the stronger the encryption protection.

Key Lifetime-Key lifetime is the time duration that a key or code for a process is valid or able to be used.

Key Management-Key management is the creation, assignment, storage, delivery/transfer and revocation of unique information (keys) by recipients or holders of information (data or media) to allow the information to be converted (modified) into a usable form.

Key Performance Indicator (KPI)-Key performance indicators are the metrics (measurements) established by upper management to determine the accomplishment of objectives set for organizational units. Examples of KPIs include net customer additions to a system, net customer loss rate (net churn) and total revenue.

Key Quality Indicators (KQI)-Key quality indicators are the metrics (measurements) established to help evaluate the Quality of service (QoS) for the desired performance and priorities of a system or services. KQI measures may include service availability, bit error rate (BER) and other measurements that are used to ensure quality communications service.

Key Renewability-Key renewability is the ability of an encryption system to issue new keys that can be used in the encoding or decoding of information.

Key Revocation-Key revocation is the process of deleting or modifying a key so a user or device is no longer able to decode content. Key revocation may be performed by sending a command from a system to a device.

Key Selection Vector (KSV)-A key selection vector is a value that is sent or used to determine the location a key in a table or set of keys.

Key Server-A key server is a computer that can create, manage, and assign key values for an encryption system.

Key Service Unit (KSU)-The central operating unit of a key telephone system (KTS) or non-PBX/ACD telephone system (small customer premises telephone switch).

Key Telephone System (KTS)-Key telephone systems are (usually small) multi-line private telephone network that allows each key telephone station to select one of several telephone lines, place a line on hold, and call via an intercom circuit between key telephones. Key systems contain a central key service unit (KSU) that coordinates status lights and lines to key telephones ("Key Sets"). Early KTS system technology was based on electromechanical relay hardware. They required all the outside telephone lines to be connected to all of the key telephone sets in the installation. In addition, two additional pairs of wire were used in conjunction with each telephone line, one pair for the A/A1 connection indicating if that line is off hook at that particular key telephone set, and another pair to operate a small light to indicate the status of that line. Consequently, each key telephone set was connected to the central KSU via a thick cable containing 50 wires (25 pairs). Newer KTS systems typically use only 4 wires to connect the electronic KSU to each electronic key telephone set, and are often called "skinny wire" key systems. Modern electronic key systems are small microprocessor controlled switching systems and have some of the same advanced call processing features such as call hold, busy status, multi-line conference, abbreviated dialing, and station-to-station intercom that are available in a larger PBX.

This figure shows a typical key telephone system. This diagram shows telephones wired to a key service unit (KSU) that is connected to the PSTN. The KSU allows the telephones to have access to the outside lines to the PSTN. The KSU controls lights on the telephone sets, intercom access, and call hold.

Keyboard-A physical device that allows a user to enter data to a computer or other electronic device. A keyboard usually consists of individual key switches that are assigned one (or several) alphanumeric character and/or function.

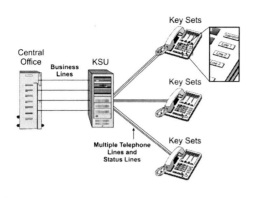

Key Telephone System (KTS) Operation

Keyboard Template-A keyboard template is a display card that fits on top of a keyboard or area of a keyboard (such as the function keys) which contains information about the functions of the keys for specific applications (such as word processors).

Keyboarding-Keyboarding is a name for entering information via a keyboard.

Keystroke Monitoring-Keystroke monitoring is the tracking of keys that are pressed on a keyboard or data entry device. Keystroke monitoring may be used to track the performance of workers or to track responses to presentations or other media (e.g. which options are selected and how long it took to select the specific options).

Keyword Advertising-Keyword advertising is a marketing process that uses key words that potential customers enter into search engines to find product or service information. Keyword advertising is usually paid for by a fixed fee or bidding process. To Keyword advertise, a list of keywords is selected and associated with a URL and a short message to accompany the listing. When the search term(s) matches the keyword, the URL and the descriptive text are displayed. These listings are called sponsored listings.

Keyword Bidding-Keyword bidding is the process of assigning bid amounts that are associated with a specific advertising message that is associated with a key word that will be used in an a search process. When the criteria is matched (such as matching a search word in an online search), the bid amounts are reviewed and the highest bids are selected and the advertising message(s) are displayed in the order of bid amount.

This figure shows the basic keyword advertising process. In this example, four companies have submitted ads to a search engines that match a keyword. When an Internet user enters a search word into the search engine, the search engine provides the user with a list of URLs found along with a list of sponsored ads. The sponsored ad presentation (impressions) is organized with the highest bid on top and ads with lower bids positioned lower on the screen.

Keyword Marketing-Keyword marketing is the process of selecting key words (adwords) and using them in a marketing program that has advertising messages appear when a user has selected or searched for an adword, portion or an adword or a related word (content match).

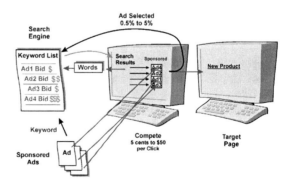

Keyword Bidding

kHz-kilohertz

Kickback-A kickback is a value that is provided for assisting in business dealings such as enabling the operation of systems or the sale of products or services.

Kill File-A kill file is a data file that is used to remove or delete records or files.

Kill Message-A kill message is an announcement that is played at the beginning of a call where the caller pays to inform the caller that charges will apply if the call is completed. This gives the caller the opportunity to hang up (disconnect the call) before call charges are applied.

Killer App-Killer Application

Killer Application (Killer App)-A software application of such great importance to the end customer that it alone motivates the customer to buy the entire system. One example is a computer game having devoted users.

kilo (k)-Prefix indicating 1000 units in the metric system (Decimal 1000). Note the use of a lower case k.

kilobit-A quantity equal to one thousand bits.

kilobits Per Second (kbps)-A measure of data transmission equal to one thousand bits per second.

kilobyte (kB)-A kilobyte is a thousand bytes.

kilobyte (kbyte)-A kilobyte is a quantity equal to one thousand bytes. When referring to computer memory, a kilobyte means 1024 rather than 1000.

kilohertz (kHz)-A kilohertz is a unit of measure of frequency equal to one thousand signal cycles that occur per second.

kilowatt (kW)-A unit of measure equal to one thousand watts.

Kilowatt Hour (kWh)-kWh is a widely used unit of energy, equal to 1000 watts of power per hour. See also joule.

Kiosk Billing System-A uniform billing system used by France Telecom in its deployment of mass-market videotex services. The system takes its name from the kiosk, or newsstand, where information is purchased on an as needed basis.

KISS-Keep It Simple Stupid

Kludge-A kludge is a solution that has been improvised or temporary solution.

Knowledge Base-Knowledge base is a system that can process information (data) into a useable form. A knowledge base has a set of rules that can assist people and systems in finding and using information.

Knowledge Density-Knowledge density is the ratio of knowledge elements within the body of other words.

Knowledge Element (KE)-A knowledge element is the component parts of information or content. A knowledge element can be the smallest common denominator of a knowledge-based system.

Knowledge Management-Knowledge management is the identification, organization and processing of knowledge elements.

Knowledge Neighborhood -Term used to define a group of related departments and disciplines. Groups within a company who share a common vocabulary, view of the world and information resources. For example, all groups involved in customer relationship management form a CRM Knowledge Neighborhood.

Knowledge Thread-Knowledge threads are relationships that exist between knowledge elements.

Knowledge Worker-Knowledge worker is a person who identifies, organizes, processes, and packages knowledge elements.

Known Device-A known device is communication assembly that has been identified to another device by its address and possibly additional information that can be used to communicate with the device.

KPI-Key Performance Indicator

KQI-Key Quality Indicators

KSU-Key Service Unit

KSV-Key Selection Vector

KTS-Key Telephone System

kW-kilowatt

kWh-Kilowatt Hour

L

L_CH-Logical Channel

Label Templates-Label templates are files, data or programs that can be used to adapt the presentation format and position of information so that it can be printed on labels.

Laboratory Testing-Laboratory testing is a process of testing a device, assembly, or system at a location that typically involves its design, prototyping or performance certification.

LAC-Location Area Code

LAMA-Local Automatic Message Accounting

Lambda-The Greek symbol for "L" that is used to represent wavelength, for instance in optical and electrical signals. Wavelength is the distance one cycle of a wave travels. Wavelength can be calculated by dividing the speed of the wave by its frequency. For example, in free space (in a vacuum) the wavelength of an electromagnetic wave is equal to: (300,000,000 meters/sec)/ frequency (Hz). The wavelength of a radio or lightwave traveling in vacuum at 300 MHz is 1 meter.

LAN-Local Area Network

LAN Segmentation-The practice of dividing a single LAN into a set of multiple LANs interconnected by bridges.

Landline-A conventional domestic or business telephone circuit. The term landline applies to telephone lines that are either buried or carried just over the ground.

Landline Network-The communications infrastructure that generally is associated with the public switched telephone network. (See also: landline.)

Language-(1-general) A set of symbols, characters, conventions, and rules used for conveying information. (2-programming) A set of commands and associated parameters that can be decoded to represent specific sequences of computer processing instructions.

Language Prompts-Language prompts are messages that alert a caller or viewer to the option of other languages that may be used for the navigation of a device or system.

Language Variations-Language variations are alternate languages that are available for selections, presentations or programs.

Laser Safety Officer (LSO)-A laser safety officer (LSO) is a person who is in charge of overall safety from exposure to laser light at a company or facility.

Lashing-Lashing is the attachment of a cable to a support strand by using helically wrapping materials such as dielectric filament or steel wire to hold the new line to another line. This figure shows a basic method of installing aerial optical cable. This diagram shows that the optical cable is being attached to a messenger wire between telephone poles. This diagram shows that the optical cable is sent through a cable guide and a cable lashing machine that is attached to the messenger cable. The optical cable is supplied from a cable reel trailer that is attached to a truck. The end of the lashing cable is attached to the messenger wire near the pole via a lashing clamp. An installer pulls the cable lasher via a pulling rope. As the cable lasher is pulled, the

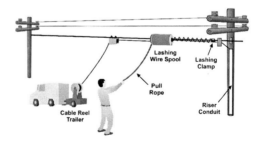

Aerial Cable Installation

lashing wire inside is spun around the cable and the messenger wire holding them (lashing) together. The truck and installer travel at approximately the same speed.

Lashing Machine-A lashing machine is an assembly that dispenses lashing wire around a cable and messenger wire as it travels along the length of the messenger cable.

Lashing Wire-Lashing wire is a material that is wound around a cable and a messenger cable to secure the cable to the messenger wire.

Last Call Return-A telephony service that allows a telephone user to automatically call back the phone number of the last received incoming call. Last call return is normally accomplished by the customer entering the service code (e.g. "*69").

Last Channel Recall-Last channel recall is a feature that allows a control device (such as a television remote control) to select the last channel that was used.

Last Choice Route-A last choice route is the last choice trunk group or series of such groups between two switching systems. This term also can refer to a final trunk group or only-route trunk group.

Last Choice Trunk Group-A final trunk group or only-route trunk group in a hierarchical network.

Last Known Good Configuration-The last known good configuration is a set of modes and parameters that existed at a certain time that can be used to restore equipment of software programs to an previously known good operational condition.

Last Mile-The last portion of the telephone access line that is installed between a local telephone company switching facility and the customer's premises.

Last Number Redial-The ability for a telephone to remember and dial the last dialed telephone number.

Last Trunk Busy-The condition in which the last available trunk in a trunk group is busy.

Last-in-First-Out (LIFO)-Last-in-first-out is the process of tracking or accounting for prod-

ucts that are produced where the last products entering into a process or area (such as into an inventory) are counted if the first products exiting the process or area.

LATA-Local Access And Transport Area

Late Bid-A late bid is a response to a bid request such as a request for proposal (RFP) that is received after the required submission date.

Late Entry-Late entry is the ability of a communication device to be added to a communication session after the communication session has been established.

Late Fees-Late fees are charges that are added to an account or invoice for failure to pay charges on a date or series of dates that have been previously agreed upon.

Late Payment Fees-Late payment fees are values or assessments that are created as a result of payments not being received by a defined payment date.

Late Payment Processing-Late payment processing is the identification and review of payment dates to determine and apply additional fees or assessments to an invoice or account.

Latency-Latency is the amount of time delay between the initiation of a service request for data transmission or when data is initially received for retransmission to the time when the data transmission service request is granted or when the retransmission of data begins.

Latency Guarantee-Latency guarantee is the commitment by a company to provide a maximum amount of time delay between transmission and reception of data. Because the latency commitment is commonly associated with financial penalties (reduced billing cost), the service provider provisions their network to ensure the bandwidth and packet processing resources are sufficient to ensure the latency does not exceed that maximum amount.

Layer-(1- LAN) A collection of related network processing functions that constitutes one level of a hierarchy of functions. (2-video) A single video image that is processed so that it can be inserted into a final composite image. There may be other layers in the image, which can be prioritized as to location. (3-computer) An overlay, which can be used to place information, so that CAD/CAM drawings can be logically subdivided for viewing or hardcopy purposes. (4-boundaries) A group of one or more entities contained within an upper and lower logical boundary. Layer (N) has boundaries to the layer (N + 1) and to the layer (N-1). (5-web page) Web page layers are unique windows or containers that can be simultaneously presented on a display with other layers.

Layered Architecture-Layered architecture divides the services or applications in a device or system into layers that perform specific functions. The precise definition of each layer allows products to be developed for specific functions by different companies. Examples of a layered architecture include physical layer, logical transport layer and logical channels.

LBS-Location Based Services

LBT-Listen Before Talk

LCA-Local Calling Area

LCC-Life Cycle Cost

LCI-Logical Channel Identifier

LCN-Logical Channel Number

LCR-Least Cost Routing

LD-Long Distance

LDAP-Lightweight Directory Access Protocol

LDS-Local Digital Switch

Lead Time-Lead time is the time between when you order or reorder a product to when the product is received or scheduled to be received.

Lead to Cash Process-A lead to cash process is an end-to-end process that captures the flow of events as well as the interaction points between the point where a prospect's details are captured and the point where the prospect (after turning into a customer) pays for the services. The interaction points are defined between the customer, service provider internal departments, as well as its partners. This process flow is often used to benchmark and compare process KPIs between peer service providers. The Lead-to-Cash Process is often decomposed into multiple sub-processes to bring in the accountability and ownership within the organization.

Lead to Order Process-A lead to order process is the capturing of the flow of events as well as the interaction points between the point where a prospect's details are captured and the point where the prospect orders the service.

Leakage-(1-rights) Rights leakage is the loss of controlled content from a system that is protected with a rights management system. (2-signal energy) Leakage is the loss of energy resulting from the flow of electricity past an insulating material, the escape of electro-magnetic radiation beyond its shielding, or the extension of magnetic lines of force beyond their intended working area. (3-revenue) Revenue leakage is the loss of revenue that occurs from a service that is provided but cannot be billed.

Lean Principles-Lean principles are the process of obtaining step-change improvements in costs, cycle times, and quality by focusing on improving the yield of business and operational parts. Lean practices define the ways and means for the systematic elimination of waste, systematic reduction in variability and systematic removal of inflexibility. The concept inherited from the automotive industry.

Lean Protocol-A lean protocol is a set of commands and processes that are used to perform a specific function and the number of commands and/or level of detail or processes is limited to the specific set of functions.

Leap Second-A leap second is a time step of one second, used to adjust coordinated universal time to ensure approximate agreement with international universal time. An inserted second is called a positive leap second, and an omitted second is called a negative leap second.

Learning Curve-A learning curve is a mathematical function or graphical representation that defines how costs of development and production decrease over time and as the volume of production increases.

Leased Line-Leased lines are telecommunication lines or links that have part or all of their transmission capacity dedicated (reserved) for the exclusive use of a single customer or company. Leased lines often come with a guaranteed level of performance for connections between two points.

Leased Network-A data network using circuits or channels leased from a telephone company (telco) or other telecommunications carrier and dedicated to use solely by the lessee.

Leased Service-The exclusive use of any channel or combination of channels designated to a subscriber.

Leasee-A leasee is a person or company who owns and provides the rights to lease products or services.

Leasing-Leasing is the assignment of the rights to use or occupy personal property (such as building or equipment) to a user or controller of the asset (the lessor).

Leassor-A leassor is a person or company who obtains the rights to use products or services through a lease agreement.

Least Cost Routing (LCR)-Least cost routing is a connection patch control feature that seeks or selects connection paths over the least expensive paths available.

LEC-Local Exchange Carrier

Legacy-Legacy is established or well known systems, technology or products.

Legacy Application-Legacy refers to a system or established technology that has been used in the past.

Legacy Phone Adapter-An adapter that allows the use of existing telephones with a new communication system.

Legacy Support-Legacy support is the ability of a new product or service to work with (support) an existing (or older) product or service.

Legacy System-A legacy system is a communication system or network that satisfies specific business needs using established technology or equipment. Legacy systems may become obsolete or is incompatible with new industry standards. To extend the life of existing investments in legacy systems, new technologies or systems are often designed to communicate with legacy systems.

Legacy Wiring-Legacy technology is commonly used to describe a previously used technology that is undergoing a change or is being replaced with a new technology to reduce costs or to satisfy new functional requirements.

Legal Rights-Legal rights are actions that are authorized to be performed by individuals or companies that are specified by governments or agencies of governments.

Length Counter-A length counter is a device that counts the length of cable that has been pulled through the counter.

Length Field-A length field is a data field within a packet header that holds a number or value that indicates the length of a data packet or a block of data.

Length Indicator-(1-General) A length indicator is a number or value that indicates the length of a data packet or a block of data. (2-CCS) In common-channel signaling, a 6-bit field that differentiates between message, link status, and fill-in signal units. When the binary value of an indicator is less than 63, it indicates the length of a signal unit.

Lessee-A lessee is a person or company who pays a person or company for the rights to use something.

Lessor-A lessor is a person or company who provides a person or company with the rights to use something in return for something of value (such as rent).

Letter Of Agency-A letter of agency is a document that authorizes a person or company to perform actions on their behalf. A letter of agency is a specialized form of power of attorney.

Letter of Credit-A letter of credit is a financial payment instrument that authorizes the transfer of funds or other valuable assets to the recipient of the letter of credit provided the terms of the letter of credit are fulfilled (such as the receipt of products at a specified location).

Level-(1-general) The strength or intensity of a given signal (2-crosstalk) The power of the crosstalk signal compared with a reference signal. (3-peak) The maximum applied sound or signal amplitude. (4-speech) The energy of speech measured in volume units (VU), and typically displayed on a VU meter. (5-speech power) The acoustic power in human speech. (6-transmission) The ratio of the power of a test signal at one point to the test signal power applied at another point in the system used as a reference. (7-video routing) An independently controllable spectrum of signals within a routing switcher. Typically, a routing switcher has a video level and one or more audio levels.

Liabilities-Liabilities are obligations or assets that are owned by other people.

Libel-In reference to any printed or written publication that may be slanderous.

Library-A library is a collection of programs, media or applications that can be accessed or used.

License-A license is a contract that grants specific rights to use of intellectual property. A license typically identifies issuers, principles, rights, resources, conditions and grants.

License Constraints-License constraints are rules or limitations that are placed on the delivery of services and/or processing of media or information.

License Exempt-License exempt systems or products do not require a license for their operation.

License Fee-License fees are an amount charged or assigned to an account for the authorization to use a product, service, or asset. License fees can be a fixed fee, percentage of sales, or a combination of the two.

License Server-A license server is a computer system that maintains a list of license holders and their associated permissions to access licensed content. The main function of a license server is to confirm or provide the necessary codes or information elements to users or systems with the ability to provide access to licensed content.

License Terms-License terms are the specific requirements and processes that must be followed as part of a licensing term agreement.

Licensee-A licensee is the holder of license that permits the user to operate a product or use a service. In telecommunications, a licensee is usually the company or person who has been given permission to provide or use a specific type of communications service within a geographic area.

Licensing-Licensing is the defining, authorizing and compensating for the rights to develop, use or sell products and services.

Licensing Collective-A licensing collective is a group or organization that represents a several or many rights holders which has the authority to negotiating and administer licenses agreements. ASCAP is an example of a licensing collectives.

Licensing Fees-Licensing fees are an amount charged or assigned to an account for the authorization to use a product, service, or asset. Licensing fees can be a combination of fixed fees or a percentage of sales.

Licensing Rules-Licensing rules are the processes and/or restrictions that are to be followed as part of a licensing agreement. Licensing rules may be entered into a digital rights management (DRM) system to allow for the automatic provisioning (enabling) of services and transfers of content.

L

Licensor-A licensor is a company or person who authorizes specific uses or rights for the use of technology, products or services.

Life Cycle-Life cycle is the time period that a particular device, assembly or a class of equipment is usable under normal working conditions.

Life Cycle Cost (LCC)-Life cycle cost is the combination (addition) of the initial acquisition cost, operational and maintenance cost, the cost of disposition (removal for upgrade or end of service) less the recovered cost from the sale of salvaged equipment (if any).

Life Safety System-A system designed to protect life and property such as emergency lighting, fire alarms, smoke exhaust and ventilating fans, and site security.

Life Span-Life span is the time duration that a product or service will operate within its defined performance characteristics.

Life Test-A test in which random samples of a product are checked to see how long they can continue to perform their functions satisfactorily. A form of stress testing is used, inducting temperature, current, voltage, and/or vibration effects, cycled at many times the rate that would apply in normal usage.

Lifeline Service-A communication service that is considered a "Lifeline" in case of emergency. Communication service that assures a person can call for assistance or be contacted.

Lifetime Revenue per Subscriber (LRS)-Lifetime revenue per subscriber is the sum of the revenues for each subscriber (customer) in a system. This value is calculated by multiplying the annual average revenue per user (ARPU) by the average time period (lifetime) that a subscriber purchases products and/or services.

Lifetime Value (LTV)-Lifetime value is a technique utilized to define how valuable a customer will be to the company, over their lifetime.

LIFO-Last-in-First-Out

Light-Electromagnetic radiation visible to the human eye. The visible wavelength of light ranges from approximately at 400 nm to 700nm. Commonly, the term is applied to electromagnetic radiation in most fiber optic communication systems. An electromagnetic radiation with wavelengths from 400 nm (violet) to 740 nm (red), propagated at a velocity of roughly 300,000 km/s (186,000 miles/s), and detected by the human eye as a visual signal in the optical communication field, the term also includes the much broader portion of the electromagnetic spectrum that can be handled-died by the basic optical techniques used for the visible spectrum. This extends the definition of light from the near-ultraviolet region of approximately 300 nm through the visible region, and into the mid-infrared region of 3.0 to 30 nm.

Lightning-A flow of current between a charged cloud and the ground resulting from an electric discharge due to large potential differences between cloud charge and ground (or the lightning strike point).

Lightning Protector-(1-general)A device that limits impulse voltages from lightning to prevent damage to people and electronic equipment. Basic spark-gap ("carbon block" or gas tube) protectors installed in buildings are inadequate to protect modern electronic equipment. Supplementary lightning protectors (surge protectors) are frequently installed at the equipment to protect against excess voltages on both signal and AC connections. (2-rod)A lightning rod system that routes lighting voltages to ground.

Lightware-Lightware is a software program and/or service that has a reduced set of features as compared to other products or services offered in other versions of the product or service.

Lightweight Directory Access Protocol (LDAP)-A standard protocol that allows users to find other devices and services in a communication network. It provides directory services for LAN and the Internet. LDAP is a subset of the X.500 protocol that operates over TCP/IP.

Limited Access-An arrangement in which only some traffic offered to a group of servers has access to all the servers in the group.

Limited Company (LTD)-A limited company is a business entity that is owned by shareholders.

Limited Liability Company (LLC)-A limited liability company is a business entity that is owned by a limited number of shareholders. LLC companies provide some of the limited liability benefits (protection from company lawsuits) while maintaining some small company tax advantages (ability to record company losses on personal income taxes).

Limited Liability Partnership (LLP)-A limited liability partnership is a business entity that is owned by partners that offers provide some of the limited liability benefits (protection from partner liabilities) while maintaining some partnership tax advantages .

Line Card-A line card is a plug-in electronic circuit card that connects telephone switching environment to telephone lines. The line card adapts signal levels, senses and inserts control commands and tones and performs other functions that allow the line card to communicate with specific types of telephone lines.

Line Item-A line item is a component or detail of an order. An example of a line item is a product that is part of an order. The line item may contain details of the product such as quantity and per item product cost.

Line Not Cutting (LNC)-A notation that itemizes the telephone numbers that will not be affected by a work order that involves changing plant facilities. This notation may be on an engineering work order that involves splicing "live" cable which has working subscribers.

Line Side Connection-Line side connections are an interconnection line between the customer's equipment and the last switch (end office) in the telephone network. The line side connection isolates the customer's equipment from network signaling requirements. Line side connections and are usually low capacity (one channel) lines.

Line Tapping-Line tapping is the connection and monitoring of a communication line.

Line Testing-Line testing is the measurements of the characteristics of a communication line (such as the dialtone level on a telephone line) to determine that the line is in service (operational) or is operating within expected performance levels (such as error or distortion levels).

Linear Television (Linear TV)-Linear television is the providing of television programs in a time sequence.

Linear TV-Linear Television

Link-(1-telecommunications) A link is a transmission facility or medium that can be used to transfer data or media between devices or applications. (2-common channel signaling) A communications path between two adjacent common channel signaling nodes. (3-computer program) The part of a computer program, in some cases a single instruction or address, that identifies the location (link) to another program or module. (4-web) Web links (Hyperlinks) are tags, icons, or images that contain a crossed reference address that allows the link to redirect the source of information to another document or file. These documents or files may be located anywhere the link address can be connected to.

Link Encryption-Link encryption is a process of a protecting voice or data information from being obtained by unauthorized users on a specific link connection.

Link Establishment-(1-general) Link establishment is the process of setting up a communication channel for use between two devices. This may involve the creation of a physical link and/or the creation of a logical channel on the physical link. (2-Bluetooth) A procedure is used to setup a physical link, specifically an Asynchronous Connectionless (ACL) link between two Bluetooth devices.

Link Exchange-A link exchange is the agreement by two (or more) companies to display web links, banner ads or images with hyperlinks on each other company's web site.

Link Identifier (LinkID)-A link identifier is a unique code or label that identifies a specific connection between communication devices.

Link Key-A link key is a word, algorithm or value that is used to encrypt and decrypt a message that is sent on a communication link.

Link Layer-The link layer facilitates the detection of and recovery from transmission errors on a specific link connection.

Link Selection-Link selection is the process of selecting a connection address or path (such as a web address). Link selection can initiate a process within an application or it may launch a new application such as a web browser.

LinkID-Link Identifier

Linux-Linux is an operating system that is used for computers and consumer electronics products. Linus Torvalds developed Linux in 1991 while at the University of Helsinki in Finland. Linux manages system services and there are various other utilities and programs that have been developed to use and interact with the Linux system. The Linux system is available for use without any fees ad defined by the Free Software Foundation's GNU Project.

List Bartering-List bartering is the providing of a list or lists in exchange for items or services of value other than cash.

List Box-An interactive screen display box that provides the user with the ability to select an item from a list of pre-defined available choices.

List Broker-A list broker is a person or company who rents or sells lists of prospects for marketing purposes (such as direct mail campaigns).

List Importing-List importing is the process of transferring list items (such as customers or sales prospects) into a file, program or system.

List Management-(1-marketing) List management is the acquiring, sending and updating of lists of people or companies that share common interests. (2-video) Video editing list management is the process that allows the system operator or manager to change the edit the lists of programs.

List Manager-A list manager is a person or company who is responsible for overseeing the gathering, updating and use of lists.

List Ownership-List ownership is the person or company who has the rights to use, maintain or sell lists.

List Price-List price is an offer value that is assigned to a product or service for general offerings. List price is commonly the highest prices referenced for a product or service.

List Rental-Lists rental is the process of obtaining a list of names, addresses and/or email addresses to be used for a purpose such as a direct mail promotional campaign. Lists that are rented are commonly sent to a 3rd party such as a mailing house so the user does not have direct access to the list.

List Rental Agreement (LRA)-A list rental agreement is a document that defines a list of information (such as a mailing list) that will be provided for use by another person or company, the terms of use of that list, and value(s) that will be exchanged in return for the use of the list.

List Transfer-List transfer is process of selecting, grouping, and moving groups of contacts or data from one system to another.

List Validation-List validation is the process of contacting or confirming some or all of the information that is contained in a list (such as a sales prospect list) is accurate.

Listen Before Talk (LBT)-A process of listening to an access channel to determine if the channel is busy before attempting access to the channel or system.

Listing Fee-A listing fee is a value that is assessed or assigned for the inclusion of a product or service in a list of products or services that are available.

Listing Services System-An interactive software system that manages customer listing information, including name, address, and telephone number. The data can be used to support customer and network services as well as to compose listings for use in white pages directories and other specialized products.

Lists-Lists are groups of information elements (list items) that may be displayed and/or selected by users. Lists may be displayed in any order (unordered) or organized (sorted) based on the list items and the specified order criteria (such as alphabetically ascending).

Live-(1-electric circuit) A device or system connected to a source of electrical potential. (2-acoustical) An area in which sound is not greatly absorbed by the walls and timings; the room, therefore, reverberates. (3-media) A media source that is transmitted when the conversion of media first occurs or within a perceived live time period (up to several seconds of delay).

Live Asset-A live asset is a media source that is being provided in real time.

Live Content-Live content is the real time or near real-time transfer of information from a non-stored content source (such as a news camera) to viewers of that information.

Live Feed-A live feed is a media connection that is used to transfer media or programs from a device that is captured in real time (such as a mobile camera) to a distributor of the media or programs.

Live Streaming-Live streaming is the process of transferring audio or video streaming for which the clients may not control the playback time of the media. That is, the clients may not control when the stream starts, pause the stream, skip to a different time within the presentation, and so on. Live streaming is often used for broadcast of an event happening in real time.

Live Support-Live support is the real time processes and communication that occurs between customers and companies to enable customers to resolve problems and successfully obtain products and services from the company. Examples of live support include call centers that communicate through telephone and interactive messaging services.

Live Television (Live TV)-Live television broadcasting is the transmission of video and audio to a geographic area or distribution network in real time or near-real time (delayed up to a few seconds).

Live TV-Live Television

Live Video-Live video is image media that is viewed immediately (or within a short delayed period such as a few seconds) when the program media is created (such as at a sports event).

LLC-Limited Liability Company

LLP-Limited Liability Partnership

LLU-Local Line Unbundling

LNC-Line Not Cutting

LNP-Local Number Portability

Load And Work Time Record (LWTR)-Load and work time record is a listing of work requests for a particular employee or contractor and the status and completion of task information.

Load Balance-Load balance is a condition where the reception of traffic (such as television or telephone signals) is controlled or adjusted to match the distributed channels from the sources of the traffic.

Load Balancing-Load balancing is a process of equalizing or redistributing the usage load of line concentrators in a switching system. Depending on the switching system, loads can be balanced simply by controlling the number of subscribers in each class of service assigned to each concentrator. In the case of a severe load, working lines can be physically rearranged to achieve load balance.

L

Load Factor-The ratio of the average load over a designated period of time to the peak load occurring during the same period.

Load Generator-A device that can be connected to a line or trunk to originate or terminate simulated telephone calls for the purpose of verifying load-handling capability. A load generator also may be called a load box.

Load Sharing-(1 computer) An operating mode of duplicated processors whereby two units share processing operations. In the event of a failure of one unit, the other can take over the entire load. (2 -telecommunications) A process by which signaling traffic is distributed over two or more signaling or message routes to equalize and efficiently handle traffic for security purposes.

Load Time-Load time is the amount of time it takes for a file or media segment to transfer from a source to a receiving device. Examples of load time is the amount of time it takes from the initial connection to a web site URL, to the time a web page begins to display on the viewing device.

Loader-A loader is a program that manages the loading of additional programs or software modules.

Loading-(1-circuit) The addition of electrical inductance to a metallic transmission line to improve the frequency characteristics of the line. Loading a line increases the distance over which a quality signal can be sent. (2-antenna) The addition of an inductance to enable an antenna to be tuned to a frequency lower than its natural frequency. (3-multichannel communications) The insertion of white noise or equivalent dummy traffic at a specified level to simulate system traffic performance. (4-system) The total signal power of a multichannel system, expressed as the total of the average power on all channels, or as the per channel load that may be carried by all channels. (5-cable) (6-wind) The total ice and wind pressure allowed for in the design of a tower, pole, or line.

LOC-Local Operations Center

Local-Applies to the users or devices within a specific portion of a network or domain.

Local Access And Transport Area (LATA)-A geographic region in the United States where a local exchange carrier (LEC) is permitted to provide interconnected telephone service. LATAs were created as a result of the division of the company AT&T by the designated by the Modification of Final Judgment (MFJ). A LATA contains one or more local exchange areas, usually with common social, economic, or other interests.

Local Ad Insertion-Local ad insertion is the process of inserting an advertising message into a media stream such as a television program at the local area of signal delivery.

Local Area Network (LAN)-Local area networks (LANs) are private data communication networks that use high-speed digital communications channels for the interconnection of computers and related equipment in a limited geographic area. LANs can use fiber optic, coaxial, twisted-pair cables, or radio transceivers to transmit and receive data signals. LAN's are networks of computers, normally personal computers, connected together in close proximity (office setting) to each other in order to share information and resources. The two predominant LAN architectures are token

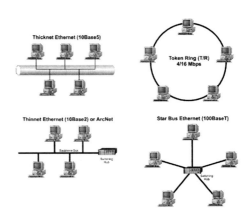

Local Area Network (LAN) Systems

ring and Ethernet. Other LAN technologies are ArcNet, AppleTalk, and This figure shows several of the most popular LAN topologies and their configurations. Some data networks are setup as bus networks (all computers share the same bus), as start networks (computers connect to a central data distribution node), or as a ring (data circles around the ring). This diagram shows for popular types of LAN networks: Thinnet, Thicknet, token ring networks, and Ethernet star network.

Local Automatic Message Accounting (LAMA)-Local automatic message accounting is an end-office accounting process that can automatically reconcile billing data for customer dialed and/or toll calls.

Local Call Accounting-Local call accounting is the process of computing and recording usage or costs for local calls.

Local Call Billing-Local call billing is the process of gathering event information created by a local telephone call and combining and computing the fees associated with the call. Local usage fees may include per call events, message units or other types of usage charges.

Local Calling Area (LCA)-Applies only to originating minutes of use and foreign carrier (OHX) account and second dialtone (OHY) accounts. The file contains the subscriber line counts by interexchange carrier (IXC) for each end office in the local access and transport area (LATA). The line counts are used to calculate ratios (factors) that are then multiplied by the IXC's OHY actual or assumed originating minutes of use (MOU) in that LATA to assign MOU to end offices for reclassification. In billing, LCA usually refers to an area within which a customer (typically residential) is not charged for usage.

Local Digital Switch (LDS)-A digital switch that is the final switching point between the end customer and the public switched telephone network.

Local Exchange-Another term for an end office (EO) telephone switching system. The local telephone company is sometimes called the local exchange.

Local Exchange Carrier (LEC)-Local exchange carriers (LECs) or post and telegraph and telecommunications (PTT) companies provide telephone services directly to residential and business customers located within a localized geographic area. Typically, these telephone companies provide services via copper lines that extend from a local carrier's switching facilities to the end customer's premises equipment (CPE). This is referred to local loop.

Until the early 1990's, most countries had a single company that provided local telephone services. This company was either owned or highly regulated by the government. To increase competition and reduce telephone service prices to consumers, some governments have begun to allow other companies to provide basic (local) telephone service. These competitive local exchange company (CLEC) or competitive access providers (CAPs) provide alternative connections to the public switched telephone networks (PSTN). The established telephone companies are now called the incumbent local exchange carriers (ILECs),

Local Feed-Local feed is a media connection that is used to transfer content from local sources. Examples of local feeds include connection from sportscasters, news crews and live studio cameras.

Local Headend-A local headend is part of a broadcast system that selects and processes video signals into local broadcast distribution system.

Local Insert-Local Insertion

Local Insertion (Local Insert)-Local insertion is the process of directing or redirecting media or content from a local source into a broadcast distribution system (such as a television system).

Local Line Unbundling (LLU)-Local loop unbundling is the regulatory process of allowing multiple telecommunications operators use of connections from the telephone exchange's central office to the customer's premises. The physical wire connection between customer and company is known as a "local loop," and is owned by the incumbent local exchange carrier.

Local Long Distance-Local long distance is the connection of calls outside the defined local service calling area but inside the local service provider's network. The local service provider may use this concept to apply toll charges to customers within their network.

Local Loop-The local loop is the connection (wired or wireless) between a customer's telephone or data equipment and a LEC or other telephone service provider. Traditionally, the local loop (also called "outside plant" or the "last mile") has been composed of copper wires that extend from the local central office, also known as the end office (EO). The EO got its name since it is part of the public switched telephone network (PSTN) that is at the edge, providing physical connections and dial tone to customers.

This diagram depicts a traditional local loop distribution system. This diagram shows a central office (CO) building that contains an EO switch. The EO switch is connected to the MDF splice box. The MDF connects the switch to bundles of cables in the "outside plant" distribution network. These bundles of cables periodically are connected to local distribution frames (LDFs). The LDFs allow connection of the final cable (called the "drop") that connects to the house or building. A NT block isolates

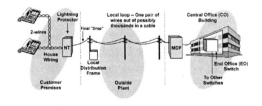

Telephone System Local Loop Operation

the inside wiring from the telephone system. Twisted pair wiring is usually looped through the home or building to provide several telephone connection points, or jacks, so telephones can connect to the telephone system.

Local Measured Service-A method of charging customers based on actual usage. Factored into local measured service are the number of local messages, the duration of those messages, the time of day, and the distance within a local exchange area.

Local Number Portability (LNP)-LNP is the process that allows a subscriber to keep their telephone number when they change service provider in their same geographic area. Local number portability requires that carriers release their control of one of their assigned telephone numbers so customers can transfer to a competitive provider without having to change their telephone number. LNP also involves providing access to databases of telephone numbers to competing companies that allow them to determine the destination of telephone calls delivered to a local service area.

This figure shows an example of the typical operation of local number portability (LNP). In this diagram, a caller in Los Angeles is calling someone in Chicago who has kept (ported) their old phone number when they connected their service to a competitive local exchange carrier (CLEC). This required the incumbent local exchange carrier (ILEC) to move (port) the telephone number to a LNP database. The line connected to the customer from the CLEC actually has a new telephone number (which the customer is not likely to be aware of). The LNP database associates the new number with the old number. This example shows how the call can be routed from an LEC in Los Angeles to the new telephone line in Chicago using the old telephone number. The call is routed from Los Angeles, through a long distance provider (IXC) who knows by the dialed area code that it needs to connect the call into a local telephone company in Chicago. Because there are several local telephone service providers in Chicago, the IXC must look first into a LNP database to see if the number has been ported to a different service provider. This LNP database (ported telephone number list) must be available to the next to last switch (called "N-1") before the call reaches the end office switch. This LNP database search instructs the last switch to the actual number used for the final connection. The call is then routed to the correct local switching office (new line) so the call can be completed.

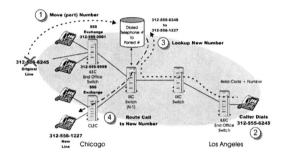

Local Number Portability (LNP) Operation

Local Operations Center (LOC)-A local operations center is a facility or system that oversees the control and distribution of services and products in a regional location.

Local Programming-Local programming is the selection of shows and programs that are offered by a local television network provider. An example of a local program is a news program that is created and broadcasted by a local broadcaster.

Local Rate-A local rate is a product or service fee that is charged to people or companies within a specific geographic area. Local rates may be less expensive than the rates charged to a national customer (such as a national advertiser).

Local Routing Number (LRN)-A local routing number is a 10 digit telephone number used for local number portability.

Local Service Management System (LSMS)-A local service management system is the information system from which local telephone companies updates from the NPAC for telephone number portability requests.

Local Service Request (LSR)-A form used by a competitive local exchange carrier (CLEC) to request local service form an Incumbent LEC (ILEC).

Local Services-Local services are the providing of work, systems or applications that benefit users where the services are created or focus on local needs or interests.

Local Spots-Local spots are advertising time periods (spots) that are available or assigned to a local broadcaster for the insertion of local advertising messages.

Local Subscriber Management-Local subscriber management is a process or system that allows for the changing of some or all of the service and features at a location that is different from the central or originating system.

Local Switching System-A switching system that connects lines to lines, and lines to trunks in an end office. The system may be located entirely in a wire center or it may be geographically disposed, as in host remote configurations

Locale-A locale is a geographic area or location.

Localization-(1-sound) The perception of sound as originating from a particular direction or distance. (2-troublshooting) The process of localizing equipment failures or below tolerance equipment.

Localized Ad Insertion-Localized ad insertion is the process of inserting an advertising message into a media stream such as a television program at a location near (local) to the receiving device. For broadcasting systems, localized ad insertions are performed at the head end near the viewing audience. For IPTV systems, localized ad insertion can be performed at locations that have access to media source and the address of the IP set top box.

Localized Pay Per Click (LPPC)-Localized pay per click advertising is an internet marketing process that displays ads in local geographic areas and only charges the advertiser only when an item is selected ("clicked").

Location Area Code (LAC)-A code assigned by the system operator to identify specific areas of operation. These LAC identifiers can be used to indicate regions that have different billing codes or the types of authorized service features, and most importantly, to limit the number of cells used to page a mobile station when setting up a mobile destination call.

Location Based Advertising-Location based advertising is the communication of a message or media content to one or more potential customers where the advertising message can vary based on the location of the recipient.

Location Based Services (LBS)-Location based services are information or advertising services that vary based on the location of the user. Mobile radio system may permit the use of different types of location information sources including the system itself or through the use of global positioning system (GPS).

Location Finder-Location finding is the process of determining the location of a person, device or service. An example of location finding is the gathering of position information from a mobile telephone so that additional functions or services can be performed.

Location Monitoring-Location monitoring is the process of continuously receiving position that identifies the location for a device or person.

Location Portability-Location portability is the ability of an end user to retain the same services or identifying numbers as they moves from one permanent physical geographic location to another.

Location Routing Number (LRN)-A telephone number (e.g. 10 digit number) that is used to route calls to and end office switch that allows for the processing of portable (assignable) telephone numbers.

Location Server (LS)-Location servers provide information regarding the location of resources that are located within a network (such as the Internet or within a SIP system). Location servers are typically databases that maintain a binding (mapping) for each registered user. This binding maps the address of the user to one or more addresses at which the user can be currently reached. The Location Service supports user mobility within a communication system. In a SIP system, the Location Service database is updated as a result of SIP User Agents performing a registration.

Location Tracking-Location tracking is the process of continuously receiving position location for a device or person.

Location Updates-Location updates are a process where a communication device informs a system as to its physical or logical location within a network. Location updates may be performed periodically to identify a specific physical location or logical address that a device is operating at so that a system can alter the routing or transfer of information so it can reach the communication device as it changes location.

Locator Page-A locator page is a web page that identifies a geographic location. A locator page can be a map with a location identified on it or it can be a text based web page with a listing of multiple companies or locations within a geographic area.

Lock Code-Wireless unit's built-in functionality which prevents unauthorized use by entering in the user-controlled lock code. It may lock out the keypad or prevent the unit from powering up altogether.

Lockout-User Lock Out

Log File-A file that contains a list of events that have occurred for a particular application or service. The log file is continually updated (added to) as new events occur. Log files are used to analyze problems that have or may occur with a particular application or service.

Log File Mining-Log file mining is the process of reviewing and analyzing log file data (such as packet routing data) for the purpose of identifying common characteristics that may be useful for other purposes (such as enhancing billing records).

Log Time-(1-general) The time at which a service or program was initiated or terminated. (2- video) new video source is placed on the program bus, usually recorded in the station log for FCC accounting and customer billing purposes.

Logarithmic Scale-(1-meter) A meter scale with displacement proportional to the logarithm of the quantity represented. (2-graph paper) A printed graph paper with one or both of the grids on a logarithmic scale rather than an arithmetic, scale.

Logging-Logging is the recording of data about events that occur in a time sequence. Logging can be used with syslog to monitor network events. Logging also applies to monitoring event, application, and system logs on Windows PC based systems. Logging is very useful in troubleshooting and correlating network and system environments and events.

Logging In (Login)-Logging in is a process that allows user to gain access to the system by the identification of the user account (login ID) and the password associated with the account.

Logic Bomb-A logic bomb is a process or event that occurs when a logical condition has been detected where the processes or events are intended to sabotage or do harm to a system or service.

Logical Address-A logical address is a grouping of numbers or codes that identifies one or more devices or services on these devices in a communication network.

Logical Channel (L_CH)-Logical channels are a portion of a physical communications channel that is used to for a particular (logical) communications purpose. The physical channel may be divided in time, frequency or digital coding to provide for these logical channels.

Logical Channel Identifier (LCI)-A logical channel identifier is a code or name that is assigned to a logical (virtual) channel.

Logical Channel Number (LCN)-A logical channel number is an identification code that is used to indicate virtual channels. For example, a LCN is used in television broadcast systems to identify a program that is part of a transport channel.

Logical Link-A logical link is a communication connection that uses codes or labels to identify and redirect data or packets that are transferred across physical connections. Several logical links can share a single physical connection.

Login-Logging In

Login ID-A login ID is a name or another form of identification given in order for a user to access a computer, site or network

Login Script-A login script is a small file that contains a set of instructions that assist or automatically submits information necessary to log a user into a system.

Login String-A login string is a sequence of parameters that is used to login a user to a system. A login string may contain a user name, destination service, and a password.

Logoff-Logoff is the process of ending a communication session that has previously been authorized by logging on.

Logon-Logon is the process of beginning a communication session that requires a user to provide identity and/or other types of validation information.

Long Distance (LD)-The connection of calls outside the local service calling area. Long distance services may be charged at a toll rate.

Long Haul System-(1-general) A communication system which includes a number of drop/add points, repeaters locations, over long distances that extend outside the local service area. (2-microwave) A microwave system that the longest radio circuit of tandem radio paths exceeds 402 km (250 miles).

This diagram shows a terrestrial microwave system-connecting IXC switches in Philadelphia and New York City. The microwave signals are moved between the two switching offices through a series of relay microwave systems located approximately 30 miles apart. Microwave is a line-of-sight technology that must take the earth's curvature into consideration. Also note that microwave towers are not limited to only facing one or two directions. A single tower can be associated with several other towers by positioning and aiming additional transceiver antennas at other microwave antennas on other towers.

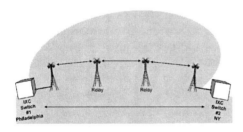

Long Haul Microwave

Long Line Adapter-A long line adapter is a circuit or assembly that is designed to extend the distance that a telephone line can be used. Long line adapters may adjust, process or enhance specific types of signals including line supervision, dialing tones or boosting ringing signals. A long line adapter also is known as a range extender.

Long Lines-Long lines were communication lines that interconnected telephone system (long distance service).

Long Term-Long term is a relative measure of an item or service that is substantially longer than other items or services. Long term financial accounts may have a measure of greater than one year.

Lookup Table-A lookup table is a data file that stores information that defines the structure of a database. A lookup table may be used in relational databases to define the names and data types of the data files (tables) and their fields (columns). A database dictionary may also be called a database dictionary.

Loop-(1-general subscriber) A loop is an insulated wire pair connecting a subscriber's telephone set (station set) to a central office switch. In historical documents dating from circa 1890 and slightly later, the loop was an innovation by John Carty of AT&T. The loop replaced the earlier practice (copied from telegraph systems of the 19th century) of using a single wire between the central office and the subscriber with earth/ground conductivity to complete the circuit. Historical single wire subscriber connections were less costly but had extremely variable total path resistance due to changes in soil moisture content, and also suffered from bad crosstalk due to use of the same shared earth conductor path for several simultaneous conversations. (2-software) The repetition of a group of instructions in a computer routine.

Loop Analyzer-A device that analyzes the performance characteristics of a local loop line.

Loop Assignment Center-An operations center that assigns customer loop facilities, telephone numbers, and central-office lines and equipment.

Loop Back Testing-Loop back testing is a process of configuring and sending test signals or information that is relayed back (looped back) to the sender for analysis. The successful reception of information indicates that system parts that the signals or information passed through are working correctly. Loop back testing commonly uses successively larger test loops (e.g. local, mid-distance, remote system) that validate which sections of a communication system are operating correctly.

Loop Checking-Loop checking is the process of determining if a communication line is operating correctly or if it is capable of providing specific types of services.

Loop Length-Loop length is amount of distance a communication circuit between its transmission origin points (e.g. central office switch) and its reception points (e.g. the customer's network termination point).

Loop Plant Improvement Evaluator (LPIE)-A system that analyzes the economics of proposed changes to facilities, such as serving area interface redesign and cable replacement.

Loop Qualification-Loop qualification is the tests or analysis that is performed to determined if a line or communication link (e.g. local loop) has the necessary characteristics to provide specific types of services (such as DSL service).

Loop Reach-Loop reach is the maximum distance that a communication line can provide service to end users.

Loop Signaling-Signaling protocols and processes used in a distribution network, or loop.

Loopback Testing-Loopback testing is a process of testing the transmission capability and functioning of equipment within a system in which a signal is transmitted through a loop that returns the signal to the source. The test verifies the capability of the source to transmit and receive signals.

This figure shows how loopback testing can be used in an IP Telephony system to progressively test, confirm, and identify failed equipments or portions of a network. In this example, the test signals is created by a test device that is connected to a local area data network. This example shows that the first test involves

programming the media gateway to loopback mode so the received test signal from the test device can be returned to the test device. The test device can report if the signal was received and what the quality of the signal is (how many errors). The second test involves programming a remote gateway to loopback mode. This test confirms that the local data network, local media gateway, and wide area network are functioning correctly. The third test in this example sets a remote test device to loopback mode. This test confirms that the local data network, local media gateway, wide area network, remote media gateway, remote data network, and remote test device are working correctly. Failure of one or more of these tests can be used to isolate and help diagnose problems with the system.

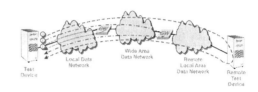

SIP System LoopBack Testing

Looping-(1-software) A programming technique by which a portion of a program is repeated until a certain result is obtained. (2-post production) The replacement of dialogue in post production. The term looping is derived from earlier film processing techniques that used loops of film and magnetic film stock to facilitate dialogue replacement.

Lot Size-A specific quantity of similar material or a collection of similar units from a common source; inspection work, the quantity offered for inspection and acceptance at any one time. The lot size may be a collection of raw material, parts, subassemblies inspected during production, or a consignment of finished products to be sent out for service.

Loudspeaker-A transducer (converter) that transforms audio electrical signal into sound waves (audible signals).

Low Impact Impairment-Low impact impairments are changes to a system or process that results in a small reduction in the ability of the system or process to fulfill the mission(s) it was developed for.

Low Level-(1-MPEG) Low level media formats is a low complexity, low bit rate version of the media. (2-Programming) Low level programming is the creation of programs using commands or instructions that are at or near the level of instructions that are used by the machine or microprocessor (e.g. assembly language).

Low Level API-Low level application programming interfaces are software programs or processes that allow other programs or services to communicate and interact with primitive (elemental) components of a system or service.

Low Level Language-A programming language that reflects the structure of a computer or that of a given class of computers. A low level language consists of instructions that are converted directly into machine code.

Low-Tier-A wireless system which uses low-power levels intended for pedestrians and other slow moving traffic.

Loyalty Index-Loyalty index is a measure of the value a customer has toward future purchases of products or services from a specific company or brand.

Loyalty Program-A loyalty program is a marketing process that identifies and rewards customers for repeated visits or purchases.

LPIE-Loop Plant Improvement Evaluator

LPPC-Localized Pay Per Click

LRA-List Rental Agreement

LRN-Local Routing Number

LRN-Location Routing Number

LRS-Lifetime Revenue per Subscriber

LS-Location Server

LSMS-Local Service Management System

LSO-Laser Safety Officer

LSR-Local Service Request

LTD-Limited Company

LTV-Lifetime Value

Lumped Loading-The use of inductors, spaced at regular intervals along a transmission line (such as loading coils), to improve the transmission characteristics of the line over a specific frequency band.

LWTR-Load And Work Time Record

M

M-Mega

m-Milli

MAC-Medium Access Control

MAC-Message Authentication Check

MAC-Message Authentication Code

MAC-Moves, Adds, And Changes

MAC Address-Medium Access Control Address

Machine Code-Machine code is the instructions for an electronic device that can be decoded and used by the hardware of a microprocessor. Machine code is the direct representation of the computer instruction in memory.

Macro-(1-computers) An abbreviation for macro-instruction, an instruction that generates a larger sequence of instructions for a computer. (2-video) A special function of some zoom lenses that permits an object to be in focus at closer than usual distances to the objective element. The function usually offers magnification of the object. (3-application) A set of stored keystroke sequences or processes that are grouped to allow the user to perform repetitive control or editing sequences of application commands.

Macro Virus-A macro command that attaches itself to application documents that are capable of running macro (multiple keystroke) commands. Macros are commonly used in Word processing or spreadsheet applications to execute repetitive commands and to open, edit, and delete files on a computer.

MAE-Metropolitan Area Exchange

Mag Stripe-Magnetic Stripe

Magalogue-A magalogue is a combination of a magazine that has content and advertising and a catalog of that lists products and services.

Magnetic Card-A magnetic card with a magnetizable layer or set of components that can be used to store information (data storage).

Magnetic Stripe (Mag Stripe)-A strip of magnetic material affixed to a badge, credit card, or other item on which data can be recorded and read.

MAH-Mobile Access Hunting

Mail House-Mailing House

Mail Merge-A mail merge is a process that merges an item to be mailed (such as an email message) with unique information for each item or message that will be sent.

Mail Robot (Mailbot)-A mailbot is a software program or function that allows the automatic processing and routing of email messages. Mailbots may be autoresponders that automatically reply to messages or mailbots may be used to automatically sort and forward messages to specific recipients.

Mail Server-A host, with its associated network software, that offers electronic mail reception and (optionally) email forwarding service. Users may send messages to, and receive messages from, any other user in the system.

Mailbot-Mail Robot

Mailbox (MBX)-A system for storage and transmission of electronic text messages. Mailboxes are often storage areas on computer hard disks that are managed by mail server computers that interconnect to data networks such as the Internet. Mailbox systems often provide notification of an incoming message and confirmation of delivery.

Mailing-The term "mailing" is a marketing term that describes the process of sending (postal mailing or email mailing) direct mail brochures to a group of customers.

Mailing Frequency-Mailing frequency is the number of media items or messages that are sent to the same recipient(s) over a period of time. An example of mailing frequency is the sending of a news update to recipients every 2 to 4 weeks.

Mailing House (Mail House)-A mailing house (also called a letter shop) is a company that provides services to perform or assist with mailing. Examples of mailing services include labeling, assembling, collating, inserting, sorting, and metering.

Mailing List-A mailing list contains the contact information of a group of people who have a common attribute. Mailing lists may contact physical address information (for direct mail) or email addresses (for email marketing).

Main Station-A telephone connected directly to a central office by either an individual or a shared line. The principal telephone of each party on a party line is a main station ion. Telephones that are connected manually or automatically to a central office through a private branch exchange (PBX) or extension telephone are not main stations, but usually are called equivalent main stations.

Mainframe-Computer systems that are used for handling large quantities of central data processing and information storage applications. Mainframe computers are used for applications including invoice creation, account reconciliation, and management information reporting.

Maintenance-Any activity intended to keep a functional unit in satisfactory working condition. The term includes the tests, measurements, replacements, adjustments, and repairs necessary to keep a device or system operating properly.

Maintenance and Repair Organization (MRO)-A maintenance and repair organization is a group or partition of a company that performs services that to keep functional units in satisfactory working condition. MROs may perform tests, measurements, replacements, adjustments, and repairs necessary to keep a device or system operating properly.

Maintenance Center-An operations center that administers all upkeep and repair work in an outside plant network.

Maintenance Contracts-Maintenance contracts are documents or recordings that define the terms for the providing of activities that are intended to keep a device or system in satisfactory working condition.

Maintenance Fees-Periodic fees which must be paid over the life of a patent in order to keep the patent in force. Most countries require the payment of maintenance fees. Failure to pay maintenance fees can result in premature expiration of a patent.

Maintenance Records-Maintenance records are the history of services and test measurements that are performed on networks, systems and transmission lines. Maintenance records help technicians to troubleshoot communication lines and systems as they provide locations and expected performance results (such as optical communication line losses) at the time the systems were installed and setup.

Major Trading Area (MTA)-A geographic region within the United States where most of the area's distribution, banking, wholesaling is performed. The United States has been divided into 51 MTAs and personal communications services (PCS) licenses were granted based on MTA.

Make Busy-(1-general) The setting of a line, trunk, or switched equipment unit to make it unavailable for service. To anyone seeking a connection, the circuit appears to be busy. (2-automatic call delivery) The marking of a customer service representative line as busy ("busy out") so the system will not transfer calls to that phone.

Malfunction-An equipment failure or a fault.

Malicious Call Trace-A process that allows the identification of the location of an undesired caller. Malicious call trace is activated after the recipient has informed the telephone company. Malicious call trace will work even if the unwanted caller's telephone number is blocked. For privacy purposes, the telephone company may only provide the unwanted caller's telephone number to the public safety authorities (such as the police) rather than directly to the recipient of the unwanted call.

MAN-Metropolitan Area Network

Man in the Middle (MITM)-Man in the middle is a process of processing and potentially modifying data by a person or device that is between a sender and receiver of information. MITM is commonly associated with the security threat of receiving, potentially changing and forwarding data as it is passed between devices without the knowledge of the sender or receiver.

Man Machine Interface (MMI)-The man machine interface is the definition of how a user will enter (input) and receive information from a device or system.

Managed Currency-A managed currency is a financial unit measurement where the exchange rate is controlled or influenced by the government that controls it.

Managed IPTV-Managed IPTV is the delivery of IP television services over a managed (controlled) broadband access network. They can control and guarantee the quality of television services. Managed IPTV systems are traditionally provided by telephone (telco) or cable service providers.

Managed Object-An atomic element of an SNMP MIB with a precisely defined syntax and meaning, representing a characteristic of a managed device.

Management Information Base (MIB)-Management information bases (MIBs) are a collection of definitions, which define the properties of the managed object within the device to be managed. Every managed device keeps a database of values for each of the definitions written in the MIB. MIBs are used in conjunction with the simple network management protocol (SNMP) as well as RMON to manage networks. MIBs (referred to now as MIB-i) were originally defined in RFC1066.

Management Information System (MIS)-A management system gathers, organizes, and processes information for a department or company. MIS systems are developed and used by companies to manage its information needs.

Management Layer-Management layer is a collection of related network processing functions that coordinate the overall operation of a system or service. A management layer may be responsible for configuring equipment or protocols in a network, setup (provisioning) of services, system maintenance, and repair (diagnostic) processes.

Management Team-A management team is a group of people who are responsible for overseeing the actions of a project or business. A management team is usually composed of leaders who have different and complementary skill sets.

Manifest-A manifest is a listing of items in a container such as the goods that are being shipped in a truck or cargo ship.

Manual Bidding-Manual bidding is the process of a person adjusting bid amounts so that your bid is larger than the next highest bidder.

Manual Intervention-Manual intervention is the process of altering the automatic or predefined flow of operations to adjust or modify data or processes.

Manual Transaction Processing-Manual transaction processing is the steps and processes taken to complete a transaction (such as an order) without the use of automation equipment. Manual transaction processing can use processing equipment that is used in a mode that is different (reduced functionality) than normal transaction processing.

Manufacturing Costs-Manufacturing costs are the resources and/or fees that are required costs to produce units of product.

MAPI-Messaging Application Programming Interface

Mapping-(1-communication) Mapping is a process of assigning information to specific time, frame or code locations on communication channels or circuits. When the information is received, the mapping process can be used to extract the channels or information from the time, frame, or code positions as needed. (2-location) Mapping is the process of gathering location information so that the locations and possibly the attributes of the items (e.g. roads or towers) can be presented in a form that is interpreted by a user.

Margin-(1-performance) The difference between the value of an operating parameter and the value that would result in unsatisfactory operation. Typical parameters include signal level, signal to noise ratio (SNR) , distortion, crosstalk coupling, and/or undesired emission level. (2-receiver) The signal power available to a receiver in excess of its design limit. (3-financial) Margin is the difference between the revenue produced by a product or service and the cost to produce or provide the product or service.

Marginal Costs-Marginal costs are the resources and/or fees that are required costs to produce additional units of product.

Market Analysis-Market analysis is a review of market information such as the sales of specific types of products or services to determine the key characteristics of a market or industry.

Market Convergence Management -A formal approach used by the managers of different telephone company (telco) product groups to combine their separate smaller customer populations into a much larger, shared pool, allowing the maximizing of revenues through cross selling, brand extension and churn proofing.

Market Development Funds (MDF)-The allocation of funds or sales credit allowances that are given by manufacturers as incentives to retailers to promote their products.

Market Price-Market prices are the values that consumers are willing to exchange for a product or service.

Market Segment-Market segment is a group of potential customers for a product or service that share common characteristics.

Market Share-Market share is a comparison of the amount of products or services that are sold by a particular company to the total number of products and services that are sold for those types of products and services.

Marketing-Marketing is the process of promoting and selling products or services. Marketing is commonly divided into product (item or service) , price (retail, wholesale) , promotion (communication) , and place (distribution) categories.

Marketing Agency-A marketing agency is a company that represents or manages the services of other companies for the purposes of providing media communication services.

Marketing Campaign-A marketing campaign is any of a broad range of marketing activities designed to send messages to customers about products, services, and options that are or will be available. Marketing campaigns can be executed via advertising, direct marketing, public relations, place, or other media.

Marketing Channels-Marketing channels are marketing materials and communication channels that are used to promote products or services to people or companies.

Marketing Communications-Marketing communications is the process or division at a company that coordinates the flow of media information to and from a company. Ideally, marketing communications coordinates and controls the brand and image of a company by ensuring all the media and communications with the company contain desired and consistent messages.

Marketing Management-Marketing management is the process that identifies, defines and tracks media communication information (such as brochures) that is sent between companies and customers.

Marketing Plan-A marketing plan contains the objectives of the marketing process, the responsibilities and incentives of those involved in the marketing process, and the resources that will be available or used for the marketing programs. The marketing plan usually includes objectives (e.g. communication targets) and it may identify the promotional companies or representatives and their responsibilities along with their territories and incentive structures.

Marketing Program-A series of related marketing campaigns assembled to accomplish a single objective (i.e. a customer retention program could be a series of advertising, sales, and direct marketing campaigns with a related set of messages, concepts, and icons).

Marketing Requirements Document (MRD)-A marketing requirements document is a set of requirements that are necessary to meet the needs or desires of typical users of a system and/or services.

Marketing Research-Marketing research is information that is gathered, analyzed and converted into a form that can be used to represent information or characteristics of companies, products or services as related to their sales or business performance.

Marketing Strategy-Marketing strategy is the planned actions and processes that will contribute to the successful marketing of products or services.

Marketing Tactics-The actions that are taken by people or companies to achieve marketing objectives.

Marketplace-A marketplace is the businesses and customers that buy specific types of products or services.

Markov Model-A statistical model of the behavior of a complex system over time in which the probabilities of the occurrence of various future states depend only on the present state of the system, and not on the path by which the present state was achieved. This term was named for the Russian mathematician Andrei Andreevich Markov (1885-1922).

Markup-Markup is the difference between the offering prices to the cost of a product or service.

Markup Schedule-A markup schedule is an itemized list or table that provides pricing adjustment information for products or services. The price markup schedule will usually include the percentage and/or amount of adjustment based on the type of items, quantity of product purchased and the types of customers that determine the amount of adjustment (e.g. retail or reseller).

Mashup-Mashup (Web Application Hybrid) is a web application that combines data from more than one source into a single integrated tool. An example of a mashup is the use of cartographic data from Google Maps to add location information to real-estate data from Craigslist, thereby creating a new and distinct web service that was not originally provided by either source.

Master Control-A master control is a device, assembly or console that has overriding authority over other controls or consoles in a system.

Master Control Room (MCR)-A master control room is the coordinating center for broadcast communication systems. The master control room coordinates the collection of programming content, the conversion of the content into a format that can be distributed, schedules the programming and monitors the overall operation of the broadcast system.

Master/Slave-(1-video editing) A system in which one or more video tape recorders (VTR) slaves are controlled by another VTR master. (2-sync generator) A system in which several video sync generators (slaves) are controlled by one main sync generator (master). (3-Bluetooth) A relationship between devices where one device coordinates communication (the master) and the other device (the slave) follows the commands of the master.

Material Release-Material release is an authorization to transfer or provide products to a person or company.

Materials Requirements Planning (MRP)-Material requirements planning is a set of processes that are used to identify, forecast and schedule the acquisition of materials that are used to produce products or services.

Matrix-(1-general) A logical network configured in a rectangular array of intersections of input/output signals. (2-disk manufacture) Nickel electroplated onto a lacquer master, forming a negative image of it, and from which the metal is produced. (3-electronics) A routing or switching array with multiple inputs and outputs. (4-mathematics) An arrangement of numbers representing the coefficients in simultaneous linear equations. (5-microphone technique) A circuit combining a unidirectional and a directional microphone into M-S stereo. (6-optical recording) A method of recording for playback channels onto two discreet optical tracks, also referred to as a 4-2A matrix. (7-TV receiver) A circuit that combines the luminance and color signals and transforms them into individual red, green, and blue signals. In a TV set, these signals then are applied to the picture tube grids.

Maximum Busy Hour-The busiest hour of the busiest day of a normal week, excluding holidays, weekends, and special event days.

MB-Megabyte

MBONE-Multicast Backbone

Mbps-Mega Bits Per Second

MBps-Mega Bytes Per Second

MBR-Multi-Bit Rate

MBR-Multiple Bit Rate

m-Business-Mobile Business

MBX-Mailbox

MC-Message Center

MC-Multicarrier Mode

MCC-Mobile Country Code

M-Commerce-Mobile Commerce

MCR-Master Control Room

MDF-Market Development Funds

MDMF-Multiple Data Message Format

MDN-Mobile Data Network

MDN-Mobile Directory Number

MDS-Media Distribution System

MDT-Mobile Data Terminal

MDU-Multiple Dwelling Unit

Meal Penalties-Meal penalties are fees that are paid to an actor by a producer for not providing the actor or talent with scheduled meal time breaks. These fees are charged when the producers violate the SAG guidelines requiring them to alot a specific amount of time to an actor for meal breaks.

Mean-In statistics, an arithmetic average in which values are added and the sum divided by the number of such values.

Mean Opinion Score (MOS)-Mean opinion score (MOS) is a measurement of the level of quality as perceived by a person. The MOS is number that is determined by a panel of viewers or listeners who subjectively rate the quality of audio on various samples. The rating level varies from 1 (bad) to 5 (excellent). Good quality telephone service (called "toll quality") has a MOS level of 4.0.

Mean Time Between Failures (MTBF)-For a particular time period (typically rated in hours) , the total functioning lifetime of an assembly or item divided by the total number of failures for that item within the measurement time interval.

Mean Time to Failure (MTTF)-Mean time to failure is an estimated time period (typically rated in hours) of the functioning time of an assembly or item divided before it is expected to fail.

Mean Time to Repair (MTTR)-Mean time to repair is an estimated time period (typically rated in hours) of the time it takes to repair or return an assembly from a failed condition to an operational condition.

Measured Load-The load that is indicated by the average number of busy servers in a group over a given time interval.

Measured Rate-Measured rate is a billing fee structure in which the monthly charge services or facilities is specified for a number of events (such as calls made) within a defined region (such as a limited geographic area).

Measurement-(1-general) A procedure for determining the amount of a quantity. (2-data) The output of a data collection system that indicates the load carried or service provided by a group of telecommunications servers.

MECABS-Multiple Exchange Carrier Access Billing

Media-Media is information that may be stored or transmitted such as voice, data or video.

Media Adapter-A media adapter is a device or assembly that can convert media that is in one format (such as analog video) into another format (such as digital video).

Media Attachments-Media attachments are media files or objects that are linked to or sent along with other messages or transmitted items.

Media Broker-A media broker is a person or company who assists advertisers to find and acquire media communication resources (such as television advertising time).

Media Buyer-A media buyer is a person or company that acquires media communication resources such as advertising space or television advertising time.

Media Components-Media components are the parts of a multimedia session or file. Media components may include multiple streaming connections or the physical media parts of a multimedia file (such as a web page that includes text, images and video).

Media Compression-Media compression is the process of transforming digital information into a form that requires a smaller amount of space for storage.

Media Device-A media device is an interface assembly that can render, store, redistribute and/or export media.

Media Distribution-Media distribution is the process of transferring information between content providers and content users. The types of media distribution include direct distribution, multilevel (superdistribution) and peer to peer distribution.

Media Distribution System (MDS)-A media distribution system is a network that can receive, route and deliver media. Media distribution systems may contain processing

systems that can encode, encrypt and add copy protection information to the media signals.

Media Encoder-A media encoder is a device or circuit that converts a signal into a media format that is suitable for transmission over a communication channel.

Media Extension-Media extension is the process of re-purposing content for use with other devices or interfaces. An example of media extension is the reformatting of HDTV media so that it can be viewed on a mobile device.

Media Format-Media format is a method of containing audio, video, and/or other digital media within a file structure. Media formats are usually associated with specific standards like MPEG video format, or software vendors like Quicktime MOV format or Windows Media WMA format. In a few cases like MP3 files, the media "format" is little more than a single codec bitstream in a file. However in most cases the media format is not to be confused with the codecs used for any compressed bitstreams within specific files of that format.

Media Gateway (MG)-A media gateway is a network component which converts one media stream to another media stream. In IP telephony this most commonly refers to a device which converts IP streams (such as audio) to the TDM or analog equivalent. A media gateway may interact with call controllers, proxies, and soft switches via proprietary or standard protocols such as MGCP, Megaco (H.248), and SIP.

This diagram shows the functional structure of a media gateway (MG) device. This diagram shows that this gateway interfaces between a public telephone network line side analog connection to a Internet packet (IP) data network connection. The overall operation of the voice gateway is controlled by a media gateway controller (MGC.) The MGC section receives and inserts signaling control messages from the input (telephone line) and output (data port). The MGC section may use separate communication channels (out-of-band) to coordinate call setup and disconnection.

Signals from the public telephone network pass through a line card to adapt the information for use within the media gateway. This line card separates (extracts) and combines (inserts) control signals from the input line from the audio signal. Because this audio signal is in analog form (another option could be an ISDN digital line side connection,) the media gateway converts the audio signal to digital form using an analog to digital converter. The digital audio signal is then passed through a data compression (speech coding) device so the data rate is reduced for more efficient communication. This diagram shows that there are several speech coder options to select from. The selection of the speech coder is negotiated on call setup based on preferences and communication capability of this media gateway and the media gateway it is communicating with. After the speech signal is compressed, the digital signal is formatted for the protocol that is used for data communication (IP packet.) This diagram shows that the call processing section of the media gateway is not part of the gateway. It is a separate controller that commands the gateway to insert mes-

sages in the media stream (in-band signaling) or it may communicate with the other gateway through another media gateway controller (MGC.)

Media Gateway Controller (MGC)-The media gateway controller is the portion of a PSTN gateway that acts as a surrogate call management system (CMS). The MGC controls the signaling gateway and the media gateway (MG). The protocols between the MGC and MG include media gateway control protocol (MGCP) , MEGACO, and H.323. The MGC acts as a call agent coordinating sessions between devices. Signaling between MGCs (agents) may use SIP or H.323 protocol.

Media Hub-A communication device that distributes or adapts multiple types of communication media to one or several devices in a network through the re-broadcasting of data that it has received from one (or more) of the devices connected to it.

Media Ingestion-Media ingestion is the process of transferring media into a storage or content management system.

Media Key-A media key is a unique code (a key) that is used to decode encrypted media. The media key may be created (such as in the AACS system) by using a unique device key in combination with a media key block (MKB).

Media Kit-A media kit is a set of information brochures, images or data (media) that describe a product or service.

Media Player-A media player is a software application and/or device that can convert media such as video, audio or images into a form that can be experienced by humans. Media players may contain support for service different media formats, compression (codec) formats as well as being able to communicate using multiple network streaming protocols.

Media Portability-Media portability is the ability to transfer media from one device to another. Media portability can range from storing media locally received from a system

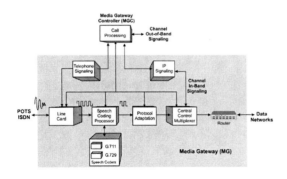

Media Gateway Operation

in a hard disk (such as a personal video recorder) to sharing media through home connections (such as a premises distribution network).

This figure shows how digital rights management (DRM) may allow for media to be transferred from one device to another. In this example, a digital movie is downloaded to an IP Set Top Box (IP STB). The IP STB can then transfer the movie to a portable media player. The user will be allowed to transfer and view

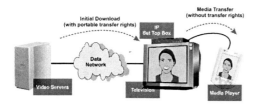

DRM Media Portability

the movie provided the DRM information is set to allow the transfer and viewing and the keys or viewing authorization information has not expired.

Media Processing-Media processing is the operations that are used to transfer, store or manipulate media (voice, data or video). The processing of media ranges from the playback of voice messages to modifying video images to wrap around graphic objects (video warping). Media processing in set top boxes includes video processing, audio processing and graphics processing.

Media Provider-A media provider is a person or company that provides content or data to companies or consumers.

Media Rights Manager-A media rights manager is a device or application that is used

to assign, monitor or control the assignment of rights for media storage, transfer or rendering.

Media Server (MS)-(1-digtial media) Media servers (sometimes called streaming servers) are computers that receive requests for media, setup a communication session to the requesting media client and provides the downloading or continuous transmission (streaming) of digital media. (2-telephone) Media servers provide common telephony features and/or specialized telephony capabilities to communication systems. The media servers' many functions are to process call connections and manage media access to media resources. Media servers can be hardware based or software-based. Hardware based media servers are specifically designed to efficiently and effectively perform call processing and media management. Software based media servers use software that operates on common computing equipment. Examples of media servers include announcement servers, conference servers, voicemail servers and CALEA server.

Media Transmission-Media transmission is the transferring of information or data symbols that represent media by wire, fiber optic or radio means. Media transmission may be in the form of point to point (1 to 1), point to multipoint (1 to many) or broadcast (1 to all).

Media Value Chain-A media value chain is the operational model that describes the core functions that are required to deliver media products or services to the end customer. The blocks in a typical media value chain include content producers, content aggregators, media servers, distribution systems and media players.

Media Workstation-A media workstation is a computing system that allows people to edit and produce programs.

Median-A median is a centralized value in a series that has as many readings or values above it as below it. A median is not influenced much by significant variations, as an average value would be affected.

Mediation-Mediation is the receiving, processes and reformatting of data or information into other formats.

Mediation Device-A mediation device is a network component in a telecommunications network that receives, processes, reformats and sends information to other formats between network elements. Mediation devices are commonly used for billing and customer care systems as these devices can take non-standard proprietary information (such as proprietary digital call detail records) from switches and other network equipment and reformat them into messages billing systems can understand.

This figure shows a mediation system that takes call detail records from several different switches and reformats them into standard call detail records that are sent to the billing system. This diagram shows the mediation device is capable of receiving and decoding proprietary data formats from three different switch manufacturers. The mediation device converts these formats into a standard call detail record (CDR) format that can be used by the billing system.

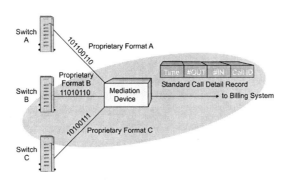

Mediation System Operation

Mediation Layer-A mediation layer adapts information or processes from one layer into a form that can be used by another layer within a device or system.

Mediation System-A mediation system adapts information or processes from one or more systems into a form that can be used by another system. These processes may include protocol conversion, message routing and store-and-forward processing.

Medium Access Control (MAC)-Medium access control is the processes used by communication devices to gain access to a shared communications medium or channel. Examples of MAC systems CSMA/CD and Token Passing.

A MAC protocol is used to control access to a shared communications media (transmission medium) which attaches multiple devices. MAC is part of the OSI model Data-Link Layer. Each networking technology, for example Ethernet, Token Ring or FDDI, have drastically different protocols which are used by devices to gain access to the network, while still providing an interface that upper layer protocols, such as TCP/IP may use without regard for the details of the technology. In short, the MAC provides an abstract service layer that allows network layer protocols to be indifferent to the underlying details of how network transmission and reception operate.

This diagram shows the key ways networks can control data transmission access: non-contention based and contention based. This diagram shows that non-contention based regularly poll or schedule data transmission access attempts before computers can begin to transmit data. This diagram shows that a token is passed between each computer in the network and computers can only transmit when they have the token. Because there is no potential for collisions, computers do not need to confirm the data was successfully transmitted through the network. This diagram also shows contention based access control systems allow data communication devices to randomly access the system through the sensing and coordination of busy status and detected collisions. These devices first listen to see if the system is not busy and then randomly transmit their data. Computers in the contention-based systems must confirm that data was successfully transmitted through the network, because there is the potential for collisions.

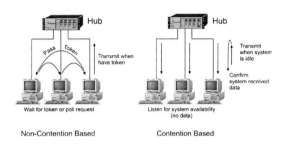

Medium Access Control (MAC) Operation

Medium Access Control Address (MAC Address)-The medium access control (MAC) address used to distinguish between units participating in a data network. MAC addresses are low-level address and are only associated with the MAC layer of the system that the data device is operating in.

Meet Me Conference-A telephone conference call arrangement, usually on a private PBX system that enables callers to use an access code to connect to a specific conference call.

Meet Point-A point, designated by two exchange carriers, at which one carrier's billing responsibility for service begins and the other's ends. There can be one or more meet points on a circuit.

Meet Point Billing (MPB)-Billing systems that must meet when unbundled network elements (UNE) or access services are provided by two or more providers, or by one provider in more than one state.

Mega (M)-A prefix meaning one million.

Mega Bits Per Second (Mbps)-A measurement of digital bandwidth where 1 Mbps =1 million bits per second (1,000,000 bits per second). The word "mega" is sometimes used to describe the nearest integral power of 2, namely 1,048,567.

Mega Bytes Per Second (MBps)-A measurement of the amount of information being transferred on a communications link in one second where 1 MBps =1 million bytes per second (1,000,000 8 bit bytes per second).

Megabyte (MB)-A megabyte is one million bytes of data. When megabyte is used to identify the amount of data storage space (such as computer memory or a hard disk) , a megabyte commonly refers to 1,048,576 bytes of information.

Megahertz (MHz)-One million hertz, or cycles per second.

Megapixel-A Megapixel is one million image elements (pixels).

Membership Activity-Membership activity is any action that is taken by a person or company who is a member of a program or subscription service.

Memorandum Of Understanding (MOU)-(1-general) A statement or agreement stating the objectives and scope of a project or plan. (2-GSM) A legal agreement between the GSM committee members to create the GSM network and its revisions. The GSM MOU was signed in 1987.

Memory-Memory is a device, circuit, or system that stores data for retrieval at a later time. The types of memory range from real time random access memory (RAM) to removable media components such as optical disks.

Memory Capacity-The total number of bits or bytes that can be stored in a device or assembly within a device.

Memory Chip-A memory chip is a component storage device that can hold data or media. A memory chip may temporarily hold its contents (volatile memory) or it may permanently hold data (non-volatile memory).

Memory Footprint-Memory footprint is the amount of memory or storage area a program requires to operate.

Memory Stick-A memory stick is a portable memory storage device. A memory stick usually can connect to a standard data connection such as a USB port.

Memory System-A memory system is a combination of storage mediums (such as computer hard disk) and information management system that can identify, retrieve and/or store data or media.

Menu Screen-A computer monitor or display screen format that lists a set of options from which users make a choice.

Merchant Account-A merchant account is a unique identifier that is used to group charges and credits associated with a specific company. The term is commonly associated with credit card merchant processing where the credit card transactions associated with specific company are posted and settled. The credit card transactions are performed by a merchant processor.

Merchant ID-Merchant Identifier

Merchant Identifier (Merchant ID)-A merchant identifier is a code or a label that uniquely identifies a company (a merchant) to financial institution (e.g. a bank) for the processing of financial transactions (such as credit card processing).

Merchant Processor-A merchant processor is a company that processes credit card transactions for companies where money from the transaction is paid to the merchant and the money to pay for the transaction is collected from the company that issues the cards.

Merging-Merging is the process of combining new data with existing data.

Message-(1-general) Any idea expressed briefly in a plain or secret language and prepared in a form suitable for transmission by any means of communication. (2-data communications) A set of information, typically digital and in a specific code (such as binary or ASCII), carried from a source to a destination.

(3- ISDN) In the integrated services digital network (ISDN), a set of layer 3 information that is passed between customer premises equipment and a stored-program control switching system for signaling. (4-telephone communications) A communication session or a successful call attempt.

Message Authentication-Message authentication is a security process that verifies the identity of the person or device that is sending a message.

Message Authentication Check (MAC)-Message authentication check is the process that is used to confirm that the message was originated from an identified sender.

Message Authentication Code (MAC)-A message authentication code is a unique number that is used to identify that a person or device created a message.

Message Body-The data information or message words that are contained in a communication message.

Message Center (MC)-The message center is a node or network function within a communications network which accommodates messages sent and received via short messaging service (SMS).

Message Format-The rules for the placement of such portions of a message as its heading, address, text, and end.

Message Investigation-A generic term used to describe the processes and group(s) responsible for investigating call detail records (CDRs) that are rejected by the rating engine portion (rate selection and cost allocation) of a billing system.

Message Minute Miles-The total number of message minutes carried on a trunk or circuit group, multiplied by the average route miles.

Message Minutes-The connection time used by a customer for transmission.

Message Processing System (MPS)-The function within a billing system that processes the events recorded in the network. MPS is sometimes referred to as the rating engine. Typically the rating engine receives events from the network, reformats each event into an internal standard, identifies the customer to be billed (see "Guiding") , and assigns a rate (see "Rating") based on parameters such as: date, day-of-week, rate period, call type, jurisdiction, an others.

Message Rate Service-Message rate service is a billing fee structure where the user pays for each event (call) based on the number of message units assigned for that particular call's destination and duration.

Message Recording-A message coding system used to distinguish customer dialed messages from operator completed messages.

Message Retrieval-The process of locating a message that has been entered in a telecommunications system.

Message Service (MS)-Message service is the transferring of small amounts of data or text (usually several hundred characters) between devices. Messaging services can be sent point-to-point (paging or email) or broadcast without acknowledgement (e.g. traffic reports).

Message Type-A message that is defined according to how it is billed and the way in which it is paid. Message types include: sent paid noncom, third number, credit card, collect, special collect, coin sent paid, and coin collect. Message type also is called message class.

Message Waiting Indicator (MWI)-A feature that informs a user that they have messages waiting in email, voice mail, or video mail. Optionally it may indicate how many mail messages are waiting without the user having to call their voice mailbox. MWI may use unique tone (rapidly changing dial time) or an indication on the telephone device (such as a light) as an indication of message waiting. MWI should not impact a subscriber's ability to originate calls or to receive calls. If the dial tone is altered to indicate a message is waiting, it will typically reset to a standard dialtone after a time period or after the user has re-established the connection. As a result, MWI may affect auto-dialers (such as modems) that sense for a dialtone signal

before sending the dialed digits. Message waiting indication is also called message waiting notification (MWN)

Messaging-A telephone system feature that alerts station users via a lighted lamp or other visual display or by an interrupted dial tone that messages are waiting.

Messaging Application Programming Interface (MAPI)-A standard program interface that was developed by Microsoft to allow the transfer of messages between software applications.

Messaging Services-Messaging services are the transfer of short information messages between two or more users in a communication system. Mobile messaging services are typically limited to a few hundred characters per message.

Meta Ad-A Meta Ad is a descriptive script about an advertising message or banner that will be displayed that can change based on certain criteria (such as the location, time of day or key word searched).

Meta Tag (Metatag)-Meta-tags are text identifiers that are used in a web page (such as a HTML or ASP page) to identify a type of information that will follow. An example of a Meta tag is <meta name="description" content="Web Marketing Dictionary, Internet Web Marketing Industry Terms and Definitions."

Meta Tag Management-Meta tag management is the process of acquiring, maintaining, distributing and the deletion of meta tags that describe program content.

Metadata-Metadata is information (data) that describes the attributes of other data. Metadata or meta tags are commonly used in databases where the attributes of a particular column of data are defined. For example: the phone number column of the employee data table contains groups of numbers 3 or 4 characters in length separated by a hyphen "-".

Metadata Coding & Transmission Standard (METS)-Metadata coding & transmission standard is an XML document format that is used to encode the metadata to manage digital library objects within and between repositories.

Metadata Normalization-Metadata normalization is the adjustment of metadata elements into standard terms and formats to allow for a more reliable organization, selection and presentation of program descriptive elements.

Metadata Server-A metadata server is a computer that can receive, process, and recommend media programs or applications that have matching or related descriptive metadata attributes.

Metafile-A metafile contains both data (content) and control (formatting) information. An example of a metafile is an HTML web page that contains text, images, and formatting information.

Metatag-Meta Tag

Meter-(1-SI unit) The meter of length is now precisely defined as the length of the path traveled by light in vacuum during the time interval 1/ 299 792 458 of a second. It can be equivalently defined as 1 650 763.73 wavelengths of the orange-red light from the isotope krypton-86, measured in a vacuum. Historically, the meter was 1/40 000 000 of a meridian of longitude passing through the poles, as determined in 1797 by a land survey of a partial meridian distance between Barcelona, Spain and Dunquerque, France. (2-ampere-hour) A device that integrates current and time to indicate the number of ampere-hours of power consumed by a load. (3-field strength) A combination radio receiver and meter (calibrated for use with a particular antenna) designed to give a direct reading of the strength of a radio signal at a given point. (4- instrument) A device for measuring the value of some quantity. (5-running time) A totaling clock that runs whenever a device is in operation. Such meters are used with various types of equipment so that maintenance work can be carried out at appropriate times.

Metering-Metering is the process of tracking a quantity of service or material over a period of time or event period.

Method of Notice-A method of notice is the communication channel or process that is used to alert people or companies of events or conditions. An example of a method of notice is the posting of a company stock offering in a news publication to alert the public of the financial changes in a company structure.

Metric System-A decimal system of measurement based on the meter, the kilogram, and the second.

Metrics-Metrics is the gathering and/or use of values that indicate the status or performance of a system or service.

Metropolitan Area Exchange (MAE)-Network points that interconnect ISP points of presence and national backbones. They provide a similar funcation as Network Access Points (NAPs).

Metropolitan Area Network (MAN)-A MAN is a data communications network or interconnected groups of data networks that have geographic boundaries of a metropolitan area. The network is totally or partially segregated from other networks, and typically links local area networks (LANs) together.

This diagram shows a five node MAN connecting that connects several LAN systems via a

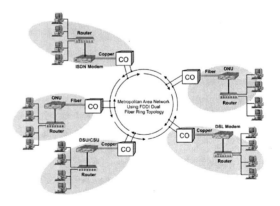

Metropolitan Area Network (MAN) System

FDDI system. This diagram shows that each LAN may be connected within the MAN using different technology such as T1/E1 copper access lines, coax, or fiber connections. In each case, a router provides a connection from each LAN to connect to the MAN.

Metropolitan Service Area (MSA)-Metropolitan service area or metropolitan statistical area. An area designated by the FCC for service to be provided for by cellular carriers. There are two service providers for each of the over 300 MSA's in the United States.

Metropolitan Statistical Area (MSA)-A geographic area in the United States that typically includes a city that has a population of at least 50,000 people that have similar economic and social characteristics. MSAs are defined by the federal government and used to gather and report statistics.

METS-Metadata Coding & Transmission Standard

MExE-Mobile Execution Environment

MG-Media Gateway

MG-Minimum Guarantee

MGC-Media Gateway Controller

MHD-Mobile Handheld Device

MHEG-Multimedia/Hypermedia Expert Group

MHP-Multimedia Home Platform

MHz-Megahertz

MIB-Management Information Base

MIB Instance-A MIB instance is a suffix identifier associated with a particular MIB object. Usually the instance has a value of "0" when the MIB object is to return 1 value. The instance suffix identifier can increment as well like in gathering interface statistics, where there can be more than 1 interface and many values for 1 object.

For example: In the case of the Octets MIB object it would look like: 1.3.6.1.2.1.2.2.1.10.<instance number> Where the instance number is the interface number for which you want the received octets for, so if you wanted the first interface then the MIB object definition would look like: 1.3.6.1.2.1.2.2.1.10.1

Micro Browser (Micro-Browser)-Micro-browser is a software application used to display web page documents in wireless devices (usually using WML formats). Various micro-browsers are available for different types of wireless devices.

Micro-Browser-Micro Browser

Microchip-A common term for an integrated circuit component.

Microcomputer-A small-scale program or machine that processes information; it generally has a single chip as its central unit and includes storage and input/output facilities in the basic unit.

Microcontent-Microcontent is data or information that conveys one primary idea or concept, is accessible through a single definitive URL or permalink, and is appropriately written and formatted for presentation in email clients, web browsers or on handheld devices as needed. A day's weather forecast, the arrival and departure times for an airplane flight, an abstract from a long publication or a single instant message can all be examples of microcontent.

Microfilter-A microfilter is a small filter device that attaches to a communications jack (such as a telephone jack) that blocks unwanted signals from entering into the communication device (such as a telephone). A DSL microfilter is a blocking filter device that attaches to a telephone jack that blocks unwanted high-speed data signals from entering into the telephone.

Microkernel-A microkernel is a software part of an operating system that is fundamental to its overall operation that has been efficiently designed to operate with more limited resources than standard operating systems (such as operating within a mobile phone). The microkernel software continually manages specific processes such as system memory, the file system and disk operations. A kernel may also run processes such as the communication between applications, coordinating the input and output of information within an application. Once loaded, the microkernel software usually remains stored and operates from computer memory and is typically not seen by the users of the system.

Micromarketing-Micromarketing is focusing of marketing programs to match small groups of users. Micromarketing may divide groups into small geographic areas or individual characteristics.

Micrometer-One-millionth of a meter. Usually abbreviated as um.

Micron-A unit of length equal to one millionth part of a meter. Also called a micrometer, written μm (not to be confused with a measuring device called a micrometer).

Micropayment-A micropayment is a transfer of value in return for the receiving of goods or services that involve a small amount of value (e.g. money).

Microprocessor-A single package (normally a single chip) electronic logic unit capable of executing from external memory a series of general-purpose instructions contained in the external memory. The unit does not contain integral user memory although memory on the chip may be present for internal use by the device in performing its logic functions.

Microsecond (usec)-One millionth of a second.

Microsoft Disk Operating System (MS-DOS)-Microsoft disk operating system is a group of software programs and routines that directs the operation of a computer in its tasks and assists programs in performing their functions. The operating system software is responsible for coordinating and allocating system resources. This includes transferring data to and from memory, processor, and peripheral devices. Software applications use the operating system to gain access to these resources as required. MS-DOS is a registered trademark of Microsoft.

Microtransaction-A microtransaction is a transaction for goods or services that involve a small amount of value (e.g. money).

Microwatt (mW)-One millionth of a Watt.

Microwave-(1-radio) The portion of the electromagnetic spectrum between approximately 1 GHz and 100 GHz. (2-heating device) A radio oven that operates at approximately 2.4 GHz..

Microwave Dish-An antenna system that uses a parabolic-shaped reflector to reflect received signals to a specific focal point (signal feed element). Because of the high signal gain and the requirement for precise positioning, dish antennas are commonly used for transmission and reception from point-to-point microwave stations and fixed position GEO communications satellites.

Microwave Radio Relay System-A point-to-point radio transmission system in which microwave radio signals are received, amplified, and retransmitted by one or more transmission radios or systems. These systems may use Plesiochronous Digital Hierarchy (PDH), Synchronous Digital Hierarchy (SDH), or other forms of communication channels.

Mid Section-In a lumped cable loading system, the middle of a load section between loading coils.

Mid-Call Teardown-Mid-call teardown is the process of ending a connection between users before either call party has elected to terminate the call. Mid-call teardown may be performed when a prepaid account balance has fallen below a specified amount or to enable the communication resources to be reassigned to other priority users (such as pubic safety officials during a crisis event).

Middleman-A middleman is a person or a company that is located between a supplier and a customer.

Middleware-Middleware is the software programs that operate between the core application layer of a system and a lower layer of the network. An example of middleware is electronic programming guide (EPGs) that resides in cable converter boxes that allow a customer to select from a list of available video programs.

his diagram shows how middleware is used on an IP Television system to link together the end user's equipment (IP Set Top box) to the media management and delivery systems. This diagram shows that the users set top box has an operating system (e.g. Linux or Windows) which controls the components of the set top box (such as infrared remote control and digital video creation). Middleware is the software that communicates with the user and the operating system to send and receive commands and media from the IP television service provider. The IP television service provider has middleware on their media management system that communicates with end customers and manages the selection and delivery of media to the end user. The middleware system also provides information to the billing system to allow the IP television service provider to track usage and create customer billing records.

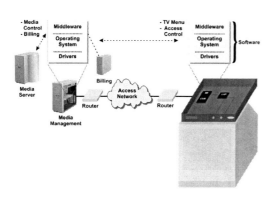

IP Television Middleware

Middleware Compatibility-Middleware compatibility is the ability of a device to accept software programs (clients) that interface the device to the system host (servers). A middleware client is a software module that is installed in a device that is configured to request and deliver media or services from a server (e.g. to request television programs from media network).

MIDP-Mobile Information Device Profile

Migration-Migration is the process of converting customers, software applications, or equipment from one technology or system to another technology or system.

Mil Specs-Military Specifications

Mileage Band-A group of individual mileage steps, such as from 0 to 50 miles, or 50 to 100 miles, measured in airline miles and used to determine billing rates for telecommunications services. This billing rating process is also called "banding."

Military Products-Military products are devices or services that are purchased or used by military organizations. Military products are commonly designed to military specifications (Mil Specs) which can have extreme operational performance and durability requirements.

Military Specifications (Mil Specs)-Military specifications are documents or data that contains physical, operational, or performance requirements for devices or services purchased or used by military organizations.

Milli (m)-A prefix used in a unit of measure meaning one thousandth.

Million Instructions Per Second (MIPS)-A unit of measure for the millions of processing steps that can be accomplished by a computer processing device in one second.

Millions of Operations Per Second (MOPS)-Millions of operations per second is the measure of a system or processor within a system to perform retrieve, manipulate and store information in units of millions.

Millisecond (msec)-One-thousandth of a second (0.001 S).

Millivolt (mV)-One-thousandth of a volt (0.001 V).

MIME-Multipurpose Internet Mail Extensions

MIN-Mobile Identification Number

Minimum Cost Routing-In automated facility planning, a circuit-routing scheme that determines a path through the network for each point-to-point demand for each year so that, when point-to-point demands are provided on these paths and the resulting capacity expansion problem is solved, the total cost of transmission facilities is minimized. Minimum cost routing is not related to least cost routing.

Minimum Guarantee (MG)-Minimum guarantees are a fixed cost that is assessed for the purchase or use of products or services when the usage or rated cost basis falls below the fixed amount.

Minimum Point Of Entry (MPOE)-A minimum point of entry is a location where communication lines first enter the customer's building or facility. An MOPE can be a network interface termination (NT) or wiring closet inside the building.

Minimums-Minimums are the lowest wages (pay) that can be paid which are assigned by an authority (such as a union or guild).

Minutes Of Use (MOU)-A measurement (usually billing related) of the number of minutes, actual or assumed, in traffic-sensitive (usage) equipment and facilities.

MIPS-Million Instructions Per Second

MIPTV-Mobile Internet Protocol Television

Mirror Site-A duplicate data or Internet Web site. Mirror sites are used to process communication traffic to local or regional areas as each mirror sites contain the same information as the other mirror site. Mirror sites are also used for mission-critical applications that allows a company or user to continue processing information in the event of system failure or network loss due to natural disasters.

Mirroring-A fault prevention architecture in which a backup data storage device maintains data identical to that on the primary device, and can replace the primary if it fails.

MIS-Management Information System

Miscellaneous Fees-Miscellaneous fees are costs or assessments of value for goods or services that cannot be assigned into a category.

Mission Critical-Mission critical are applications or services (missions) that are essential to the operation, performance or safety of individuals or companies.

Mission Statement-A statement of the primary mission or objectives of a company.

MITM-Man in the Middle

Mixed Media Services-Mixed media services are combinations of multiple media types that may be combined to enhance or produce new services.

Mixer-(1-general) A circuit used to combine two or more signals to produce a third signal that is a function of the input waveforms. (2-audio) An audio console used to switch and combine various audio sources to produce a finished output (3- broadcast) The studio control console or other unit used to combine or "mix" the various program elements into a final program that is sent to the transmitter. (4- receiver) The stage in a superheterodyne radio re receiver at which the incoming signal is modulated with the signal from the local oscillator to produce an intermediate frequency signal. (5-video) A European term for production switcher. The complete term is vision mixer.

This diagram shows how a mixer combines two signals to produce a sum or difference fre-

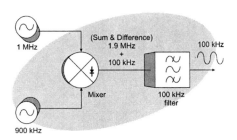

Mixer Operation

quency. This diagram shows this mixer contains a diode (non-linear device) that allows the two-incoming signals to interact with each other to produce the difference (subtractive) frequency and sum (additive) frequencies. The output of this mixer circuit contains a tuned circuit (resonant circuit) that only allows the difference frequency to transfer out of the mixer.

MM-Mobility Management

MMI-Man Machine Interface

MMP-Mobile Media Player

MMS-Multimedia Messaging Service

MNC-Mobile Network Code

Mnemonic-A memory aid in which an abbreviation or arrangement of symbols has an easily remembered relationship to the subject.

Mnemonic Address-A simple address code with some easily remembered relationship to the actual name of the destination, often using initials or other letters from the name to make up a pronounceable word.

MNP-Mobile Number Portability

Mobile Access Gateway-Mobile access gateways are communications devices that transforms and control data that is received from one network into a format that can be used by a mobile communication devices. A mobile access gateway usually has more intelligence (processing functions) that include authentication, secure encoding of media, and adapting of media into formats suitable that match the capabilities of mobile devices.

Mobile Access Hunting (MAH)-A telephone call-handling feature that causes a transferred call to "hunt" through a predetermined group of telephones numbers until it finds an available ("non busy") line. The mobile systems search a list of termination addresses sequentially for one that is idle and able to be alerted. If a particular termination address is busy, inactive, fails to respond to a paging request, or does not answer alerting before a time-out, then the next termination address in the list is

tried. Only one termination address is alerted at a time. The mobile telephones in the group may be alerted using distinctive alerting. Additional calls may be delivered to the MAH Pilot Directory Number at any time.

Mobile Ad Insertion-Mobile ad insertion is the process of inserting an advertising message into a media stream such as a multimedia mobile device. For mobile broadcasting systems, ad inserts are typically inserted on a national or geographic basis that is determined by the distribution network. For two-way mobile systems, ad inserts can be directed to specific users based on the viewer's profile. This figure shows how a mobile advertising system can work. In this example, a company creates and submits a small banner ad to a mobile advertising network and selects location, time and category as the ad promotion criteria. The mobile advertising network submits these ads to multiple mobile networks and keeps track of the transmission, selection and response to these ads. The mobile system operators review the capabilities of the mobile devices in their network to determine which devices can receive and respond to mobile ads.

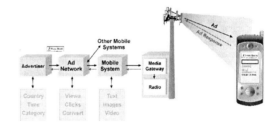

Mobile Advertising

Mobile Advertising Networks-Mobile advertising network is a group of advertising systems (such as multiple web sites) that

M

allow product or service sellers to reach a broad audience of mobile devices. Mobile advertising networks typically provide discounts to companies or people for advertising to multiple wireless networks.

Mobile Business (m-Business)-Electronic commerce (e-commerce) that is performed through the use of mobile communication devices.

Mobile Carriers-(1-service provider) Companies that provide mobile communication (e.g. cellular or PCS) services. (2-electrons) Free electrons and holes that move through a conductor or semiconductor.

Mobile Channel Type-Mobile channel type is a set of characteristics associated with how products are communicated from the manufacturer or source of the products to the customers who use or consume them.

Mobile Commerce (M-Commerce)-A shopping medium that allows wireless devices in a telecommunications network to present products to the customer and process orders.

Mobile Communities-Mobile communities are web portals that allow people to use their mobile devices to communicate and share information. Mobile communities are commonly made of people who share common interests.

Mobile Cookies-Mobile cookies are a small amount of information that is stored on a user's mobile device (a client) that is used by a web site (web server) to help control the content and format of information to the user during future visits to the web site.

Mobile Country Code (MCC)-An identity code that is assigned to identify the country of a GSM system.

Mobile Coverage Area-A geographical area that has a sufficient level of radio signal strength to allow two-way communication with mobile radios.

Mobile Data-Mobile data is the transmission of digital information through a wireless network. The term mobile data is typically applied to the combination of radio transmission devices and computing devices (e.g. computers electronic assemblies) that can transmit data through a mobile communication system (such as a wireless data system or cellular network).

Mobile Data Network (MDN)-A mobile data network is a communication system that allows users to connect to the data network at different locations.

Mobile Data Terminal (MDT)-Mobile data terminals are data input and output devices that are used to communicate with a remotely located computer or other data communication device. Mobile data terminals frequently consist of a keyboard, video display monitor, and radio communication circuitry that can connect the data terminal with the remotely located computer.

Mobile Directory Number (MDN)-A mobile directory number (MDN) is a number that can be dialed through the public telephone network. A full mobile directory number contains the country code, national designation code, and subscriber number and it can be up to 15 digits.

Mobile Downloadable-Mobile downloadable is the capability to transfer data or a program from a source (such as a web server) to a mobile device (such as a mobile telephone that is connected to the Internet through a wireless connection).

Mobile Equipment SIM Lock-A mobile phone feature that ensures a mobile phone will only work with one or a group of subscriber identity module (SIM) cards. The mobile phone is programmed with information that must match the information on the SIM card to enable operation. This information may include one or more of the following; IMSI, Group identifier (Service Provider identity) or PLMN identity (MCC+MNC). If a SIM card is inserted that does not match the stored information, the phone will be inoperative.

Mobile Execution Environment (MExE)- Mobile station application execution environment (MExE) is a group of defined application processes and data structures that allow application service providers to develop software programs that will perform specific operations on GSM or 3G UMTS handsets. Mobile telephones are classified into types of applications that are based on wireless application protocol (WAP), JavaPhone and personal Java (class 2) and Java mobile environment (J2ME). MExE is defined in GSM specification 02.57.

Mobile Gaming- Mobile gaming is an experience or the actions of a person that are taken on a skill testing or entertainment application with the objective of winning or achieving a measurable level of success on a mobile device (such as a mobile telephone).

Mobile Handheld Device (MHD)- A mobile handheld device is a portable communication assembly that can convert media such as video, audio or images into a form that can be experienced by humans. Mobile handheld devices for home networks is defined by the Digital Living Network Alliance (DLNA).

Mobile Identification Number (MIN)- The mobile identification number (MIN) is a 10-digit number that represents a mobile telephones (mobile station) identity. It is divided into MIN1 and MIN2. MIN1 is the 7 digit portion of the number. MIN2 is the 3 digit area code portion of the number.

Mobile Information Device Profile (MIDP)- Mobile information device profile is a defined implementation of processes and protocols (such as Java 2 mobile environment) that ensures that mobile devices will operate in a specific way for mobile data applications.

Mobile Internet Protocol (Mobile IP)- Mobile IP is a protocol that allows IP communication devices to use the same IP address as

it moves between locations and even different types of networks (e.g. Cellular to Ethernet).

Mobile Internet Protocol Television (MIPTV)- Mobile Internet protocol television is the process of providing television (video and/or audio) services through the Internet protocol (IP) that are sent through mobile systems. MIPTV systems initiate, process, and receive voice or multimedia communications using IP protocols. MIPTV systems may be a combination of public and private communication systems (e.g. Internet and Mobile Data Networks).

Mobile Internet Television- Mobile Internet television is the ability for a user or device to connect to digital television provided through the Internet via mobile connections. An example of mobile Internet television is the requesting and connection to a television gateway or media server via a data connection on a mobile telephone.

Mobile IP- Mobile Internet Protocol

Mobile Landing Page- A mobile landing page is a web page that a customer has landed on as a result of selecting a link on a mobile device. A mobile landing page is usually customized for a specific product and the type of device that is accessing the landing page (such as a mobile telephone with a small display screen).

Mobile Media Player (MMP)- A mobile media player is a self-contained mobile device (e.g. a mobile telephone) and associated software application that can convert media such as video, audio or images into a form that can be experienced by humans.

Mobile Network Code (MNC)- A mobile network code is a unique identifier that is assigned to identify a mobile network operator in a specific country.

Mobile Number Portability (MNP)- Mobile number portability is the process that allows a subscriber to keep their telephone number when they change telephone service providers.

Mobile Originated Short Message Service (MOSMS)-The sending of a message from a mobile telephone to a system or to another user in a system.

Mobile Party Pays (MPP)-Mobile party pays is a billing process where the mobile telephone user pays for the acceptance of delivery of a call or services to their mobile device.

Mobile Portal-A mobile portal is an Internet web site that acts as an interface between a mobile device user of a product or services and a system that provides the product or services. Mobile portals are designed by using graphics and protocols (such as WAP) that provide a better user experiences on mobile devices.

Mobile Position Center (MPC)-A mobile position center is the process or service within a communication system that selects a position determination entity (the location hardware and software) that determines the position of a mobile device.

Mobile Satellite Service (MSS)-Mobile satellite service is a form of wireless service that employs satellites as part of the wireless infrastructure which communicates directly with mobile satellite receivers. MSS systems are capable of serving very large geographic areas. The use of MSS may be appropriate for areas that are economically not viable for land based radio towers or to provide wide area group call (dispatch type) services.

Mobile Satellite Television-Mobile satellite television is the broadcasting of television signals through satellites to viewers who have portable viewing devices that can receive satellite signals.

Mobile Service-A service of radio communication between mobile and land stations, or between mobile stations.

Mobile Service Area (MSA)-A geographic area authorized for mobile radio service coverage that is usually divided into a number of smaller radio coverage areas (cells.)

Mobile Service Provider-A generic term used to describe companies, including cellular carriers, radio common carriers, and private carriers that provide mobile telecommunications service.

Mobile Station (MS)-A mobile radio telephone operating within a wireless system (typically cellular or PCS). This includes hand held units as well as transceivers installed in vehicles.

Mobile Station Identity (MSID)-A unique number used by a mobile system to enables the current networking and authentication mechanisms to validate the identity of mobile devices.

Mobile Storefront-A mobile storefront is the online store that permits person or companies to purchase products and/or services via mobile devices (such as via mobile telephones).

Mobile Subscriber ISDN (MSISDN)-The telephone number assigned to mobile telephones. This number is compatible with the North American number plan.

Mobile Switching Center (MSC)-Switching system that are used for mobile communication networks (cellular, PCS, and 3G.) The MSC was formerly called the mobile telephone switching office (MTSO).

This figure shows that a MSC consists of controllers, switching assemblies, communications links, operator terminals, subscriber databases, and backup energy sources. The controllers are computers which control communications and call processing. Controllers help the MSC to understand and create commands to and from the base stations. In addition to the main controller, secondary controllers devoted specifically to control of the cell sites (base stations) and to handling of the signaling messages between the MSC and the PTSN are also provided. A switching assembly routes voice connections from the cell sites to each other or to the public telephone network. Communications links between cell sites and the MSC may be copper wire, microwave, or fiber optic. An operator terminal allows operations, administration and maintenance of the system. A subscriber database contains features the customer has requested along with billing records. Backup energy sources provide power when primary power is interrupted. As with the base station, the MSC has many standby duplicate circuits and backup power sources to allow system operation to be maintained when a failure occurs.

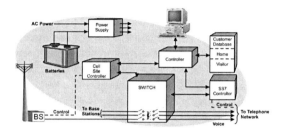

Cellular Mobile Switching Center (MSC)

Mobile Targeting-Mobile targeting is the process of selecting which mobile devices advertising messages can be sent to. Mobile targeting criteria may include country, context, carrier, device type, device capabilities, device manufacturer, and device platform.

Mobile Telephone-A mobile telephone is a wireless telephone that operates within a wireless communication system (typically cellular or PCS). This includes hand held units as well as transceivers installed in vehicles.

Mobile Television-Mobile television is the delivery of digital television services to portable devices over wireless connections. Mobile television may be provided by mobile telephone radio channels, satellite channels or other types of wireless channels.

Mobile Television (Mobile TV)-Mobile television is the transferring of signals that carry moving picture information to mobile devices. Mobile TV systems commonly refer to mobile devices (such as multimedia telephones) that can connect to and view broadcasted or live television programs.

Mobile Terminals-Transceivers (radios) that are mobile (as opposed to fixed position terminals.)

Mobile TV-Mobile Television

Mobile TV Bypass-Mobile TV bypass is using mobile systems to deliver high-quality television services and bypassing traditional television systems such as cable, satellite and terrestrial broadcast television.

Mobile Video (MV)-Mobile video is the transferring of signals that carry moving picture information to mobile devices. Mobile video is commonly associated with supplying video signals to mobile telephones.

Mobile Virtual Network Enabler (MVNE)-A mobile virtual network enabler is a company that provides infrastructure and services to enable mobile network operators to offer services and have a relationship with end-user customers.

Mobile Virtual Network Operator (MVNO)-A Mobile Virtual Network Operator (MVNO) is a mobile communications service provider that resells the communication services of other wireless communication network operators. MVNO providers purchase airtime (minutes of use) in quantity and resell the airtime to customers they obtain and manage.

MVNO's may provide value added services such as information services, brand labeling, special sales support, and support of unique distribution channels. MVNO's attempt to position their services so customers do not recognize that the operator does not own a network. To provide advanced services, some MVNO operators may own network equipment that interfaces with wireless networks. This may allow MVNO's to have more control over customer databases and SIM cards.

Mobile Virus-A mobile virus is a software program that is spread by automatic copying from memory or computer networks and intended to interrupt or destroy the functioning of a mobile device.

Mobile Web-Mobile web is the connection of computer services that can receive and process requests for services from mobile communication devices. The mobile web may be associated with providing services to mobile telephones that have multimedia capabilities.

Mobile Web Best Practices (MWBP)-Mobile web best practices is a set of guidelines that assist mobile web application developers to design systems that provide a better and more consistent web interactive experience for mobile device users.

Mobilecasting-Mobilecasting is the providing (transmitting) of television programs or other media on a mobile communication system.

Mobility-Mobility is the capability of a device to operate at different locations. Mobility commonly refers to the ability of communication devices to operate in different geographic areas.

Mobility Management (MM)-Mobility management is the processes of continually tracking the location of mobile telephones or devices that are connected to a communication system. Mobility management typically involves regularly registering telephones or communication access devices. Mobile telephones typically automatically register when they are first turned on (attach) and when they are turned off (detach).

Mobility Services-Mobility services use communication to successfully satisfy the information requirements of the end customer that may be moving throughout a geographic area. While mobility services typically involve the use of wireless communication, mobility services may be applied to other technologies such as Internet access and IP Telephony that can be used at many geographic connection points. Mobility services involve interaction between the mobile device and the communication system to facility registration (location updates), authorization of services (authentication) and delivery (distribution) of information.

Mock Up (Mock-up)-A mock up is a representation of a proposed or new product or brochure in a physical form that closely resembles the appearance of the product.

Mock-up-Mock Up

MoDem-Modulator/Demodulator

Modem Pool-A grouping of modems that are shared by users and other network devices to allow different types of communication devices (such as mobile telephones and PSTN telephone lines) to communicate with each other.

Moderator-A moderator is a person who coordinates the communication between people. Moderators may be used to help control the communication between members of newsgroups to ensure the communication conforms to specific rules and requirements.

Modification of Bid-A modification of bid is changes to a bid proposal. A modification of bid may be only allowed if the change is provided to the bidding office prior to the final submission date.

Modular-(1-general) A unit or program that is independent or may be defined as a part of a larger system. (2-telephone cord connector) Trade name for small molded plastic electrical connectors used on handset and mounting (wall) cords. The RJ11 connector is one example of a modular connector. Called Teledapt in Canada."

Modulator/Demodulator (MoDem)-Modems are devices that convert signals between analog and digital formats for transfer to other lines. Data modems are used to transfer data signals over conventional analog telephone lines. The term modem also may refer to a device or circuit that converts analog signals from one frequency band to another.

This figure shows how a data modem converts digital information into analog signals that can be transmitted on an analog communications network. In this example, the data signal comes from a computer (called the data terminal equipment (DTE)), via an RS-232 serial data interface. The RS-232 data interface uses pre-defined signaling commands and data transmission rates to communicate with the data modem. The modem performs a digital-to-analog conversion and from the line to the DTE an analog-to-digital conversion.

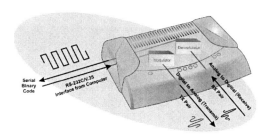

Data Modem Functional Operation

Module-(1-circuit) An assembly replaceable as an entity, often as an interchangeable plug-in item. A module is not normally capable of being disassembled. (2-software) A program unit that is discrete and identifiable with respect to compiling, combining with other modules, and loading.

Monetary Unit-Monetary units are the financial basis of measurement that is defined by a country or other authorized entity.

Money Order-Money order is a document or item that is issued by a bank or other authority that can be used by a recipient to obtain money.

Monitization-Monitization is the conversion of content, services or products into a valuable commodity (such as money) that can be exchanged for other assets or services.

Monitoring-The process of listening to or viewing a communication service for the purpose of determining its quality or whether it is free from trouble or interference.

Monopoly-A monopoly is an industry where a single company supplies or controls all the customers. Monopoly companies may be owned or controlled by government regulations.

Monopsony-A monopsony is a marketplace where only one buyer exists.

Monthly Fee-A monthly fee is an assessment or charge for a service that paid for on a monthly.

Monthly Rate-Monthly rate is the fee structure for a service that is purchased on a month to month basis.

Monthly Recurring Cost (MRC)-A cost for service or equipment usage that is continuously charged on a monthly basis.

Monthly Service Charge-A recurring fee that is charged to a customer for the monthly maintenance of their communication service.

Moore's Law-Named after Gordon Moore who in 1965 predicted that the number of electronic devices that can be made upon a circuit chip would double every year.

MOPS-Millions of Operations Per Second

Moratorium-A moratorium is an action or requirement to delay a process. An example of a moratorium is a delay on all payments that are processed during the settlement of a dispute.

MOS-Mean Opinion Score

MOSMS-Mobile Originated Short Message Service

Motion Picture Experts Group (MPEG)-Motion picture experts group (MPEG) standards are digital video encoding processes that coordinate the transmission of multiple forms of media (multimedia). Motion picture experts group (MPEG) is a working committee that defines and develops industry standards for digital video systems. These standards specify the data compression and decompression processes and how they are delivered on digital broadcast systems. MPEG is part of International Standards Organization (ISO).

MOU-Memorandum Of Understanding

MOU-Minutes Of Use

Mouse-A handheld device that translates movement or click-button instructions into corresponding movement of a cursor on a video display. Mouse movement, measured in mickeys, relates the on-screen distance a cursor moves to the distance the mouse moves across the desk.

Mouse Potato-A mouse potato is a person who spends an above average amount of time in front of a computer running programs or surfing the Internet.

MOV-Movie Format

Moves, Adds, And Changes (MAC)-These define a system administration activity responsible for reconfiguring telephone sets and computer workstations on an existing switch or host system. This also includes the reconfiguration of local area network (LAN) devices.

Movie Format (MOV)-Movie format (MOV) is Apple's Quicktime multimedia digital video format that interleaves digital audio and digital video frames into a common file. MOV files have the ability to synchronize digital audio and digital video along with managing other forms of media.

Movie Studio (Studio)-A movie studio is a room or company that is used to produce movies. A move studio is composed of one or more rooms or sets that are designed for the recording of video and audio.

Movies-Moving Pictures

Moving Pictures (Movies)-A movie is a set of images that are played in rapid sequence to create the illusion of movement.

MP4-MPEG-4

MPB-Meet Point Billing

MPC-Mobile Position Center

MPEG-Motion Picture Experts Group

MPEG (.MPG)-MPEG is a digital media container file format that identifies the digital audio and digital video components into a common MPEG stream or file format.

MPEG Level-MPEG levels are the amount of capability that a MPEG profile can offer. MPEG levels can range from low level (low resolution) to high level (high resolution).

MPEG-1-MPEG-1 is a multimedia transmission system that allows the combining and synchronizing of multiple media types (e.g. digital audio and digital video). MPEG-1 was primarily developed for CDROM multimedia applications.

MPEG-2-MPEG-2 is a frame oriented multimedia transmission system that allows the combining and synchronizing of multiple media types. MPEG-2 is the current choice of video compression for digital television broadcasters as it can provide digital video quality that is similar to NTSC with a data rate of approximately 3.8 Mbps.

MPEG-21-MPEG-21 is a multimedia specification that adds rights management capability to MPEG systems.

MPEG-4 (MP4)-MPEG-4 is a digital multimedia transmission standard that was designed to allow for interactive digital television and it can have more efficient compression capability than MPEG-2 (more than 200:1 compression).

MPEG-7 Multimedia Content Description Interface-The MPEG-7 multimedia content description interface is a multimedia standard that adds descriptive content to media objects.

MPOE-Minimum Point Of Entry

MPP-Mobile Party Pays

MPS-Message Processing System

MRC-Monthly Recurring Cost

MRD-Marketing Requirements Document

MRO-Maintenance and Repair Organization

MRP-Materials Requirements Planning

MS-Media Server

MS-Message Service

MS-Mobile Station

MSA-Metropolitan Service Area

MSA-Metropolitan Statistical Area

MSA-Mobile Service Area

MSAN-Multiservice Access Network

MSC-Mobile Switching Center

MS-DOS-Microsoft Disk Operating System
msec-Millisecond
MSID-Mobile Station Identity
MSISDN-Mobile Subscriber ISDN
MSS-Mobile Satellite Service
MTA-Major Trading Area
MTA-Multimedia Terminal Adapter
MTBF-Mean Time Between Failures
MT-RJ Connector-An MT-RJ is a two-fiber modular connector similar to an 8 pin RJ-45 wired connector. Because MT-RJ connectors are the same physical size as the RJ-45 connector, this allows for existing modular wall plates to be easily updated with MT-RJ sockets.
MTTF-Mean Time to Failure
MTTR-Mean Time to Repair
Muckraker-A muckraker is a person who performs actions that are intended to disturb or result in conflict between people or groups.
Multi Tasking (Multitasking)-Multi tasking is a processing or computing system that can only run more than one process or task at a time. An example of a multi tasking computing system is Windows where more than program could be running at a time.
Multi User Software-An application designed for simultaneous access by two or more network nodes, typically employing file and/or record locking.
Multi Vendor Integration Protocol (MVIP)-A standard protocol used for telephony and data switching that allows the design of advanced telephony applications such as voice mail, PBX, faxback service and others that are interoperable with other hardware equipment and software programs.
Multi-Bit Rate (MBR)-Multi-bit rate is the capability of a communications service or transmission lines to operate at different data transmission rates. Multi-bit rate services usually require some real-time interactivity with bursts of data transmission. An example of a MBR application is a video server that can provide digital video signals at different transmission rates.

Multicarrier Mode (MC)-Multicarrier mode (MC) is a communication system that combines or binds together two or more communication carrier signals (carrier channels) to produce a single communication channel. This single communication channel has capabilities beyond any of the individual carriers that have been combined.
Multicast-Multicast service is a one-to-many media delivery process that sends a single message or information transmission that contains a multicast address (code) that is shared by several devices (nodes) in a network. Each device that is part of a multicast group needs to connect to a router (node) in the network that is part of the multicast distribution tree. This means that the multicast media (such as an IPTV channel) is only sent to the users (viewers) who have requested it. The benefit of multicasting is the network infrastructure near the user (e.g. a home) only needs to provide one or two channels at once, drastically reducing the bandwidth requirements.
This figure shows examples of how multicast services can be implemented. The first method uses encoded video broadcast transmission and encoded messaging to allow only a select group to view the received information. While all the television broadcast receivers all receive the same radio signal, only the receivers with the correct code will be able to descramble (decode) the television signal. The second method uses multicast routing in the Internet to store and forward data to an authorized group of recipients that are connected to its router. When a router in the Internet that is capable of multicast service receives a multicast message, it will store the message for forwarding. It then uses the multicast address to lookup a list of authorized recipients in its routing table. The stored message is then forwarded to the authorized receiving device or next router that is part of the multicast service.

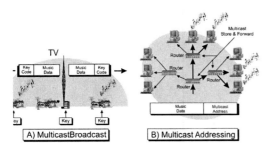

A) MulticastBroadcast

B) Multicast Addressing

Multicast Operation

Multicast Address-A multicast address is a unique identification code that is used by a group of routers that receive and forward packets for a multicast group. The multicast address is stored in a multicast routing table.

Multicast Backbone (MBONE)-Multicast backbone is a high-speed data communications system that interconnects the Internet that allows multicast services. The MBONE network is composed of interconnected multicast LANs.

Multicast Stream-A multicast stream is a flow of data or information that is a one-to-many media delivery process that sends a single message or information transmission that contains a multicast address (code) that is shared by several devices (nodes) in a network sent to devices that are part of a multicast group. Each part of the multicast group connects to a router (node) in the network that is part of the multicast distribution tree. A stream may be continuous (circuit based) or bursty (packetized).

Multichannel Audio-Multichannel audio is the use of multiple sound channels to produce an enhanced listening experience. Examples of multichannel audio include stereo, surround sound and low frequency enhancement (LFE).

Multiformat-Multiformat is the capability of a piece of equipment, service or program to access (as inputs) and/or provide (as outputs) multiple signal types, such as digital, analog component or analog composite.

Multi-Function Subscriber Identity Module (SIM) Cards-SIM cards that are capable of performing more functions than the providing the identity of a wireless telephone subscriber. These functions may include electronic cash, prepaid calling cards, security access card, medial record storage, electronic airline travel ticket and many other functions.

Multilevel Access Privileges-Multilevel access privileges is the assignment of access credentials that allow different users to connect, process, use or transfer groups of products or services which can have different access requirements..

Multilevel Distribution-Multilevel distribution is the transferring of products, services or content through multiple types or levels of distribution. An example of multilevel distribution is the selling of books from a publisher to a wholesaler who sells books to retail stores who sells to book readers.

Multilingual Content-Multilingual content is media (such as television programs or web sites) that include information in multiple languages.

Multilingual Training-Multilingual training is the providing of information materials in a multiple language formats that helps a person to gain knowledge or skills.

Multimedia-Multimedia is a term that is used to describe the delivery of different types of information such as voice, data or video. Because Internet service is often used with broadband (high-speed) data services, it is possible to send multiple types of information at the same time.

Multimedia Computer-A multimedia computer is a data processing device that is capable of using and processing multiple forms of media such as audio, data and video. Because many computers are already multimedia and Internet ready, it is often possible to use a multimedia computer to watch IP television through the addition or use of media player software. The media player must be able to find and connect to IP television media servers, process compressed media signals, maintain a connection, and process television control features.

Multimedia Home Platform (MHP)- Multimedia home platform is an industry middleware standard used in the digital video broadcasting (DVB) system that defines the communication of the DVB system with the integrated receiver device (IRD) and other set top box receivers that have MHP capability. MHP is designed to allow the user to access additional interactive services such as Internet browsing and electronic programming guides.

Multimedia Messaging Service (MMS)- Multimedia messaging service is a system that allows for the sending of messages to include graphics, audio or video components. Multimedia messages can be sent using WSP or HTTP protocols.

Multimedia Terminal Adapter (MTA)-A customer premises device that connects the subscriber's telephone to a managed broadband IP network (HFC cable, ADSL, fiber, wireless) and call control elements in the network to deliver high-quality telephony services. Multimedia Terminal Adapters (MTAs) provide the codecs and all signaling and encapsulation functions required for media transport and call signaling.

Multimedia/Hypermedia Expert Group (MHEG)-The multimedia/Hypermedia expert group oversees the creation of industry standards for multimedia hypermedia information objects that can be exchanged between different services and applications. Information can be found about MHEG at www.MHEG.org.

Multipart Virus-A multipart virus is a program that can spread itself to other programs or data files through the use of multiple delivery types such as embedding itself into the boot sector or into executable files. Multipart viruses may be difficult to locate because they can reappear in other formats to avoid detection and deletion.

Multi-Party Call Conference-An enhanced telecommunications service that allows the (or more) users to be connected to the same call.

Multiple Access-The capability of a communications system to allow more than one user to access to one ore more channels in the system.

Multiple Award-A multiple award is the assignment of a contract or order to more than one supplier.

Multiple Bit Rate (MBR)-Multiple bit rate is the storing and/or streaming of media using different bit rates to represent the media (such as low or high resolution digital video formats).

Multiple Bookings-Multiple bookings is the scheduling of a program playout or reservations for resources that will be required to provide services (such as broadcasting television programs) for a specific company or project for multiple events.

Multiple Cycle Billing-Multiple cycle billing is the dividing of customers into groups that allows their bills to be processed in specific cycles (or "billing cycles").

Multiple Data Message Format (MDMF)- The caller identification format that transfers the calling telephone number along with additional text (name) information.

Multiple Dwelling Unit (MDU)-Multiple dwelling units are facilities or buildings that have multiple living areas. An example of a MDU is an apartment building.

Multiple Exchange Carrier Access Billing (MECABS)-Multiple exchange carrier access billing is a billing standard that was developed by Alliance for Telecommunications Industry Solutions (ATIS) to enable billing records to be exchanged between multiple service providers.

Multiple Platform Access-Multiple platform access is the ability of a user or device to connect to and communicate with multiple types of hardware and software that programs or services operate.

Multiple Unit Pricing-Multiple unit pricing is the assigning of prices that depends on the quantity of product purchased. Multiple unit pricing may define the discounts based on each purchase order or on the number of quantity purchased over a period of time.

Multiplex-(1-general) The use of a common channel to make two or more channels. This is accomplished either by the splitting of the common-channel frequency band into narrower bands, each of which is used to constitute a distinct channel (frequency-division multiplex), by allotting this common channel to multiple users in turn, to constitute different intermittent channels (time-division multiplex), or by allowing the simultaneous transmission of channels using unique identification codes (code-division multiplex). (2-frequency division) A multiplexing system in which different frequency bands are used by different channels, enabling many different channels to be carried by a single frequency bearer channel. (3-time division) A multiplexing system in which the original analog signals are converted into digital form. The digital signals (for each of many channels) are transmitted sequentially at different time instants. (3-code division) A multiplexing system in which the original signals are converted into digital form and multiplied by a unique identification code. The digital signals (for each of many channels) are transmitted in parallel using different code identifiers.

Multiplexer-A multiplexer is a device or assembly that accepts a number of input signals and reformats/combines these signals into a single transmission format.

Multiplexing-Multiplexing is a process that divides a single transmission path to parts that carry multiple communication (voice and/or data) channels. Multiplexing may be time division (dividing into time slots), frequency division (dividing into frequency bands) or code division (dividing into coded data that randomly overlap).

This diagram shows how multiplexing can combine two or more low speed channels into one higher speed communication channel. In this diagram, there are eight 8 kbps communication channels that are supplied to a multiplexer. The multiplexer stores and sends 8 bits of each slow speed communication channel during each 125 usec time slot on the 64 kbps channel.

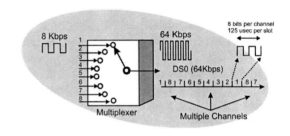

Multiplexing

Multipoint-Multipoint systems can transfer information from one device (or point) to multiple devices (multiple receiving points).

Multiprocessing-Multiprocessing is the concurrent or interleaved execution of two or more programs or sequences of instructions by one or more computers by using parallel processing, multiprogramming or a combination of both.

Multipurpose Internet Mail Extensions (MIME)-A data communication format this allows information blocks (such as binary images and multimedia data) to be sent with email messages that may be developed primarily for text (7-bit) characters. MIME is defined in RFC 1521.

Multiservice Access Network (MSAN)-A multiservice access network is a portion of a communication network (such as wired or wireless networks) that allows individual subscribers or devices to connect to the core network and receive multiple types of services (such as IPTV, IP Telephony and Internet web access).

Multi-Site Enterprise-A business or corporation that has multiple locations that are part of its overall business.

Multitasking-Multitasking is a process of providing services (such as computer processing) to multiple processes. For sequential computer processors, multitasking involves the switching from one task to another on a computer without losing track of either. Multitasking usually is accomplished by time slicing shared resources.

Multitasking-Multi Tasking

Multi-Tenant Service-The sharing of centralized equipment, facilities, and telecommunication services by occupants in a building or office complex.

Multithreading-Multithreading is the processing of identifying and organizing multiple threads that link elements or tasks (such as computing tasks).

Multi-Tier Pricing-Multi-tier pricing is the assignment of a selling price to services or items where the selling price changes based on the quantity of products or services purchased.

Multivendor Compatibility-Multivendor compatibility is the ability for electronics systems or equipment to connect and/or exchange signals or data directly between them.

Multi-Vendor Interoperability-Multi-vendor interoperability is the ability of devices produced by different manufacturers that are within a similar category to work or interact with each other.

Multi-Vendor Network-A multi-vendor network is a group of interconnected equipment (networks) that is composed of assemblies and/or services that are produced by different manufacturers.

Must-Carry Rules-Must-carry rules are requirements for non-radio broadcasters (e.g. cable TV operators) to retransmit (carry) programs transmitted by local broadcasting companies.

Mutual Authentication-Mutual authentication is a process where each device in a communication session authenticates each other before the session can start.

mV-Millivolt

MV-Mobile Video

MVIP-Multi Vendor Integration Protocol

MVNE-Mobile Virtual Network Enabler

MVNO-Mobile Virtual Network Operator

mW-Microwatt

MWBP-Mobile Web Best Practices

MWI-Message Waiting Indicator

N

n-Nano

NACD-Network Automatic Call Distribution

Nailed Up Connection-The assignment of a long term (permanent) dedicated path that created a network.

Naked DSL-Naked DSL is digital subscriber line service that does not include dialtone service.

Name Resolution-A process of translating a text based name into a numeric address, such as Internet Protocol (IP) address. See domain name server (DNS).

Name Server-The data processing device that translates text names to address information (such as an Internet addresses).

Name Suppression-Name suppression is the process of stopping the sending of mailing or other promotional information to people who have their name on a suppression list.

Namespace-A namespace is the combination of domain names that exist on a system below the root system.

Naming Convention-Naming convention is the structure of the labels or codes that are assigned to devices, services, files, applications and/or the locations (e.g. the end points in a system).

Nano (n)-A metric preface representing 0.000 000 001.

Nanometer (nm)-A measurement of signal using wavelength of 0.000 000 001 (10^{-9}) meter.

Nanosecond (nsec)-A measurement of time using 0.000 000 001 of a second.

NANP-North American Numbering Plan

NAP-Network Access Point

NAS-Network Attached Storage

National Change of Address (NCOA)-National change of address is a database that is maintained by the United States postal service that identifies the addresses of people and companies that have moved or are moving to a new address. NCOA can be used by mailing list companies to reduce undeliverable mail.

National Direct Dialing (NDD)-National direct dialing is the access codes that are used to make call within a country from one city location to another city.

National Number-The telephone number identifying a calling subscriber station within an area designated by a country code.

National Programming-National programming is the selection of shows and programs that are offered by a television network provider that provides media channels over national geographic area. An example of a national program is a television program that is created and a network operator that is broadcasted by many local television systems throughout a nation.

NCOA-National Change of Address

NDA-Non-Disclosure Agreement

NDCBU-Neighborhood Delivery and Collection Box Unit

NDD-National Direct Dialing

NDM-U-Network Data Management - Usage

NDS-Network Directory Server

NE-Network Element

Near Real Time-Actions that occur within a short time period that is perceived or used (such as within a few minutes) to perform or record events when they are required or used.

Near Real time Roaming Data Exchange (NRTRDE)-Near real time roaming data exchange is a standard developed for GSM operators for the exchange of information about calls being made by subscribers on networks other than their home network. It is designed to replace the existing High Usage Report process with an industry-wide procedure . The procedure has two essential functions: it reduces the required timeframe for delivery of fraud related roaming information from 36 hours to a maximum of 4 hours and it provides individual call detail record information in sufficient form to allow operators to cost effectively manage roaming fraud

The GSM Association (GSMA) is recommending all mobile service providers across the world should implement NRTRDE by October 1, 2008 to combat fast-growing incidents of global roaming fraud.

Near Video On Demand (NVOD)-Near video on demand is a video delivery service that allows a customer to select from a limited number of broadcast video channels when they are broadcast. NVOD channels have pre-designated schedule times and are used for pay-per-view services.

This diagram shows a near video on demand (NVOD) system. This NVOD system allows a customer to select from a limited number of broadcast video channels. These video channels are typically movie channels that have pre-designated schedule times. This system allows the user to unblock an encoded channel during pre-scheduled play times.

Nearline Asset-Nearline assets are files or data that represents a valuable form of media that are not immediately available for use but can be made available for use in a short period of time.

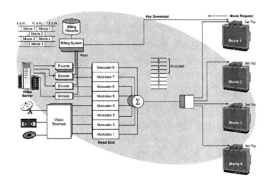

Near Video on Demand (NVOD) Operation

Negative Cash Flow-Negative cash flow is a financial condition where the cash or liquid assets that are being received are less than the cash or liquid assets that are being expended.

Negotiation-The fourth phase of the marketing process in the Costa Model that motivates the customer to engage in negotiation with a company for the sale of products or services. Negotiation can be explicit (talking with the sales rep about options) or implicit (the customer chooses between options without direct interaction).

Neighborhood Delivery and Collection Box Unit (NDCBU)-A neighborhood delivery and collection box unit is a device that is authorized to receive mail on behalf of people that live in a neighborhood.

Net-Net is the difference between an amount (such as sales receipts) and a cost associated with that amount (such as the cost of goods).

Net-Internet

Net Invoice Price-Net invoice price is the total price for an invoice's less defined deductions such as taxes, shipping, insurance and non-product or service related charges.

Net Negative-Occurs when the number of gross adds for a given period is less than the number of existing customers who cancelled service for the same period.

Net Neutrality-Net neutrality is the providing of services on a network (such as the Internet) without regard for the characteristics of data or content that is sent through the network.

Net Presence-Net Presence is the activity or perceived activity and availability on the Internet.

Net Price-A net price is the cost of a product or service after all the additional fees and credits have been applied.

Net Proceeds-Term often used in contract communicating direction stating the amounts it will take to be paid.

Net Sales-Net sales are the recorded revenue for products or services minus discounts, allowances, or returns. Some net sales deductions may include transaction fees, marketing costs, and catalog management fees.

Net Settlement-Net settlement is the financial difference between revenues generated from charging for products or services provided by a company minus the fees that other companies have charged for their products or services.

NetComp-Network Compensation

Netpreneur-A netpreneur is an entrepreneur who focuses on developing online Internet business.

Netting-Netting is the process of matching revenue and cost items.

Network-Networks are a series of points that are interconnected by communications channels, often on a switched basis. Networks are either common to all users or privately leased by a customer for some specific application.

Network Access-Electronic circuitry that determines which, when, and how a communication device may transmit and communicate with the system. This circuitry may be centrally located or may be located in each of the network interface controllers.

Network Access Charge-A fee paid by an operator of a system for access to other network systems.

Network Access Point (NAP)-A network access point is a key physical (layer 2) interconnection point that interconnects regional Internet systems and sub-systems. NAPs are public network exchange facility where Internet Service Providers (ISPs) can connect with one another in peering arrangements. The NAPs are a key component of the Internet backbone because the connections within them determine how traffic is routed. At the end of the 20th century, they were one of the points of Internet congestion. As a result of congestion of NAPs, several ISP have invested in private NAPs to interconnect to each other's network.

Network Access Revenue-Service revenue that results from charge for service access between carriers, such as when inter-exchange (IXC) carriers pay to connect through the local telephone access network infrastructure.

Network Adapter-A device or assembly that converts the format of information that is transferred between one network or device to the format used by another network or device. An example of a network adapter is an Ethernet network adapter card that converts information in an Ethernet packet into a format that can be transferred into a computer's internal communication bus.

Network Address-(1-general) A unique number associated with a network host that identifies it to other hosts and devices during network communication. (2-SS7) The Signaling System 7 signaling code that contains a network identification, a network duster, and network cluster member fields.

Network Administration Center-An operations center with administrative responsibility for local and tandem switching systems.

Network Administrator-(1-coordinator) A person responsible for managing the day-to-day operations of a network. (2-humor) The person blamed for all computing problems, whether they are related to the network or not.

Network Advertising-Network advertising is sending of advertising messages (adverts) to media communication systems (such as television networks or terrestrial broadcasters).

Network Architecture-Network architecture is the design, physical structure, functional organization, data formats, operational procedures, components, and configuration of a network. Network architectures usually divide network functions into layers of software and hardware. Each network layer serves a specific purpose. There are often specific relationships between network layers that allows different manufacturers or equipment that operate at different layers to interoperate with each other (e.g. a router interfacing with a network hub.)

Network Assurance-Network assurance is reviewing of the ability of a network to operate or perform as expected.

Network Attached Storage (NAS)-(1-product) A collection of mass-storage devices contained in a single chassis with a built-in operating system. Typically connected to a local area network, these devices usually support Network File System (NFS) and common Internet file system (CIFS) as a means to share data in a departmental or enterprise environment. NAS products are marketed to small and medium businesses as self-contained, plug-and-play, easy to operate storage expansions.

(2-architecture) Network Attached Storage is an architecture in which traditionally LAN-oriented technologies, such as Ethernet and TCP/IP are used to connect storage. NAS utilizes LAN technology in place of traditional storage protocols such as Fibre Channel and parallel-SCSI to produce large arrays of disk drives with a virtualized interface. One NAS disk array may by configured to appear as a single disk drive or as multiple volumes of varying sizes.

Network Automatic Call Distribution (NACD)-NACD is a is a call processing system that routes (distributes) incoming telephone calls to specific telephone sets or stations calls based on the characteristics of the call or network settings. These characteristics can include an routing on network congestion, time of day routing, and other criteria.

Network Availability-Network availability is the ability of a network to perform the services or operate within the parameters defined for the services for features.

Network Backbone-A network backbone is the core infrastructure of a network that connects several major network components together. A backbone system is usually a high-speed communications network such as ATM or FDDI.

Network Busy Hour-The hour in a given 24-hour period during which the total load carried by all trunks in a network is greater than the total load carried during any other hour.

Network Call Center-A network call center is a system that processes calls between a company and a customer. Network call centers typically assist customers with requests for new products and service along with providing information about product and service features. A network call center usually has many stations for call center agents that communicate with customers.

Network Centric-Network centric is the primary use of a network system or function to provide features or services.

Network Channel Code-An encoded description of a channel provided by a local exchange carrier from the point of termination at an interexchange customer's premises to a central office, or from that customer's premises to an end-user location or Centrex system.

Network Compensation (NetComp)-Network compensation is the amount of money a broadcast station receives from a network for transmitting network programs.

Network Configuration-Network configuration is a set of network conditions and parameters that are used to allow one or more types of services and applications.

Network Connection-A network connection is the point of connection between equipment and a communication network. A network connection is the physical and electrical boundary between two separate communication systems.

Network Connectivity-Network connectivity is the ability of a device to setup a connection and communicate with a network.

Network Control-Network control is the transmission of signals or messages that perform call control, equipment configuration, or information management functions. Network control can be centralized or distributed. The control of public telecommunications networks is a centralized system as call processing is coordinated through a common channel signaling (CCS) network. The Internet uses distributed control as the switching information dynamically changes in packet switching centers (routers) throughout the Internet network.

Network Cost-Network cost is the charges and fees associated with the setup and operation of networks.

Network Data Collection Center-An operations center that administers network data collection and supervises the operation and maintenance of the Engineering and Administrative Data Acquisition System.

Network Data Management - Usage (NDM-U)-The network data management-usage (NDM-U) is a standard messaging format that allows the recording of usage in a communication network, primarily in Internet networks. The NMD-U defines an Internet Protocol detail record (IPDR) as the standard measurement record.

The IPDR record structure is very flexible and new billing attributes (fields) are being added because Internet services are now offered in almost all communications systems. The NMD-U standard is managed by IPDR organization at www.IPDR.org.

Network Directory-A network directory is a listing of items (such as data files) and their location within a network.

Network Directory Server (NDS)-A network directory server is a processing system that can receive and respond to requests for the identification and location of items (such as data files) within a network.

Network Element (NE)-A network element is a facility or the equipment used in the provision of a communications service. The term includes subscriber numbers, databases, signaling systems, and information sufficient for billing and collection or used in the transmission, routing, or other provision of a communications service.

Network Equipment-The telecommunications equipment and facilities owned, installed, and maintained by a telephone company or service provider and that are part of a telecommunications network.

Network Feeds-A network feed is a media connection that is used to transfer media or programs from a network to a distributor of the media or programs.

Network Gateway (NGW)-A media and signaling adapter (gateway) used in a network to interface between different types of networks. A network gateway can convert both the media and signaling control messages between the systems.

Network ID-Network Identifier

Network Identifier (Network ID)-A network identifier is a unique identifier assigned by a system to the communication device that identifies unique flows that may be concurrently sent on the network.

Network Integration-The joint provision of telecommunications services and the joint assumption of risk through a partnership arrangement among telephone companies. The expression often is used to describe both technical and economic integration.

Network Interface (NI)-A network interface is the point of connection between equipment and a communication network. A network interface is the physical and electrical boundary between two separate communication systems.

Network Interface Device (NID)-A connection point between the end customers equipment and the telecommunications network. This is also called the demarcation point.

Network Maintenance Center (NMC)-A facility that allows monitoring, testing and maintenance of a telecommunications network. The NMC is typically operational 24 hours a day, 7 days per week.

Network Management (NM)-Network management is the process of configuring equipment in the network, the setup (provisioning) of services, system maintenance, and repair (diagnostic) processes. Network management systems are commonly composed of a network management server computer and network management software.

Network management systems usually include a set of procedures, equipment, and operations that keep a telecommunications network operating near maximum efficiency despite unusual loads or equipment failures.

Network Management Center (NMC)-An operations center that monitors and controls traffic flow to help ensure the most efficient and economical use of available network capacity. The center plans strategies and works to minimize the effects of disasters, abnormal traffic loads, and switching system or facility failures.

Network Management System (NMS)-A network system is a combination of equipment and software that is used to setup, control, monitor and manage the operation of a communication network.

Network Neutrality-Network neutrality is the selection of network or providing of services on a network (such as the Internet) without regard for the network type or characteristics of data or content that is sent through the network.

Network Operating System (NOS)-A software program that manages communication between devices within a network. The NOS oversees resource sharing and often provides security and administrative tools.

Network Operations Center (NOC)-(1-Surveillance) A center responsible for the surveillance and control of telecommunications traffic flow in a service area. (2-Service) A facility or organization responsible for maintaining, monitoring, and troubleshooting a network infrastructure.

Network Operator-A company that manages the network equipment parts of a communications system. A network operator does not have to be the service provider. Also see Service Provider and Reseller.

Network Planning System (NPS)-An interactive computer program that assists strategic planners in the development of interoffice facilities and wire centers, and aids in the planning of traffic and distribution routes.

Network Port-A communication input/output access point to a network. The network port usually has specific network access protocols and security levels associated with it.

Network Probe-A network probe is a device or process that is inserted into a network to monitor for specific characteristics or conditions. Probes are can be non-intrusive and simply monitor and report on information that passes through or by the probe.

Network Program-(1-computer system) A program that operates on a server within a network. (2-distribution) Any program delivered simultaneously to more than one broadcast station regional or national, commercial or noncommercial.

Network Security-The processes used within a network to validate the identity of users (authentication), access control of services (authorization), and information privacy protection (encryption).

Network Selection-Network selection is the process of identifying and connecting to a network line or system. IP STB may have one or more network connection options that include Ethernet, DSL modem, cable, Wi-Fi, satellite, DTT or other network system. The choice of network connection may include availability of a network signal (is it attached to the network), preferred connection type (higher speed connections) and cost considerations.

Network Servers-Hosts, and sometimes personal computers, may function as specialized types of nodes called network servers. These specialized nodes serve the other nodes by storing many of their files and running much of their common software.

Network Service Provider (NSP)-Any company that provides network services to customers or devices.

Network Subsystem (NSS)-(1-general) The network parts of a communication system. network. (2-GSM) The system parts of a GSM network this includes the mobile switching center (MSC), home location register (HLR), visitor location register (VLR) and equipment identity register (EIR).

Network Termination (NT)-A final end point in a network that is usually owned by the network service provider. After the network termination, the equipment is commonly owned by the customer (called customer premises equipment-CPE). When the network termination (NT) is an active device, it typically has standard communications parameters such as protocols, timing and voltages to allow specific types of equipment to correctly communicate with the network.

Network Termination Equipment (NTE)-Network termination equipment (NTE) are the devices used by end-user to access the network. In the traditional PSTN world, network termination equipment was generally confined to the telephone, headset or conference phone. The cellular industry expanded this to include cell phones and pagers. In IP Communications network termination equipment can include all these traditional devices as well as computers and PDAs.

Network Utilization-Network utilization is a comparison of how many network resources are being used as compared to the total amount of availability network resources.

Network Utilization Charge-A network utilization charge is a fee assessed for the use of network resources. Examples of network utilization charges is bandwidth, storage or network access fees.

Network View-Network view is a graphical display of network elements in a system. A network view may include information or graphics about the status of each of the elements or the flow of information or data through the elements.

Networking-The connection of geographically separate computers and communication devices using transmission line facilities.

Never Aware Applications-Never aware applications are software programs that can perform operations using commands or information from other sources (such as a user at a keyboard) independently from other systems or networks.

Newbie-A newbie is a person who is new to the use of the Internet.

News Tickers-News tickers are devices that can display news information on a continuous basis.

Next Generation Operations Support System (NGOSS)-Next generation operations support system is a set of commands, processes and procedures that is used by a network to allow a network operator to perform the administrative portions of business operations such as billing and customer care.

NGOSS-Next Generation Operations Support System

NGW-Network Gateway

NI-Network Interface

NID-Network Interface Device

Night Answer-A telephone system feature that redirects in-coming calls during designated times of the day, such as after business hours.

Night Service-Night service is a processing state of a telephone system (such as a PBX) during hours of operation when the company is closed or in a different state of business operation. Night service usually provides a different greeting messaging and call routing (transfer) capability. Night service may prompt callers to leave messages instead of being routed to an operator.

This diagram shows how a telephone system can change its basic operation for daytime and nighttime telephone service. In this example, during the day, all the incoming calls are routed to (received by) a receptionist at extension 1001. At night (between 5 pm and 8 am), the calls are automatically redirected to an automated telephony call processing system that is connected to extension 1014. When the automated attendant detects a ring signal, answers the phone (off-hook signal) and plays a pre-recorded messaging informing the caller of options they may choose to direct the call to a specific extension. In this example, the auto-

N

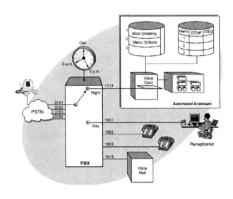

Night Service Operation

mated call attendant software decodes DTMF tones or limited list of voice commands to determine the routing of the call. The automated call attendant software then determines if the destination choice is within the option list and if the extension is available. If the extension is available, the automated attendant will send a command to the computer telephony board (voice card) that can switch the call to the selected extension. If the extension is not valid or not available, the automated attendant will provide a new voice prompt with updated information and additional options.

NIH-Not Invented Here

nm-Nanometer

NM-Network Management

NMC-Network Maintenance Center

NMC-Network Management Center

NMS-Network Management System

NNX-A term used for qualifying dialing digits. N stands for any number from 2 through 9 and X stands for any number 0 through 9. An example of a valid NNX is 230.

No Bid-No bid is a response to a request for proposal that indicates the recipient has received the request but is not interested in responding. Companies may submit a no bid response to ensure that they continue to receive bid requests in the future.

No Charge Traffic-Traffic, such as 611 service request and 911 emergency calls, classified as "no-charge" in a tariff on file with an appropriate regulatory agency.

No Circuit Tone-A low tone, interrupted 120 times per minute (02 5 on and 0.3 5 off) indicating that no trunk is available. This term also is known as a reorder, all-circuits busy, or fast-busy tone.

NOC-Network Operations Center

Node-(1-network) In network topology, a terminal of any branch of a network, or a terminal common to two or more branches of a network. (2-ascending) The point where a satellite crosses the plane of the equator when moving north. (3-current) The points at which the current is at minimum in a transmission system in which standing waves are present. (4-descending) The point where a satellite crosses the plane of the equator when moving south. (5-network) A terminal on any branch of a network. (6-switching) The switching points in a switched communications network, including patching and control facilities. (7-transmission line) A point of interconnection on a transmission line. (8-tree structure) A point where subordinate data originates. (9-telephony) A switching office or facility junction. (10-test facility) A remote test facility. (11-voltage) The points at which the voltage is at a minimum in a transmission system in which standing waves are present.

Nomadic Service-Nomadic service is the providing of communication services to more than one location. While nomadic service may be provided to many locations, nomadic service typically requires the transportable communication device to be fixed in location during the usage of communication service.

Nominal-The most common value for a component or parameter that falls between the maximum and minimum limits of a tolerance range.

Nominal Value-A nominal value is a specified or anticipated value.

Non-Chargeable-Non-chargeable is a billing record or service that is not or cannot be charged.

Non-Compete-Non-Competition Agreement

Non-Competition Agreement (Non-Compete)-A non-competition agreement or clause is commitment of a company or person that they will not work or assist other people or companies in a competing business or industry. A non-compete agreement usually restricts competitive actions for a limited period of time.

Non-Disclosure Agreement (NDA)-A non-disclosure agreement is a binding contract between parties not to disclose and to keep confidential information shared among the parties from being spread to other parties. Commonly referred to as an NDA, in certain circumstances such an agreement can preserve the novelty of an invention. Documents exchanged under a properly written and executed NDA may not be considered as a publication, or public dissemination of an invention and can preserve the right of the inventor to apply for patent protection.

Non-facilities Based Carrier-Refers to carriers that do not operate their own switches and networks

Non-Payment Churn-Non-payment churn is the disconnection or loss of customers from a telecommunications service provider due to non-payment of services.

Nonprofit-Any corporation, foundation, or association that is not operated to benefit any private shareholder or individual.

Non-Recurring Charge (NRC)-A cost for a facility or product that only occurs one time or is not periodically charged.

Non-Recurring Engineering Costs (NRE)-Costs associated with a product or service that are associated with the development of the product and not associated with the marketing or support of the product sales.

Non-Refundable Retainer-A non-refundable retainer is a value that is provided to a person or company in advance of the receiving of products of services which is not refunded even if the services are not performed or charged.

Non-Renewal-Non-renewal is the termination or confirmation that a license providing services will end on its pre-defined date or termination criteria.

Non-Repudiation-Non-repudiation is the use of information or evidence that confirms that a person or company has received or used the products or services that they have ordered.

Nonresponsive Proposal-Nonresponsive proposals are proposal submissions that have omitted required portions or that have been structured in a format other than the format that was required in the request for proposal. Nonresponsive proposals may be excluded from consideration for the awarding of projects.

Nonusage Charges-Nonusage charges are fees that are assessed for the providing of services that are measured by time periods (e.g. monthly access charges) or other non-varying fees.

Normal Business Hours-Hours during which most similar businesses in the community are open to serve customers. In all cases, "normal business hours" must include some evening hours at least one night per week and/or some weekend hours.

Normal Routing-The routing of a given signaling traffic flow under normal conditions, that is, in the absence of failures.

North American Numbering Plan (NANP)-The NANP is an 11 digit-dialing plan that is used within North America. It contains 5 parts: international code, optional intersystem code (1 +), geographic numbering plan area (NPA), central office code (NXX), and station number (XXXX). The NPA code defines a geographic area for the serving telephone system (such as a city). The NXX defines a particular switch that is located within the telephone system. Finally, the station code identifies a particular line (station) that the switch provides service to.

NOS-Network Operating System

Not Invented Here (NIH)-Not invented here is a concept or a viewpoint that new products or services should be invented within a company rather than being developed outside the company.

Notices-Notices are declarations or messages that are provided or sent to people or companies to alert them of events or conditions. Notices may be required for certain business or legal transactions.

Notification-Notification is the process of sending a message that indicates the status of an event or action that will occur. An example of notification is the alerting of mobile telephones that a group or broadcast call is occurring.

Notification Server-A computing device (typically a computer with communications software) that provides notification to users or devices when specific events occur.

NPA-Numbering Plan Area

NPA Codes-Interchangeable Numbering Plan Area

NPDB-Number Portability Database

NPS-Network Planning System

NRC-Non-Recurring Charge

NRE-Non-Recurring Engineering Costs

nsec-Nanosecond

NSP-Network Service Provider

NSS-Network Subsystem

NT-Network Termination

NTE-Network Termination Equipment

NTS-Number Translation Service

Number Pooling-The grouping or issuing of telephone numbers in block sizes. It is common to pool telephone numbers into thousand block groups.

Number Portability-Number portability involves the ability for a telephone number to be transferred between different service providers. This allows customers to change service providers without having to change telephone numbers. Number portability involves three key elements: local number portability, service portability and geographic portability.

The first part of the telephone number (NPA-NXX) usually identifies a specific geographic area and specific switch where the customer subscribes to telephone service. If a telephone number is assigned to another system (different NXX) in the same geographic area (same NPA), the interconnecting carriers (IXCs) connecting to that system must know which local system to route the calls based on the selected local service providers. In this case, the IXC must look up the local telephone number in a database (called a database dip) prior to delivering the call to the end customer.

Number Portability Database (NPDB)-Number portability database enables number portability by providing information that helps to route calls to their destination which may have an assigned telephone number that is different (number has been ported) than the destination phone number.

Number Translation Service (NTS)-A process of converting one number (such as a telephone number in one country) into another telephone number (such as a telephone number in a call center of another country). NTS service is commonly used to provide local telephone numbers that are automatically connected (re-routed) to call centers in other areas.

Numbering Plan-A numbering plan is a system that identifies communication points within a communications network through the structured use of numbers. The structure of the numbers is divided to indicate specific regions or groups of users. It is important that all users connected to a telephone network agree on a specific numbering plan to be able to identify and route calls from one point to another.

Telephone numbering plans throughout the world and systems vary dramatically. In some countries, it is possible to dial using 5 digits and others require 10 digits. To uniquely identify every device that is connected to public telephone networks, the Consultative Committee for International Telephony And Telegraphy (CCITT) devised a world numbering plan that provides codes for telephone access to each country. These are called country codes. Coupled with the national telephone number assigned to each subscriber in a country, the country code telephone makes that subscribers number unique worldwide. The International Telecommunications Union (ITU) administers the World Numbering Plan standard E.164 and publishes any new standards or modifications to existing standards on the Internet.

Numbering Plan Area (NPA)-A 3-digit code that designates one of the numbering plan areas in the North American Numbering Plan for direct distance dialing. Originally, the format was NO/IX, where N is any digit 2 through 9 and X is any digit. From 1995 on, the acceptable format is NXX.

Numbering Scheme-British English synonym for North American "Numbering Plan."

Numeric-A display, message or readout that contains numerals only, such as paging.

Numeric Paging-A paging application in which information is sent and displayed as numeric characters. Usage is typically to display phone numbers which user call back to This figure shows basic operation of a numeric paging system. This diagram shows how a numeric paging system receives incoming calls and creates a numeric message that can be sent to a numeric pager to alert the numeric pager of the telephone number that the caller wants the recipient to call. In this example, a caller dials the numeric pagers assigned telephone number. When the numeric paging system receives the incoming call, it plays an interactive voice response message to the caller to leave their number by entering the

telephone number via the touch-tone buttons. This example shows that an operator may be used to receive the telephone number verbally as a phone with touch-tone may not be available. The paging system then converts the telephone number to the pagers identifying address and sends the message to the paging transmitters. This diagram shows that the numeric pager uses paging groups and this pager is assigned to paging group 4. When paging group 4 starts, the pager wakes up and looks for messages with its address and the data that follows the message (usually the caller's telephone number). When it finds this message, it displays the paging user to the pager user and stores the message in temporary memory. When the user sees the number, the user will go to a telephone and call the displayed number (usually the caller's telephone number).

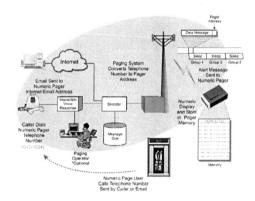

Numeric Paging Operation

NVOD-Near Video On Demand
NXX-A term used for qualifying dialing digits. N stands for any number 2 through 9 and X stands for any number 0 through 9. An example of a valid NXX is 201.

370

O

O&O-Owned and Operated

OA&M-Operations Administration And Maintenance

OASIS-Organization for Structured Information Standards

OBE-Out of Box Experience

OBF-Ordering And Billing Forum

Object Oriented Programming-Object oriented programming is a process of developing an assembly of instructions for a computer through the use of independent blocks of programming code (objects) that are designed to receive inputs and outputs from other objects to perform specific functional processes.

Object Request Broker (ORB)-Object request broker is a communication processing function is an object oriented distributed computing environment, which adapts information or communication from one object to another. The ORB standards are defined by the Object Management Group. More information about ORB can be found at www.OMG.org.

Object Rights-Object rights are the authorizations of use for specific portions of programming code (objects) that have designed to receive inputs and outputs from other objects to perform specific functional processes. Object rights may include the ability to use the objects with other programs or to modify the objects to perform different or new functions.

Obsolete Technology-Obsolete technology are processes, components or systems that have been replaced by better solutions or are no longer needed to perform the functions they provide.

OC-Operations Center

OC-Optical Carrier

OC&C-Other Charges & Credits

OC&C-Other Charges and Credits

OCA-Outside Collections Agency

OCC-Other Common Carrier

Occupancy-The fraction of the time that a circuit or equipment is in use, expressed as a decimal. Occupancy is the erlangs carried and is equal to the hundred call seconds (CCS) carried divided by 36. It includes both message time and setup time.

OC-n-Optical Carrier Hierarchy

OCR-Optical Character Recognition

OCS-Online Charging System

ODBC-Open Database Connectivity

Odd Even Pricing-Odd even pricing is the process of changing price values from even numbers to odd numbers to influence the perceived value of the price. An example of odd even pricing is the reducing of a $100.00 price to $99.99 which makes it appear significantly lower in cost to the buyer.

ODM-Original Design Manufacturer

ODS-Operational Data Store

ODU-Outdoor Data Unit

OEM-Original Equipment Manufacturer

OEM Account-Original equipment manufacturer accounts are companies or organizations that provide a product, which uses components, assemblies, or completed products that are produced for their customers. OEM account sales support needs may include technical support, design services and skilled negotiations.

OEM Products-Original equipment manufacturer products are objects or assemblies that are designed, modified or labeled so they can be sold by other companies under their brands.

Off Catalog Item-An off catalog item is a product or service that is not listed in a catalog. Customers may request off catalog items if they do not find exactly what they are looking for.

Off Line (Offline)-(1-general) A condition of devices or subsystems not connected into, not forming a part of, and not subject to the same

O

controls as an operational system. (2-computer system) A circuit or device that is disconnected from a system, usually a remote computer, and not available for use.

Off Net-Off Network

Off Network (Off Net)-Off network is the providing or controlling of services within networks that other companies own, manage or control.

Off Peak-A time period where a system usage level is lower, typically after normal business hours. Some telecommunications service providers charge a reduced rate for the use of services during off-peak hours.

Off Premises Extension (OPX)-A call processing feature that allows a call to be forwarded to a telephone at a secondary location that is located off the premises of the phone system that is transferring the telephone call.

Off Shore Development-Off shore development is the use of companies and/or people that are located outside a country or continent to assist in the development of applications, products or services.

Off Site Training-Off site training is the providing of lectures or a presentation of learning materials at a location (site) other than the company that is receiving the training.

Off the Shelf (OTS)-Off the shelf is an equipment or service that is ready to use or to install directly after it is purchased.

Offer-An offer is an item and/or characteristic of a product or service that is offered to a customer in return for money or something the customer provides in exchange for that product or service.

Offer Letter-An offer letter is a document that defines goods, services or other valuable assets (such as an job at a company) along with the related terms and what is required to receive the offer if the offer (such as an acceptance letter).

Offer Management-Offer management is the process of assigning and tracking specific product and service offers from people and companies.

Off-Hook-An electrical signal that occurs when a customer typically removes a telephone receiver off of its cradle, thus releasing the hook switch. When the hook switch is released (off-hook), this typically causes a drop in telephone line voltage due to connecting of the local loop telephone wires together. Automatic devices such as a computer modem can also initiate an off-hook signal.

Office Expenses-Office expenses are the fee or assessments that occur as a result purchasing or using of equipment or services for office (activity supporting) activities. Examples of office expenses include administrative software, paper supplies and telephone services.

Official Rate-An official rate is a pricing or valuation system that is defined or managed by an authority or perceived authority. An example of an official rate is a currency exchange rate that is assigned by a government.

Offline-Offline is a condition where a device is unable or inhibited from communicating with a network (such as when a computer dta connection with the Internet has terminated).

Offline-Off Line

Offline Asset-Offline assets are files or data that represents a valuable form of media that is not directly or immediately available for use.

Offline Behavior-Offline behavior is the actions that programs perform when they are not connected to the Internet. An example of offline behavior is the disabling of features and services that require connection to the Internet for authorization purposes.

Offline Charging-Offline charging is the process of gathering and processing billing records and invoices after the service is provided.

Offshore Company-An offshore company is a business entity, which has its principle offices in another continent.

OI-Operator Interrupt

Oligopoly-An oligopoly is an industry where a limited number of companies supply or control all the customers. An example of an oligopoly is the licensing of a limited number of companies to provide wireless services such as wireless broadband Internet.

OLTP-Online Transaction Processing

OMA-Open Mobile Alliance

OMAP-Operation, Maintenance and Administration Part

OMC-Operations And Maintenance Center

On Approval-On approval is a term or condition that allows a buyer to take possession or control of goods or services to determine if they want to accept (approve) of the order.

On Demand Ad Insertion-On demand ad insertion is the process of dynamically inserting (on request) an advertising message into a media stream such as a television program or during the replay of a stored program. For video on demand systems, on demand ad insertion offers the ability to insert different advertising messages for the viewer depending on a variety of factors including location, time and viewer profile.

On Demand Billing-On demand billing is the process of grouping service or product usage information for specific accounts or customers and displaying or producing invoices or billing statements when requested.

On Demand Programming-On demand programming is providing or making available programs that users can interactively request and receive.

On Demand Services-On demand services interactive programming services that provide or entitle a customer to receive or gain access to specific services after they request (demand) the service. On demand services are typically provided for a single use or session with a fixed termination event (download complete) or time (viewing period). On demand services users are often billed per event for the specific service that is requested. On demand services through the television may include Internet access, videoconferencing, instant messaging and a variety of other interactive services. On demand services may be made be simplified and made more valuable through the use of sophisticated Electronic Program Guides (EPGs) that can be dynamically changed or updated.

On Line (Online)-(1-general) A device or system that is energized and operational, and ready to perform useful work. (2-computer system) A circuit or device that is connected to a system, usually a remote computer, and available for use.

On Net-On Network

On Network (On Net)-On network is the providing or controlling of services within the network that the operator owners, manages or controls.

On Order-On order is a term or condition that requires the buyer to take possession agreeing to own the goods or services when they initiate the order.

On Peak-A time period where a telecommunication system usage is higher, typically during normal business hours. Some telecommunications service providers charge a premium rate for the use of services during peak hours.

On Site Training-On site training is the providing of lectures or presentation of learning materials at a location (site) that is defined by the recipient (such as in a building that is used by the company that is receiving the training).

On the Fly (OTF)-On the fly are processes that occur instantly or within a short time period that is perceived or used (such as within a few milliseconds) to perform or record events when they are required or used.

On The Spot Packages-On the spot packages is the ability to offer a combination of items as part of a package offer. An example of an on the spot package is the combining of two products and a service at a discount price that were not previously offered or available in the order processing system.

Oncost-Ongoing Cost

One Click Signup-One click signup is a process that allows a user or viewer to select and accept the purchase of products or services without the need to complete additional information.

One Flat Business Line (1FB)-A telephone line used by a business that is charged a single monthly fee regardless of how many calls that are originated or received during each month.

One Off Charge-A one off charge is the billing for a service that is provided as a result of a single event or service (such as purchasing the 24 hour right to view a pay per view movie).

One Off Payment-A one off payment is the processing of a payment each time one order is processed.

One Time Usage-One time usage is the authorization to use a product or service only one time.

One Way Encryption-One way encryption is the process of converting data or information elements into a new format in a way that some of the conversion information is lost in the process so that it is not possible to recreate the original information from the coded data. An example of a one way encryption process is password hashing where a password that has many characters (ASCII characters) can be summed and only the last few digits of the sum are used. This makes it impossible to exactly determine the original password from the hashed password code.

One-Way Paging-A paging technology whereby the signal is sent from the base station to the paging unit only, without a return verification signal or other 2-way capabilities.

Ongoing Cost (Oncost)-Ongoing cost is continuing accumulation or addition of costs. Examples of oncosts include repetitive operations costs such as rent and utilities.

On-Hook-An electrical signal that occurs when a customer typically replaces a telephone receiver onto its cradle, thus opening the hook switch. When the hook switch is opened (on-hook), this typically causes an increase in telephone line voltage due to removal of the connection between the telephone wires on the local loop line.

ONI-Operator Number Identification

Online-Online is a condition where a communication device is ready and able to communicate with a network (such as when a computer is connected to the Internet).

Online-On Line

Online Access-Online access is the ability of a device, system or service to be controlled via a web page or Internet access device.

Online Account-Online accounts are financial instrument can be processed (charged) by companies for the payment for products or services (which can process online transactions) and the charges accumulated on the account is paid for by the owner or user of the card. An online account has a unique identifier that designates a customer or company that is used to associate billing charges for products and services that are usually available for viewing via a web portal.

Online Account Management-Online account management is the authorizing, recording, and assignment of costs to users and groups through the use of a web portal.

Online Assets-Online assets are files or data that represents a valuable form of media that is directly or immediately available for use.

Online Bank-An online bank is a financial institution that receives, processes and lends money through the Internet.

Online Behavior-Online behavior is the actions that programs perform when it is connected to the Internet. An example of online behavior is the providing of authorization for features and services that would not be available to the user if the computer was not connected to the Internet.

Online Billing-Online billing is the process of grouping service cost or product usage information for specific accounts or customers and where the usage records, invoices and payments may be viewed and possibly controlled via a Internet web portal.

Online Catalog-An online catalog is the presenting of items available for selecting or ordering in a web page format. Online catalogs formats range from a linear display of products to an interactive dynamic listing of items that match users search criteria.

Online Charging System (OCS)-An online charging system is a set of interconnected network elements that enable the identification, rating and posting of charges in real time (or near real time).

Online Check-An online check is order directing a bank to pay money when can be transferred through the Internet.

Online Commerce-A shopping medium that uses a Internet to present products and process orders.

Online Conversion-Online conversion is the ratio of people or companies that respond or take action as a result of a promotional campaign via an Internet process (such as email, web registration or instant messaging).

Online Database-An online database is a collection of information items (records) and their associated elements (fields) that can be accessed through the Internet.23

Online Editing-Online editing is the modification of media programs through the use of network connections (such as sending program change commands through the Internet).

Online Help-Online help is the presenting of information through the Intenret that can help users to answer questions, discover solutions or solve problems.

Online Mall-An online mall is an online store hosting system that is shared by several merchants. An online mall may be promoted by both the online mall host (such as Yahoo! stores) and by the merchants that are located as part of the online mall.

Online Order-An online order is an authorization processed via an Internet connection to send and bill for products or services. An online order typically contains details of whom ordered, where it is shipped, what is ordered, order details (such as items ordered) and the method of payment.

Online Payment System-An online payment system is the process(es) used to collect payments from the buyer of products and services through the Internet. Online payment systems may involve the use of money instruments such as credit cards, electronic checks, credit memos, coupons, or other form of compensation used to pay for one or more order invoices.

Online Reports-Online reports are summaries of transactions or data that can be viewed through a web portal.

Online Retailer-An online retailer is a company that primarily sells products on or through the Internet.

Online Signature-An online signature is a graphic image that is created using a mouse or other types of drawing devices that are stored with IP addresses, email and possibly other unique identifying information.

Online Statistics-Online statistics is the ability to view the statistical analysis of data (such as online store visitor and order activity) through a web portal.

Online Storage-Online storage is a device or system that stores data that is directly and immediately accessible by other devices or systems. Online storage types can vary from disk drives to electronic memory modules.

Online Store Interfaces-An online interface is a defined set of messages, processes and/or hardware equipment that allow online store information (such as order and product details) to be exchanged to other programs or systems. Online store interfaces may allow database and accounting system connectivity.

Online Store Manager-An online store manager is a program or web portal that allows online storeowners or administrators to setup, configure, edit and obtain reports about online stores. An online store manager may contain functional areas including catalogs, shipping options, payment methods, affiliate links, and reports.

Online Storefront-An online storefront is a web site that enables visitors to find, order and pay for products and services. Online storefronts typically include catalogs, shopping cars and payment processing systems. There are many storefront options that can influence the effectiveness of a user experience and web analytics can be used to determine which mix of web pages, content choices, and product offers are most effective.

Online Tracking-Online tracking is the monitoring or following of the flow or usage of information, transmission or services via the Internet. An example of online tracking is the providing of a tracking number to a customer that they can use to see the status of their order or service.

Online Transaction Processing (OLTP)-Online transaction processing (OLTP) systems operate on or through the Internet. OLTP systems usually operate in real time (at the time the customer submits their order). Examples of OLTP systems electronic bookstores, banking systems, and travel reservation systems.

On-Net-On-The-Net Calls

ONT-Optical Network Termination

On-The-Net Calls (On-Net)-Calls that are connected on a single network (such as calls through the Internet).

Ontrepreneurs-An ontrepreneur is a person who identifies business opportunities that can be operated through the Internet and takes actions (such as establishing and running a company) to develop business transactions for that business.

ONU-Optical Network Unit

Opaque Transport-Opaque transport is the transmission of data or information through a transmission line or network where the quality of service (QoS) is defined but may not be able to be guaranteed.

Open Architecture-Open architecture is a design that permits the interconnection of system elements provided by many vendors. The system elements must conform to interface standards.

Open Database Connectivity (ODBC)-An interface for accessing data in a environment of relational and non-relational database management systems. ODBC provides a vendor-neutral way of accessing data in a variety of personal computers, minicomputer and mainframe databases.

Open Interface-A connection or access point between two assemblies or systems that is well defined and is readily available to manufacturers or users of the interface. Open interfaces are usually defined to encourage competition as multiple manufacturers can compete to produce products that have open interfaces.

Open Mobile Alliance (OMA)-The open mobile alliance is a group that is analyzing the needs and producing industry standards for the inter working of mobile services and applications.

Open Network-An open network is a communication system that is accessible by public users.

Open Numbering Plan-A numbering plan in which local numbers comprise a different quantity of digits, even in the same city, and each area or zone code typically comprises a different quantity of digits. The national telephone numbering plans of many European countries are open plans. For example, there are both 6 digit and 7 digit telephone numbers in the same city in some countries. Some area codes for small towns have more digits than the area codes for larger towns in the same country, etc. The ITU international numbering plan is an open plan, with different national telephone systems being reached via "country codes" having different numbers of digits. The country code for North America is 1, a single digit. The country code for the United Kingdom is 44, a pair of digits. The country code for Israel is 972, comprising three digits. In each case the quantity of digits that must follow the country code is also different for each destination country, and may vary among different cities in the same country.

Please note that an assembly of closed numbering plans may comprise an overall open numbering plan!

In most local numbering plans the quantity of digits in a number is tied to the leading digits of that number. This is called a "deterministic" open numbering plan. For example, in such a plan, all numbers beginning with the digits 23, 24 or 25 are 5 digits in length, while all numbers beginning with any other two digits are 6 digits in length. When an open local numbering plan is not deterministic, as in some cities in Austria, the originating telephone switch must use a "time out" method to determine when the originator has dialed the last digit. An open numbering plan has the advantage of allowing residents of small towns to dial a minimum quantity of digits actually required to distinguish the small quantity of local telephones in their local dialing area. However, it also increases the complexity of determining accurately when an originator has dialed the final digit of a non-local call. Most systems that handle open numbering plans use the "time out" method. That is, they assume that the originator has completed dialing when an interval of typically 6 seconds elapses without the originator dialing any further digits. Some systems will wait as long as 20 seconds to ensure that no further digits are dialed. This either prolongs the time to set up the call, if the waiting time is very long, or occasionally causes incorrect number connection attempts if the waiting time is too short. In some systems such as in North America, the "time out" method is used for international calls, but an originator dialing an international call can also indicate the end of the dialed digits by using the # key, but this is only possible for a originator who has a touch-tone dial. See also Closed Numbering Plan.

Open Order-An open order is a purchase authorization that has not been completely fulfilled.

Open PLC European Research Alliance (OPERA)-Open PLC European research alliance is a European commission that assist in the creation and deployment of PLC services and products in Europe.

Open Pricing-Open pricing is the making of price lists and intended prices changes are make available to all that want to see them.

Open Settlement Protocol (OSP)-A standard protocol that is designed to transfer billing information to allow inter-carrier billing between voice and data communication systems. The OSP format is approved by the European Telecommunications Standards Institute (ETSI). OSP allows communication gateways to transfer call routing and accounting information to clearinghouses for account settlements between carriers (service providers).

Open Source Software (OSS)-Open source is software that includes the original source code from which the product was developed to allow other developers to make changes to the software to meet their specific application needs.

Open Standards-Open standards are operational descriptions, procedures or tests that are part of an industry standard document or series of documents that is recognized and available to all people or companies as having validity or acceptance in a particular industry. Open standards are commonly created through the participation of multiple companies that are part of a professional association, government agency or private group.

Open System-A system whose characteristics comply with specified standards and that therefore can be connected to other systems that comply with these same standards.

Opening Ticket-An initial work order that is requested by outside plant personnel from the network operations center (NOC) or other maintenance center prior to opening an underground splice closure for either repair or splicing activity. The ticket is closed out at the end of the days activities and a new ticket issued upon request for subsequent work activity.

OPERA-Open PLC European Research Alliance

Operating Information-Operating information is data that relates the ability of a business to provide products or services. Operating information may include production capabilities, licenses and key business relationships.

Operating License-An operating license is a contract or recognized authorization (such as from a government agency) that grants specific rights to operate a system and/or to provide specific types of services.

Operating Platform-The operating platform is the combination of system hardware and operating system software that run software programs or applications.

Operating Resource Management (ORM)-Operating resource management is the process of identifying and coordinating the available resources for operation systems. An example of operating resource management is the coordination of office systems and supplies.

Operating Statistics-Operating statistics are data and associated mathematical analysis and representation of information that relates to the operation of a system or business. For communication systems, operating statistics may include the amount of usage (loading), number of customers that subscribed to services (take rate) and churn (disconnect) rate.

Operating System (OS)-An operating system is a group of software programs and routines that directs the operation of a microprocessor (a computing chip) in a device in its tasks and assists programs in performing their functions. The operating system software is responsible for coordinating and allocating system resources. This includes transferring data to and from memory, processor, and peripheral devices. Software applications use the operating system to gain access to these resources as required.

Operation, Maintenance and Administration Part (OMAP)-The application entity that is dedicated to the communications aspects of the operation, administration and maintenance of the signaling system network

Operational Data Store (ODS)-A specialized data base system that is created to serve as a storage and holding area for data extracted from operational systems and staged for passive access by other systems and for loading into data warehouses and data marts.

Operational Expenses (OpEx)-The term OpEx is used to define the day-to-day short term expenses paid to a telephone company (telco) to support continued business operations (e.g. salaries, rents, commission fees).

Operational Requirements-Operational requirements are a set of resources (such as equipment, labor and software processing) that are necessary to meet the operational needs of a business or system.

Operational Testing-Operational testing is the configuring of system equipment, application of test signals (if required) and measurements or observations of signals and test responses that ensure a system is operating correctly.

Operations-The term denoting the general classifications of services rendered to the public for which separate tariffs are filed, namely exchange, state toll and interstate toll.

Operations Administration And Maintenance (OA&M)-The functions that are necessary to operate, perform administration functions and maintain a communications network.

Operations And Maintenance Center (OMC)-The OMC includes alarms and monitoring equipment to help a network operator diagnose and repair a communications network.

Operations Center (OC)-An operations center is a facility and the associated equipment that is responsible for the operations and monitoring of communication services and system operations.

Operations Cost-Operations cost is the charges and fees associated with the administration, provisioning and management of a business or system.

Operations Department-An operations department is a group within a company or organization that is responsible for operating and maintaining the equipment and systems within the company.

Operations Support System (OSS)-Operations support systems are combinations of equipment and software that are used to allow a network operator to perform the administrative portions of the business. These functions include customer care, inventory management and billing. Originally, OSS referred to the systems that only supported the operation of the network. Recent definition includes all systems required to support the communications company including network systems, billing, customer care, etc.

Operations System-A general-purpose software system that supports the operations of a telecommunications company. Operations supported include planning, engineering, ordering, inventory tracking, automated designs, provisioning, assignment, installation, maintenance, and testing.

Operator-(1-general) A person who assists customers with the operation or use of a system or service (2-carrier) In telecommunications, this is the company that provides communication services.

Operator Assisted Call-A telephone call made by a customer who dials for an access code for assistance (such as 0 +) or is automatically connected to an operator for assistance in placing person-to-person, collect, coin, third-panty-billed, or credit card calls.

Operator Interrupt (OI)-An operator service whereby the operator may interrupt an ongoing conversation. Sometimes called an emergency interrupt (EI)

Operator Number Identification (ONI)-The manual identification of the calling num-ber of a customer dialed toll call and its entry by an operator onto an automatic message accounting tape for billing.

Operator Relay Services-A program to assist those with hearing and/or speech disabilities to communicate over telephone networks through the use of a relay operator who translates written text into speech, and spoken replies into text.

Operator Responsiveness-Operator responsiveness is the ability or performance of an operator to respond to orders, service requests, operational problems or other demands on the capabilities of the operator.

Operator Services-Operator services use an operator to assist in the handling of a processing of a call. These special handling services include collect calling (billing to a called number), third party charging (billing to another phone or calling card), identification of a person who has called (call trace services), call information services (assistance with directory number location), rate information services (call charge rates), or any other service that requires an operator for special call processing services.

Operator Services System-A service system that allows for any special handling of the calls. These special handling services include collect calling (billing to the called number), third party charging (billing to a calling card), identification of the person called (call trace services), call information services (assistance with directory number location), rate information services (call charge rates), or any other service that requires special call processing services.

Operator Trunks-A term that generally refers to multi-channel communication lines (trunks) that are located between manually operated switchboard positions and local dial central offices.

OpEx-Operational Expenses

Opportunity Costs-Opportunity costs are potential amounts of revenue or profit that is lost if a project or business acquisition is not performed. The opportunity cost is the difference between the result of the results of an investment in a project or business and the costs of resources that could have been achieved without that investment.

OPS-Order Processing System

Optical Cable-A cable that contains fibers or bundles of fiber lines that is designed for physical and optical (e.g. optical signal loss at specific wavelengths) specifications that allow it to be used in specific types of optical communication applications (e.g. undersea or in-building.)

Optical Carrier (OC)-An optical carrier is an optical signal that is modified (modulated) in amplitude, frequency or phase for the purpose of carrying information.

Optical Carrier Hierarchy (OC-n)-Optical carrier (OC-n) transmission is a hierarchy of optical communication channels and lines that range from 51 Mbps to tens of Gbps (and continues to increase). The "n" is an integer (typically 1, 3, 12, 48, 192, or 768) representing the data rate. Lower level OC structures are combined to produce higher-speed communication lines. There are different structures of OC. The North American standard is called synchronous optical network (SONET) and the European (world standard) is synchronous digital hierarchy (SDH).

Optical Character Recognition (OCR)-The recognition of printed or handwritten characters by automatic systems, often laser- and photoelectric-based.

Optical Disk-A form of data storage using a laser to optically record the data on a disk which is read with a low-power laser pickup. The primary types of optical discs are: read only (RO), write once read many (WORM), erasable/record-able (thermo-magneto-optical TMO) and phase change (PC).

Optical Link-Any optical transmission channel designed to connect two end terminals or to be connected in series with other links. Terminal hardware also may be considered within the bounds of this term.

Optical Network-Optical networks are a series of points that are interconnected by optical communications channels or systems. Optical networks are either common to all users or privately leased by a customer for some specific application.

Optical Network Termination (ONT)-The optical network termination converts optical signals into another format and operates as the end network termination point in a communications system. ONTs are typically located in the customer's premises.

This figure shows an optical network termination (ONT) is a special form of ONU. This diagram shows ONT receives signals from the optical system (the OLT) and converts the signals into other forms of information. This diagram shows an ONT that includes a data router that has a television RF port, 4 Ethernet ports (RJ-45 connectors) and two standard POTS voice ports (RJ-11 connectors).

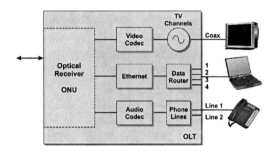

Optical Network Termination Operation

Optical Network Unit (ONU)-An optical networking unit is a device that can receive, multiplex and demultiplexes optical signals and converts the signals to a format suitable for distribution to other systems or to a customer's equipment such as copper lines.

This figure shows the basic operation of an optical network unit (ONU). This diagram shows that an ONU receives optical signals from OLT on multiple wavelengths. The ONU can receive and process one or more optical wavelengths and convert these optical signals into an electrical form. This diagram shows an ONU that can process electrical signals into video (television), data (Internet browsing) and digital audio (telephone) formats.

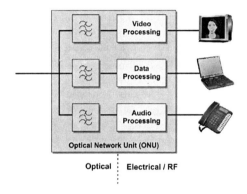

Optical Network Unit (ONU) Operation

Optical System-Optical systems are the combination of optical equipment, protocols and transmission lines that are used to provide communication services.

Optimal Solution-Optimal solutions are the selection and implementation of equipment, software and human resources that provide services or benefits that closest match the objectives of a company or person.

Optimize-The process of adjusting for the best output or maximum response from a circuit or system.

Optimized Network-A network in which each trunk group has been sized to operate at its specified economic or service objective when traffic is routed according to a specified plan.

Optimum Layered Pricing Strategy-A pricing plan that allocates different prices to different customers for the same service levels.

Option to Renew-An option to renew is a term or condition that authorizes a person or company to renew an agreement or portions of an agreement when certain conditions exist.

Optional Calling Plan-A service offering that gives customers the choice of expanding a local calling area for an additional monthly charge, or selecting a smaller calling area and paying toll charges for all calls outside that area. The plan includes Extended Area Service (EAS).

Options-Options are the rights to perform an action (such as to purchase stock or equity in a company) when certain conditions exist (such as when the buyer is vested and pays a pre-defined price).

OPX-Off Premises Extension

ORB-Object Request Broker

Order-An order is an authorization to perform an action such as to send products and to invoice for the contents of the order. An order contains details of whom ordered, where it is shipped, what is ordered, the order details (such as items ordered), and the method of payment.

Order Approval-Order approval is an authorization to accept the terms of an offer to purchase goods or services.

Order Confirmation-Order confirmation is a notice or message that confirms an order has been received and processed. Order confirmation may be sent to the customer and/or sales representative who processed the order.

Order Entry-Order entry is the process of capturing or transferring order information into (data entry) a billing or operations system.

Order Entry Life Cycle-Order entry life cycle is the time period that a sequence of order entry related events occurs between the initiation of a new order and when the order has been entered into an order processing system.

Order Form-An order form is a set of information areas that a user (such as a web site visitor) can enter the product item and customer data that is stored or transferred when the user selects the submit button on the form.

Order Fulfillment-Order fulfillment is the process of gathering the products and materials to complete an order and shipping the products or initiating the services that were ordered.

Order Handling-Order Handling is the process of entering the orders gathered by the sales organization into the billing and network management systems.

Order Management-Order Management is a function in the provisioning cycle. Order management handles the orders for the service the customer requests, supports changes when necessary, provides status updates to customer on the progress of their order, and tracks and manages till successful and on-time completion. Core processes in order management includes accepting orders, pre-order activity and credit check, price estimates, order plan development, request customer deposit, reserve number, initiate service installation, notifying the customer, and initiating the billing process.

Order Manager-An order manager is a program or system that coordinates the review, selection and tracking of orders.

Order Processing-Order processing are the steps involved in selecting the products and agreeing to the terms that are required for a person or company to obtain products or services.

Order Processing System (OPS)-(1-general) Order processing systems gather information related to orders, process the information into specific orders, and create actionable information that allows the fulfillment of the orders. (2-IP Telephony) IP telephone systems can be integrated with order processing systems to allow interactive control with customers to allow the capturing of order information directly from customers and to assist in fulfillment of the order.

Order Status-Order status is the condition of the progress of an order through is processing cycle. Order states may include received, accepted, processed, and fulfilled.

Order Submission-Order submission is the process of transferring of the details that authorize an order to be processed. Order submission methods can range from a verbal order to detailed orders that are transferred via electronic data interchange (EDI) systems.

Order to Bill Process-An order to bill process is the flow of events as well as the interaction points between the point where a prospect orders the service and the point where the bill for the service is generated and dispatched to the customer.

Order Tracking-Order tracking is the process of identifying the status of an order.

Order Types-Order types are categories of orders that can be processed by an order management system. Types of orders may include create (new) orders, change (modify) orders, cancel orders, suspend, and resume orders.

Order Wire-A dedicated voice grade line for communications between maintenance and repair personnel. In digital carrier systems, the order wire is a communication (talk) channel that allows near and far end telephone company (telco) personnel to communicate using telephone sets.

Order Worksheet-An order worksheet is a template or form that contains information entry areas for a salesperson to enter information that will be used by to complete and order from a prospective customer. Order worksheets are used to identify and gather the key terms or items that will be part of the sale.

Orderable-Orderable is a condition or capability of a product listing that allows the product to be ordered.

Ordering And Billing Forum (OBF)-The ordering and billing forum is a subcommittee of alliance for telecommunications industry solutions (ATIS) which recommends and provides direction on the methods and procedures in telecommunications billing.

Organization for Structured Information Standards (OASIS)-Organization for structured information standards a non-profit organization that helps to develop standards and guidelines for the information systems and services. OASIS standards include extensible markup language (XML) and hypertext markup language (HTML).

Organizational Learning-Process whereby the entire organization is able to receive feedback from the environment (the market, competition, new technology providers) and dynamically adapt itself accordingly.

Organizational Structure-An organizational structure is the key functions or people within an organization and the relationships that exist between them.

Original Design Manufacturer (ODM)-An original design manufacture is a company which develops and products a product or assembly for another company.

Original Equipment Manufacturer (OEM)-The original manufacturer of equipment regardless of who sells the equipment or the name marked or associated with the equipment. The term OEM is sometimes used to refer to companies that use other manufacturers to produce products for them. These companies sell, name, and/or use their distribution system for the product that was produced by the other manufacturer. When a company adds assemblies, software, or documentation to products produced by OEMs. These are referred to as a value-added reseller (VAR).

Originating Number-In identifying number of a device (such as a telephone number) that originated a call or request for service. For telephone systems, the originating number is often provided as an ANI.

Originating Rate Center-The office in a geographic area that is designated as a rate center for a service originating within that area.

Originating Zone-An originating zone is a geographic area, region or a group of points that shares a common identifier or characteristic where a service or process starts.

Origination Gateway-A gateway that is used as an origination (starting) point for a communication service. An example of an origination gateway is an audio gateway that is used to connection public telephone calls (the origination of a VoIP connection) to a data network (such as the Internet).

ORM-Operating Resource Management

OS-Operating System

OSI Reference Model-This reference model was created in 1982 by the International Standards Organization (ISO) to standardize communication systems. The model standardized nomenclature across existing protocols and provided guidelines for new protocols using 7 layers. Each successively higher layer builds on the functions of the layers below, as follows:

Application layer 7 The highest level of the model. It defines the manner in which applications interact with the network, including database management, e-mail, and terminal-emulation programs.

Presentation layer 6 Defines the way in which data is formatted, presented, converted, and encoded.

Session layer 5 Coordinates communications and maintains the session for as long as it is needed, performing security, logging, and administrative functions.

Transport layer 4 Defines protocols for structuring messages and supervises the validity of the transmission by performing some error checking.

Network layer 3 Defines protocols for data routing to ensure that the information arrives at the correct destination node.

Data-link layer 2 Validates the integrity of the flow of data from one node to another by synchronizing blocks of data and controlling the flow of data.

Physical layer 1 Defines the mechanism for communicating with the transmission medium and interface hardware.

OSP-Open Settlement Protocol

OSP-Out Side Plant

OSS-Open Source Software

OSS-Operations Support System

OTA-Over the Air

OTAP-Over The Air Programming

OTF-On the Fly

Other Charges & Credits (OC&C)-Charges and credits which do not fall under any other billing category. An example of OC&C: One time waiver of charge.

Other Charges and Credits (OC&C)-Other charges and credits are fees or assessments for services or products that cannot be classified in other categories (e.g. miscellaneous charges).

Other Common Carrier (OCC)-A pre-divestiture term for a telecommunications common carrier, other than a former Bell operating company, authorized to provide a variety of private line services. This term has been replaced by interexchange carrier. The Federal Communications Commission (FCC) also uses the terms miscellaneous, or specialized common carrier.

Other Income-Other income is revenue amounts that are received that cannot be assigned to other defined income categories.

OTS-Off the Shelf

Out of Box Experience (OBE)-Out of box experience is the perceived interactions that a person has with a product or service when it is first opened or used.

Out Side Plant (OSP)-All telephone company (telco) facilities that are located from the main distribution frame (MDF) outward toward the subscriber, interoffice or toll facility. This includes all toll, trunk, exchange grade facilities whether copper, fiber, or wireless.

Outage-Any disruption of service that persists for more than a specified time period.

Outage Probability-The probability that the outage state will occur within a specified time period. In the absence of specific known causes of outages, the outage probability is the sum of all outage durations divided by the time period of measurement.

Outage Threshold-A defined value for a supported performance parameter that establishes the minimum operational service performance level for that parameter.

Outbound Call Center-A call center (group of customer service agents) that originate telephone calls from customers. Outbound call centers (telemarketing centers) are often used by companies to solicit new business or to obtain statistical or other business related information.

Outbound Customer Contact-Outbound customer contact is a communication session with a customer that is initiated by a company or its agent.

OutCollects-Outcollects are charges a network operator sends to other telecommunication companies for services they provided to customers that are not registered in the local network (such as completing their calls in local networks). Outcollects are call detail records (CDRs) that are sent by service provider A to service provider B for services provided by A to B's customers. An example of an Outcollect is a roaming billing record that is sent to the home service provider of a customer that details the usage of the customer for services that were provided by the visited system.

Outdoor Data Unit (ODU)-An outdoor data unit is part of a communication system that is located outside environmentally controlled areas. ODUs are typically constructed of more rugged materials than indoor units.

Outgoing Call Restriction-In telephone call-processing feature that restricts telephone use to specific authorized dialing patterns (typically local phone number).

Outside Collections Agency (OCA)-An external organization that attempts to collect past due money from delinquent customers. OCAs generally perform collection services for a fee (commission) that is based on the success of their collection activity.

Outsourced Management-The use of an outside company to perform system management functions that would or could be conducted internally.

Outsourcing-The use of an outside firm to produce products, assemblies or to perform specific business functions that would or could be conducted internally. When a communication carrier contracts to another company to provide teleservices or facilities management is an example of outsourcing.

Over Specification-Over specification is a standard or operations procedure that contains detailed information that limits the ability of a device or system to be designed or produced with innovative and new features.

Over Subscription-Over subscription is the assignment of more connections, services or users to system or link can it accommodate or provide service to at the same time.

Over the Air (OTA)-Over the air is the ability for a wireless service provider to send program information that can be stored in a mobile telephone or radio device. OTA allows over the air activation after the customer initially purchases a mobile telephone.

Over The Air Activation-The ability for a wireless service provider to program or activate service features for a mobile telephone or radio receiver after the product has been purchased.

Over The Air Programming (OTAP)-The ability for a service provider to directly program information stored in a mobile telephone or radio device. This allows over the air activation after the customer initially purchases a mobile telephone.

Over the Top Services-Over the top services are applications and their associated communication transmission provided for the benefit of the user through the use of underlying services. An example of an over the top service is a telephone service that is provided over an Internet connection.

Overage Charges-Overage charges are fees that are imposed or added to an account when a user exceeds the allocated amount of usage for a defined level or quantity of service.

Overage Fees-Overage fees are amounts that are assessed for usage or services that are provided beyond the pre-defined account limits.

Overbought-Overbought is the purchasing of more goods or services that was requested or required.

Overbuild-Overbuild is the installation or use of systems or equipment that have capabilities that are significantly beyond the maximum needs or requirements of an application or service.

Overhead-Overhead is the combined costs are the fees or bills that are associated with the operational support systems of a company.

Overprovisioning-Overprovisioning is the providing more capacity than is actually required for a given application. Overprovisioning may include reserving excessive switching capacity or using a higher speed communications link. Overprovisioning is sometimes performed to reduce the transmission delay of data transmission.

Override-An override process or condition that allows an action to occur that would normally have been restricted or unable to process.

Override Commission-An override commission is a value that is assigned or awarded to people or companies on products or services that are paid in addition to or instead of other commissions. An example of an override commission is a sales percentage that is paid to a division leader for all the products that are sold in their department regardless of who sold the products or services.

Overstock-Overstock is a condition where the inventory level for a product exceeds its expected demand.

Oversubscription-Oversubscription is a situation that occurs when a service provider sells more capacity to end customers than a communications network can provide at a specific time period. This provides a benefit of reduced network equipment and operational (reduced leased line) cost.

Over-subscription is a common practice in communications networks as customers do not continuously use the maximum capacity assigned to them and customers access the networks at different time periods. Unfortunately, over-subscription in telecommunications can cause problems when customers do attempt to access the network at the same time. For example, when customers open their presents at a holiday event (e.g. Christmas) and attempt to access the Internet at the same time.

Overwrite-Overwriting is the process of replacing existing data with new data.

OWL-Web Onthology Language

Owned and Operated (O&O)-Owned and operated is a company or system that is owned and operated by the same entity.

Ownership Dilution-Ownership dilution is the reduction of existing shareholder's fractional ownership that results from the issuance of additional shares of common stock.

P

PABX-Private Automatic Branch Exchange

Package-A package is a combination of items. For media broadcasters, it is a set of programs offered by a network as a discounted package. For publishers, it is a package of multiple items such as brochures.

Package Deal-A package deal is on offer to a buyer that includes additional discounts or promotional items in return for agreeing to purchase multiple products or services.

Package Tracking-Package tracking is the monitoring or following of the fulfillment of an order or the shipment of a package. Package tracking may involve the providing of a tracking number(s) to a customer, which they can use to inquire about the status of their order or package.

Packaging-Packaging is the process of combining multiple items into a group or container. Packaging may involve the inclusion of other supporting or related items such as inserting product literature into an order package.

Packet Billing-Packet billing involves the authorizing, gathering, rating, and posting of account information for the transmission of data packets. Packet billing may be based on the number of packets, amount of packet data transferred, the time packet data session was in progress or other measurable service information.

Packet Buffering-Packet buffering is the process of temporarily storing (buffering) packets during the transmission of information to create a reserve of packets that can be used during packet transmission delays or retransmission requests. While a packet buffer is commonly located in the receiving device, a packet buffer may also be used in the sending device to allow the rapid selection and retransmission of packets when they are requested by the receiving device.

Packet Count-The total number of packets that have been sent since a session began.

Packet Data-The sending of data through a network in small packets (typically under 100 bytes of information). A packet data system divides large quantities of data into small packets for transmission through a switching network that uses the addresses of the packets to dynamically route these packets through a switching network to their ultimate destination. When a data block is divided, the packets are given sequence numbers so that a packet assembler/disassembler (PAD) device can recombine the packets to the original data block after they have been transmitted through the network.

Packet Data Session-A packet data session is the setup, packet transfer and disconnection of data flow between two or more communications points or devices.

Packet Data Switched Network (PDSN)-The packet data serving node (PDSN) is used to control (route) data packets to and from the PCF functions (BS packet controllers) that communicate with access terminals. The PDSN is responsible for originating, maintaining, and terminating data interfaces between the access terminals (via the PCF and base stations) and packet data networks (such as the Internet).

Packet Error Rate (PER)-Packet error rate is calculated by dividing the number of packets received in error by the total number of bits transmitted. It is generally used to denote the quality of a digital transmission channel.

Packet Structure-A packet structure is the division of sequence of digital information into different fields (information) parts. Packet structure fields typically include a header that contains a synchronization code, destination address field, control flags, user data, and error detection.

Packet Switch-A packet switch is a device in a data transmission network that receives and forwards packets of data. The packet switch receives the packet of data, reads its address, searches in its database for its forwarding address, and sends the packet toward its next destination.

Packet switching is different than circuit switching because circuit switching makes continuous path connections based on a signal's time of arrival (TDM) port of arrival (cross-connect) or frequency of arrival. In a packet switch, each transmission is packetized and individually addressed, much like a letter in the mail. At each post office along the way to the destination, the address is inspected and the letter forwarded to the next closest post office facility. A packet switch works much the same way.

Packet Switched Public Data Network (PSPDN)-Public data networks interconnect data communication devices (e.g. computers) with each other through a network that is accessible by many users (the pubic). To allow many different users to communicate with each other, standard communication messages and processes are used. The Internet is an example of a public data network (there are other public data networks) that uses standard Internet protocol (IP) to allow anyone to transfer data from point to point by using data packets. Each transmitted packet in the Internet finds its way through the network switching through nodes (computers). Each node in the Internet forwards received packets to another location (another node) that is closer to its destination. Each node contains routing tables that provide packet-forwarding information.

Packet Video-Packet video is the transfer of video information in packet data format.

Packet Voice-Packet voice is the process of sending voice communication by converting voice signals (audio) into digital signals and sending the digital signal through a data communication network (such as the Internet) using data packet transmission.

Packing Slips-A packing slip is a document that contains information about the contents of a package. Packing slips may not contain prices of the items on the list.

PADS-Product Acquisition and Development System

Page Views (PV)-Page views are the number of times a web page has been requested and displayed.

Paging-Paging is a process used to deliver a message, via a public communications system or radio signal, to a person whose exact whereabouts are unknown by the sender of the message. Paging can be a dedicated service (such as numeric pagers) or it may be a general process of alerting devices that they are receiving a call, command, or message.

For paging service, users typically carry a small paging receiver that displays a numeric or alphanumeric message displayed on an electronic readout or it could be sent and received as voice message or other data.

Paging Service-Transmission of coded radio signals for the purpose of activating specific pagers; such transmissions may include text, voice, or data messages.

This diagram shows how paging systems can provide for one-way and/or two-way paging operation. This diagram shows that paging sytsems typically use high-power transmitters to broadcast paging message to a relatively large geographic area. All pagers that operate on this system listen to all the pages sent, paying close attention for their specific address message. Paging messages are received and processed by a paging center. The paging center receives pages from the local telephone company or it may receive messages from a satellite network. After it receives these messages, they are processed to be sent to the high power paging transmitter by an encoder. The encoder converts the pagers telephone number or identification code entered by the caller to the necessary tones or digital signal to be sent by the paging transmitter.

Two-way paging systems allow the paging device to acknowledge and sometimes respond to messages sent by a nearby paging tower. The two-way pager's low power transmitter necessitates many receiving antennas being located close together to receive the low power signal. This Figure shows a high power transmitter (200-500 Watts) that broadcasts a paging message to a relatively large geographic area and several receiving antennas. The reason for having multiple receiving antennas is that the transmit power level of pagers are much lower than the transmit power level of the paging radio tower. The receiving antennas are very sensitive, capable of receiving the signal from pagers transmitting only 1 watt.

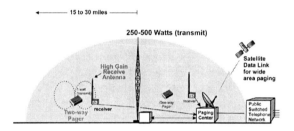

Paging System

Paid Download-A paid download is a data file or program that requires a user to pay for the transfer (downloading) of the information. While the user may pay for a file or media download, the rights to play (render), transfer (copy) or alter may be limited.

Paid Inclusion-Paid inclusion is the process used to allow companies or people pay to have their advertisements or web addresses included in the presentation of materials such as a search engine listing.

Paid Listing-A paid listing is a web address that is included in a listing output of a search engine.

Paid Placement-Paid placement is a marketing program where companies or people pay a fee for a specific location on a media page (such as a web page). The use of a paid placement program for keyword advertising assures the position (usually on top) of a specific listing regardless of its actual popularity ranking.

Paid Subscription-Paid subscription is the distribution of a magazine or publication where the recipient pays for the subscription. While paid subscriptions may not require qualification from the reader, the willingness of the reader to pay for the subscription indicates the subscriber is likely to read and is qualified to read the publication.

Pairing-Pairing is an initialization procedure whereby two devices communicating for the first time create an initial secret link key that will be used for subsequent authentication. For first-time connection, pairing requires the user to enter a Bluetooth security code or PIN.

PAN-Personal Area Network

PAP-Password Authentication Protocol

Paradigm-An example or pattern that is commonly associated with thought processes or business operations.

Paradigm Shift-A significant change in the thought patterns or business operations that have been previously established. An example of paradigm shift includes sending mail through the Internet instead of sending mail through the postal service.

Parameter-(1-component or circuit) A specific value of some variable as applied to a particular component or circuit. Examples include the resistance of a resistor or the operating frequency of a transmitter. (2-machine language) A variable that identifies and contains information needed to execute a command. (3-variable) A variable that must have a constant value for a specified application.

Parameter Negotiation (PN)-Parameter negotiation is the process of requesting and agreeing on the preferred characteristics for a communication session.

This diagram shows how two data communication devices negotiate for data transmission rates and protocols selection in a data network using the preferences assigned by a user along with the options determined by equipment availability. In this example, data terminal 1 sends a connection request message to data terminal 2. This connection request indicates that the data terminal prefers to use a 56 kbps data transmission rate because it has enough bandwidth. Unfortunately, the data terminal cannot accept the request for 56 kbps because it's access bandwidth is low speed (28 kbps). The receiving data terminal sends back a request to use 28.8 kbps data transmission rate. When the originating data terminal receives this request, it accepts the request because it has that data transmission rate and protocol capability available. It then confirms the request and both devices use a data transmission rate of 28.8 kbps.

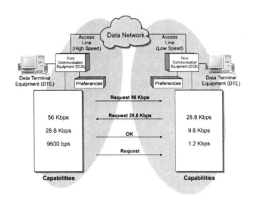

Parameter Negotiation (PN) Operation

Parameter Set-A parameter set is a group of parameters (values) that are associated or transferred to a device or software module.

Parameters-Parameters are values that are associated with an item, service or software program.

Parent Directory-A parent directory is a group of files or file pointers that share a common identifier (a directory) in a hierarchical directory system that is immediately above the current directory.

Parental Controls-Parental control are processes that allow a person (such as a parent) to control the access rights of another person (such as a child).

Parental Rating-Parental rating is the evaluation of the characteristics of media programs by another person or company to provide a rate value that indicates the maturity level that is recommended for the viewing of the program.

Pareto Rule-The Pareto rule is that 80% of business or wealth is created or controlled by 20% of the customers or population. The Pareto rule was created by Vilfredo Pareto who discovered that 80% of the wealth in Italy was controlled by 20% of the population.

Parking Customers-Parking customers is the process of assigning customers to an alternative service while a new service is being made available or optimized.

Parsers-Parsers are dividers of media, data or information.

Parsing-Parsing is the process of dividing information or data into more manageable component parts.

Part Failure Rate (PFR)-Part failure rate is a comparision of how many parts fail to perform with their defined characteristics compared to the total number of parts that are received or used.

Partner Integration-Partner integration is the linking of systems and processes together between companies with the objective of working together (partnering) on projects or promotions.

Partner Management-Partner management is the process of identifying, assigning terms and tracking performance of companies that have a collaborative (partnering) relationship.

Partner Settlement-Partner settlement is the financial difference between revenues generated from charging for services provided to customers by a business partner (such as a customer purchases of content) less than the fees that other partner charges (such as a content provider) for their portion of the fulfillment of the purchase.

Partnership-A partnership is a form of company where each partner is responsible and liable for the performance and obligations of a business.

Partnership Agreement-A partnership agreement is a document that defines the roles and responsibilities between partners in a company or between companies.

Pass Along Rate-Pass along rate is a measure of how many times an item is read, viewed or used by people. For magazines, pass along is a measure that is used to indicate how many total people read a magazine as opposed to how many people receive the magazine.

Passive Language-Passive language is the choice of words in a message (such as a news story or advertising message) that inform in a non-offensive and non-abusive way.

Passive Monitoring-Passive monitoring is the process of listening to or viewing a communication service for the purpose of determining its quality or whether it is free from trouble or interference without modifying or altering the information or processes that it is monitoring.

Passive Optical Network (PON)-A passive optical network (PON) combines, routes, and separates optical signals through the use of passive optical filters that separate and combine channels of different optical wavelengths (different colors). The PON distributes and routes signals without the need to convert them to electrical signals for routing through switches.

PON networks are constructed of optical line termination (OLT), optical splitters and optical network units (ONUs). OLTs interface the telephone network to allow multiple channels to be combined to different optical wavelengths for distribution through the PON. Optical splitters are passive devices that redi-

rect optical signals to different locations. ONU's terminate or sample optical signals so they can be converted to electrical signals in a format suitable for distribution to a customer's equipment. When used for residential use, a single ONU can server 128 to 500 dwellings. In 2001, most PON's use ATM cell architecture for their transport between the provider EO or POP and the ONU (in some case even to the user workstation). When ATM protocol is combined with PON system, it is called ATM passive optical network (APON).

This figure shows an ATM passive optical network (APON) system that locates optical network units (ONUs) near residential and business locations. This passive optical network routes different optical signals (different wavelengths) to different areas in the network by using optical splitters instead of switching devices. In this example, the optical distribution system uses ATM protocol to coordinate the PON. ONU interfaces are connected via fiber to an OLT located at the provider's EO or POP. Each ONU multiplexes user channels (between 12 and 40) into an optical frequency

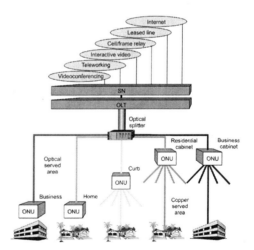

Passive Optical Network (PON) System

spectrum allocated to that ONU. Up 32 ONU's can share access to a single PON using the features of dense wave division multiplexing (DWDM). Some newer PON's use high-density wave division multiplexing (HDWDM). Use of HDWDM increases the number of ONU's per PON from 32 to 64. This diagram shows that a PON that uses HDWDM can support approximately 2500 residential customers.

Passkey-A passkey is an identification number that is entered by a user into the communication device to enable it to operate or to enable a service.

Password-Passwords are a string of characters or an equivalent combination of characters that is associated with a specific user ID. Generally, passwords are not directly stored on a computer network. The passwords selected by users are usually processed by a mathematical formula to produce a result code (a number) that is stored. When the user enters a password or the equivalent value of a password (other character sequences may also work), the result is calculated and compared to the stored result. If the result matches for the specific user ID, access to the system or resources is usually granted.

Password Authentication Protocol (PAP)-A security protocol that prompts the user to enter a user name and password before that is transferred to the system before access to service is granted.

Password Hacking-Password hacking is the repeating of attempts to access a system. Password hacking can be a systematic process such as sequentially trying password characters in sequence (e.g. A followed by B) or it can by a random process that attempts to use additional information (e.g. birthdays, names of family members).

Password Hashing-Password hashing is a computational process that converting a password or information element into a fixed length code. Password hashing can be used to convert variable length text based passwords into fixed length digital password codes.

Password Management-Password management is the creation, assignment, storage, delivery/transfer, and revocation of unique information (passwords) by recipients or holders of information (data or media) to allow access to systems, services or data.

Password Protection-Password protection is the restriction to access or use information or media by requiring the entry of one or more identifying codes (passwords) before allowing access.

Patch-(1-circuit) A temporary circuit rearrangement made by using jacks to bypass faulty circuit components or transmission facilities. (2-software) A temporary software program update that is used to correct software problems between official software release updates.

Patching-Creating business units with overlapping business functionality in order to encourage competition and excellence in delivery.

Patent-A patent is a document that grants a monopoly for a limited time to an invention described and claimed in the body of the document. A patent describes the invention and defines the specific aspects (claims) of the invention that are new and unique. There are several forms of patents including mechanical patents (processes) and design patents (appearances).

Patent Claim-A description of a specific use of technology that is new and described in the patent. There are usually many patent claims that are part of a patent.

Patent License-A license that is given for the production, use, or sale of products that may use technologies or processes defined by the claims in a patent.

Patent Licensing-The process of obtaining the rights to use technology that is described in one or more patents.

Patent Pools-Patent pools are collections of patents into a group that assist in the licensing of multiple patents by companies that want to use the patents. Patent pools are commonly established to cover a specific area of technology such as digital video or wireless communication.

Patent Review Board-A patent review board is a group of people (typically within a company) that review invention applications to determine if they are worthy of submission to the patent office.

Patent Rights-Patent rights are intellectual property rights which give the owner, or assignee, the right to prevent others from making, using, or selling the invention described and claimed in the patent. Patent rights must be applied for in the country or region where they are desired. Most countries have a national patent office, and an increasing number of countries grant patents through a single regional patent office.

Patent Royalty-A fee paid for the use of technology that is defined in the claims of a patent.

Patent Statement-A patent statement is a disclosure by a company or person for the patents they believe relate to an industry specification or system which may include their intended licensing requirements or restrictions.

Patent Troll-A patent troll is a company or person who forces a company to pay excessive fees for technologies that are covered by their patents.

Path-(1-general) The route a signal travels through a channel, circuit, or network. (2-communications system) In radio and optical communications systems, the route that an electromagnetic wave travels through space. (3-data) A logical connection between the point where a signal is assembled into a data format and the point at which the signal is disassembled. (4-file name) The complete name location for a file or information element in a computing system.

Path Cost-Path cost is the combination of the link cost(s) between entry and exit points of a path on a communications network along with equipment and data processing costs.

Path Establishment Delay-The amount of time it takes to establish a communication path or circuit.

Patriot Act-The Patriot Act is a U.S. law that was enacted on 26 October 2001 that requires companies to store and provide information to government agencies to help deter terrorism.

Pay for Inclusion (PFI)-Pay for inclusion is a marketing program that allows advertisers to pay a search engine company to ensure that search result listings include their search results.

Pay for Performance (PFP)-Pay for performance is the process or concept of exchanging value for actions or other measurable results for a marketing campaign.

Pay for Placement (PFP)-Pay for placement is a marketing program that allows advertisers to pay for listings in search engines. Companies or people bid or pay a fixed fee to get listed in the result of a search.

Pay for Presentation-Pay for presentation is the process or concept of exchanging value for impressions or other activities that present messages as part of a marketing campaign.

Pay Per Call (PPC)-Pay per call is a marketing process (commonly an Internet marketing program) that charges the advertiser only when a customer calls a telephone number listed in an advertising program. Pay per call may use a unique telephone number that can be tracked or it may use a "call me back" or "click to call" feature that allows the viewer to directly call the company through their internet connection.

Pay Per Click (PPC)-Pay per click is an internet marketing process that charges the advertiser only when an item is selected "clicked."

Pay Per Day (PPD)-Pay per day is a service that allows a viewer to select a movie or program group for viewing that allows the viewer to access the program(s) for a day period.

Pay Per Inclusion (PPI)-Pay per inclusion is an internet marketing process that charges the advertiser only when an URL is returned in the results of a web search.

Pay Per Lead (PPL)-Pay per lead is a marketing process that charges the advertiser only when contact information for a potential sales lead is obtained.

Pay Per Listen (PPL)-Pay per listen (PPL) is the providing of audio programming such as music, news, and other education audio that customers view for a fee. PPL services may be ordered by telephone or via an Internet data radio.

Pay Per Sale (PPS)-Pay per sale is an internet marketing process that charges the advertiser only when an item is sold.

Pay Per Use-Pay per use is the charging for usage of products or services for each use or usage period for that item.

Pay Per View (PPV)-Pay per view (PPV) is the providing of television programming such as sports, movies, and other entertainment video that customers view for a fee. PPV services may be ordered by telephone or via an interactive set top box.

Pay Phone-A telephone that requires coins to be inserted or a calling card swiped or inserted before a toll call can be placed. Some pay telephones allow toll free or freephone calls to be originated without the need for coins or calling cards.

Payee-A payee is the person or company who receives a payment.

Payer-A payer is the person or company who pays a bill or fee.

Payment Association-Payment association is the linking or assigning of a payment to an account or transaction.

Payment Authorization-A payment authorization is the indication code or information that identifies that a payment has been successfully accepted into a billing system.

Payment Center-A payment center is a program or system that coordinates the selection, entry and configuration of payment options for an order processing system.

Payment Conditions-Payment conditions are a set of rules or requirements that define how payments can be received and processed.

Payment Gateway-A payment gateway is a device or software program that adapts or transfers payments from one system to another system.

Payment in Advance-Payment in advance is a value provided in return for a product or service that will be provided in the future.

Payment in Kind-Payment in kind is an exchange of items, media or services which are used as value instead of the use of money.

Payment Instrument-A payment instrument is a financial processing tool or method that allows value to be transferred from a person or company to another person or company. Examples of payment methods include credit card, debit card, cash, check, credit memo, coupon, and electronic transfer.

Payment Interface-A payment interface is the connection point (possibly only a software interface) between systems that allow payment to be requested from one system and posted on another system.

Payment Method-Payment method is the financial instrument that is used to pay for products or services. Common payment methods include credit cards, online accounts, checks, cash, cash on delivery and vouchers. The choice of payment method may influence the payment terms or costs because some payment methods are preferred (e.g. have lower costs) by vendors.

Payment Options-Payment options are the financial instruments that can be chosen to pay for products or services. Common payment options include credit cards, online accounts, checks, cash, cash on delivery and vouchers.

Payment Processing-Payment processing is the tasks and functions that are used to collect payments from the buyer of products and services. Payment systems may involve the use of money instruments, credit memos, coupons, or other form of compensation used to pay for one or more order invoices.

This figure shows some of the different options available for bill payment. This example shows that customers can make cash payments to a company office of authorized agent, can send checks to the company, can pay via credit cards, via a 3rd party financial processor (e.g. paypal) or through electronic funds transfer (transfer fund). This example shows that payments on account are applied to specific invoices and when the invoices are paid in full, they are marked as pay and are not available for additional payments to be applied.

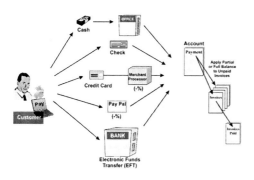

Bill Payment Options

Payment Processor-A payment processor is a company or financial system that receives payment orders and collects payments from the buyer of products and services. Payment processors may process payments multiple money instruments such as credit cards and checks.

Payment Receipt-A payment receipt is a document or data that indicates that a payment or value has been received.

Payment Schedule-A payment schedule is an itemized list or table that provides payment amounts that is provided when specific actions or payments will be processed.

Payment Services-Payment services are financial transaction applications that enable people or companies to exchange assets (such as money) for goods or services. Examples of payment services include PayPal, Visa, Mastercard or other financial applications.

Payment System-A payment system is the process(es) used to collect payments from the buyer of products and services. Payment systems may involve the use of money instruments, credit memos, coupons, or other form of compensation used to pay for one or more order invoices.

Payment Threshold-A payment threshold is an assigned value that when the value is exceeded a payment is made. Payment thresholds are commonly used to eliminate the need to process commission payments when the amount of the payment would be so small that the cost of issuing the payment would exceeded the paid amount.

Payment Types-Payment types are financial instruments that can be used to transfer value from a buyer to a seller. Common payment types include company accounts, credit cards, online accounts, checks, cash, cash on delivery and vouchers.

Payment Voiding-Payment voiding is the process of ending and possibly reversing the tasks that were associated with collection of a payment. Payment voiding may be performed before a transaction is posted or batched to financial centers to avoid processing fees.

Payments-Payments are value transfers from one person or company to accounts or recipients.

Paypal-Paypal is a financial instrument can be processed (charged) by companies for the payment for products or services (which can process Paypal transactions). The charges accumulated on the Paypal account is paid for by the owner or user of the card.

PB-Petabyte

PBX-Private Branch Exchange

PC-Personal Computer

PCMCIA-Personal Computer Memory Card International Association

PD-Portable Device

PDA-Personal Digital Assistant

PDD-Post Dial Delay
PDF-Policy Decision Function
PDF-Portable Document Format
PDN-Public Data Network
PDP-Policy Decision Point
PDS-Premises Distribution System
PDSN-Packet Data Switched Network
Peak-(1-signal) The maximum amplitude or value, usually of a voltage or current. In a periodic function, the peak is the instantaneous maximum value, either positive or negative. (2-category of usage) A category of usage that indicates that the request or usage of service is at its highest level. (3-billing) A rating time period that may assign a different billing rate (generally a higher rate) when customers use services during period of higher network usage.

Peak Bandwidth-Expressed in bytes per second, this limits how fast packets may be sent back-to-back from applications. Some intermediate systems can take advantage of this information, so that more efficient resource allocation results.

Peak Hour-The one or two hours of the day where the number of calls to a call center is at its highest level.

Peak Load-The load that results from higher-than-average traffic volume. Peak load usually is expressed as the load during a 1-hour period. It also can be expressed in terms of any of several functions of an observing interval, such as a peak hour during the day.

Peak Time-The hours designated as the heaviest usage of a system. In cellular systems, peak time usually is defined as being between 7 a.m. and 7 p.m., Monday through Friday. Usage rates are sometimes higher during peak time than during off-peak time.

Pedestrian Service-Pedestrian services are communication services provided to slow moving users (pedestrians).

Peer-A peer is a device or person who can request and provide information between other peers.

Peer Communication-Peer communication is the transferring of information between users or devices that operate on the same hierarchical level (usage types that have similar characteristics).

Peer to Peer Distribution-Peer to peer distribution is the process of directly transferring information, services or products between users or devices that operate on the same hierarchical level (usage types that have similar characteristics).

Peer-To-Peer Network-Peer-to-peer networks or communication systems that allow users within the communication network to directly interact and exchange data with other users (peers) who are connected to other users in the network.

Peg Count-The total number of any traffic event that occurs during a given period. An example is the number of times a switching-system component functions during one hour.

Pegging-Pegging is the setting of a value such as setting the exchange rate of a foreign currency.

Penalties-Penalties are fees that are charged for actions or service usages that fall outside the agreed limits of service usage. Penalties can include early contract termination fees, charges for lost equipment or the usage of services by unapproved devices (e.g. media storage devices).

Penetration-(1-skin effect) A measure of the depth of current in a conductor. As the frequency through a conductor is increased, the current tends to travel only on the outer surface (skin) of the conductor. (2-broadcasting) The extent to which the population within a service area can receive a broadcast signal, or the extent to which receivers in a service area are available. Penetration may be measured as a percentage of all households, or of the total population. (3-cable TV) The ratio of the number of subscribers to the total number of households passed by a cable TV system.

PER-Packet Error Rate

Per Call Block-A feature of calling number identification that allows the originator of a call to disable the display of their calling number. This feature may be used when a caller wishes to place an anonymous complaint.

Per Diem Rate-A per diem rate is the amount of value or charges that may be provided per day.

Per Line Block-A feature of calling number identification that allows the user of the telephone device to disable the display of their calling number when making calls. This feature may be used when a caller wishes to place calls from a private number.

Perceived Value-Perceived value is the worth a person or customer associates with a promotion or item.

Percent Local Usage (PLU)-The percent of allocation between IntraLATA and Local traffic.

Per-CPU Rate-A per-CPU rate is an amount charged for a service that is based on the number of central processing units (CPUs) that are used to provide the services.

Per-Employee Rate-A per-employee rate is an amount charged for a service per a specific time period that is based on the number of employees or people that will use the system or service.

Perfect Degree of Security-Perfect degree of security is the amount of risk that a security system has if private keys are only distributed to the security communication participants.

Performance Bond-A performance bond is a financial instrument that is used by a company that is purchasing products or services to ensure compensation is paid in the event a product, system or service does not achieve the guaranteed performance objectives.

Performance Counter-A counter that increases each time a performance threshold has been reached.

Performance Gateway (PG)-A performance gateway is a communications device or assembly that monitors the performance of data communication that is received from one network into a format that can be used by a different network. A performance gateway usually has more intelligence (processing function) that can change system configurations to adjust the performance of a system to provide established performance objectives or levels.

Performance Guarantees-Performance guarantees are commitments from providers of products or services on the performance levels that will be achieved. Performance guarantees usually define what will happen if a performance level is not achieved such as financial penalties. System performance levels may be defined for different types of operational modes including normal and loaded (stressed) conditions. To ensure that financial penalties will be paid in the event of inability to achieve desired performance, a supplier may be required to provide (pay for) a performance bond.

Performance Monitoring-Performance monitoring is the process of obtaining information from products or services that can be used to determine the performance of systems or services.

Performance Reporting-Performance reporting is the creation of media (such as a printed report) that identifies the performance of systems or services.

Performance Specifications-Performance specifications are documents that describe the essential technical or application requirements for items, materials or services including the procedures by which it will be determined that the requirements have been met.

Performance Tests-Performance tests are measurements of operational parameters during specific modes of operation. Performance tests are used to determine if the device or service is operating within its designed operational parameters. Performance tests can be performed over time to determine if a system is developing operational problems.

Period-The time required for one complete cycle of a regular, repeating series of events.

Periodic-A recurring function or signal that repeats at regular time intervals.

Periodic Charges-Periodic charges are fees that are associated with a product or service that is assessed on a regular interval (i.e. monthly, quarterly, annually).

Peripheral-An auxiliary device, such as a printer, mouse or graphics pad, that is connected to a computer and provides service to it but does not participate in principal computer functions.

PERL-Practical Extraction and Report Language

Permanent Virtual Circuit (PVC)-A permanent virtual circuit is a logical communication path that is manually created for a continuous communication connection. After a permanent communications circuit is established, a data path (logical connection) is maintained.

This diagram shows how a permanent virtual circuit (PVC) is used to allow the transfer of data through a communications network through a pre-established logical (virtual) path. In this example, a PVC is created by programming routing tables in 4 switches before any data is sent. These routing tables assign data transfer connections between input and output channel on each switch. For example, as data from the sending computer (portable computer) is sent into input channel 3 of the first switch, it is transferred to the output channel 5. This process will repeat for any data that is sent from the sending computer to the destination computer. This example also shows that the PVC path remains active even if the portable computer is disconnected for a period of time.

Permissions Department-A permissions department is a group or groups within a company that manage the assignment of usage rights to content or portions of content.

Perpetual Inventory-Perpetual inventory is the process of maintaining a continual availability of products in stock at all times.

Per-Seat Rate-A per-seat rate is an amount charged for a service (such as a training program) that is based on the number of employees or people that will use the system or service.

Personal Agent-Software that allows a communication user to program specific telephone or communication device responses so that it can perform operations without the need of the communication user. Personal agent software may be used to provide call screening or intelligent call forwarding services.

Personal Area Network (PAN)-Personal area networks (PANs) are short-range communication networks that typically have a range of 5 to 25 feet. A PAN is used to connect terminal equipment (mice and keyboards) to computing and data network equipment.

Personal Computer (PC)-A computer that is primarily used for home, non-business applications.

Personal Computer Memory Card International Association (PCMCIA)-A standard physical and electrical interface that is used to connect memory and communication devices to computers, typically laptops. The physical card sizes are similar to the size of a credit card 2.126 inches (51.46 mm) by 3.37 inches (69.2 mm) long. There are 4 different card thickness dimensions: 3.3 (type 1), 5.0

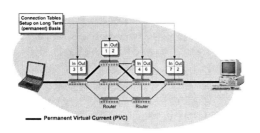

Permanent Virtual Circuit (PVC) Operation

(type 2), 10.5 (type 3), and 16 mm (type 4-unofficial size). The number of connections (pins) is 68. The first PCMCIA standard was approved in 1991 with the concept of standardizing computer memory cards. Because the interface defined how to address, control, and transfer data in standard formats, the PCMCIA standard was quickly used for other types of communication and storage accessories such as modems, network interface cards (NICs), hard disk storage, and controller adapters.

Personal Digital Assistant (PDA)-Personal digital assistants are small computing devices that contain their own software operating system that allows the user to run software processing applications. Personal digital assistants are often used to organize personal activities and may provide access to communication services (such as web browsing and email).

Personal Guarantee-A personal guarantee is a pledge by a person to accept responsibility of a debt or other type of liability where the persons personal assets are used to reduce the risk to a lender or other providing entity.

Personal Identification Number (PIN)-A personal identification number is a code or value that is provided to or created by an individual subscriber which is used to gain access to specified services, such as credit card calling or prepaid wireless services.

Personal Identification Number Blocking (PIN Blocking)-The use of personal identification numbers (PINs) to control access to 976 information access service. In practice, such blocking allows customer access only after a valid PIN is entered.

Personal Investment-Personal investment is the providing of assets owned or controlled by a person into a company or business where some or all of the assets are at risk.

Personal Media Server (PMS)-A personal media server is a computer host that can provide a very limited number of communication sessions (such as a personal computer).

Personal Operable Device Slurping (POD Slurping)-POD slurping is the unauthorized gathering media and data files from another computer or data storage system using a portable media player or storage system.

Personal Publishing-Personal publishing is the process of making available personal media property (content) to distributors and/or users of the content.

Personal Video Recorder (PVR)-A personal video recorder is a device that stores video images in digital format for personal use.

Personalization-Personalization is the process of modifying information (such as an email message) to better match the personal interests of the intended recipient.

Personalized Letter-A personalized letter is a communication document that is provided to a reader where some of the information within the document has been altered to better match the interests or needs of the recipient.

Personalized Ringing-The ability for a user to select and setup the audio sound that is used to indicate an incoming call. Personalized ringing helps individuals to identify their telephone is ringing when there are several telephones located within the same area.

Personalized Speed Dialing-The ability for a user to setup speed dial numbers on their telephone or communication device.

Persons of Population (POPs)-The population within a geographic area that can have access to the service of a communications service provider. This is commonly used for evaluating the worth of a system. The number of POPs is multiplied by the price to be paid per POP. One POP is generally equal to one person.

PERT-Program Evaluation and Review Technique

Pert Chart-A pert chart is a diagram that shows project tasks as functional blocks with interconnecting lines that show the planned progression of tasks.

Petabyte (PB)-A petabyte is one million billion (quadrillion) bytes of data. When petabyte is used to identify the amount of data storage space (such as computer memory or a hard disk), a petabyte commonly refers to 1,125,899,906,842,624 bytes.

Petty Cash-Petty cash is an amount of cash that is used for small purchases such as office supplies.

PFI-Pay for Inclusion

PFP-Pay for Performance

PFP-Pay for Placement

PFR-Part Failure Rate

PG-Performance Gateway

PGP-Pretty Good Privacy

Phantom Call-A phone that rings inadvertently without a calling party online.

Phased Invoice-A phased invoice is a set of billing charges that is assessed as a project progresses between different phases.

Phasing-(1-signal) Phasing is the relative time shift of a signal as referenced to another signal. (2-process) Phasing is the coordination of activities such as when media is presented to an Internet web user.

Phishing Fraud-Phishing fraud is the use of a fake billing issue announcement email to trick email recipients to disclose private security information.

Phone Blog-A phone log is a shared media resource on a web page that allows a blogger to contribute and view media that is related to topics or questions on their mobile telephone.

Phonelet-A phonelet is a small program that runs on a server to provide specific types of functions (such as message waiting light control) for particular models of telephones.

Phoneline Network-A phoneline network is a local area data network technology that allows standard telephone wiring to be used as network cabling without the need to disconnect standard telephones. The phoneline network uses high frequency signals that are above standard telephone and DSL frequency bands. To install a phoneline network, end users install Phoneline NICs in a similar method to adding an Ethernet card. The Phoneline networking system allows computers to be connected to each other without the use of a hub (daisy chain).

PHP-Hypertext Preprocessor

Physical Access-Physical access is the ability of a user or unauthorized user to physically send or receive information with a communication system or device.

Physical Address-A grouping of numbers that identifies a particular piece of computer hardware connected to a local area network or other data communications system. Contrast with logical address.

Physical Good-A physical product or item that is offered for sale.

Physical Link-A physical link is a connection between two or more devices which can send and receive information or data between devices using a physical medium such as wired, radio or optical transmission.

PIC-Preferred Interexchange Carrier

PICC-Primary Interexchange Carrier Charge

Pick Group-A hunt group is a list of telephone numbers that are candidates for use in the delivery of an incoming call. When any of the numbers of the hunt group are called, the telephone network sequentially searches through the hunt group list to find an inactive (idle) line. When the systems finds an idle line, the line will be alerted (ringing) of the incoming call. Hunt lines are sometimes called rollover lines.

Pico-A prefix of a unit of measure meaning one-trillionth.

Picosecond (ps)-One trillionth of a second. In one picosecond, light travels about one-third of a millimeter.

PICS-Protocol Implementation Conformance Statement

Piece Rate-Piece rate is the assignment, pricing or paying for goods or services on a per item basis.

Pilot-(1-general) A signal, usually a single frequency, transmitted over a system for supervisory, control, synchronization, or reference purposes (2- FM) A signal at 19kHz transmitted with the FM broadcast carrier. The FM stereo subcarrier at 38kHz, modulated with a L-R difference signal, is generated from the pilot, leaving the two in-phase. In FM receivers, a pilot detector circuit serves the present of the 19kHz signal, which is used to decode the stereo information. (3-business) A pilot is a small scale test of a system or service.

PIN-Personal Identification Number

PIN Blocking-Personal Identification Number Blocking

Pinout-The pin configuration for a connector or system cabling.

Pipe-A software interface or a hardware device that acts as an interface or buffer between network applications and devices.

Piracy-(1-transmission) The operation of unauthorized commercial stations, usually in international waters near target regions. Pirates violate flu frequency regulations, copyright laws, and national broadcasting laws. (2-programming) The duplication and sale of program materials, particularly TV programs on videotape, in violation of copyright laws. (3-signals) The unauthorized use of cable TV or satellite signals.

PKE-Public Key Encryption

PKI-Public Key Infrastructure

PKZip-A file compression program, which uses the Lempel-Ziv-Welch algorithm. WinZip is a similar program having a Windows-compatible graphic user interface.

PL-Preferred Language

Place Shifting-Place shifting is the viewing of a program or information at a location other than that at which it was originally received. Place shifting allows for the viewing of media programs (such as movies or television channels) at any location that has a multimedia computer and a broadband connection.

Plain Old Telephone Service (POTS)-Plain old telephone service is the providing of basic telephone service without any enhanced features. It is the common term for ordinary residential telephone service. The POTS system uses in-band signaling tones and currents to determine call status (e.g. call request). Because POTS allow for the transfer of audio signals below 3.3 kHz, POTS systems are also used for modems that allow data transmission (called dial up connection). Whenever a new service or feature is described, the author may refer to the previous available package of features and services as POTS, even when the previous package included several very sophisticated capabilities.

Plain Text (Plaintext)-Plaintext is data that can be directly used or displayed (without any additional processing such as decryption).

Plaintext-Plain Text

Plant-A general term applied to all of the physical property and facilities of a telephone company that contribute to the provision of communications services.

Plant Location Record-Records that show the placement of distribution terminals and the location, length, size, date, and nomenclature of cables.

Plant Retired-A plant that has been removed, sold, abandoned, destroyed, or otherwise withdrawn from service.

Platform-A platform is the combination of system hardware and software that programs or services operate.

Platform Access-Platform access is the ability of a user or device to connect to and communicate with a combination of system hardware and software that programs or services operate.

Platform Dependent-Platform dependent refers to devices or systems that can only communicate with specific platforms that contain specific hardware types and software programs.

Platform Independent-Platform independent is the ability of a program or system to operate on or in other hardware and software devices or systems.

Platform Neutral-Platform neutral is the ability to use software programs or services on systems that use different combinations of hardware and operating systems.

Platform Specific-Platform specific is the combination software that programs or services operate that can only be used on specific combinations of hardware and operating systems.

Playout System-A playout system is an equipment or application that can initiate, manage and terminate the streaming or transferring of media to users or distributors of the media at a predetermined time schedule or when specific criteria have been met (such as when user registration and payment).

PLC-Public Limited Company

PLM-Product Life Cycle Management

PLS-Position Location System

PLU-Percent Local Usage

Plug And Play (PnP)-Plug and play is a compatibility system that simplifies the installation and removal of hardware devices into a personal computer. PnP includes the automatic recognition of hardware installation and removal, activation and deactivation of software drivers, and system accessory management functions.

Plug In (Plug-in)-A plug-in is a software program that works with another software application to enhance its capabilities. An example of a plug-in is a media player for a web browser application. The media player decodes and reformats the incoming media so it can be displayed on the web browser.

Plug-in-Plug In

PM-Portable Media

PMP-Point to Multipoint

PMP-Portable Media Player

PMS-Personal Media Server

PN-Parameter Negotiation

PnP-Plug And Play

PNP-Private Numbering Plan

PO-Purchase Order

POC-Push to Talk Over Cellular

POD-Push On Demand

POD Slurping-Personal Operable Device Slurping

POE-Point of Entry

POI-Point Of Interconnection

Point of Entry (POE)-A point of entry is a location where communication lines first enter the customer's building or facility.

Point Of Interconnection (POI)-The point of connection between exchange carriers.

Point of Origin-Point of origin is the location where is shipment is initially sent from.

Point Of Presence (POP)-(1-interconnection) A point of presence physical location that allows an interexchange carrier (IXC) to connect to a local exchange company (LEC) within a LATA. The point of presence (POP) equipment is usually located in a building that houses switching and/or transmission equipment for the LEC. (2-authentication) A processing function in a network that is capable of looking up information in databases.

Point of Sale (PoS)-Point of sale is a process or location that is used to capture or complete a transaction.

Point Of Termination (POT)-The point within a local access and transport area (LATA) at which a local exchange carrier's responsibility for access service ends and an inter-exchange carrier's responsibility begins.

Point to Multipoint (PMP)-Point to multipoint communication is transferring information from one device (or point) to multiple points or devices (multiple receiving points). PMP services are broadcast or multicast services.

Point to Point (PTP)-Point-to-point communication is the transmission of signals from one specific point to another. Point-to-point communication uses addressing to deliver information to a specific receiver of the information. It is possible to implement point-to-point communication through a broadcast network by using device addressing or through a network using network routing.

Point-To-Point Communication-The transmission of signals from one specific point to another, as distinguished from broadcast transmission which blankets the general public. Radio relay links are a common type of point-to-point communication.

Point-To-Point Protocol (PPP)-Point to point protocol (PPP) is a connection oriented protocol that is established between two communication devices that encapsulates data packets (such as Internet packets) for transfer between two communication points. PPP allows end users (end points) to setup a logical connection and transfer data between communication points regardless of the underlying physical connection (such as Ethernet, ATM, or ISDN). PPP is described in the IETF RFC 1661.

This diagram show how a point-to-point protocol (PPP) connection allows a computer connection to be established, verified, and maintained. This example shows that PPP protocol can be used on any type of access connection. The PPP protocol allows for the separation of the link protocol on the PPP link and the network control protocol that operate above the high level data link control (HDLC) protocol that is used to coordinate the PPP link. The PPP protocol also includes security features such as password authenticated protocol (PAP) and a more secure challenge handshake authentication protocol (CHAP).

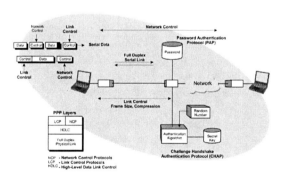

Point to Point Protocol (PPP) Operation

Pole Attachment-Any attachment by a line (such as a cable television system) to a pole, duct, conduit, or right-of-way owned or controlled by a utility.

Pole Attachment Contracts-Agreements that are government-regulated that affect utility pole owners, joint use contracts, and companies that enter into agreements to lease space on poles to attach facility cabling, wiring, terminals, etc.

Pole Rental-Pole rental is fees that are charged for the authorization to use a pole (e.g. telephone pole).

Policies-Policies are the rules are processes that are applied to a device, service or user.

Policing-Policing is the monitoring of data transmission or service usage to ensure it meets user and/or network objectives or requirements.

Policy Control-Policy control is the processes that are used to modify the configurations and/or parameters of network elements or devices that control the providing of services to ensure the performance or operation attempts to confirm to defined service levels.

Policy Decision Function (PDF)-A policy decision function is a server that uses a set of rules to makes decisions as to whether connections or service flows should be admitted or added to a network.

Policy Decision Point (PDP)-A policy decision point is a server (such as in a COPS system) that makes decisions as to whether connections or service flows should be admitted or added to a network.

Policy Platform-A policy platform is the combination of system hardware and software that is used to monitor and control the rules or management (policies) of a system.

Policy Server-A policy server is a communications server (computer with a software application) that coordinates the allocation of network resources based on predetermined policies on the priorities and resources required by communication services and applications within the network. A policy server is used to help manage network operation in the event of loss of resources. The pre-set policies define which communication services are critical (such as voice) and how many resources should be allocated to these critical services at the expense of other communication services (such as web browsing).

Polling-Polling is the process of sending a request message (usually periodically) for the purpose of collecting events or information from a network device. The receipt of a polling message by a device starts an information transfer operation for a specific time period.

Polyphonic Tones-Multiple tones that are used to indicate an event (such as an incoming telephone call).

PON-Passive Optical Network

Pooling-Pooling is the making available of a product or service to multiple devices, accounts or services.

POP-Point Of Presence

POP3-Post Office Protocol Version 3

POPs-Persons of Population

Popup-Popups (pop-over) are windows that are opened on an Internet browser when the user opens a browser window or opens a window at a new web address.

Port-(1-general) A port is a physical or logical connection point between a computer or computer-based machine and other hardware devices. (2-network) A place of access to a device or network where energy may be supplied or withdrawn, or where the device or network variables may be measured. (3-software) The process of moving source code and executable programs from one computing system to another of a different type without substantive changes to the source code.

Port Charges-Port charges are fees or assessments for the providing of access points (ports) into a communication network.

Port Growth-Port growth is the number of telephone stations that a telephone system can expand to. In some cases, additional switching modules or cards can be added to accommodate port growth.

Port Identifier-A value assigned to a port that uniquely identifies it within a switch. Port Identifiers are used by both the spanning tree and link aggregation control protocols.

Port Multiplier-A port multiplier is a device or software program that allows for the creation of additional physical and/or logical connections for devices or networks.

Port Sharing-The process of allowing multiple virtual connections (logical channels) to share the same port connection in a data network.

Port Solver Protocol-Port solver protocol is a set of commands and procedures that can be used to setup and manage logical connections (ports) to devices that are connected to shared networks (such as Ethernet networks).

Portability-This means that application software can be dragged across different computing platforms and operating systems. With the TAO spec, developers can write an application that will run on a PC, Alpha Server or Tandem host. SCSA's operating-system-independent APIs give developers a uniform method for supporting multiple operating systems.

Portability Rights-Portability rights are the permissions granted from an owner or distributor of content to transfer the content to other devices (such as from a set top box to a portable video player) and other formats (such as low bit rate versions).

Portable Device (PD)-A portable device is a small computing device that contains it's own software operating system that allows the user to run software processing and media player applications.

Portable Document Format (PDF)-Portable document format (pdf) is a file format industry standard that allows for the transfer, presentation, and printing of electronic books and/or documents independent of the types of computing or communication systems that are used.

Portable Media (PM)-Portable media is information content (e.g. digital audio) that can be transferred to one or more media players.

Portable Media Player (PMP)-A portable media player is a self-contained device and associated software application that can convert media such as video, audio or images into a form that can be experienced by humans.

Portal-A portal is an Internet web site that acts as an interface between a user and an information service.

Portal Service-A service that provides an interface between a user and specific types of information services.

Ported Number-A ported number is a telephone number that has been setup to be redirected to a new switch port (new destination number).

Porting-Porting is the converting of programs or services into a form that can be used on different systems or platforms.

PoS-Point of Sale

Position Location System (PLS)-A position location system that gathers and processes information to determine the geographic location of devices or equipment. The position location information may be gathered by the network sensing the position of the device relative to one or several antennas in a network (such as the Teletrac system) or by the vehicle reporting its location using external position locating devices (such as the Global Positioning System).

Position Tracking-Position tracking is the process of a continuously receiving position that identifies the location for a device or person. Position tracking services include fleet management, workforce location tracking, automatic vehicle location (AVL), and asset tracking.

Positive Cash Flow-Positive cash flow is a financial condition where the cash or liquid assets that are being received are more than the cash or liquid assets that are being expended.

Post Billing-Post billing is the process of gathering and processing billing records and invoices after the service is provided.

Post Dial Delay (PDD)-The time period between when a user dials the last digit of a phone number and hears the phone at the other end begin to ring.

Post Dialing-Refers to the ability of a terminal to send dialing information after the outgoing call request setup message is sent.

Post Mission-Post mission is the processing or using of information after a mission (such as measuring location information) has been performed.

Post Office Protocol Version 3 (POP3)-Post office protocol is an advanced set of commands and processes that are used with simple mail transfer protocol (SMTP) to reliably and securely store and transfer e-mail messages from one email application to another.

Post Payment Processing-Post payment processing is the tasks and functions that are used to collect payments from the buyer of products and services after they process their order.

Post Processing (Postprocessing)-Post processing is performing operations on information or data that occurs after the event that created the information or data has occurred.

Post Production-Post production is the editing process for data (e.g. billing records) media programs (e.g. movies) that is performed after the service or process (e.g. filming) has been performed.

Post, Telephone And Telegraph (PTT)-A term used for a government agency in many countries that supplies and maintains the infrastructure and provides basic telecommunication services.

Postal Code-A postal code is sequence of numbers or characters that are used to assist in the routing of letters or packages through the postal distribution system.

Posting-(1-Internet) Posting is the process of sending a message or information to a newsgroup, blog or bulletin board. (2-accounting) Posting is the recording of accounting data into accounting documents or systems.

Postpaid Service-Postpaid services is a process that allows customers to pay for services after they receive or use the products or services.

Postpay Service-A billing or coin telephone service that allows for payment of service after communication has started (e.g. called party answers.) Postpay, now rarely used, is the simplest form of coin service but requires additional operator attention on toll calls.

Postprocessing-Post Processing

PostScript-PostScript is a text and graphics description language. It allows the universal creation of documents which various style and position features.

POT-Point Of Termination

Potential Entrant-A potential entrant is a company that has qualities or ambitions that make it a likely candidate to enter into a new industry or business segment.

POTS-Plain Old Telephone Service

Power-(1-general) Power is the time rate at which energy is generated, consumed, or converted from one form to another. The electrical unit of power is the watt. One watt is the product of one volt of voltage and one ampere of current. In mechanical units, one watt is the product of one Newton of force with one meter per second of velocity. In non-technical discourse, the four words force, energy, power and momentum are often unfortunately used as synonyms for each other. In science and technology, each of these words describes a distinct and separate physical quantity. (2-alternating current) The product of the effective voltage, the effective current, and the cosine of the phase angle between them. (3-apparent) The product of the voltage and the RMS current. (4-available) The maximum power available from a source by a suitable adjustment of the load. (5-average) In a pulsed laser, the energy per pulse (joules) multiplied by the pulse repetition rate (in Hertz). (6-average speech) The total speech energy over a period of time, divided by the length of the period. (7-carrier) The average power supplied to the antenna transmission line by a radio transmitter during one radio frequency cycle under conditions of no modulation. For each class of emission, the condition of no modulation should be specified. (8-direct current) The product of the voltage and the current, or of the resistance and the square of the current. (9-effective radiated, ERP) The product of the power supplied to an antenna and the antenna gain relative to a half-wave dipole. (10-mean) The power supplied to an antenna transmission-mission line by a transmitter during normal operation averaged over a time sufficiently long compared with the period of the lowest frequency encountered in the modulation. A time of 0.1 seconds during which the mean power is greatest normally will be selected. (11-peak) The maximum power emitted by an optical source or a radio frequency transmitter. (12-peak envelope) The average power supplied to the antenna transmission line by a transmitter during one radio frequency cycle at the highest crest of the modulation envelope taken under conditions of normal operation. (13-real) The power in an alternating current circuit that is used in doing work. It is the product of the RMS voltage times the RMS current times the power factor (the cosine of the angle between voltage and current).

Power Consumption-Power consumption is the amount of energy that a device or system uses over a period of time.

Power Dialing-The process of dialing calls in lists, connecting the call to a customer service representative if it is answered by a person, remembering unanswered calls for a later callback, and updating records to indicate the calls that have been completed to avoid repeat calling. Power dialing systems are sometimes linked to customer information databases to allow the customer service representative to see the call recipients account information when the call is connected.

Power of Attorney-Power of attorney is an authorization to perform legal actions on behalf of another person or company. A power of attorney authorization is typically limited to a specific set of actions that are authorized to perform (such as purchasing a property or obtaining documents).

PPC-Pay Per Call

PPC-Pay Per Click

PPD-Pay Per Day

PPDN-Public Packet Data Network

PPI-Pay Per Inclusion

PPL-Pay Per Lead

PPL-Pay Per Listen

PPP-Point-To-Point Protocol

PPS-Pay Per Sale

PPSN-Public Packet Switched Network

PPV-Pay Per View

Practical Extraction and Report Language (PERL)-Practical extraction and report language is set of commands and processes that can be used to analyze, process and display data.

PRD-Program Reference Document

Predatory Pricing-A predatory pricing is the adjusting prices of products or services to reduce or disable a competitor's ability to compete in a marketplace. Predatory pricing may lower the prices below cost to keep a competitor from entering into or staying in a market segment.

Predictive Analytics-Predictive analytics are the processes that are used to evaluate the likely future actions or performance of programs or services. Predictive analytics can be used to modify marketing programs or offers based on changes in predicted consumer behavior.

Predictive Dialing-Predictive dialing is an automated method for making outbound telephone calls in which a scheduling (pacing) algorithm determines the number of calls placed in advance of actual operator availability.

Preemptible Rate-Preemptible rate is a discounted fee structure that is used for advertising time where the advertising insertion is not guaranteed (can be preempted by other ads).

Preemption-Preemption is the inhibiting of a service so another service may be provided that has a higher priority level.

Preferential Numbers-Preferential numbers are telephone numbers or other labels that have desirable characteristics or have been selected by users as an option.

Preferred Access Method-A preferred access format is the selection of one of several available access systems that can be used to receive and send signals.

Preferred Interexchange Carrier (PIC)-The assignment or use of an inter-exchange carrier to complete calls from a customer outside their systems calling area. The PIC code is obtained when a customer dials a number that requires inter-exchange carrier (IXC) service. The PIC code is used to route the call to the IXC carriers point of presence (POP) switching center.

Preferred Language (PL)-Provides a telephone service subscriber with the ability to specify the language for network services. This service allows the subscriber to specify service in English, Spanish, French or Portuguese.

Preferred Media Formats-A preferred media format is the selection of one of several available formats that can be transmitted, received and decoded.

Prefix-Any digit dialed before a destination address. Prefixes indicate service options. For example, a prefix of "1" indicates a 10-digit call address in some areas, and a "0" indicates a request for the services of an operator.

Premise-Premise is the space occupied or owned by a user or customer or business within a building facility.

Premises Distribution System (PDS)-Premises distribution systems are the combination of equipment, protocols and transmission lines that are used to distribute communication services in a home or building.

Premium-Premium Offer

Premium Channels-Premium channels are media services (such as television programs) that are offered as a supplement or upgrade to other services. Premium channels may be provided as a bundled package (multiple channels), on a per channel subscription or on a pay per view (PPV) basis.

Premium Offer (Premium)-A premium is an item or other valuable object that is offered as an added incentive to a potential customer to help motivate them to take action or to increase the amount of a product or service purchased by that potential customer.

Prepaid Account-A prepaid account is a unique identifier that designates a billing entity that is used to associate billing charges for products and services where services are provided only when there is value stored in the prepaid account. A prepaid account may only have value transactions and a limited amount of owner or user identity information.

Prepaid Calling Card-A card that is issued by a telecommunications service provider that contains coded identification information that permits the card holder to initiate a call or request and receive an information service. Calling cards contain a number or code contained on a magnetic stripe that uniquely identifies the card and authorized services to the system.

Prepaid Service-Prepaid communication services is a process that is used by service providers (such as communication providers) to be paid for services they provide in the future. This allows the provider to obtain revenue for services without the risk of bad debt and it eliminates the need and cost for billing operations. Prepaid service is often associated with customers that may be credit challenged or who want more control over bills.

Prepaid System-A prepaid system is a combination of software and hardware that receives service usage information, identifies the specific account that the usage amounts are associated with, calculates the charges and deducts the charges from prepaid amounts on the accounts.

Prepaid Wireless-Wireless connection whereby service is prepaid before usage is accumulated. Typical users hear an announcement prior to the call noting how many minutes or dollars or units they have remaining on their account.

Prepay Service-Coin telephone service in which an initial coin deposit is required before a connection is established on chargeable calls. Prepay service is provided either by coin-first or dial-tone-first coin service.

Preprint-A preprint is a production of a media item (such as an advertisement) that can be viewed or used before actual production of the media item.

Pre-Proposal Conference-A pre-proposal conference is a meeting where the anticipated requirements for a proposal document are described and discussed.

Pre-Proposal Review-A pre-proposal review is a meeting where the anticipated requirements for a request proposal document are described and discussed. A pre-proposal review can be an informal environment that allows for explanations and discussion of what is needed and some of the available options that may meet those needs. A pre-proposal review can help to shape and clarify an RFP before it is released.

Pre-Provisioning-Pre-provisioning is the process of assigning and/or programming information into a system or device before it is sold or provided to a user to setup or assist in the setup in the providing service to the system or device.

Pre-Qualification Application-A prequalification application is a form that a person or company uses to identify that they are qualified to participate in a project (such as the issuance of a request for proposal). A pre-qualification application may be used to limit the people or companies who can participate in a RFP or tender offer to those that have specific attributes such as financial strength and technical abilities.

Presence-(1-communication) Presence is sensing and triggering the sending of information alert messages on specific activities or status changes that have occurred by a user or device. Presence is one of many features that can be offered as part of multimedia services bundle (such as IPTV). Based on a common feature in most of today's PC-based instant messaging applications, presence refers to the ability to view what friends and family are on the network through a "buddy list" that can be graphically displayed on-screen. Advanced presence features include the ability to alert others when you are on the phone or what program you are watching. (2-noise level) The natural noise level of a room or environment without dialogue or other produced sounds. Presence also is known as room tone. Presence is used to fill holes created during editing or footage recorded without sound.

Presence Constraints-Presence constraints are the bounds or restrictions for the use of a product or service in a particular instance are state of being (presence).

Presence Services-Presence services are the providing of information transfer, processes or authorizations that enable users, devices or systems to perform actions that they desire that is based on the detection the status of a person or device at a location or connection point.

Presentation Control-Presentation control is the ability to setup and control the display and operation of media on a display such as a television along with the placement and operation windows on the users display.

Pressure Sensitive Label (PSL)-A pressure sensitive label is a sticker that can be peeled off a low-stick backing material and affixed to another material (such as an envelope) through the use of pressure.

Presubscription-Presubscription is the preselection of service that enables a customer served by an equal-access end office to automatically route all interLATA and international calls to a long distance carrier of the customer's choice. Carrier access codes, such as 1OXxx, where Xxx is a unique code identifying a long-distance carrier, are not required to dial a call. (See also: primary interexchange carrier.)

Pretty Good Privacy (PGP)-Pretty good privacy is an open source public-key encryption and certificate program that is used to provide enhanced security for data communication. It was originally written by Phil Zimmermann and it uses Diffie-Hellman public-key algorithms.

PRI-Primary Rate Interface

Price-A price is an offer value that is assigned to a product or service.

Price Agreement-A price agreement is a set of terms agreed to by vendor(s) and customer(s) that defines pricing levels that depend on volume commitments and/or other criteria (such as promotional support).

Price Analysis-Price analysis is the evaluation of price points and the results they generate. Price analysis of related products and services can be used to determine more optimal prices that will be set for new or updated products.

Price Cap-A price cap is a value or maximum number that a price can increase to during an agreed period.

Price Cap Index-A price cap index is a list of maximum price point requirements that is calculated for specific types of products or services. A price cap index may be applied to companies by government regulations.

Price Cap Regulation-Price cap regulation is a set of pricing rules imposed by governments that companies must follow for providing certain types of products or services.

Price Competition-Price competition is the use of price points to influence buying patterns as compared to other companies.

Price Differentiation-Price differentiation is the presentation of cost attributes of a product or service where the recipient or user of the products or services perceives a different price or value as compared to similar products or services that would be available from other companies.

Price Discrimination-Price discrimination is the providing of the same types of products or services at different prices to different customers due to unjustified preferences.

Price Elasticity -Economic indicator used to define how sensitive consumers will be to changes in prices. A high elasticity means that customers will change their demand drastically with small changes in price.

Price Engine Web Site-A price engine web site is a web portal that allows users to search for a product or service and displays the prices and/or stores that offer those products or services.

Price Erosion-Price erosion is decline in the market price that results from the addition of products offered by competitors or price reductions of similar related goods or services.

Price Leadership-Price leadership is the offering of a product or service in the marketplace at the lowest price. Other companies may not reduce their prices below the price leader to avoid a price war.

Price Per MHz Per Pop-The price per MHz per POP is a measure of the value of radio system or radio license that is based on the amount of radio bandwidth in MHz and population covered by the geographic area assigned to the radio license holder.

Price Plan-Pricing Plan

Price Point-A price point is a value or an amount that is presented to a buyer where there is a definable change in the buying or usage patterns.

Price Testing-Price testing is the offering of the same product at different price levels to determine an optimal price level for selling a product or service.

Price War-A price war is the continual reduction of prices by multiple companies so that each can attempt to gain a market share increases.

Pricing-Pricing is the assignment of value to products or services that are offered for sale.

Pricing Models -Pricing models are mathematical formulas or processes that are developed to help determine what prices should be charged to what customers for different services.

Pricing Options-Pricing options are a set of price plans or rates that are based on qualification or selection criteria. An example of pricing options is the difference in prices that are charged for wholesale and retail customers.

Pricing Plan (Price Plan)-A price plan is a document that defines the objectives of product or service pricing, what rules to use to calculate prices and how it expects prices will change when certain events occur (such as competitive price erosion).

Pricing Scheme-A pricing scheme is a structure that is used for the assignment of value to products or services that are offered for sale.

Pricing Strategy-A pricing strategy is a set of tactics that may be used to set or adjust prices of products or services.

Primage-Primage is an extra cost or fee that is assessed for handing goods when they are being loaded or unloaded onto a transportation carrier.

Primary Event-Primary events are media programs in a playout system that is scheduled to be broadcasted.

Primary Interexchange Carrier Charge (PICC)-A recurring fee that is added to the bill of the local telephone service customer that charges them for access to inter-exchange carrier (IXC) services. This fee is charged in addition to other fees or tariffs (such as a percentage of billed long distance usage) that are paid by the IXC to the local exchange carrier (LEC). The purpose of the PICC charge is to help recover the cost of providing local loop access service in a marketplace that has declining long distance revenue charges.

Primary Key-A primary key is a label or code that is used to uniquely identify an object in a list or set.

Primary Rate Interface (PRI)-Primary rate interface (PRI) is a standard high-speed data communications interface. This interface provides a standard data rates for T1 1.544 Mbps and E1 2.048 Mbps. The PRI interface can be divided into several channels (channelized). These channels can be combinations of 64 kbps (B) 384 kbps (H) channels or other channels. PRI connections must include at least one 64 kbps (D) control channel.

Prioritization-Frame and packet prioritization assigns different priority codes to packet that are transmitted through a communication network. This allows some frames or packets to receive a higher transmission priority for time sensitive data communications (such as packetized voice).

Priority Ring-A service that provides a telephone customer with distinctive ring sounds when a call is received from a telephone number that the customer has previously identified as a priority number (e.g. family or office). See also distinctive ring.

Privacy-(1-encryption) A term used with regard to encryption to indicate a level of protection that is minimal, corresponding to a moderate amount of effort on the part of the eavesdropper to understand the private communication, but not so good as the better levels designated by the words "secret" or "secure."
(2-channel separation) An electrical capability to prevent other extensions on a multi-extension analog telephone line or key telephone system from connecting during a conversation, typically via diodes or other devices actuated by the decrease in subscriber loop voltage when one set is off hook. Also the name of an equivalent capability using an electronic key system.

Privacy Breach-A privacy breach is the transferring of information to recipients or through a process that violates privacy rules or terms.

Privacy Policy-A privacy policy is the self proclaimed rules a receiver of information claims to follow when a customer or visitor sends or provides information. Privacy policy rules typically state how the information may be used and who the information may be distributed to.

Privacy Statement-A privacy statement is a declaration of rules or policies that a receiver of information claims to follow when a customer or visitor sends or provides information. Privacy policy rules typically state how the information may be used and whom the information may be distributed to.

Private-(1-general) Private means secure or not shared with others. (2-Bluetooth) A mode of operation whereby a device can only be found via Bluetooth baseband pages, it only enters into page scan.

Private Automatic Branch Exchange (PABX)-A telephone switch that is generally located on a customer premise. Often referred to as a PBX, CBX, EPABX. This provides for the transmission of calls internally as well as to and from the public network.

Private Branch Exchange (PBX)-PBX systems are private local telephone systems that are used to provide telephone service within a building or group of buildings in a small geographic area. PBX systems contain small switches and advanced call processing features such as speed dialing, call transfer, and voice mail. PBX systems connect local telephones ("stations") with each other and to the public switched telephone network (PSTN).
This diagram shows a private branch exchange (PBX) system. This diagram shows a PBX with telephone sets, voice mail system, and trunk connections to PSTN. The PBX switch calls between telephone sets and also provides them switched access to the PSTN. The voice mail depends on the PBX to switch all calls needing access to it along with the appropriate information to process the call.

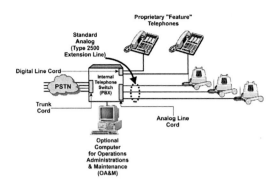

Private Branch Exchange (PBX) System

Private Carrier-An entity licensed in the private services and authorized to provide communications service to other private services on a commercial basis.

Private Internet Address-An Internet address that can be transferred by routers on a private data network.

This figure shows how private Internet address numbers are used to uniquely identify devices connected within a private network that uses Internet protocol transmission. This diagram shows that packets that are designated for the Internet must be translated to an Internet address that can be routed through the Internet.

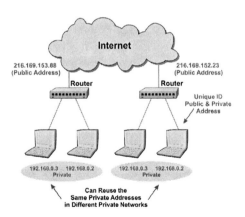

Private Internet Address

Private IPTV-Private IPTV systems are IP television systems that have been designed or setup to be used in private facilities such as in hotels, cruise ships or college campuses.

Private Key-A key used to decrypt a message encrypted with the owner's public key or to sign a message from the key's owner. Sometimes called a secret key.

Private Line-A dedicated communications circuit that is leased by a customer from a telephone service provider for voice, data, or video services. While a private line may be connected through a switching facility, the connection resources are constantly dedicated to the customer who is leasing the line.

Private Network-A network designed for the exclusive use of one customer. Often, such a network is nationwide in scope and serves large corporations or government agencies.

Private Numbering Plan (PNP)-A feature that enables subscribers to call defined private-network extensions using an abbreviated dialing pattern. When a PNP subscriber dials a private-network extension, the cellular network translates the dialed extension to a number in the North American numbering plan (NANP) and routes the call.

Private Telephone System-Private telephone systems are independent telephone systems that are owned or leased by a company or individual. Private telephone networks include key telephone systems (KTS), private branch exchange (PBX) and computer telephone integration (CTI).

This figure shows the different types of private telephone systems. This diagram shows that key telephone systems (KTS) allow each telephone station to access some or all of the lines available to the company. Private branch exchange (PBX) systems include a local telephone line switching with call processing software to allow simple stations to connect to many different telephone lines. Computer telephony (CT) systems merge computer intelligence with telecommunications devices and often link telephone systems with company information systems. Private telephone sys-

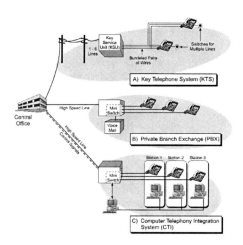

Private Telephone Systems

tems that use Internet protocol data networks to interconnect telephone stations are called IPBX systems. This diagram shows an IPBX system that shares a data network for both telephone calls and computer workstations.

Private Television (Private TV)-Private television systems are audio and visual media distribution systems that have been designed or setup to be used in private facilities such as in hotels, cruise ships or college campuses.

Private TV-Private Television

Privatization-The process of selling or reallocating government owned assets or resources to companies or investors. Privatization in the telecommunications industry often includes the issuance or re-allocation of licenses to provide a telecommunications services.

Proactive Monitoring-Proactive monitoring is the process of gathering information from a system or other source that will assist in the diagnosis of a problem that may occur or to ensure that a process operates as expected.

Probe-(1-test) A test prod used to check components for the presence of signals. (2-cavity) A wire loop inserted in a cavity for coupling energy to an external circuit. (3-communication) A process of sending a message or alert signal to another device or group of devices

that discovers if there is information to be gathered.

Problem Handling-Problem handling is the process of entering in trouble reports from customers and making sure that the problems are resolved.

Process Server-A process server is a computer system performs specific processes or functions related to a specific user, device or service.

Processing Delay-The time required for processing information. Processing delay for coding and decoding of voice or data information is usually measured in msec.

Processor-(1-computing) A circuit or device that systematically executes specific operations on digital data as directed by a program stored in memory. (2-finance) A processor is a financial institution that processes credit card transactions.

Procurement-Procurement is the processes that are involved in the acquisition of products or services.

Product Acquisition and Development System (PADS)-A process that is used for the identification and coordination of product development activities. The PADS system is divided into 5 phases; concept, feasibility, planning, development, and introduction.

This diagram shows a basic product development process to help ensure the successful

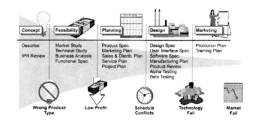

Product Acquisition and Development System (PADS) Operation

P

development of communication products and services. This example shows that this product development process evolves through a series of gates (called toll gates) that are used to determine if a product should continue in the normal development process or eliminated (discarded) so additional resources can be allocated to other products. In this example, the product development cycle begins with product ideas (concept), business evaluation (feasibility), resource scheduling (planning), technology design and production testing (design), and market introduction and distribution support (marketing).

Product Activation-Product activation is the process of enabling a product to begin operation by entering information into the product through the use of either local operations (e.g. user keypad) or via an external connection (downloading the information into the product). Product activation usually requires that certain customer financial criteria must also be met before the product activation is performed or before the necessary product entry information is provided to the customer.

Product Adaptation-Product adaptation is the conversion or alteration of a product so that it can be used for a different purpose or sold to different regions. An example of product adaptation is the conversion of a product voltage rating so that it can work in other countries that use different voltage levels.

Product Bundling-Product bundling is the marketing strategy that involves offering several products for sale as one product. Communication service providers (CSPs) may use product bundling of voice, data, video, and wireless services to provide a higher perceived value to end users (consumers or enterprise). Some CSPs may offer the capability for users to select or setup their own product bundles.

Product Catalog-A product catalog is a central repository of products managed within the scope of the service delivery organization with a catalog of associated components to be deployed for enabling the service. Product catalogs can provide multiple views including a commercial product catalog and technical product catalog. Commercial catalogs cover the product representation from customer, sales, financial perspective while technical catalog provides the technical realization view.

Product Code-A product code is a unique number or label that identifies a product or service.

Product Creation-Product creation is the development of devices, applications or services that can be sold to people, companies or devices.

Product Development-Product development is a process identifying, defining, developing, and producing a new or improved product.

Product Differentiation-Product differentiation is the modification of a products characteristics or the use of promotional materials to differentiate its' products or services from each other or from other competitor's products.

Product Fulfillment-Product fulfillment is the process of gathering the products and materials to complete an order and shipping the products or initiating the services that were ordered.

Product Key-A product key is a code or process that is used to enable the ability of device or product to begin operation of to obtain access to a service.

Product Launch-A product launch is the process of making a product available for purchase and distribution with significant promotional efforts. Companies may perform a hard product launch strategy after a soft launch or it may use a hard launch campaign to maximize the effects of new product publicity.

Product Life Cycle-Product life cycle is sequence of events and time period that occurs from the origination of a product idea until when the product or support for the product is no longer available to users or buyers. Product life cycle stages can be classified into introduction, growth, maturity, and sales decline stages.

Product Life Cycle Management (PLM)-Product life cycle management is the sequence of events and time period that occurs from the launching or introduction of a product or service until when the product or support for the product is no longer available to users or buyers. Product life cycle stages can be classified into introduction, growth, maturity, and sales decline stages. PLM for the communication industry include the coordination of services, and products. Their variants groom the product launch to its retirement.

Product Management-Product management is the process of assigning and tracking specific tasks and functions related to ensuring the success of products or services.

Product Matching-Product matching is the process of identifying a product or service that has characteristics or features that match the requirements or desires of a prospective customer.

Product Packaging-Product packaging is the combining of a product and its supporting elements (accessories or files) into a form or container that is used to deliver the product to the consumer or user of the product.

Product Placement-Product placement is the providing of a product within a media program (such as a movie or television program) as a way of promoting the product and/or branding of the product.

Product Profile-A product profile is a set of characteristics that are associated with a product or service. These characteristics may include a basic description, service capabilities, feature sets, and device capabilities along with many other characteristics.

Product Sample-A product sample is an item or service that is provided to a customer or prospective customer for free or at reduced cost to help them decide or evaluate the purchase of additional products or services.

Product Specific Promotion-Product specific promotions are messages that provide offer(s) for specific products.

Product Specification-A document that defines the key features and characteristics of a product.

Product Support-Product support is the communication, system and processes that are used to help customers or users of products to purchase, use, return or solve problems related to the product they purchased.

Product Trainers-Product trainers are people who provide information and educational support to users or sellers of products or services. Product trainers may be used to help users or retailers to become more familiar with the capabilities and features of a product or service. This makes them more likely to sell or use the product.

Product Types-Product types are the defining of products into groups that are based on characteristics such as size, function, brand, durability, or cost. Examples of product types include consumer, original equipment manufacturer (OEM), military, and industrial. While it is desirable that product types are unique, in practice, some product types may overlap.

Production-Production is the process of creating or physically producing products, media programs or services.

Production Bonus-Sum of money due paid to contracted writer or other staff members upon the occurrence of an initiation of a production.

Production Department-A production department is a group within a company or organization that is responsible for producing products, media programs or services.

Production Facilities-Production facilities are equipment and software applications that are used in the production of media content or programs.

Production House-A facility that typically does everything to generate final video productions except shooting the original video-tape. Services typically include editing raw master tapes, modifying recorded material, and creating new effects. Projects include advertising, training, promotion, music videos, and TV shows.

Production Schedule-A time schedule that identifies what products or assemblies will be produced and when they will be produced.

Production Switcher-A production switcher is a device that is capable of connecting media communication paths or signals between different sources (such as cameras) and receivers (recorders). A production switcher usually contains a special effects generator.

Production Testing-Production testing is the measurements and adjustments that are performed on products during the production process.

Production Yield-Production yield is the percentage of acceptable products that are produced compared to the total number of products that are manufactured.

Productivity Application-A productivity application is a software program that performs specific operations to enable a user to improve or more efficiently perform business or work related functions.

Professional Licenses-A professional license is a certificate of authorization for a person in a particular industry or job type to perform specific types of tasks or projects.

Professional Piracy-Professional piracy is the acquiring, duplication and sale of information or program materials, particularly TV programs on videotape, in violation of copyright laws by skilled professionals.

Profile-A profile is a particular implementation or instantiation of a more general protocol. Many protocols are extremely general and allow one to specify a restricted set of messages and their actions for a particular purpose. Such a set is known as a profile. For example, NCS is a profile of MGCP with a few extensions.

Profile Privacy-Profile privacy is the level of anonymity that a user or device has from com-

panies or people finding out about the features and/or characteristics associated with their usage of a product or service.

Profiling-Profiling is the process of monitoring, analyzing and characterizing a user of a product or service as to their usage patterns or preferences.

Profit Forecasting-Profit forecasting is the process of estimating the future profitable performance for a product or service.

Profit Margin-Profit margin is the percentage of the revenue generated from sales that is above the total costs of selling or providing products and services.

Profitability-Profitability is a measure of the efficiency of a person or company to generate earnings.

Proforma Invoice-A proforma invoice is a grouping of billing charges that would occur provided an order was placed. A proforma invoice may be requested so that product and cost details are provided before an order is initiated.

Program-A sequence of instructions used to tell a computer how to receive, process, store, and transfer information.

Program Development-Program development is the writing, entering data, translating and debugging a software program.

Program Evaluation and Review Technique (PERT)-Program evaluation is a review technique that is a management procedure that is often computerized to provide close control of all operations needed to complete a project on time and within budget. The procedure enables actual progress to be compared with the progress originally planned. Control charts prepared under this process are called PERT diagrams.

Program Guide-A program guide is a listing or an interface (portal) that allows a customer to preview and select from possible lists of available content media. Program guides can vary from simple printed directory to interactive filters that dynamically allow the user to filter through program guides by theme, time period, or other criteria.

Program Reference Document (PRD)-Program reference documents are a set reference materials that are used to develop or submit products for specific programs.

Programmed Telephony-The use of programming to control the setup, routing, and termination of telephone calls. See computer telephony integration (CTI).

Programmer-A programmer is a person who prepares sequences of software instructions for a computer.

Programming-(1-software) Programming is the process of developing an assembly of instructions for a computer to enable it to carry out a particular job. (2-television) Television programming is the selection of shows and programs that are offered by a television service provider.

Programming Department-A programming department is a group within a company or organization that is responsible for the selection and scheduling of media programs (such as a television program schedule).

Programming Expenses-Programming expenses are the costs of acquiring and using media content.

Programming Language-A language designed to be under-stood by a computer. A high-level programming language is converted into the required machine code by a program called a compiler.

Projection Ratio-A projection ratio is a scaling value (multiplier) that can be used to estimate future sales using the ratios from existing sales multiplied by expected market changes (e.g. growth factors).

Promo-Advertising Promotion

Promoter Group-The promoter group is the companies that initially started a consortium (such as Bluetooth).

Promotion Program-A promotion program is a campaign that is designed to communicate (promote) a product or service.

Promotion Rules-Promotion rules are a set of conditional processes that are used to select and offer incentives to potential customers for products or services.

Promotion Tool-A promotion tool is the processes and media channels that are used to communicate messages to potential consumers of products or services.

Promotional Allowances-Promotional allowances are amounts of value that are given to offset a price or previously agreed upon term in return for promotional consideration (such as including a product in an advertising message).

Promotional Campaign-Promotional campaign is a set of marketing activities that are designed to send messages to customers about products, services, and options to meet specific marketing (promotional) objectives.

Promotional Item-A promotional item is an object or service that is given to a person or company to help build awareness, interest or provide convenient contact information. Examples of promotional items include pens, coffee mugs, or memory sticks.

Promotional Materials-Promotional materials are items or literature that can be provided to people or potential customers to promote action (such as buying products or services).

Promotional Product-A promotional product is an item or service that is offered or provided as part of a promotion program.

Promotional Schemes-Promotional schemes are a structure that is used for creating and managing offers for products or services.

Prompt-A cue to help the operator choose the next action.

Proof-A proof is a sample of an item that will be produced or distributed that is usually given as a final check to determine errors or to review how the item looks before it is released to production or broadcasting.

Proof of Performance-Proof of performance is the information or records that are provided that can be used to validate that services (such as broadcasting a television commercial) are correctly performed and billed.

Proof Of Performance Measurements-Proof of performance measurements is metrics that is gathered to validate the performance of a system or service. An example of proof of performance measurement is the gathering of radio field strength measurements at locations that can determine the radiation pattern characteristics of a radio transmission system.

Proof of Purchase-Proof of purchase is some item or detailed information received from the customer of a product that reasonably proves that they were the purchaser or owner of the product.

Proposal-A proposal is a presentation of an offer to provide products or services under defined terms and conditions.

Proprietary-Proprietary means owned or controlled by another person or company.

Proprietary Information-Proprietary information is data or materials that are owned or controlled by another person or company that may be harmful or perceived to be harmful if it is obtained by unauthorized recipients. Proprietary information may include customer lists, business processes, operational capabilities or technologies.

Proprietary Software-Proprietary software is files, programs, or applications that are privately owned and controlled by a person or company. Proprietary software is usually created or used for a specific purpose or company. Applications for specific devices or operating systems used for communications equipment (such as a mobile phone) may be considered proprietary even though many devices and applications may be available for them.

Proprietary Systems-Proprietary systems are equipment, assemblies, or networks that are unique to a specific manufacturer or company. While proprietary systems may contain components that are non-proprietary (industry standard), the interoperation of the components or systems may be unique and proprietary.

Prorating-The process of fractionalizing charges for a partial period. In order to determine the number of days for which to charge, a "multiplier" is used (See "Multiplier").

Prospect-A prospect is a person or company that is likely to be qualified to purchase products or services or to perform desired actions (such as volunteer for a group cause).

Prospect List-A prospect list is a group of contacts that have qualities or characteristics that indicate they are candidates to purchase products or services from a company or person.

Prospect Universe-A prospect universe is the total number of potential customers for a product or service.

Prospective Suppliers-Prospective suppliers are a group of companies or people who have characteristics that indicate that they are likely to offer or they are qualified to produce certain types of products or services.

Prosumer-A prosumer is a person or company who both uses and creates products or services. An example of a prosumer is a person who produces and consumes video clips.

Protest of Award-Protest of award is the process of requesting that an RFP issuer review an award selection for various causes such as asserted biased evaluation, errors and omissions or other criteria with basis. The process for protesting awards may be defined in an RFP and/or procedures may be defined by government regulations (such as when a contract is awarded from a government). The original RFP may define a time period that protest of awards must be filed after an award selection has been made.

Protocol-Protocols are the languages, processes and procedures that perform functions used to send control messages and coordinate the transfer of data. Protocols define the format, timing, sequence, and error checking used on a network or computing system. While there may be several different protocol languages that can be used for communications services, the underlying processes (setup and disconnection of communication sessions) are fundamentally the same.

Protocol Implementation Conformance Statement (PICS)-A protocol implementation conformance statement is a document that is provided by a company or testing facility that states that the product or system provides and/or supports a specific set of protocols.

Protocols-(1-rules) Protocols are a precise set of rules, timing, and a syntax that govern the accurate transfer of information between devices or software applications. Key protocols in data transmission networks include access protocols, handshaking, line discipline, and session protocols. (2-connection) A procedure for connecting to a communications system to establish, carry out, and terminate communications.

Prototype-A prototype is a device or product that has been produced for the purposes of demonstration and/or testing but is not made commercially available. Prototype units may use temporary components and/or parts that are not fully designed or are not durable.

Provisioning-Provisioning is a process that is used by a service provider or network operating to establish new accounts, activation and termination of features within these accounts and coordinating and dispatching the resources necessary to fill those service orders.

Provisioning Server-A server that coordinates the activation setup, authorization of features, and elimination of users from a communications system.

Proximity Services-Proximity services are information services that indicate the proximity of people or items to a specific location or locations.

Proxy-A proxy is a type of server (computer with specific application software) that is used to communicate with other devices or users on behalf of another user or device (a proxy).

Proxy Server-Proxy servers are computing devices (typically a server) that interface between data processing devices (e.g. computers) and other devices within a communications network. These devices may be located on the same local area network or an external network (e.g. the Internet). A proxy server usually has access to at least two communication interfaces. One interface communicates with a device requesting services (e.g. a client) and a device that is being requested for a service (the server).

ps-Picosecond

PSAP-Public Safety Answering Point

PSDN-Public Switched Digital Network

PSL-Pressure Sensitive Label

PSPDN-Packet Switched Public Data Network

PSTN-Public Switched Telephone Network

PSTN Fallback-The ability to use the public switched telephone network as a backup connection in the event that another network (such as a data network or the Internet) are not able to provide communication services. PSTN fallback is commonly used in voice over data communication systems.

PSTN Gateway-A PSTN gateway is a communications device or that transforms data that is received from one network (such as the Internet or DSL network) into a format that can be used by the PSTN network. The PSTN gateway may be a simple device that performs simple call completion and adaptation of digital audio into a compatible signals for the PSTN. Or, it may be a more complex device that is capable of advanced services such as conference calling, call waiting, call forward and other PSTN like services. The PSTN gateway must create signaling protocols and compensate for timing differences between a end users computer and the public switched telephone network (PSTN).

PTP-Point to Point

PTT-Post, Telephone And Telegraph

PTT-Push To Talk

Public--(1-general) Available to everyone (2-Bluetooth) A mode of operation whereby a device can be found via Bluetooth baseband inquiries; that is, it enters into inquiry scans. A public device also enters into page scans.

Public Data Network (PDN)-Public data networks interconnect data communication devices (e.g. computers) with each other through a network that is accessible by many users (the pubic). To allow many different users to communicate with each other, standard communication messages and processes are used. The Internet is an example of a public data network (there are other public data networks) that uses standard messages and message transfer processes (Internet protocol) to allow anyone to transfer data from point to point by using data packets. Each transmitted packet in the Internet finds its way through the network switching through nodes (computers). Each node in the Internet forwards received packets to another location (another node) that is closer to its destination. Each node contains routing tables that provide packet-forwarding information.

Public domain-Products which may be expired or unable to be copyrighted, therefore making them accessible to the public.

Public Internet Address-An Internet address that can be recognized and transferred (routed) on the public Internet.

This figure shows that Internet address number formats are divided into large networks,

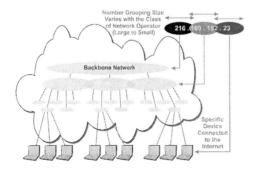

Public Internet Address

smaller networks, and local networks. This example shows that an Internet address only identifies a specific device (data connection point) within the Internet. Because devices can be connected to the Internet anywhere in the world, the Internet address does not identify a specific location.

Public key-A public key is a cryptographic key that is made available to the public for the purpose of encrypting messages to add privacy (security) to messages that are sent and received.

Public Key Certificate-A public key certificate is information that is encapsulated in a file or media stream that identifies that the media has been originated by a specific person or device by binding them (validating them) using public key information from a certificate authority (CA).

Public Key Encryption (PKE)-Public key encryption is an authentication and encryption process that uses two keys, a public key and a private key, to setup and perform encryption between communication devices. The public key and private keys can be combined to increase the key length provider and a more secure encryption system.

Public Key Infrastructure (PKI)-Public key infrastructure is the verifying and authenticating users through the use of systems that can be accessed and used by the public through a system of trusted entities. The PKI system uses trusted certificate authorities (CAs) that publish public keys and security certificates that can be validated by multiple parties.

Public Limited Company (PLC)-A public limited company is a business entity that can be formed in the UK that is owned by shareholders.

Public Network-A public network is a communication system that is operated by common carriers, local and long-distance, that is made available for private users and the public.

Public Notice-Public notice is a message that

is communicated to the pubic in general (usually through advertising or public media channels) that alerts interested parties that a change in regulation is occurring.

Public Packet Data Network (PPDN)-A packet data network that is generally available for commercial users (the public). An example of a PPDN is the Internet.

Public Packet Switched Network (PPSN)- An exchange access capability that provides packet-switched data transport for terminals, value-added networks, host computers and interexchange carriers.

Public Safety Answering Point (PSAP)- Public safety answering points (PSAPs) are facilities that receive and process emergency calls. The PSAP usually receives the calling number identification information that can be used to determine the location of the caller. The PSAP operator will then initiate and/or route calls to assist with the emergency situation.

Public Switched Digital Network (PSDN)-A high-speed data network that uses a public switched telephone system to link users over greater distances than can be provided by local area networks.

Public Switched Telephone Network (PSTN)-Public switched telephone networks are communication systems that are available for public to allow users to interconnect communication devices. Public telephone networks within countries and regions are standard integrated systems of transmission and switching facilities, signaling processors, and associated operations support systems that allow communication devices to communicate with each other when they operate.

This diagram shows a basic overview of the Public Switched Telephone Network (PSTN) as deployed in a typical metropolitan area. PSTN customers connect to the end-office (EO) for telecommunications services. The EO processes the customer service request locally or passes it off to the appropriate end or tandem office. As different levels of switches

interconnect the parts of the PSTN system, lower-level switches are used to connect end-users (telephones) directly to other end-users in a specific geographic area. Higher-level switches are used to interconnect lower level switches.

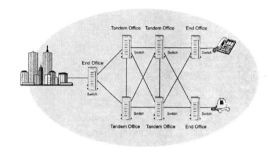

Public Switched Telephone Network (PSTN)

Public Telephone-A telephone provided by a telephone company through which an end user may originate interstate or foreign telecommunications for which he pays with coins or by credit card, collect or third party billing procedures.

Public Utilities Commission (PUC)-A state regulatory body that sets rates and rules for local exchange carriers and interexchange carriers that provide long distance (interLATA) service within a state boundary. In some states, it is called a public service commission (PSC), board of public utilities (BPU), or corporation commission (CC).

Public Warehouse-A public warehouse is a building or facility where goods from different people or companies can be received, stored and shipped.

PUC-Public Utilities Commission

Pull Notification-Pull notification is when a device (such as a wireless telephone) polls the server for any new events or information. This figure shows how pull notification works with a WAP server. This example shows a

P

WAP push proxy gateway that receives email messages that are addressed to the WAP client. The push proxy gateway stores these messages until it receives a request from the WAP client for the delivery of messages. The WAP client will then download (pull) the messages from the push proxy gateway so the messages can be displayed on the users phone.

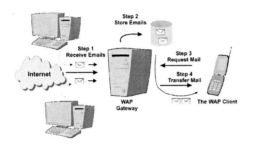

WAP Pull Notification

Pull Video On Demand (Pull VOD)-Pull video on demand is the downloading of video content after the consumer has requested it.

Pull VOD-Pull Video On Demand

Pulse Dialing-A dialing process that uses momentary pulses to transfer a telephone number from a telephone (rotary telephone) to a telephone system.

Purchase Manager-A purchase manager is a program or service that coordinates how users and their devices are used to purchase products or services. A purchase manager may be responsible for selecting and ensuring secure payment methods and transactions.

Purchase Order (PO)-A purchase order is a document or record from a person or company that authorizes the sale of products or services to that person or company.

Purchasing System-A purchasing system is a set of accounts, rules, processes, software programs, equipment, and other mechanisms that are used to identify suppliers, materials and services that need to be ordered, issuing purchase orders, tracking receipts of materials and services, and issuing payments for these materials and services.

Purge-Purging

Purging (Purge)-Purging is the process of removing multiple data entries or erroneous information items from a file or list. Purging mailing lists may include finding and removing invalid or old contact information to ensure the list is current and more accurate.

Push Notification-Push notification is when the server contacts the device (such as a wireless telephone) without any request coming from the device, and then pushes the information down to the device. One of the most common examples of push notification is when the server notifies the wireless device that a new email has arrived.

Push On Demand (POD)-Push on demand is a process of initiating the transfer of content or control to another communication that is typically started when an event occurs or when a specific condition exists. An example of push on demand is the sending of program media from a television content provider (e.g. a television show) to networks so the networks can either immediately broadcast the media or they can store it for broadcasting at a later time.

Push Technology-The ability of a communication user to push content or control to another communication user. Push technology is commonly used in web seminar (webinar) systems where an instructor can push presentations to others who are participating in a webinar session.

Push To Talk (PTT)-Push to talk (PTT) is a process of initiating transmission through the use of a push-to-talk button. The push to talk process involves the talker pressing a talk button (usually part of a handheld microphone) that must be pushed before the user can transmit. If the system is available for PTT service (other users in the group not talking), the talker will be alerted (possibly with an acknowledgement tone) and the talker can transmit their voice by holing the talk button. If the system is not available, the user will not be able to transmit/talk.

Push to Talk Over Cellular (POC)-Push to talk over cellular is the providing of group calling services over a cellular system. PoC allows a user to request to talk to members of a group by pushing a "Talk" button. If the system is available to allow the PoC service (e.g. if it is not busy serving other PoC customers), the "talker" will begin to transmit to the system and the system will copy and distribute the audio signal to all members of the group.

Push Video on Demand (Push VOD)-Push video on demand is the downloading of video content that is not previously requested by the consumer so that it can be immediately available if the user selects to watch the program.

Push VOD-Push Video on Demand

PV-Page Views

PVC-Permanent Virtual Circuit

PVR-Personal Video Recorder

Pyramiding Late Fees-Pyramiding late fees is the charging of late fees from a previous invoice period so that the payments applied are not enough to cover the current late fees resulting in the assessment of additional late fees.

Q

QA-Quality Assurance
QBE-Query By Example
QC-Quality Control
QoE-Quality of Experience
QoR-Quality of Restitution
QoS-Quality Of Service
QoS Policy-Quality of Service Policy
QoSM-Quality of Service Metrics

Quadruple Play-Quadruple play refers to providing four main services such as data, voice, video and electrical service by one company. For telephone companies, this sometimes refers to the providing of fixed telephone, broadband Internet, television and mobile voice.

Qualification-Qualification is the validation of characteristics that indicate a person, company or other item meets certain requirements.

Qualification Requirements-Qualification requirements are a set of criteria that is used to determine if a person or company is likely to be able to perform or provide services or products.

Qualification Test-A qualification test is one or more tests that are performed on products or services to ensure they qualify to be used with and/or attached to a network or system.

Qualification Worksheet-A qualification worksheet is a data gathering form that is used to gather information that determines if a person or company is qualified (or likely) to purchase a product or service.

Qualified Vendor-A qualified vendor is a company or supplier who has met certain criteria to authorize them to be a supplier. Examples of vendor qualification include ISO-9000 compliance, financial performance and support systems.

Qualifier-A qualifier is a condition or value that is used to determine if or how an action or process can proceed.

Qualifying-Qualifying is the process of researching and analyzing of prospects to determine if they have characteristics that indicate a need for your product or services.

Quality Assurance (QA)-All those activities, including surveillance, inspection, control, and documentation, aimed at ensuring that a given product will meet its performance specifications.

Quality Control (QC)-A function whereby management exercises control over the quality of raw material or intermediate products to prevent the production of defective devices or systems.

Quality of Experience (QoE)-Quality of experience (QoE) is one or more measurement of the total communications experience or the entertainment satisfaction from the perspective of the end user. QoE measures may include service availability, audio and video fidelity, types of programming and the ability to use and the value of interactive services.

Quality of Restitution (QoR)-Quality of restitution (QoR) is one or more measurement of ability of a system to adjust or continue to operate at a specific performance level for a range of impairments.

Quality Of Service (QoS)-Quality of service (QoS) is one or more measurement of desired performance and priorities of a communications system. QoS measures may include service availability, maximum bit error rate (BER), minimum committed bit rate (CBR) and other measurements that are used to ensure quality communications service.

Quality of Service Metrics (QoSM)-Quality of service metrics is one or more parameters that are used to define acceptable or desired performance for systems and services. Examples of QoSM include latency, jitter, peak throughput and error rate.

Quality of Service Policy (QoS Policy)- Quality of service policy is a set of rules that are used to coordinate the allocation of network resources based on predetermined policies on the priorities and resources required by communication services and applications within the network. A QoS policy server is used to help manage network operation in the event of loss of resources. The pre-set policies define which communication services are critical (such as video or voice) and how many resources should be allocated to these critical services at the expense of other communication services (such as web browsing).

Quantity Discounts- Quantity discounts are cost reductions that are provided to a purchaser for goods or services in return for ordering higher quantities of products or services. Quantity discounts may be defined in a discount schedule.

Quantization- Quantization is the process of representing sample signals with a range if defined values. In analog-to-digital conversion, a continuous analog value is represented by one of a finite number of quantized values. In lossy signal compression (such as an A-law encoding) one digital value is represented by another one which is usually not precisely the same. Except in lucky cases where the quantized value is exactly the same as the original, quantization introduces error (or noise).

This figure shows how the audio digitization process can be divided into sampling rate and quantization level. This diagram shows that an analog signal that is sampled at periodic time intervals and that the each sample gets a quantization level. This diagram shows that the quantization level may not be exactly the same value of the actual sample level. The closest quantization level is used and this causes quantization distortion.

Quarantined Connection- A quarantined connection is a communication link (such as an Internet connection) that can only access a limited number of connection points.

Query- A request for information that matches or approximates a specific criteria (e.g. search word).

Query By Example (QBE)- A database front-end routine that requests the user to supply an example of the type of data to be retrieved.

Query Language- A query language is a software programming language containing commands that allow users or software applications to retrieve and/or process information from databases.

Question Mark (?)- A question mark is a character that is used as a wildcard (substitution for any character) when searching a computer for a filename or filename extension.

Queue- A queue is a list of people, tasks or software instructions that are waiting to be processed or served.

Queue Delay- Queue delay is an amount of time it takes for tasks or people waiting to be processed or served to when they are provided with services or actions.

Queuing- Queuing is a process of delaying or sequencing messages. Queuing involves receiving requests for service, prioritizing these requests, storing them in appropriate order and transferring the messages when the facilities (channels) are available to send them.

Queuing systems may change the order of messages or services to be provided based on priority access. For example, communication requests from a public safety official may be given priority over a communication request from a consumer.

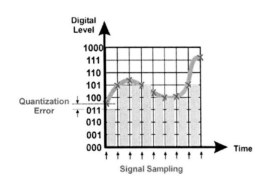

Audio Sampling and Quantization

Queuing Scheme-Queuing scheme is the set of rules and process that are used to identify and priorities tasks, processes or transfer of data or information.

Quid Pro Que-Quid pro que is the providing of something in response to receiving something.

Quotation (Quote)-A quotation is a set price or value that is required for a person or company as compensation for products or services that are offered for sale.

Quote-Quotation

Quoting-(1-general) Quoting is the copying and use of other materials. (2-business) The providing of a cost estimate for providing a product or service.

R

RA-Registration Agency

Rack-An equipment rack, usually measuring 19 inches (48.26 cm) wide at the front mounting rails.

RAD-Rapid Application Development

Radio Access-A method or technology used to coordinate access to radio channels or portions of radio channels.

Radio Access Network (RAN)-The radio access network is a portion of a communication network that allows individual subscribers or devices to connect to the network via radio signals.

Radio Advertising-Radio advertising is the sending of promotional messages or media content to one or more potential program listeners.

Radio Coverage-Radio coverage is a geographic area that receives a radio signal above a specified minimum level.

Radio Local Loop (RLL)-Radio local loop (RLL) is the providing of local telephone service via radio transmission. Radio local loop systems often use a radio conversion device located at the home or business to allow the use of standard telephones. Although RLL systems may provide for traditional dialtone service, RLL systems commonly provide for multiple types of services such as telephone service, Internet access, and video programming.

Radio Telephone Messages Telecommunications Service Tariff-A tariff that applies to a radiotelephone message telecommunications service is furnished through land radiotelephone base stations of an AP or in conjunction with other carriers and to customer, otherwise known as FCC Tariff No.3.

Radio Tower-Radio towers are poles, guided towers or free standing constructed grids that raise one or more antennas to a height that increases the range of a transmitted signal.

Radio towers typically vary in height from about 20 feet to more than 300 feet.

Radio Transceiver-A radio transceiver is a combination of a radio transmitter and receiver into one radio device or assembly. A portable cellular phone is a radio transceiver.

RADIUS-Remote Access Dial In User Service

RAID-Redundant Array of Inexpensive Disks

Raising Capital-Raising capital is the process of finding potential sources of resources (investors), providing with them with information on why they provide money and getting them to sign loan and/or stock purchase agreements.

RAM-Random Access Memory

RAN-Radio Access Network

RAND-Reasonable and Non-Discriminatory

Random Access Memory (RAM)-A data storage device, often an integrated circuit, from which a word of data can be removed without regard for the time it was stored or its location. Internal semiconductor RAM, static or dynamic, is volatile in nature and does not retain the data stored within if power is removed from the device.

Random Errors-Random errors are bits in a received digital signal that are received in error that occur in such a way that each error can be considered statistically independent from any other error.

Ranking-The position of items within a list based on specific criteria such as popularity of an item or relevance to the terms in a specific query.

RAO-Revenue Accounting Office

RAP-Returned Account Procedure

Rapid Application Development (RAD)-Rapid application development is the use of tools or processes that can speed up the creating of software programs or applications.

RAS-Remote Access Server

Rate-The price charged a customer for a particular service, as specified and defined in a tariff.

Rate Adapter-A rate adapter is a unit or process that converts the data transmission rate from a system or communication line to the data transmission rate of another system or communication line.

Rate Band-A rate band is comprised of the combination pair of originating and terminating numbers where the "banded rating" amount (rate) is fixed within the same band. International calling is an example of banded rating (e.g. all calls from anywhere in the US to anywhere in France will carry the same rate.)

Rate Card-A rate card is a fee schedule for services. For publishing, a rate card is the fees that are charged for the insertion of ads into magazines, journals or other media publication.

Rate Center-As defined by rate map coordinates, the point within an exchange area that is used as the primary basis for determining toll rates. Rate centers also can be used for determining selected local rates. A rate center also is called a rate point.

Rate Class-A rate classification determined by the type of message, station, or customer, and the rate in effect at the originating rate center at the time a telephone call begins.

Rate Increase-Any change in a tariff which results in an increased rate or charge to any of the filing carrier's customer.

Rate List-A rate list is a schedule (listing) of items and their associated (assigned) rates (costs).

Rate Management-Rate management is the assignment, monitoring and adjustment of equipment configurations and resources to achieve or to change in the direction of desired rates. An example of rate management is the adjustment of lossy data compression (higher or lower compression) in a digital video system to maintain or limit the maximum data trans-

mission rate so it can be sent on a transmission channel that has a fixed transmission rate.

Rate of Exchange-Rate of exchange is a value that is used to convert from one unit to another unit.

Rate Plan-A rate plan is the structure of service fees that a user will pay to use services. Rate plans are typically divided into monthly fees and usage fees.

Rate Shaping-Rate shaping is the identification, categorization and prioritization of the transfer of data or information through a system or a network to match user requirements with network capacity and service capabilities.

Rate Table-A rate table is a group of billing codes or product identifiers and their associated billing rates and characteristics that are used to calculate the charges for billing events or billing records.

Ratification-Ratification is the process of reviewing and approving a proposal or contract.

Rating-Rating is a function within the billing system that assigns a rate (cost parameter) to a usage record. Rating typically involves using the originating number or network address, terminating number or destination network address, date the service was used, amount of usage or time period, usage type and tax jurisdiction to determine the initial charge assigned to the usage record. The actual cost of the usage record may be adjusted based on volume discounts or other rate plan considerations.

Rating Elements-Rating elements are the functional parts of a billing system that enhance usage records with by adding or modifying usage rates.

Rating Engine-A rating engine is a function within a billing system that assigns the charging rates to a usage event or call detail record (CDR).

This figure shows the how a rating engine of a billing system can calculate the fees associated with specific usage events. This diagram

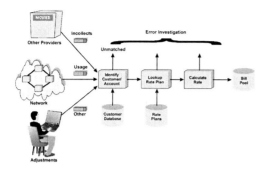

IPTV Billing Record Rating

shows that the rating engine receives usage records from network usage, incollects from other companies and other billing records (such as adjustments). The rating engine first identifies the account associated with the usage records and checks for duplicate records. The rate table for the customer account is selected for these records and the appropriate fee is calculated for each record. After the calculated rate is added to the usage detail record, it is sent to the bill pool to await the processing of invoices.

Rating Points-Rating points are values assigned to evaluation criteria (such as credit score). Each element that may be evaluated is assigned points and the total of these points is used to in the evaluation of a process (such as a credit application).

Rating Scheme-A rating scheme is a structure that is used for the assignment of values to services that are used or offered for sale.

Rating System-Ratings systems measure usage or viewer habits. Ratings systems such as Neilson ratings identify the percentages of households that watch specific programs.

Ratings Based Sales-Ratings based sales is the offering of services (such as advertising) where the value or cost of the services is determined by the popularity or subscription to the services.

Raw Materials-Raw materials are the fundamental elements or products that are used to produce a product or service.

RC-Recurring Charge

RCF-Remote Call Forwarding

RDBMS-Relational Database Management System

RDC-Regional Data Center

Reach Extended Asymmetric Digital Subscriber Line (RE-ADSL)-Reach extended ADSL is an improvement to ADSL systems that optimizes the allocation of DMT transmission channels to extend the reach of ADSL systems. This increases the maximum distance that an ADSL system can operate.

Reaction Time-Reaction time is the elapsed time between the occurrence of an event (such as a service interruption problem) and when the processes to begin correction activities begin to occur.

Reactivation-Reactivation is the process of inputting specific information into a system or communication network to re-authorize account usage or to re-initiate a service.

Read Only Memory (ROM)-A memory circuit or device in which any address can be read from, but not written to, after initial programming. The ROM is an asynchronous device with an access time dictated by internal circuit time delays. Semiconductor ROM storage is nonvolatile and retains data when power is removed.

Readout-A visual display of the output of a device or system.

RE-ADSL-Reach Extended Asymmetric Digital Subscriber Line

Real Time-Real time actions are processes (such as transmission) that occur instantly or within a time period that is perceived or used (such as within a few seconds) to perform or record events when they are required or used.

Real Time Billing-Real time billing involves the authorizing, gathering, rating, and posting of account information either at the time of service request or within a short time afterwards (may be several minutes). Real time billing is primarily used for prepaid services such as calling cards or prepaid wireless.

This figure shows how real time billing for Internet telephone service commonly allows the customer to display their billing records immediately after they are created (in real time). This example shows how the ITSP keeps track of each call as it helps to connect the call and to process advanced features. It uses the call setup and termination information to adjust your bill and in this example, these charges can be displayed immediately through an Internet web page.

This information is sent back to the gateway and the gateway completes (connects) the call. During the call progress, the gateway maintains a timer so the caller cannot exceed the maximum amount of time. After the call is complete (either caller hangs up), the gateway sends a message to the real time rating system that contains the actual amount of time that is used. The real time rating system uses the time and rate information to calculate the actual charge for the call. The system then updates the account balance (decreases by the charge for the call).

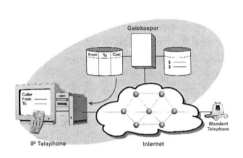

ITSP Real Time Billing Operation

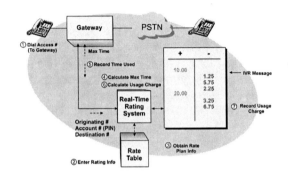

Real Time Billing Operation

This figure shows a real time prepaid billing system. In this example, the customer initiates a call to a prepaid switching gateway. The gateway gathers the account information by either prompting the user to enter information or by gathering information from the incoming call (e.g. prepaid wireless telephone number). The gateway sends the account information (dialed digits and account number) to the real time rating system. The real time rating system identifies the correct rate table (e.g. peak time or off peak time) and inquires the account determine the balance of the account. Using the rate information and balance available, the real time rating system determines the maximum available time for the call duration.

Real Time Payment Processing-Real time payment processing is the tasks and functions that are used to collect payments from the buyer of products and services when they process their order or when they submit their payment for it.

Real Time Rating-Real time rating is a function within the billing system that assigns a rate (cost parameter) to a usage record where the cost of the usage is calculated immediately or within a short time after the usage occurs.

Reason Code-A code that indicates a reason that the status of a message or communiation session has changed. An example of a reason code is a message that indicates that a communication session has ended due to a user (caller) disconnecting by hanging up a telephone.

Reasonable and Non-Discriminatory (RAND)-Reasonable and non-discriminatory are licensing terms (fees and restrictions) offered by owners of intellectual property that allows other companies to build products that incorporate their technology while providing the potential for sustainable profits. Companies or standardization bodies, such as ETSI, may require that members who have patented inventions, which are incorporated into published standards agree to make licenses available to anyone who wants to use the standard on fair, reasonable, and non-discriminatory terms.

Rebate-A rebate is a value provided in return for purchasing a product or service that meets certain conditions. Rebates often require the submission of a coupon or rebate submission form so the buyer can qualify and receive the rebate.

Reboot (Reset)-To restart an electronic assembly or computer.

Receipt-A receipt is a document or data that indicates that a product or service has been received. Receipts may indicate the payment status such as balance due or paid in full.

Receivables-Receivables are value amounts that have been earned but payment has not been received.

Receiver Off-Hook Tone-A tone that is provided when the phone is off-hook.

Recency-Recency is a measure or indication of the amount of time that has elapsed since information has been gathered or validated.

Recharge-(1-battery) To store energy in a battery. (2-calling card) A process of adding additional time to a calling card or prepaid wireless account using either a credit card or debit card.

Reciprocal Compensation-Reciprocal compensation is the process of accessing similar costs for similar services. The 1996 Telecommunications Act mandated that local telecommunications companies exchange rev-

enue for the cost of terminating calls that originated on the wireline network. Previously, only wireless companies were obligated to pay compensation for calls originated on their networks but terminated on the wireline network.

Reciprocal Link-A reciprocal link is a link on one web page that is provided in return for the insertion of a link on another page. The benefits of reciprocal links is the increasing of web traffic between pages and a potential increase the ranking of search engines for the linked web pages.

Recomputation-Recomputation is the process of performing calculations on information or data (such as accounting records) to correct errors or to apply new or updated rules or formulas.

Recon-Reconnect

Reconciliation-Reconciliation is the process of comparing records or data from multiple sources to confirm the completeness or accuracy of the data. An example of reconciliation is the comparison of checks written on an account to transaction statements provided by the bank that manages the account.

Reconnect (Recon)-In outside plant construction, the moving of subscriber dropwires onto new cable facilities as engineered on a work order. This often relieves congestion at poles and terminals in addition to improving quality of service for customers.

Reconnection Charges-Reconnection charges are fees that are assessed for the connection of a service that was previously activated. Reconnection charges may be lower than new service activation charges because much of the customer information is already available which reduces the amount of resources that are required to reactivate service.

Reconstruction-Reconstruction is the repairing or processing of data or information so that it is converted back into its original form. An example of reconstruction is the replacement of a shortened temporary address with a complete address.

Record-An assemblage of data elements, all of which are in some way related and handled as a unit.

Record Handling-Record handling is the process of filtering, guiding (matching) and rating of billing records.

Record Validation-Record validation is the process of reviewing the data that is stored in a data record (such as the billing record) to determine if conforms to rules, formats or correctly relates to other information.

Recording Time-Recording time is the duration a measurement (or series of measurements) is made.

Recovery-Recovery is a process of reconnecting a communication session, restarting a system or reusing previously discarded or unused equipment or facilities.

Recovery Detection-Recovery detection is the process of gathering information that can confirm that a process or communication function that has been previously disabled has been re-enabled.

Recovery Plan-A recovery is set of actions or processes that may be used to assist in the reconnecting or restarting of an operation, system or service in the event of trouble or failures.

Recurring Charge (RC)-A predetermined charge associated with a product or service that is assessed on a regular interval (i.e. monthly, quarterly, annually).

Redundancy-Redundancy is a system design that includes additional equipment for the backup of key systems or components in the event of an equipment or system failure. While redundancy improves the overall reliability of a system, it also increases the number of equipment assemblies that are contained within a network. Redundancy usually increases cost.

Redundancy Plan-A network structure plan that defines the alternate equipment configurations and routes that are used when specific failures occur.

Redundant Array of Inexpensive Disks (RAID)-Redundant array of independent disks (RAID) is a computer information storage system that uses multiple independent disk storage devices (hard disks). RAID systems were first defined in 1988 and they were originally called redundant array of inexpensive disks. RAID systems can use standard hard disk drive interfaces such as SCSI or Integrated Drive Electronics (IDE).

RAID systems can be configured in various ways to ensure data integrity and to provide the ability to remove and replace disks while the system continues to operate in the event of equipment failure ("hot-swap".) The different types of configurations are called RAID levels. There were six original levels (RAID 0 through RAID 5). Several manufacturers have combined the RAID levels to produce new unique levels above RAID 5.

For RAID 0, the data is distributed (striped) over several drives but there is no redundant storage of data. RAID 1 combines two hard disks of equal storage capacity that simultaneously store the same information (mirror). RAID 2 allows data to be corrected on one drive from another drive by interleaving (distributing) the information across multiple drives and using a data protection formula (algorithm) that relates the information. RAID 3 stores information on several disk drives where only one drive is used for parity bits that are used for error detection and correction. RAID 4 systems store data in multiple drives without the need to mirror information at the bit level. RAID 5 writes data and parity information on the same disk but to different sectors to increase the data transfer performance.

Redundant Server-The inclusion of a second communication server that allows the automatic transfer of service (reconfigure) in the event a failure occurs on part or all of the other server. In a redundant server, an equipment failure should cause no loss of data or information.

Redundant System-A redundant system is a duplicate configuration of equipment in communication systems that allows some of the equipment to automatically reconfigure in the event a failure occurs on part or all of the duplicated system. In a redundant system, a failure should cause no loss of data or information.

Reference Model-A reference model is a representation of a product, system or process in a form that can simulate or assist in the understanding of the operation and estimation of likely results provided specific inputs or events that occur to the product, system or process.

Reference Point (RP)-A reference point is a conceptual point in a system or network that defines information or processes that flow through it.

Referral Premium-A referral premium is a value incentive (such as a gift or discount) that is provided to customers who refer a friend or acquaintance to purchase a product or service.

Refund-Refunds are credits that are applied to an account or invoice to reverse a previously received payment.

Refundable Retainer-A refundable retainer is a value that is provided to a person or company in advance of the receiving of products of services which is refunded (or the unearned balance is refunded) if the services charged are less than the retainer fee.

Refurbished-Refurbished is a device or system that has been repaired, modified or upgraded. Refurbished equipment may have been returned for repair or refund.

Regional Data Center (RDC)-A regional data center is a collocation facility that allows a carrier interconnection point-of-presence (POP) within a rate center to provide information services.

Regional Enterprise Network (REN)-A regional enterprise network is a subset of the enterprise network (EN) that interconnects corporate offices located within the same city.

Registered User-A registered user is a person or company that is on a list or in a database that is used to authorize the use products or services.

Registrar Operator-A registrar operator is a company, organization or person who operates a system that receives processes and manages registration requests (such as domain name registrations).

Registration-(1- FCC) A legally required procedure whereby vendors must submit their telephone equipment for testing and certification before it can be directly connected to a public telephone network. (2-wireless system notification) A process where a mobile radio transmits information to a wireless system that informs it that it is available and operating in the system. This allows the system to send paging alerts and command messages to the mobile radio. (3-VoIP) The process of an Internet telephone terminal registering with a gatekeeper. (4-Internet) The process of a web site visitor entering information into a web site to record their contact information (name, address, and email address).

Registration Agency (RA)-A registration agency is a company or entity that receives, stores and manages registrations for products or services.

Registration of Interest-Registration of interest is the providing of a request to a potential supplier or facilitator that they are interested in certain types of products or services.

Registries-A registry is a company, entity or agency (such as an industry association or government agency) that receives, stores and manages registrations for products or services.

Registry-(1-computer) A database of information that contains computer's hardware and software information that is used by it's operating system to manage communication between hardware and software applications. Information store in the registry is usually updated automatically as configuration information changes through the addition, removal, or modification of hardware and software controlled by the system. The registry database holds information that was commonly stored in initialization files (.INI) files. (2-security) A registry is a company, entity or agency (such as an industry association or government agency) that receives, stores and manages registrations for products or services.

Regression Analysis-A forecasting method used where the factor to be predicted may be expressed as a function of one or more variables. The variables themselves may have to be forecasted.

Regulation-(1-electrical quantity) A process of adjusting the parameters of some signal or system (such as circuit gain). (2-government) Rules established by a government agency that are designed to maintain the service public communications systems.

Regulatory Fees-Regulatory fees are taxes, assessments or other amounts levied by regulatory authorities for the authorization to develop, operate or change systems or services that are overseen by the government.

Reinstatement-Reinstatement is the reactivation of the provisioning of a system or service to enable a user or device to have access to systems, services or authorization levels they have been previously assigned.

Reject Processing-Reject processing is the selection, organization and routing of records or data that could not be processed.

Relation-A relation is a link, process and/or criteria that exist between elements or data tables. A relation is created by specifying the properties and processes that are used to link one type of object to another.

Relational Database-A relational database is a set of record files (tables) which have their records and/or fields within the records linked (related) to other tables, records or values by logical connections or formulas.

Relational Database Management System (RDBMS)-A relational database management system is a collection of hardware and software that enables users to setup controls (relationships) that can access, organize and process information or data.

Relational Model-Relational model is a set of links, formulas or processes that are defined between database elements that are used to organize and process data or information.

Relational Systems-Relational systems is the combination of equipment and software that allows for the gathering, accessing and updating of sets of record files (tables), which where their records and/or fields within the records can be linked (related) to other tables, records or values by logical connections or formulas.

Relaunch-Relaunching is the releasing of a previously released product or service. Relaunching may be performed when changes have been made to the previous product or service.

Release-(1-communications) A signal that indicates that a line or circuit has been released from use. (2-product) A release is the process of distributing or the making a product or service available.

Release Form-A release form is an agreement that authorizes the use of media, data or content (release of rights). Release forms usually define the information that will be used and how it will be used.

Release Management-Release management is the process of identifying, distributing and maintaining the products or services that have multiple versions.

Release Window-A release window is the scheduled time period that a product or service will be released. An example of a release window is the making available a movie for video rental distribution a specific time period after the movie has been initially released.

Remediation-Remediation is the correction of a condition or problem. For communication systems, remediation may involve the dispatching of a qualified technician or field service representative to validate, configure or correct faulty installations.

Remote Access-Remote access is the ability of a device, user or system to communicate or control devices, services or systems at locations outside the boundaries of the device or system they are controlling.

Remote Access Dial In User Service (RADIUS)-Radius is a network protocol that can receive identification information from a potential user of a network service, authenticates the identity of the user, validates the authorization to use the requested service and creates event information for accounting purposes. RADIUS is specified in RFC's 2138 and 2139, RADIUS is a client/server protocol that uses UDP.

Remote Access Server (RAS)-Remote access server (RAS) is software that is a part of Microsoft Windows NT Server that allows remote users or devices to connect to a server and access resources through a data network connection (such as through the Internet or through a data modem).

Remote Administration-Remote administration is the processes and tasks that are performed by a system administrator to add, change, or disconnect devices, features and equipment from locations other than the general location of the system they are controlling. Remote system administration tasks may include monitoring the performance of the network and making configuration adjustments to the network as necessary.

Remote Call Forwarding (RCF)-A service offering in which calls to a given directory number are redirected to another number, often in another city. Also, a service that enables a customer to invoke call forwarding whether the customer is at home or away. Control normally is limited to direct local access to a line from which forwarding is to be authorized. Remote control is invoked by calling a shared number at a home office, entering a PIN (personal identification number), then processing as if local control applied.

Remote Diagnostics-Remote diagnostics is a process and/or a program that is built into a device or network component that allows users or devices from remote locations to access information or control operations that can be used to discover the functionality or performance of a device or system connection.

Remote Line-A service that forwards calls from a distant location (such as a well known city) to a local (remote) telephone number. Remote line service is often used for companies to indicate a presence in a leading city while having their facilities in a remote location.

Remote Procedure Call (RPC)-A remote procedure call is a process that is used to request and control processes an on remote or distributed computer systems. Remote procedure calls are described in RFC-1057.

Remote Purchase Record Collection-Remote purchase record collection is the process of storing billing record transactions at remote locations and transferring these records back to a billing system for further processing. The initiation of the transfer process may be periodically (sent at regular intervals) transferred on request by the billing system (e.g. polling) or sent when communication links become available (such as when a dial connection is setup).

Remote Station-A remote station is a terminal or device that is physically located at a remote location from a main system or computer that uses a communication channel to access the system or computer.

Remote Support System (RSS)-Remote support systems are the services and/or equipment that can be provided by a supplier, person or company to maintain the operation of a product or service via a remote connection or location.

REN-Regional Enterprise Network

Rendering-Rendering is the process of converting media into a form that a human can view, hear or sense. An example of rendering is the conversion of a data file into an image that is displayed on a computer monitor.

Renewability-Renewability is a process or ability of a system to update or adding value to a device, card or system.

Renewal License-A renewal license is authorization for an extended period of time that a company can operate and/or provide services on an existing system (such as a cable television system).

Renewing Certificate-A renewing certificate is an authorization command and necessary information (such as decryption keys) that enables or continues the allowance of use of media or information.

Rental Window-A rental window is a scheduled time period that a product or service will available for viewing or use.

Renumeration-Renumeration is the paying or providing of value for the receipt of products or services.

Reorder Points-Reorder points is a quantity level of inventory that indicates that the product should be reordered. When the inventory quantity falls below the reorder point, it may trigger an automatic ordering of additional products.

Reorder Tone-A tone applied 120 times per minute indicating that all switching paths or trunks are busy. The reorder tone also is called a channel busy or fast busy tone.

Rep Firm-Sales Representative Firm

Repair Service Bureau-A repair service bureau is a group that coordinates the repair of products or services for customers.

Repeat Customer-A repeat customer is a person or company who orders of products and services again from the same company or supplier.

Repeat Dialing-A service that automatically dials the last dialed number a repeated number of times or until an event occurs.

Reply Messages-Reply messages are commands or data that is sent from a device or user in response to the reception of a message (such as a banner ad). Reply messages can range from simple acknowledgement commands to opening a voice communication channel (initiating a telephone call).

Report Program Generator (RPG)-A report program generator is a software application or service that can create output data according to user specified formats. RPG options may include heading, style and data location.

Reported Minutes-Reported minutes are the amount of time that a communication session occurs between users. Reported minutes may be rounded to the next higher or lower number.

Reporting System-A system that provides information about specific events (such as call or service usage). The reporting system usually provides details on the cause of the event (user identification), type of event (telephone call type), and amount of event (minutes or amount of data transferred).

Reporting Template-A reporting template is a set of information elements (data) that are formatted into a structure (such as a tables or graphs).

Reports-A report is a display of information that represents specific aspects of data.

Repository-A repository is a storage system that is used to hold media or data.

Reprint-Reprinting

Reprinting (Reprint)-(1-media) Reprinting is a service that provides media items (such as a magazine) that are reprinted for a particular use. Reprints are commonly segments of the original media item (such as a magazine cover and a single article). (2-billing) The reprinting of a bill or billing records.

Repudiation-Repudiation is the refusal to pay or provide value for a product or service. Repudiation may be performed when a product or service does not meet the requirements or expectations of the purchaser.

Repudiation Attack-A repudiation attack is processes that are used to deny services to others that are authorized and should be enabled to access services. An example of a repudiation attack is the blocking or redirecting of information transfer that is necessary to enable access to services (such as login messages or access to supporting data files).

Request-A request is an instruction from a device or user to select, enable or to receive a product or service.

Request For Budgetary Proposal (RFBP)- A statement of requirements and technical specifications designed to provide information on the costs of products or services from suppliers.

Request For Comments (RFC)-A requirement or draft standard document created by a standards body that solicits comments from manufacturers, carriers and industry experts to finalize the standard. When used by the Internet engineering task force (IETF) every major Internet Protocol is specified first by an RFC. There are many RFC documents available and they are a significant method used to define Internet protocols and technical standards. Because RFCs are commonly updated or revised, they are sometimes called "Requiring Further Correction."

Request For Information (RFI)-A request for information is a formal statement of information needs that may be used to determine the feasibility products, services or systems. An RFI may be issued prior to an RFP to help determine if and what a company will request in an RFP.

Request for Offer (RFO)-A request for offer is a formal statement that requests a vendor send a quotation or offer options for products or services.

Request For Proposal (RFP)-A Request for Proposal (RFP) is a formal statement of requirements and technical specifications that constitute a company's needs for the evaluation, recommendation and bid proposal of systems, services or products from vendors.

This figure shows an outline of key elements of an RFP. This diagram shows that an RFP is typically starts with a general overview and the scope of what is being requested. This is usually followed by several pages that detail key requirements such as product types, services, and support needs. The RFP also includes the terms & conditions that define how the vendor should respond such as who to respond to (contact information), what format to respond (printed and/or electronic file formats), and key response submission dates.

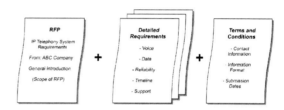

RFP Elements

This figure shows the general process used for the initiation, development, and completion of the RFP process. This diagram shows that the process generally begins with the RFP issuer and the RFP responder does not become involved until they are invited to attend an RFP conference. The RFP issuer spends a substantial amount of time and effort in determining the requirements that will be defined in the RFP document. This process shows that the RFP creation process is typically interactive between the issuer and responder with questions, responses, and clarification of information being exchanged during the RFP process. This diagram shows the RFP process typically ends with an announcement of the vendor (or vendors) whom has been selected as a winner of the RFP.

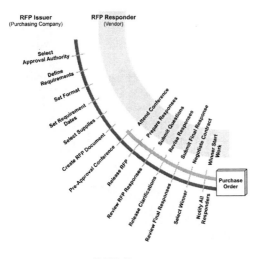

RFP Issuer
(Purchasing Company)

RFP Responder
(Vendor)

RFP Process

This figure shows the many requirements that are commonly considered when creating a request for proposal (RFP) for an IP Telephony system. This diagram shows that some of the requirements include services such as telecom voice and data communication. The requirements also include installation, testing, and support requirements along with disaster recovery and security considerations.

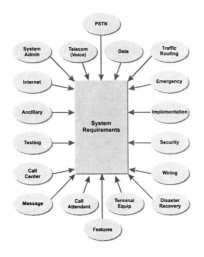

IP Telephony RFP Requirements

Request For Quotation (RFQ)-Request for Quotation (RFQ) is a document that is prepared by a manufacturer, a carrier or a company that finds the needs of a company and solicits a response for quotation for specific equipment and services that are specifically defined in the RFQ. The terms RFQ and RFP are often confused. The purpose of a RFQ is to receive a financial proposal on a specific product or service where as an RFP is used to first gather information that may be used to better define the product or service before purchasing products and/or services.

Required Availability-Required availability is a measure of the amount of time that a service must be available provided specific conditions exist.

Requirement Dates-Requirement dates are the time periods when events should occur such as submitting specific items or documents that are required to be transferred or received for an RFP or when specific implementation actions (such a "system cutover") should occur.

Requirements Phase-Requirements phase is a portion of a development process that is used to determine the technical and/or business requirements for a project to progress.

Requisition-A requisition is a request to order products or services.

Re-Rating-Re-rating is the process of evaluating and updating the rates in usage records. Re-rating may be performed to adjust charge amounts as a result of rate plan allowances (such as free minutes or usage).

Reseller-A reseller buys network services in bulk from an existing carrier for resale to the public or other customers. The reseller provides sales and support services to the customer and the customer usually pays the reseller for the communication services it receives.

Reserve Price-Reserve price is a minimum price level that is set in a bid auction.

Reserved Agents-Reserved agents are customer service representatives that are reserved to answer questions of certain categories or situations.

Reset-The act of restoring a device to its default or original state. Reset also may refer to restoring a counter or logic device to a known state, often a zero output.

Reset-Reboot

Residential Gateway (RG)-Residential gateways facilitate the circuit-to-packet conversion from analog endpoints in the home out to an IP network. Typical devices include IADs, cable modems, DSL modems, and wireless broadband units.

Residual Interconnection Charge (RIC)-An additional telecommunication charge that allows a telecommunications carrier to recover revenue losses (shortfalls) that were created by the implementation of the FCC's local transport restructure.

Residual Payments-Residual payments are fees that are paid to a seller or sales agent for the continued purchase of products or services.

Residuals-Residuals are fees that are paid to a seller or sales agent for the continued purchase of products or services.

Resource Availability-The amount of available system capacity to provide services to additional users. Resource availability may be measured by amount of data transfer (Mbps), number of users, or number of calls in a specific time period.

Resource Grouping-Resource grouping is the assigning of assets or resources to a group code or project.

Resource Management-Resource management is the process of identifying and coordinating the available resources in a product, service or system. An example of resource management is the coordination of available memory in a communication device (such as a television set top box). The resource manager will assign and remove resources from applications that use the limited memory and system resources in the device.

Resource Manager-A resource manager is a controller in an assembly that is responsible for identifying and coordinating the available resources in a product, service or system. An example of resource management is the coordination of available memory in a communication device (such as a Bluetooth transmitter). The resource manager will assign and remove resources from applications that use the limited memory and system resources in the device.

Resource Partitioning-Resource partitioning is the dividing of resources or system processing functions into parts that are used or dedicated for particular services.

Responder-A responder is a person or company who replies to a request for proposal.

Response Capabilities-Response capabilities are the processes that a device or service can use to respond to actions or requests.

Response Rate-A response rate is the ratio (percentage) of recipients who respond to a media promotion, request, or query.

Response Time-The elapsed time between the generation of an inquiry and the receipt of a reply in a communications system. The response time includes the transmission time, processing time, time for searching records and files to obtain relevant data, and transmission time back to the inquirer. In a data system, it is the elapsed time between the end of transmission of an inquiry message and the beginning of the receipt of a response message, measured at the inquiry originating station.

Rest of World (ROW)-Rest of world is an assignment of geographic areas that are outside other assigned territories.

Restoration-The repair or returning to service of one or more telecommunication services that have experienced a service outage or are unusable for any reason, including a damaged or impaired telecommunications facility. Such repair or returning to service may be done by patching, rerouting, substitution of component parts or pathways, and other means, as determined necessary by a service vendor.

Restricted Data-Restricted data is information or media that requires an authorization or privilege to identify, access, process, or store.

Retail Account-Retail accounts are customers that own or operate stores that sell products directly to consumers. Retail account sales support needs may include product training, point of sale, materials and displays, and checking and restocking of inventory.

Retail Channel-Distribution channel that makes use of retail stores to provide goods and services to customers.

Retail Price-A retail price is an offer value that is assigned to a product or service for sale of products to end users.

Retail Rating-Retail rating is the assigning of retail prices to usage records.

Retainer-A retainer is a value that is provided to a person or company in advance of the receiving of products of services. A retainer may be requested or given as a gesture of good faith or to ensure a person or company will be paid for their services.

Retainer Letter-A retainer letter is a document that confirms that a company or person has agreed to retain the services of another company or person along with the conditions or limitations associated with the retention of services.

Retainer Schedule-A retainer schedule is an itemized list or table that provides payment amounts that is provided when specific actions or objectives are reached.

Retention-Retention is the process of actively or passively preventing customers from leaving and switching to other service providers or products. A measurement of retention is churn rate.

Retention Rate-Retention rates are the percentages of customers or potential customers who maintain their subscription for an application or service.

Retention Tool-A retention tool is a program, item or system that can be used to influence potential customers to keep buying or using products or services.

Retransmission Policy-Retransmission policy is the set of rules or processes used by networks to define if, when and how retransmissions of data or information will occur.

Retrial-Any subsequent attempt by a customer, operator, or switching system to complete a call within a measurement period.

Retrieval Services-A service that allows authorized users of the service to retrieve information from an information center.

Retrofitting-Retrofitting is the conversion of a product or system to allow it to have new capabilities.

Return Call-A feature that allows a telephone user to program the telephone system to return a call (usually to the originating telephone) when a specific telephone number that is busy becomes available for call delivery.

Return Channel-A return channel is a communication path from a device to the controlling or central point in a system.

Return Days-Return days are a period of time that an affiliate person or company can continue to receive sales commissions when an affiliate customer returns to a vendor to purchase additional products or services.

Return of Investment (ROI)-Return of investment is the recovery of an investment or a portion of an investment. Investors sometimes joke that they hope business owners will have high projections of return on investment, which actually means return of investment.

Return on Advertising (ROA)-Return on advertising is the recovery or value received for the fees paid for media communication services.

Return on Advertising Spend (ROAS)-Return on advertising spending (ROAS) is the recovery or value received for the fees paid for media communication services.

Return on Investment (ROI)-Return on investment is a financial measurement that compares the profit with the original investment. ROI is to evaluate the impact of an investment on the telephone company's profitability or operational efficiency. Return on Investment is reported in terms of dollars spent compared to benefits gained.

Return on Marketing Activity (RoMA)-A financial measurement that is utilized to evaluate the effectiveness of a marketing activity. Return on Marketing Activity is reported in terms of dollars spent / market share realized.

Return Policy-Return policy is the set of rules that define how products must be handled and processed so that a return payment or credit is issued.

Returned Account Procedure (RAP)-Returned account procedure is a GSM-industry billing standard for rejecting and returning invalid call records. The overall GSM billing standard uses two formats: the Transferred Account Procedure (TAP) for roaming call events and RAP. TAP and RAP files facilitate inter-company settlements and bill presentation to the customer of all call activity.

Returns-(1-sales) Returns are products or items that have been sent back or provided to the seller, distributor or retailer. (2-email) Email messages that could not be delivered.

Returns and Allowances-Returns and allowances are amounts of value that are given to offset a price or previously agreed upon term.

Returns Management-Returns management is the process of identifying, receiving and managing the return of products or data.

Revenue-Revenue is any form of income that is received.

Revenue Accounting Office (RAO)-An accounting office that handles billing and related data processing activities for telecommunications services provided to customers.

Revenue Assurance-Revenue assurance is reviewing of systems and records that are associated with revenue streams to ensure billable services are correctly recorded and collected.

Revenue Diversity-Revenue diversity is the process of generating sales through unrelated processes.

Revenue Model-A revenue model is a representation of the processes of a business that generates revenue. These representations may be in mathematical formulas or as functional descriptions of processes that are used to define how a business will generate sales.

Revenue Operations Center (ROC)-A revenue operations center is a facility, system or operational function that can capture, analyze and use financial parameters of a service provider to capture, maintain and adjust operations and functions to control and optimize revenue generation. ROC brings the information together, from the traditionally separate systems that exist in the service provider's landscape and provides a visual representation of this information from a business perspective and in a manner that enables issues affecting business performance to be easily identified, investigated, diagnosed and corrected. Modeled on the NOC concept, ROC is intended to provide an equivalent view to the financial community of the operational effectiveness of a telecom operator's revenue network, as the NOC itself does for network operations.

Revenue Protection-Revenue protection is the processes that are used to ensure that purchases and services that are used are recorded, re used within their defined authorizations and the fees associated with the purchase or use of these services are posted and collected.

Revenue Sharing-Revenue sharing is the process of transferring revenues generated by one or more companies to one or more other companies for the exchange of products or services. The products or services that are provided by the recipient(s) of revenue sharing may or may not be related to the creation of the revenues that are shared. Revenue sharing is commonly performed in the communications industry between companies that provide revenue generating services to customers and companies that provide underlying services necessary to provide for these services. An example of revenue sharing is the access fees that are paid to local telephone companies (LECs) by long distance inter-exchange companies (IXCs). The long distance revenues of the IXCs are shared with the LECs in return for the LECs providing telephone access between the end-user and the IXC.

Revenue Split Advertising-Revenue split advertising is the process of sharing the revenue generated as the result of an advertising program with the advertiser and the company that provides the product or service.

Revenue Stream-A revenue stream is a source of money or value received that is associated with the sale or providing of services or products or can be associated with a specific type of sales or marketing process.

Revenue Target-Revenue targets are sales values that are desired along with the associated time periods that the sales should be achieved. Revenue targets may defied in gross sales or net sales amounts.

Revenue Volume Pricing Plan-Revenue volume pricing plans apply discounts to billing charges or usage fees that are based on volume of revenue amounts.

Revenues-Revenue is the receipt of cash or other assets that occurs as a result of actions. These actions can be business operations or they can be the transference of assets as a result of legal actions (such as inheritance).

Reversal-(1-financial) A reversal is a transaction that reverses the effects of a previous transaction. (2-electrical) A reversal is a change in magnetic field direction or electrostatic charge to the opposite direction or value.

Reverse Auction-A reverse auction is a pricing system process of allowing the price to be reduced until the product is sold.

Reverse Billing-Reverse billing is the process of charging the recipient for the services that are initiated or used by another person or company.

Reverse Charging-Reverse charging is the applying of costs to the account or device that received a service from another user or device.

Reverse DNS Lookup-A reverse DNS lookup is a process that obtains a web address (URL) from a DNS server by providing an IP address.

Reverse Engineering-Reverse engineering is the process of studying, testing, evaluating and analyzing a product to decipher how it works.

Reverse Pricing Model-A reverse pricing model is the use of mathematical formulas or processes that are developed by customers to help determine what prices should be set or offered.

Revocation Check-Revocation checking is a process that is used to determine if a certificate or authorization to use a product or service has been removed or revoked.

Rewards Based Invitation-Rewards based invitation is an invitation to a free product or special offer that is the result of qualifying for a reward.

Rework-Rework is the process of repairing or modifying of assemblies that have already passed through a production process with failures or poor performance.

RFBP-Request For Budgetary Proposal

RFC-Request For Comments

RFI-Request For Information

RFO-Request For Offer

RFP-Request For Proposal

RFP Invite-An RFP invite is a request from an RFP issuer that is sent to a company to invite them to participate in an RFP process. An RFP invite may define the products and/or services desired along with some of the basic terms (e.g. non-disclosure agreement) and/or qualifications (company size, technologies and support capability) for a company to participate in the RFP process.

RFP Issuance-RFP issuance is the copying and distributing of an RFP to a list of authorized recipients. This distribution may be in the form of paper and/or electronic copies. Potential recipients of RFPs may have received and responded to an RFP Invite. For some RFP issues, only vendors that have been invited to participate may respond. The issuance of an RFP may also involve the signing of non-disclosure to ensure that confidential information about the company issuing the RFP (such as new service and expansion plans) is not made available to the public or to competitors.

RFP Issuer-An RFP issuer is the company that issues a request for proposal.

RFP Objectives-RFP objectives are statements that identify the desired results that should be achieved as a result of the completion of the project defined by the RFP.

RFP Release Date-The RFP release date is the date that an RFP is shipped or made available to participating companies.

RFP Requirement Dates-RFP requirement dates are the specific dates that specific actions must be taken to successfully participate in an RFP process. Failure to complete a task by one or more requirement dates typically disqualifies the vendor from continuing in the RFP bidding process.

RFP Requirements-RFP requirements are the specific are set of actions, documents and data that a company must perform and describe when submitting an RFP response. The RFP requirements may include information types, timing of when the information is to be provided, format of the information provided (printed and or electronic documents) and additional details about actions that may occur between the RFP issuance and response processes.

RFP Respondent-An RFP respondent is a person or company that responds to a request for proposal.

RFP Response Evaluation-RFP response evaluation is the process of judging the RFP responses. RFP response evaluation is usually based on a pre-defined set of criteria. RFP evaluation typically includes weighting of criteria such as vendor's experience with systems they are proposing, their financial and business stability, and how well their proposal satisfies the requirements of the RFP.

RFP Response Procedure-An RFP response procedure is the instructions to the RFP responder that defines the steps that the responder must perform to comply with the terms of the RFP submission process. The response procedure may define levels of compliance and how variations from compliance will be handled.

RFP Response Review-RFP response review is process of receiving, reviewing and evaluating responses to RFPs. As RFPs are received from responders, they are marked and/or logged as received with the date they are received.

RFP Responses-Request for proposal (RFP) responses are documents and supporting materials that are provided by companies that have received and reviewed an RFP that define how can satisfy the objectives and requirements defined in the RFP.

RFP Submission Procedure-The RFP submission procedure is the sequence of tasks, delivery components and dates that a company or person must use to submit a response to an RFP. The submission procedure section of an RFP should include the method of response and how interactions between an RFP responder and RFP issuer should occur. The submission procedures usually include details on the expected preparation and format of the RFP response, the methods the company will use for receiving responses to the RFP, how RFP responses will be evaluated and how RFP respondents will be notified of the winner of the RFP.

RFP Transmittal-RFP transmitter is the process used to submit an RFP. RFP transmittal requirements include who the RFP is submitted to, when it should be received, the contents of an RFP, supporting documents and the format. RFP submittal formats may include electronic and/or printed copies.

RFQ-Request For Quotation

RFTP-Right First Time Provisioning

RG-Residential Gateway

RIC-Residual Interconnection Charge

Right-A right is a legally recognized authorization or entitlement to use, transfer or display content.

Right First Time Provisioning (RFTP)-Right first time provisioning is a process within a company or network that allows for establishment of new accounts, activation and termination of features within these accounts, and coordinating and dispatching the resources necessary to fill those service orders without errors or unexpected operations.

Right of Resale-Right of resale is the authorization to purchase and resell products or services.

Right of Way (ROW)-Right of way is the authorization provided by a government or a property owner to install, build, or maintain communication lines and/or facilities.

Right Sizing (Rightsizing)-Right sizing is the adjusting of the number of jobs at a company to better match the business needs of the company. The term right sizing is commonly used when reducing the number of jobs.

Rights Clauses-Rights clauses are specific terms and their requirements (such as financial consideration) that are associated with the rights (ability to use, distribute or sell) associated with a product, service or intellectual property (content).

Rights Consideration-Rights consideration is the authorization or transfer of value in return for the assignment of rights. Rights consideration may be in the form of money or any other quantifiable item or service.

Rights Extents-Rights extents are the amount of usage of a content item that is authorized. Rights extents may be defined in units of time, the number of uses and what places (geographic regions) that the rights apply.

Rights Management-Rights management is a process of organization, access control and assignment of authorized uses (rights) of content. Rights management may involve the control of physical access to information, identity validation (authentication), service authorization, and media protection (encryption). Rights management systems are typically incorporated or integrated with other systems such as content management system, billing systems, and royalty management.

This figure shows basic rights management processes. This example shows that rights management can involve rendering, transferring and derivative rights. Rendering rights may include displaying, printing or listening to media. Transfer rights may include copying, moving or loaning media. Derivative rights can include extraction, inserting (embedding) or editing media.

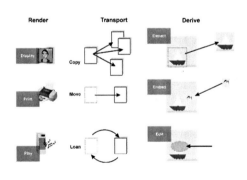

Rights Management

Rights Model-Rights model is a representation of how the content rights are transferred and managed between the creator or owner of content and the users or consumers of content.

Rights Transaction-Rights transaction is the transfer of rights from a content owner to a content user or distributor. A rights transaction may involve the use of a formal agreement (e.g. a publishing agreement) or it may occur through an action (e.g. a customer buying a book).

Rightsholder-A rightsholder is a company or person that has been given authorization to a set of rights for a specific amount or form of content.

Rightsizing-Right Sizing

Ring-(1-wire) One of the two subscriber loop/line wires. It is connected to the so-called ring conductor on a manual switchboard plug, hence its name. In North America, red insulation color is used to identify it. Corresponds to the European "B" subscriber wire. (2-audio) Audible alerting signal at subscriber set indicating incoming telephone call. May be a ringing bell or other sound. (3-business) A linking of business relationships where one helps another who helps another which eventually links back to the original business relationship.

Risk-Risk is a measure of uncertainty or possibility that losses or unwanted events may occur.

Risk Assessment-Risk assessment is the identification, qualification, quantification and analysis of risks that exist for performing projects, processes or developing businesses.

Risk Capital-Risk capital is resources that are provided to companies or business entities where the capital is has the potential of less than desired rate of return or an uncertainty of repayment.

Risk Management-Risk management is the processes and steps taken to identify, minimize and correct the potential losses or unauthorized uses of products or services.

Risk Tools-Risk tools are software programs or processes that can be used to help identify, evaluate and assess the probable outcomes of taking actions (such as developing and launching a new product).

RLL-Radio Local Loop

ROA-Return on Advertising

Road Maps-Road maps are graphic illustrations that provide information about the location of roads and other objects of interest in a defined geographic area.

Road Warrior-A worker or contractor who continuously moves to many locations to perform their job or assignments.

Roaming-Roaming is the capability to move from one carrier's system area to another carrier's service area and obtain service. While it is desirable to roam without loosing functionality of the phone or device, some communication systems offer advanced features (such as high speed data) that other systems may not have installed. This may limit the operation of advanced features.

Roaming Agreement-A roaming agreement is a service agreement between service providers that defines the services that will be provided to subscribers who are visiting in each others systems and the charges that will be charged to each carrier for providing these services.

Roaming Fees-Roaming fees are the billing charges to a customer for the usage of a product or service while operating in another service providers' communication system. Roaming fees may be composed of daily access charges and usage charges.

Roaming Indicator-A roaming indicator is an icon or other alerting mechanism on a device that indicates that it is operating on a system other than its home service provider.

Roaming Partner-A roaming partner is a company or system that allows visiting users to gain access to their systems or services.

ROAS-Return on Advertising Spend

ROC-Revenue Operations Center

Rogue Employees-A rogue employee is a person who performs actions that not authorized or desired by their employer. An example of a rouge employee action is the selling of media (such as television programs) to unauthorized users or distributors.

Rogue Standard-A rogue standard is a specification for a program, product or service that is created outside the traditional standard development processes.

ROI-Return of Investment

ROI-Return on Investment

Rollback-Rollback is the process of restoring previous data or values.

Rollout-Rollout is the introduction of a product or service into the marketplace.

Rollover Lines-Telephone lines that are placed in a hunt sequence where calls will move to the next available line when those higher in the hunt group are found busy. This also refers to the movement of inbound call traffic from one telephone switch or ACD queue to another based on current busy conditions or programmed instructions within the telephone switch.

ROM-Read Only Memory

RoMA-Return on Marketing Activity

Room Terminal Monitoring-Room terminal monitoring is the sensing, monitoring or gathering of information about the status or use of a communication device that is located in a room (such as a hotel room).

Root-A root is the main focal point in a system or the base of a tree structure or hierarchy system.

Root Cause Analysis-Root cause analysis is the processes that are used to find the source (the root) of a problem or degradation in a system or service performance.

Root Certificate-A root certificate is an authenticated document or electronic key that is signed by the highest level authority that issues certificates.

Root Directory-A root directory is the highest level directory from which all other sub directories must branch from.

Rotary Dial-Rotary dialing is a process of sending dialed digits through the use of a spring-loaded mechanical switch that produces pulses as it rotates through 10 positions (1 through 9, and 0). As the rotary dial turns, a switch briefly interrupts the loop current. The number of pulses per rotation is counted to determine the number dialed. A time pause between rotary dials is used to determine when additional digits are dialed or when the caller is finished dialing.

This diagram shows how a mechanical rotary switch can be used to gather dialed digit information. In this example, a spring-loaded rotary dial is turned to produces pulses that represent the numbers on the dial. As the rotary dial turns, a contact switch briefly interrupts the loop current. The line card in the telephone system counts number of pulses to determine the number dialed. This diagram shows that after a specified time pause occurs after a series of pulses, the counter resets and is ready to count another dialed digit.

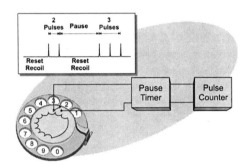

Rotary Dial Operation

Round Robin-Round robin is the selection of tasks or items in a progression where the selection repeats to the beginning location.

Rounding Rules-Rounding rules are conditions and processes that determine when and how partial unit amounts will be rounded up to the next unit amount. An example of rounding rules is to adjust the amount of time used on a telephone call up to the next higher minute when partial minute usage periods occur.

Router-A router is a device that directs (routes) content (data, voice or video) from one path to another on a network. Routers base their switching information on one or more parameters contained in the packet of content. These parameters may include availability of a transmission path or communications channel, destination address contained within a packet, maximum allowable amount of transmission delay a packet can accept, along with other key parameters.

Routers forward data packets between multiple interfaces based on the network layer. Most modern day routers support one or more of the following protocols: Internet Protocol (IP), Novell IPX or AppleTalk. Routing occurs at layer 3 of the OSI reference model and can be used to limit the broadcast domain of a bridged network.

Routine-A group of instructions for carrying out a specific processing procedure. Routine usually refers to part of a larger program. Routine and subroutine have essentially the same meaning, but a subroutine could be interpreted as a self contained routine nested within another routine or program.

Routing Options-Routing options are a set of alternative connection paths that can be used to connect two or more points with each other.

Routing Tables-A routing table is a list (a database table) that is located within a router that is used to determine the forwarding path (route) for incoming packets.

ROW-Rest of World

ROW-Right of Way

Royalties-Royalties are compensation for the assignment or use of intellectual property rights.

Royalty-A royalty is a fee that is paid to an author or composer for the right to use each copy of a work that is sold, performed, or produced under license of an exclusive right (such as patent rights). When patents were issued by English Queens and Kings, it was common to provide the monarch with a small (or large) payment in return for the grant to use the patent.

Royalty Free-Royalty free means that a patent is licensed without cost to the user. Royalty free licensing terms are mandatory for patents incorporated into Open Source software such as Linux, and W3C standards.

Royalty Import-Royalty importing is the recognizing of royalty costs or assignments and the capturing of these costs into financial systems.

Royalty Validation-Royalty validation is the reviewing of systems and records that are associated with royalty fees to ensure royalty amounts are correctly recorded and paid.

RP-Reference Point

RPC-Remote Procedure Call

RPG-Report Program Generator

RSA-Rural Service Area

RSA-Rural Statistical Area

RSS-Remote Support System

Runtime-Runtime is the operation of a program in a computer or data processing device.

Rural Service-Rural services are communication services provided to users located in rural areas. For 3rd generation systems, rural services may obtain data transmission rates of 144 kbps.

Rural Service Area (RSA)-A cellular service area in rural (small population) regions.

Rural Statistical Area (RSA)-A geographic area designated by the FCC for service to be provided for by cellular carriers that falls outside the metropolitan statistical area (MSA) regions. There are over 400 RSAs in the United States.

s-Second

SA-Security Association

SAARC-South Asian Association for Regional Cooperation

SAC-Service Access Code

SAC-Subscriber Acquisition Cost

Safe Harbor Statement-A safe harbor statement is a declaration that provides qualifying and/or limiting terms that relates to claims or assertions that are made in a promotion or advertisement. A safe harbor statement may be included to limit the potential liability of the claims or assertions by demonstrating that good faith disclosure effort was made.

Safe Mode-Safe mode is a condition of a system or device where it continues to operate with some reduced or restricted capabilities. An example of safe mode is a computer operating system that remains in local mode to enable a user or technician to help diagnose problems.

Safety-Safety is the ability of the effort or system operation to reduce the possibility of unwanted potentially dangerous (such as human injury) from happening.

SAID-Security Association Identifier

Salary-Salary is a compensation amount that is provided to an employee over interval periods such as weekly, bi-weekly or monthly.

Sales-(1-operations) The operational department responsible for identifying potential new customers, negotiating offers, and closing contracts. (2-Revenue) The amount of products or services that are sold or committed to be sold.

Sales Agency-A sales agency is a company or organization that performs sales activities for other companies. Sales agencies typically focus on a specific type of industry and obtain non-competing companies as clients.

Sales Bonus-Sales bonuses are additional sales incentives that are provided when some criteria are achieved. An example of a sales bonus is the paying of an additional $1,000 is sales commission for the obtaining of a new corporate sales customer.

Sales Campaign-Any of a broad range of sales activities designed to identify prospective customers, qualify customers, assess customer interest levels, gather sales related decision information, and to obtain new orders.

Sales Channels-Sales channels are people, companies or systems that can communicate offers and process orders. Sales channels include direct sales, sales representatives, distribution, and online.

Sales Commissions-Sales commission is a value assigned or awarded to a salesperson or other entity for obtaining or servicing a customer. The sales commission is typically paid after a sale to a customer is completed.

Sales Cycle-A sales cycle is a sequence of phases and tasks that a buyer goes through when making the decision to purchase a product or service. Some of the typical phases in a sales cycle include need awareness, research, evaluation and purchasing.

Sales Engineer-A sales engineer is a person who performs technical sales analysis and support functions.

Sales Invoice-A sales invoice is a grouping of billing charges for a time period or events that are associated with a specific sale.

Sales Leads-A sales lead is contact information about a person or company that has expressed some form of interest in a product or service. Sales leads may be qualified (some validated information or interest) or they may be unqualified (no validated).

Sales List-A sales list is a group of contacts that have a common set of characteristics such as customers that have already purchased products or are a group of contacts that are potential customers (sales prospects).

Sales Literature-Sales literature is a media item (such as a pamphlet) that is used by a sales person to help communicate or motive a prospective customer to buy a product or service.

Sales Management-Sales management is the process that records, coordinates and processes information that is used for prospecting, qualifying, interest level assessment, data gathering and order tracking.

Sales Objectives-Sales objectives are statements that identify targets that should be achieved through sales efforts that may include items such as revenue, new customer types, new channel types, profit margins or quantity targets.

Sales Person (Salesperson)-A sales person is someone who can accept an offer to purchase a product or service. The role of a salesperson can range from an order taker (a processor of orders) to an order getter (prospecting and proposing).

Sales Plan-A sales plan contains the objectives of the sales process, the responsibilities and incentives of those involved in the sales process, and the resources that will be available or used for the sales process. The sales plan usually includes objectives (sales targets), assigned sales representatives, what products they are authorized to sell, list of sales roles, responsibilities, and territories.

This diagram shows how a sales process can be managed. In this example, the sales process is divided key steps that can be defined and managed. The prospecting step is used to identify new customers or expanded needs for existing customers. The qualification step is used to determine how many of the prospects are qualified (real candidates) for the product or service. The interest assessment step is used to determine how motivated the prospect is to take action to satisfy their need for the product or service. The fact-finding stage is used to determine who are the decision makers and what steps are necessary to complete

the sale. The close involves the consolidation of the previous steps (coordination of motivated decision makers) so purchases can result. The progress between each step can be tracked and optimized. Included in the chart is an example activity time sheet showing that step may require different levels of time commitment and that the allocation of resources (time for each salesperson) is usually distributed along each step.

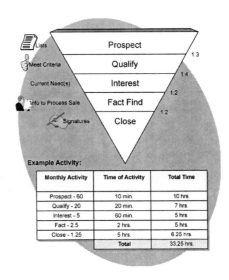

Monthly Activity	Time of Activity	Total Time
Prospect - 60	10 min.	10 hrs.
Qualify - 20	20 min.	7 hrs.
Interest - 5	60 min.	5 hrs.
Fact - 2.5	2 hrs.	5 hrs.
Close - 1.25	5 hrs.	6.25 hrs.
	Total	33.25 hrs.

Sales Management Diagram

Sales Process-The sales process involves identifying likely prospects, qualifying these prospects for your products or services, assessing the interest level of your prospects, gathering the information (fact finding) that is necessary to process the order, and developing an agreement (sales order) that both the prospect and your company agree with.

Sales Program-A series of related sales campaigns targeting the same objectives.

Sales Projections-Sales projections are the forecasting of future sales results through the use of ratios or assumptions.

Sales Rep-Sales Representative

Sales Reports-Sales reports are groupings of data that indicate sales activities that are related to specific criteria such as the particular sales channel or person who is responsible for the sales.

Sales Representative (Sales Rep)-A sales representative is a person who performs sales functions for one or more companies.

Sales Representative Agreement-A sales representative agreement is a set of terms and obligations that have been agreed to between a provider of products or services and a person or company who sells the products or services. This diagram shows a sample page from a

<Company>
Manufacturers Representative Agreement

This instrument, hereafter referred to as the AGREEMENT, executed this <Agreement Date>, between <Company>, located at
(hereafter referred to as <Company>) and _____, located at <Representative Address>
(hereafter referred to as REPRESENTATIVE). It is the intention of both parties of this Manufacturers Representative Agreement to promote the sale of <Company> products (hereafter referred to as PRODUCTS).

1. <Company> hereby appoints REPRESENTATIVE as its independent representative to solicit orders and otherwise assist in obtaining orders for products sold, licensed or otherwise distributed by <Company>, for the accounts as set forth in Addendum A here to, and accepts such appointment and agrees to solicit orders and promote the PRODUCTS in accordance with the terms of this AGREEMENT.
2. REPRESENTATIVE agrees to promote the sale of <Company> products in accordance with the instructional materials provided to REPRESENTATIVE by <Company>.
3. REPRESENTATIVE shall submit to <Company> periodic reports relating to its activities directed to introduce, sell and support PRODUCTS.
4. <Company> agrees to make available to REPRESENTATIVE sufficient samples of products, brochures, price lists and promotional aids. All aforementioned materials shall be used for the express purpose of promoting sales of PRODUCT, never intended for individual re-sale <Company> shall have the right to limit the number of samples furnished, when determined by <Company> that such samples are not necessary for REPRESENTATIVE to effectively promote the PRODUCT.
5. REPRESENTATIVE agrees to indemnify <Company> against losses from claims arising from infringement as may be made, should REPRESENTATIVE attempt to sell or market PRODUCT beyond the agreed upon account list or territory ascribed.
6. REPRESENTATIVE agrees to use only professional conduct in selling PRODUCTS, within the general guidelines provided by <Company> defining the nature and image of the PRODUCTS and trademarks. Under no circumstances shall REPRESENTATIVE act in any official capacity as a direct agent for <Company>.
7. <Company> agrees to pay commissions on the gross sales less freight to customer for product orders delivered and paid by customers, which qualify under the terms and conditions of this AGREEMENT. Commission is earned only when the transaction is complete, i.e. the goods have been shipped and paid for, regardless of the time of payment. Earned commissions shall be paid on the 15th day following receipt of payment by <Company>.

Sales Representative Agreement Sample Page

sales representative agreement.

Sales Representative Firm (Rep Firm)-A sales representative firm is a company who performs sales functions for one or more companies.

Sales Representative Support-Sales representative support is the resources that will be provided to the sales representative or to oth-

ers designated by the sales representative. Sales representative support includes literature and technical support.

Sales Resources-Sales resources are the materials and services that are provided or authorized for use to enable or assist the sales process.

Sales Spiff-A sales spiff is the providing of marketing incentives from one company directly to the sales representatives of another company (usually a retail sales company). Spiffs are often used to focus sales representatives on demonstrating and giving preference to products or services from a specific manufacturer.

Sales Support-Sales support is the resources that are dedicated or available for sales related projects or tasks.

Sales Targets-Sales targets are quantities of companies, products or revenue values that are assigned to be achieved or used to develop a list of actions to achieve them. Sales targets may be associated with a commission structure or bonus system to help motivate and reward people for helping to achieve sales objectives.

Sales Tax-A sales tax is a fee or assessment that is imposed (levied) by a government or authorized entity on the sale of a product, activity or any other asset that can be valued and transferred.

Sales Tax Computation-Sales tax computation is the calculation of a sales tax amount using order information and its associated sales tax rules.

Sales Territory-Sales territories are regions or geographic boundaries that are assigned to salespeople or sales departments. Sales territories may be referenced to regions such as countries, states, counties or areas on maps.

Sales Tracking-Sales tracking is the process of determining where in the sales cycles each customer is located.

Salesperson-Sales Person

Sample-(1-signal) A sample is the value of a signal, such as voltage, selected at a specific point in time. (2-quality assurance) One or more units selected at random from a quantity of product to represent that product for inspection purposes.

Sample Code-Sample code is a group of software instructions or segments of instructions that are used within a computer program that is provided as a reference to assist in the development of other software programs.

Sample Script-A sample script is a written example of a communication session between people. Sample scripts may be used literally (exactly as written) or they may be used as guidelines.

SAN-Storage Area Network

Sandbox-(1-software program) A sandbox is an area or set boundary limits. Sandbox limits for Java defines how downloaded software programs (applets) may operate and interact with other resources on a computer. The program can only process and interact with other objects in the sandbox area. (2-search engine) A search engine sandbox places a domain listing into a restricted access area (possibly when a domain name is initially activated) to allow verification of the validity of the web site.

Sarbanes Oxley Act (SOX)-The Sarbanes Oxley Act is a United States law that was passed in 2002 that defines new requirements and responsibilities for company financial reporting. This law was created in response to multiple scandals that resulted from misrepresentative accounting records from companies including Enron and MCI.

SAS-Single Award Schedule

SAS-Subscriber Authorization System

Satellite Distributor-A satellite distributor is a system or company that distributes information through a satellite transmission system.

Satellite Operator-A satellite operator is a company that transmits or provides information to users that are connected or able to access signals through a satellite network.

SBC-Subsequent Billing Company

SBLP-Service Based Local Policy

SC-Smart Card

SC-SMS Service Centre

Scalability-Scalability is the ability of a system to increase the number of users or amount of services it can provide without significant changes to the hardware or technology used.

Scalable-Scalable is the capability of a system or service to be expanded to produce more products or services.

Scalable Vector Graphics (SVG)-Scalable vector graphics is a media structure that describes two-dimensional graphics and graphical applications using XML format.

Scalable Vector Graphics 1.1 (SVG 1.1)-Scalable vector graphics version 1.1 is a media structure that describes two-dimensional graphics and graphical applications using XML format.

SCE-Service Creation Environment

Scenario-A scenario is the using of a particular set of conditions that might or could exist in the future to assist in determining the likely outcomes of operations or actions.

Schedule-A schedule is a listing of tasks or events along with the date or times they are planned or expected to occur.

Scheduled Conferencing-A conference bridge that allows callers to become members of a conference group during a predetermined time period.

Scheduled Multicast-A scheduled multicast is a one-to-many media distribution process that is scheduled to start or operate during a pre-defined time.

Scheduled Transmission-Scheduled transmission is a process in a communication system that assigns or coordinates the transmission time to or from a communication device to a specific time period or after an event (e.g. a start trigger signal) has occurred. When devices have received their scheduled infor-

S

mation, the can transmit at their assigned period without any concern that other devices may transmit at the same time (contention free).

Scheduling Services-Scheduling services are the medium access control functions (data flow control) that define how and when devices will receive and transmit on a communication system.

Schema-A schema is a structure or plan for data or process elements and how they relate to each other.

SCI-Supply Chain Integration

SCM-Supply Chain Management

Scope of Work (SOW)-Scope of work is the range or boundaries of the assignment of projects or tasks.

Score Threshold-A score threshold is a value that is compared to a scoring result that determines if a process or event should be initiated. A score threshold may be used to determine if the value of a search listing should be included in the display of a search results.

Scrambler-A scrambler is a device that transposes or inverts signals, or otherwise encodes a message at the transmitter to make it unintelligible at a receiver that is not equipped with a descrambling device.

Scrambling-Scrambling is a process of altering or changing an electrical signal (often the encoding or distortion of a video signal) to prevent interpretation of the signals by users that can receive the signal but are unauthorized to receive the signal. Scrambling involves the changing of a signal according to a known process so that the received signal can reverse the process to decode the signal back into its original (or close to original) form.

Screen-(1-shield) A screen is an isolating metallic shield that attenuates or blocks signals from entering an area (such as into an electronics assembly). (2-display) A display screen is an area that can present text and/or graphic images that can be perceived by a viewer.

Screen Based Telephony-The use of computer screens to provide a telephone user with information about a call in progress and/or the status of their telephone calls. Screen based telephony may allow the telephone user to dial, answer, and control their telephone calls.

Screen Management System (SMS)-A screen management system coordinates the selection and playout of digital media within a digital cinema.

Screen Name-A screen name is an identifying label that is used for instant messaging systems to identify a instant messaging person or device. The instant messaging system provides the relationship between a screen name and an email or IP address. The use of a screen name allows a person to avoid providing an email address (for privacy) and it allows the use of dynamically changing IP addresses (for systems that dynamically assign IP addresses when the user logs onto the system).

Screen Pop-Screen pops are the display of an information screen on a computer monitor that is automatically triggered by an event such as an incoming call or customer selected feature request.

Script-A script (or a Servlet) is a small program or sequence of operations (macro) that is written in a predetermined language that can be understood by the calling program to allow automatic interaction between programs or devices. Examples of scripts include login scripts that are used to provide identification information when accessing a system or Javascripts that provide advanced features on Internet web pages.

Scripting-Scripting is the creation of instructions and information to use by a software program into a file called a script. Scripts may be composed of commands from a variety of languages including Java, Perl, Microsoft Visual Basic or in a form of unique (proprietary) commands and text that only the software program can interpret.

Scripting Language-Scripting language is the specific syntax and processes that are used by a software program to execute or interpret commands in a script file.

Scroll Bar-A scroll bar is a graphic display that shows the relative position of a display area within the entire scrollable area.

Scrolling-Scrolling is the process of changing the position of the cursor or display within an allowable graphic area.

SCRP-Service Control Reference Point

SD-Service Discovery

SDMF-Single Data Message Format

SDO-Standards Development Organizations

SDP-Service Delivery Platform

SDR-Service Detail Record

SDSL-Symmetrical Digital Subscriber Line

SDU-Service Data Unit

Sealed Bid-A sealed bid is a product or service proposal that contains offer information that is not shared with other bidders during the bidding process.

Search Engine-A search engine is a web portal (web site) or software that searches through web pages or data records to find matches to specific words or items. Web page search engines allow users to enter key words (search words) to find web pages that contain the key words or links that are associated with the key words.

Search Services-Search services are companies or systems that can find and search through information sources (such as web pages or data records) to find matches to specific words or items. Companies that provide search services may be able to limit their search services to specific information sources such as specific web page (such as a web site search engine).

Seasonal Rate-A seasonal rate is a price charged to a customer for a particular service during a defined time interval (e.g. during a season).

Seat-A seat is a position in a call center, terminal for a person or a reserved spot in a training class.

SECABS-Small Exchange Carrier Access Billing

Second (s)-(1-time unit) Unit of time. It is defined as 9,192,631,770 cycles of the radiation associated with the transition between the two hyperfine levels of the ground state of the cesium-133 atom. Originally historically defined as 1/ 31 557 600 of a year. (2-order) The item in a sequence following the first item.

Secondary Event-Secondary events are the scheduling of media item broadcasts that will be combined or used with a primary event.

Secondary Route-The circuit to be used when the primary route is congested. In manual and semi-automatic operations, a secondary route also may be used when transmission on the primary route is not of sufficiently good quality, or if traffic is to be handled outside the normal hours of service on the primary route.

Secure Clock-A secure clock is a time reference that can't be modified by a user or end user device.

Secure Electronic Transaction (SET)-A secure electronic transaction is an exchange of assets or other quantifiable information represented by data (electronic media) that is encoded in such a way to be private (not viewable by others) and unaltered (not able to be changed by others).

Secure Hypertext Transfer Protocol (S-HTTP)-Secure hypertext transfer protocol is a secure version of HTTP protocol that is used to transmit hypertext documents through the Internet. It controls and manages communications between a Web browser and a Web server. S-HTTP is designed to privately send and receive messages without the need to setup and maintain a security session.

Secure Server-A secure server is a computer that can receive, process, and respond to an end user's (client's) request in a secure mode. Secure servers can use secure socket layer (SSL) protocol to establish and maintain

S

authentication (identity) and encryption privacy.

Secure Shell (SSH)-Secure shell is a secure alternative to remote shell that allows a computing device that is connected to a network to perform commands on a remote computing device. SSH creates an encrypted and authenticated channel between hosts for all communication. SSH may also be used to securely tunnel TCP traffic between two hosts' networks.

Secure Sockets Layer (SSL)-A secure socket layer (SSL) is a security protocol that is used to protect/encrypt information that is sent between end user (client) and a server so eavesdroppers (such as sniffers on a router) cannot understand (cannot decode) the data. SSL version 2 provides security by allowing applications to encrypt data that goes from a client, such as a Web browser, to a matching server (encrypting your data means converting it to a secret code) SSL version 3 allows the server to authenticate (validate the authenticity) the client.

Secure Transaction-A secure transaction is an exchange of assets or other quantifiable information that is encoded in such a way to be private (not viewable by others) and unaltered (not able to be changed by others).

Security-Security is the ability of a person, system or service to maintain its desired well being or operation without damage, theft or compromise of its resources from unwanted people or events.

Security Association (SA)-Security association is the setup of a relationship between two network elements that ensures that traffic passing through the interface is authentic, unchanged and/or cryptographically secure (typically, through encryption.)

Security Association Identifier (SAID)-A security association identifier is a label or code that uniquely identifies a security association between devices (such as between a base station and a subscriber station). For DOCSIS, a SAID is a 14-bit number that is used to identify security associations in BPI+.

Security Deposit-Security deposits are asset collections (typically cash) that are controlled by a service provider or company for the assurance that a person or company will fulfill their obligations such as performing an action or payment for services.

Security Dongle-A security dongle is a small communication assembly or device that attaches to another object or system (usually through a USB port). Security dongles contain active electronic circuits that allow them to store, use and process authentication and encryption data.

Security Framework-Security framework is a hypothetical or desired structure that is used to plan or describe the security aspects of a system or business.

Security Hole-A security hole is a weakness or potential access process that can be used to bypass or disable security controls and processes.

Security Layer-A security layer is a functional process in a communication system that controls the security procedures of a device or system.

Security Management-(1-General) Security management of a network involves identity validation (authentication), service authorization, and information privacy protection. Authentication processes identifies the device or person that is requesting the use of the telecommunications device or network services. Authorization is the process of determining what services devices are customers are permitted to use. Privacy or encryption services are used to help ensure that the information transmitted or received is not available to unauthorized recipients. (2-FCAPS) Security is one of the five functions defined in the FCAPS model for network management. Security is responsible for protecting the network against hackers, unauthorized users, and physical or electronic sabotage.

Security Manager (SM)-A security manager is a system or process that controls the security aspects of communication or operations with devices or services.

Security Modes-Security modes are the states that a communication device must be set to access specific types of processes or services. For the Bluetooth system, there are three security modes. When in security mode 1 (no security), the device never initiates a security procedure. When in security mode 2, the device does not initiate any security procedure before a channel establishment request has been received or it has initiated a channel establishment procedure. When in security mode 3, the device initiates security procedures before it sends a message indicating that link setup is complete.

Security Protocol-Security protocols are the languages, processes, and procedures that perform security functions used to send control messages and coordinate the transfer of data. Security protocols may be used to authenticate users authorize services and encrypt information.

Security Proxy-A proxy server that receives and filters information designated for a user it is serving. The security proxy restricts information based on rules it has received from the network operator and/or user of the information. A security proxy is also called a firewall.

Security Risks-Security risks are the methods or measure of uncertainty or possibility that unauthorized access, use, losses or unwanted events may occur in an operation or system.

Security Sublayer-A security sublayer is the functional process within a communication device or system that performs access controls, identify validation and/or encryption of data.

Security Threats-Potential people or processes that may allow access to services or information that is not desired or authorized by the owner or controller of the systems or information.

Security Through Obscurity-Security through obscurity is the process of keeping the security algorithms and functions secret to increase the difficulty in cracking the system.

SED-Shipper Export Declaration

Selective Call Acceptance-A service feature that only delivers calls to their dialed destination if they are on a previously specified selective call acceptance telephone number list. Calls that are received by other numbers are provided with a pre-recorded announcement that states the number is not accepting their call or the call may be routed to an alternate directory number.

This diagram shows how selective call acceptance can be used to only deliver calls from a specific list of callers. This diagram shows that regional managers from a service center can call from numbers that are pre-defined (their office numbers) and that when their calls are received, the will be connected to the specified telephone or extension. Call that are received from numbers that are not on the selective call acceptance list are transferred to an automated message unit that plays a pre-recorded announcement that states the number is not accepting their call.

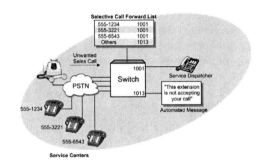

Selective Call Acceptance Operation

Selective Call Forwarding-A service feature that forwards calls to one (or multiple) telephone number dependent on the incoming call forwarding criteria. Selective call forwarding can be used to redirect calls of a specific type (such as fax calls) to a pre-designated number (such as an office fax machine.) This diagram shows how selective call forwarding can be used to deliver calls to alternate number based on a specific criteria type. This diagram shows a selective call forwarding service that routes fax calls to different telephone number or extension after it detects the call is a fax call. After the system detects that the incoming call is a fax (by the fax tones), the switch call processing software transfers the call to the destination number that is connected to a fax machine.

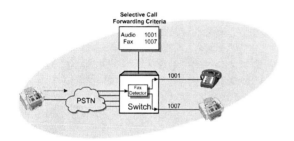

Selective Call Forwarding Operation

Selective Call Rejection-A service feature that restricts the delivery of calls to their dialed destination if they are on a previously specified call rejection telephone number list. Selective call rejection is used to block calls from undesired callers such as prank callers or harassing bill collectors. Calls that are received by numbers on the call rejection list are provided with a pre-recorded announcement that states the number is not accepting their call or the call may be routed to an alternate directory number.

Selective Ringing-A technique for ringing the telephone of only one customer on a multiparty line.

Selects-Category Selection

Self Certification-Self certification is the process of testing and declaring that products

or services conform to specifications or other requirements. Self certification may be performed by companies or people when systems or technologies are initially released and testing and validation facilities are not yet established.

Self Configuration-Self-configuration is the ability of a device or system to detect, discover, initialize and update its program features and parameters without the direct assistance of a user or technician.

Self Dispute-A self dispute is a dispute that is entered by a person or company who has a dispute with another person or company without the assistance of the other parties affected by the dispute.

Self Install-Self Installation

Self Installation (Self Install)-Self installation is the process of allowing a user of a product or service to install and possibly configure (program optional features and parameters).

Self Provisioning-Self provisioning is the ability of an end user or a device to select, initiate and configure their own services without manual intervention from the service provider. This feature may be available in the self care web portal.

Self Signing-Self signing is a process where the owner of the software adds their own identifying signature to software code to help ensure that it is not malicious or that it has not been altered from its original source.

Sell Through (Sellthrough)-Sell through is a measure of the amount of products that are sold to customers as compared to the amount of products that are shipped for sale.

Seller's Market-A seller's market is a condition of an industry or geographic region that the demand for products or services is high. This provides sellers with additional negotiating value.

Selling-Selling is the process of identifying, communicating and assisting people or companies for the acquisition of products or services that they need or want.

Sellthrough-Sell Through

Semantics-Semantics are the meanings of words, codes and sequences and their relationship to other words, codes and sequences.

Semaphore-(1-computing) A semaphore is a process of coordinating programs that share the same processing resource. (2-signaling) The use of a flags and their position to send information.

Separator Character-A separator character is a symbol or code that is designated to be used to define the separation point between data or information.

Serial-An arrangement whereby one element of data is linked to the next so that progress must proceed from the first element through the next, then the next, and so on.

Serial Communication-Serial communication is the transmission of information from a communication device to another communication device (such as from a computer to an accessory device) that is performed one bit at a time.

Serialization-Serialization is the modification of a device or assembly in a way (such as adding a serial number) so that it can be uniquely identified from similar devices or assemblies.

Series Sales Bonus-In reference to a paid sum of money paid by a studio/producer that a network uses to purchase a pilot series, thereby ordering production of a series.

Server-A computer that can receive, process, and respond to an end user's (client's) request for information or information processing.

Server Application-A server application is a software program that runs (operates) on a computer server that is designed to perform operations using commands or information from other sources (such as an Internet user) where multiple users may access or use the application at the same time.

Server Clustering-Server clustering is the grouping of resources or servers within a network to increase system reliability and/or to distribute the processing requirements of the system.

Server Co-Location-Server co-location is the installation of a server that is owned by a company at the location of another company.

Server Farm-A server farm is a group of interconnected computers that share the processing requirements for systems and services.

Server Side-Server side is the sender (provider) side of a system or service.

Service-A service is the providing of information transfer, processes or authorizations that enable users, devices or systems to perform actions that they desire.

Service Access Code (SAC)-The 3-digit codes in the NPA (N 0/1 X) format which are used as the first three digits of a 10-digit address in a North American Numbering Plan dialing sequence. Although NPA codes are normally used for the purpose of identifying specific geographical areas, certain of these NPA codes have been allocated to identifying generic services or to provide access capability, and these are known as SACs. The common trait, which is in contrast to an NPA code, is that SACs are non-geographic.

Service Accessibility-Service accessibility is the ability to obtain services when requested or authorized.

Service Activation-Service activation is the processes used to enable a service to become operable.

Service Address-A service address is a physical or logical address where a service is provided.

Service Agreement-Service agreements are documents or recordings that define the terms for the providing of services from a person or company and the person or company that will use or have the rights to use the services.

Service Area-The region covered by a given cellular carrier. It is also a landline term which means a geographical area in which local exchange carriers provide local exchange service to end users as well as network access to interexchange carriers.

Service Assurance-Service assurance is reviewing of the ability to access and obtain services that have been authorized for use.

Service Availability-Service availability is the ratio of the amount of time an authorized user is able to access the services compared to the total time service is supposed to be available.

Service Based Local Policy (SBLP)-Service based local policy is a network element that is used to control the configurations and/or parameters of local network elements or devices that control the providing of services to ensure the performance or operation attempts to confirm to defined service levels.

Service Boundary-Service boundary is the interaction or interconnection points between systems, processes or systems.

Service Bundles-Service bundles are the combining of different products and services into a "package" offer, and then offering it to a customer at a separate, combined price.

Service Bureau-A service bureau is a company that owns facilities or applications that it makes available to other users for a service usage fee. An example of a service bureau is a company that provides communication-switching services to a reseller of telecommunication services.

Service Class-Service classes are sets of communication parameters that are used or assigned to provide transmission flows that can provide services that meet specific quality of service (QoS) requirements.

Service Class Name-Service class name is a label or handle that identifies a group of characteristics that are associated with a specific service class.

Service Classifier-A service classifier is a device, assembly and/or software processes that can identify and categorize a connection or media type. After the service classifier identifies the underlying service (e.g. on connection system or by packet analysis), a service classifier may change information or append data of the header of a packet to enable alternative routing or to change the packet routing process (e.g. prioritization).

Service Context-Service context is the combination of resources and processes allocated or associated with an application.

Service Control Architecture-Service control architecture is the types of functions or processes in a system (logical elements) and how they are used to setup, manage and terminate services.

Service Control Point 800-A service control point database system that provides call processing information required for 800 number inquiries.

Service Control Reference Point (SCRP)-A service control reference point is a conceptual point in a system or network that defines service control commands and how they will interact with a service.

Service Creation-Service creation is the development of applications or processes that can be used by people, companies or devices to transfer, processes or to perform actions that they desire.

Service Creation Environment (SCE)-A development toolkit that allows the creation of services for advanced intelligent network (AIN) that is used as part of the signaling system 7 (SS7) network.

Service Data Unit (SDU)-Service data units are packets of data that contain the cumulative headers and trailers that are appended by the higher layers necessary to provide a particular type of service.

Service Degradation-Service degradation is a reduction in the ability of a system or operator to provide services to users or devices.

Service Delivery Platform (SDP)-Service delivery platform refers to hardware types and software programs that are required to deliver specific types of services.

Service Detail Record (SDR)-A service detail record holds information related to a service or group of service usage events related to a communication session. This information usually contains the origination address of the user, time of day the service or services were requested, the types and quantities of services used, and charges that may be added from supporting vendors throughout the duration of the service.

Service Discovery (SD)-Service discovery is the process of communicating with a device and obtaining information about its available services.

Service Duration-Service duration is the amount of time between when a service is setup or started to when the service ends or the process is terminated.

Service Element-A service element is a device or functional part of a system that is used to provide services.

Service Establishment-Service establishment is the process of requesting, negotiating and/or beginning to perform a service.

Service Facilitators-Service facilitators are individuals or companies who provide part of the elements needed by system operators to provide services.

Service Flexibility-Service flexibility is the ability of a communications medium to modify or add additional services to an existing platform or system.

Service Identifier (SID)-(1-cable television) A unique identifier code that is assigned by a cable modem termination system (CMTS) to a cable modem (CM). While each SID assigned by a CMTS is unique, a CM may have more than one SID. (2-billing) A unique number assigned by a billing system for a service type (such as a telephone number of circuit identification number).

Service Implementation Fee-A service implementation fee is a one-time fee that is charged for the addition or initial setup of a communication service.

Service Instance-A service instance is a component of a service that is distinct from the services provided.

Service Launch-A service launch is the process of making a product available for purchase and distribution with promotional efforts.

Service Level Agreement (SLA)-An agreement between a customer and a service provider that defines the services provided by the carrier and the performance requirements of the customer. The SLA usually includes fees and discounts for the services based on the actual performance level received by the customer.

Service Life-The period of time that equipment is or can be expected to be in active use.

Service Loop-A service loop is a bundle (loop) of cable that is part of a communication line (such as a pole mounted cable TV or telephone lines) that provides additional cable length that may be necessary to perform a splice or cable path reconfiguration at a later time. Outdoor cable loops can store 100 to 200 feet of cable for each 1,000 feet of installed cable line while indoor service loops typically store 10 to 20 feet of additional cable.

Service Management System (SMS)-A computer system that administers service between service developers and signal control point databases in the SS7 network. The SMS system supports the development of intelligent database services. The system contains routing instructions and other call processing information.

Service Mix-Service mix is the different types of services that an operator or service provider may provide to customers.

Service Model-A service model is a representation of how services will operate and what they can provide to users. These representations may be in mathematical formulas or as functional descriptions of processes that are used to define how a service will generate revenues or profits. Service models are used to simulate what a service will accomplish when it is provided with various conditions or opportunities.

Service Multiplexing-The multiplexing of different types of services such as voice, data and video onto one type of communications channel. Service multiplexing allows one piece of equipment and/or communication channel to provide services that have different delay, bandwidth and other quality of service (QoS) requirements.

Service Name-A service name is a label or a tag that is used to identify a particular service flow and its associated characteristics.

Service Number-(1-telephone number) A service number is a dialing code that is used to allow calls to be connected to a service facility. (2-repair) A service number is a trouble ticket identification number.

Service Number Portability (SNP)-Service number portability allows a customer to take their telephone number to a different type of service provider. Service number portability involves determination of the type of service provider (e.g., wireless or wired) who is responsible for completing the call using the telephone number (e.g. area code and NXX.) Service number portability may differ from local number portability as the interconnection and call processing for different types of service providers may vary.

Service Option (SO)-A service option is a feature or characteristic of a service that can be selected or changed.

Service Order (SO)-A service order is a record that describes a customer request to establish, change, or terminate a service. The service order contains all information required to meet a customer's needs.

Service Order Code (SOC)-Service order code is a unique identifier that is assigned to a service order. The SOC does not change over time as the service order is completed.

Service Oriented Architecture (SOA)-Service oriented architecture is a set of functions and their relationships that guides all aspects of creating and using business processes packaged as services. SOA can be applied throughout a service lifecycle, as well as defining and provisioning the IT infrastructure that allows different applications to exchange data and participate in business processes loosely coupled from the operating systems and programming languages underlying those applications.

Service Portability-Service portability is the capability of a communications customer to obtain their services through different types of access devices or systems.

Service Precedence-Service precedence is the hierarchy of the activation and operation of services in a communication system.

Service Profile-A service profile is a set of authorized services for a specific user, device or account and their associated characteristics or parameters.

Service Profile Identifier (SPID)-A service profile identifier is an 8- to 14-character identifier that is associated with an ISDN connection that is used to identify the services available on that connection. SPIDs are stored at the ISDN service provider company.

Service Protection-Service protection is the monitoring and control of access to services to prevent services from being improperly used or pirated in a communication network (such as in a mobile telephone system). Service protection involves uniquely identifying services, assigning the usage rights, requiring authentication and conditional access before being granted access the services.

Service Provider-A service provider is a person or company that provides information and/or performs actions (services) to customers.

Service Provider Portability-Service provider portability is the capability of a communications customer to change their selected service provider.

Service Provisioning-Service provisioning is the process of an authorized agent or process that processes and submits the necessary information to enable the activation of a service. For communication systems, this may include transmission lines, wiring, and equipment configuration.

Service Quality Churn-Voluntary churn that occurs when customers are dissatisfied with their service quality change carriers to get better quality.

Service Rates-Service rates are the fees that a user will pay for the transfer, processing, or authorizations to obtain benefits they desire. Service rates are typically divided into monthly fees and usage fees.

Service Record-A database record in a communication system that contains a description of a service and the attributes that are necessary (e.g. protocols) to use that service.

Service Regions-Service regions are the geographic areas that provide services. Examples of service regions for optical networks include metropolitan enterprise, metropolitan access, Backbone, terrestrial long-haul, and submarine long-haul.

Service Requirements-Service requirements are the communication capabilities and processes that are necessary to provide a communication service.

Service Specific Charge-A service specific charge is a fee or value assessment that is related to a characteristic or measured value of a service.

Service Subscriptions-Service subscriptions are registrations that entitle a recipient to receive and/or have access to services or multiple editions of a product (such as a magazine) over a period of time.

Service Template-A service template is a set of parameters and services that are likely to be used to setup and operate a service for a user or a device.

Servicemark-A unique symbol, word, name, picture, or design, or combination thereof, used by firms to identify their own services and distinguish them from the services sold by others. Servicemarks are Intellectual Property Rights which give the owner the right to prevent others from using a similar mark which

could create confusion among consumers as to the source of the services. Servicemark protection must be registered in the country, or region, where it is desired. In most countries and regions, servicemarks are administered by the same government agency which administers trademarks.

Services-(1-telecommunications) The provision of telecommunications to customers by a common carrier, administration, or private operating agency using voice, data, and/or video technologies. (2-performance) The overall quality of telephone system performance, sometimes stated in terms of blocking or delay.

Services Aggregator-A services aggregator is a company that obtains the rights to provide services from two or more companies and offers these services as a combined services package to customers.

Services URL-An address that is preprogrammed into or used by devices (such as an IP Telephone) where information is kept regarding services that may be used by the device. The services URL may be associated with a menu button or an icon on the display of an IP telephone.

Serving Area-(1-central office) A geographic area served by a central office or exchange. (2-outside plant) In outside plant, a distribution area that connects with a feeder route through a serving area interface.

Servlet-A Servlet (or a script) is a small program or sequence of operations (macro) that is written in a predetermined language that can be understood by the calling program to allow automatic interaction between programs or devices. Examples of servlets include login scripts that are used to provide identification information when accessing a system or Javascripts that provide advanced features on Internet web pages.

Session-Sessions are the time and activity between the operation of a software program or logical connection between two communications devices. In communications systems, the session involves the establishment of a logical channel with configuration transmission parameters, operation of higher level applications, and termination of the session when the application is complete. During a session, many processes or message transmissions may occur.

Session Based Charge-Session based charge is the measurement of a service that is provided during a communication session.

Session Identification Code (SessionID)-A session identification code is a unique number or character string that is assigned to a user or device that has initiated a communication session. The sessionID stays the same during the time the user initiates the communication until the session ends. The sessionID may be transferred along with web clicks to track the selections of the user as they navigate through web pages.

SessionID-Session Identification Code

SET-Secure Electronic Transaction

Set Aside-Set aside is the reserving or assigning of resources for a specific use or function.

Set Top Box (STB)-A set top box is an electronic device that adapts a communications medium to a format that is accessible by the end user. Set top boxes are commonly located in a customer's home to allow the reception of video signals on a television or computer.

Set Up Bonus-In reference to, on occurrence that a buyer makes an agreement with a studio in order to begin production of a property, the sum of additional money received by the owner in exchange for that property.

Settlement-A settlement is the financial difference between revenues generated from charging for services provided by a carrier to visiting customers less the fees that other carriers have charged for services provided to their customer that have visited other systems.

Settlement Agreement-A settlement agreement is a document or recordings that define the terms for two or more entities to agree on an action or resolution to a conflict.

Settlement Code-A code that identifies a billing message type and the preferred settlement procedures. Some sample settlement codes include interstate, intrastate, domestic, and overseas.

Settlement Reports-A settlement report is a statement that summarizes the transactions for services provided between two or more companies and the value(s) that were or are to be transferred to settle the overall costs of the transactions.

Settlements-Settlements are the processes or amounts of transfer of property or resources that settle the usage of products or services between companies.

Settlements Procedures-A process for distributing revenues among carriers in proportion to services provided or assets used.

Setup Routine-A software program or process that coordinates the installation of a program or system.

Setup Screen-A setup screen is a menu or series of menus that allow users to configure equipment or service features.

SGML-Standard Generalized Markup Language

Shadow Network Capacity-An amount of network capacity that the network service provider has that can be utilized to produce revenue that is unaccounted for in the current financial statements. Shadow network capacity exists because it may not be readily measurable or it has already been depreciated and amortized.

Shared Bandwidth-A characteristic of a communications channel in which the available capacity is shared among all of the attached stations.

Shared Folder-A shared folder is a directory area on a computer that can be viewed and possibly modified by other users. A shared folder may be available to all users or its access may be restricted or through use of a password.

Shared Hosting-Shared hosting is the providing of application services (such as virtual telephone service or the providing of web service) for the benefit of a client or customer where multiple companies or customers share the same resources.

Shared Key Authentication-An authentication process that shares a key between two or more users.

Shared Network-Shared networks are systems that allow companies to share and control the resources of other networks or systems.

Shared Revenue-Shared revenue is a process where a company shares revenue for the providing of services or content. Shared revenue services allow a service provider to provide a portal (access point) and billing for the delivery of media content or information services and to split (share) the revenue with the content provider.

Shared Server-A shared server is a computer or processing device that can be used by multiple users over the same time period.

Shared Tenant Service-The sharing of centralized equipment, facilities, and telecommunication services by occupants in a building or office complex.

Shareholder-A shareholder is a person, company or entity that owns a portion (as share) of a company.

Shareholder Agreement-A shareholder agreement is a contract between some or all of the shareholders of a corporation that defines the rights and obligations of the shareholders. Shareholder agreements are commonly used in tightly controlled corporations where the actions of shareholders could dramatically affect the value of the other shareholders (such as selling stock to competing companies).

Shareware-Shareware is files, programs, or applications that provide a potential user to obtain the program(s) and trial the programs before deciding to keep or purchase the software. After shareware is installed, it is usually setup to operate for a predetermined period of time (e.g. 30 days) before it deactivates itself. If the user conforms to the terms of the shareware (usually through the payment of a fee), an access code is provided that activates the shareware program.

Shawdowing-Shadowing is the process of maintaining a duplicate information or a copy of a data file and making modifications to that file in step with the modifications begin made by the user to the main version of the file.

Sheath Miles-The number of route miles installed (excluding pending installations) along a telecommunications cable path multiplied by the number of wires or fibers existing within cabling along the same path.

Sheet Engine-A sheet engine is a software program that can receive, process and respond to an end user's (client's) data entries into a spreadsheet.

Shell-A shell is a portion of a software system that allows users or programs to interact with the system without directly communicating with the underlying functions or processes. The use of a shell protects the overall operation of a system and the incorrect activation of a function or procedure can stop a program from operating and a shell can restrict the options or operations available to the users to help ensure more reliable operation.

Ship Weight-Shipping Weight

Shipper Export Declaration (SED)-A shipper export declaration is a listing of items that are being exported along with destination shipping details.

Shipping and Handling-Shipping and handling is a fee to cover packing and postage to get the item to you.

Shipping Manager-A shipping manager is a program or system that coordinates the setup, addition and configuring of shipping options and rates for an online store order processing system.

Shipping Method-Shipping method is the choice of which type of package or service provider will be used to transfer a product or service. Examples of shipping methods include ground, 2nd day and overnight.

S

Shipping Options-Shipping options are the methods of shipment that are offered or presented to a customer of a product or service.

Shipping Rate-Shipping rate is the amount that will be charged or assessed for the shipping of a product or item. Shipping rates are typically determined using a shipping rate schedule.

Shipping Rates-Shipping rates are a list of charges or fees that are associated with shipping methods and package characteristics (such as weight).

Shipping Rules-Shipping rules are the conditions and associated processes (such as calculations) that are used to select and calculate shipping fees and processes.

Shipping Weight (Ship Weight)-Shipping weight is the combined weight of a product and its packaging.

Shopping Cart-Shopping carts are the electronic containers that hold online store items while the user is shopping. The online shopper is typically allowed to view and change items in their shopping cart until they purchase. Once they have completed the purchase, the items are removed from their shopping cart until they start shopping again.

This diagram shows the basic operation of a shopping cart. This example shows that the shopping cart places the customers order information into a cookie (small memory area) on their own computer along with a user identification code assigned by the online store. This allows the customer's order information to remain available if they exit the store and renter at a later time.

Short Delivery-A short delivery is providing of some but not all of the items listed on an invoice or purchase order.

Short Message Service (SMS)-A messaging service that typically transfers small amounts of text (several hundred characters). Short messaging services can be broadcast without acknowledgement (e.g. traffic reports) or sent point-to-point (paging or email). Most digital cellular systems have SMS services. Short messaging for mobile telephones may include: numeric pages (dialed in by a caller), messages that are entered by a live operator via keyboard, an automatic message service that sends a predefined message when an event occurs (such as a fire alarm or system equipment failure), network operator announcements to customers, to and from other message capable devices in the system, from the Internet, advertisers or other information providers.

Short Message Service Center (SMSC)-A short message service center is a system or facility that processes and routes short messages to telecommunications devices (such as a mobile telephone.)

Shovelware-Shovelware is a sarcastic name for a software program that has been created through the use of readily available programs or information. Shovelware is created by using (shoveling) multiple resources to create a new product that may not have significant value to its users.

Show Stopper-A show stopper is an event or action that results or would result in the inability to continue operations as planned.

Shrinkage-Shrinkage is the amount of goods from an inventory that are lost or unaccountable. A significant cause for shrinkage may be theft.

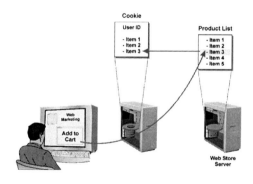

Online Store Shopping Cart

S-HTTP-Secure Hypertext Transfer Protocol

SI-Système International

SIC-Standard Industry Classification

SID-Service Identifier

SIG-Special Interest Group

Sig File-Signature File

Sign a Message-The process of adding a digital signature to a message. The digital signature is calculated from the contents of the message using a private key and appended or embedded within the message. The signing of a message allows a recipient to check the validity of a file or data by decoding the signature to verify the identity of the sender.

Sign Off-Sign off is the process of approving a project or contract.

Signal Quality-Signal quality is a measure of how well a signal represents characteristics or attributes of the information it transports (voice, data or video).

Signaling-Signaling is the process of transferring control information such as address, call supervision, or other connection information between communication equipment and other equipment or systems.

Signaling Network-A signaling network is a system that receives, processes, and distributes messages that control a system. A signaling network may be an independent communication system or it may share some or all of the resources of the communication system it controls. An example of a signaling network is the Signaling System 7 (SS7) network. The SS7 network is a packet data communication network that is used to control public switched telephone networks. The SS7 system consists of signaling points that create or receive control commands (such as switches), signaling transfer points (for distribution of control messages), and control points that can receive and process requests for more advanced control features (such as toll free call routing information).

Signaling System-A system that receives, processes and sends control information. Signaling may be in-band (replaces voice or data information) or out-of-band (is sent separately from voice or data information).

Signaling Tone-A tone that is used to indicate a status change or to transfer a signaling message on a communication system. The signaling tone is mixed with or replaces an audio signal in a communication system. Signaling tones are used on analog mobile communication system between the mobile station and the base station to indicate event changes such as handoff, the end of a call, or a special service request (e.g. hookflash).

Signature-(1-identifier) A pattern or image that uniquely identifies a person or process. (2-digital) A number calculated from the contents of a file or message using a private key and appended or embedded within the file or message. The inclusion of a digital signature allow a recipient to check the validity of the file or data by decoding the signature to verify the identity of the sender. (3-email) A short amount of information that is added at the end of each email message a person sends.

Signature File (Sig File)-A signature file is a media file that holds information that can be automatically attached to e-mail messages that are sent.

Signing Hierarchy-A signing hierarchy is a system of authenticated documents or electronic keys that are signed and are arranged in successive levels of security trust.

Silent Churn-The process where customers change their usage patterns in preparation for conversion (disconnection) from one carrier (service provider) to another. Silent churn can involve the complete or partial disconnection of service(s) by the customer. Silent churn is usually indicated by a reduction in usage of a particular service.

Silent Point Security-Silent point security is the unauthorized use, duplication or distribution of media which is performed or assisted by people who are employees, contractors or supporting personnel of companies that produce, distribute or manage the program materials.

Silo-A storage of information that is not accessible by one or other information systems.

SIM Applet-Subscriber Identity Module Applet

SIM Cards-Multi-Function Subscriber Identity Module

Simple Interest-Simple interest is the loan fee that is calculated on the outstanding principle of the loan.

Simple Mail Transfer Protocol (SMTP)-Simple mail transfer protocol is a set of text commands and processes (text based protocol) that is used to move e-mail messages from one e-mail server to another. SMTP provides a direct end-to-end mail delivery, rather than a store-and-forward protocol. The e-mail servers run either Post Office Protocol (POP) or Internet Mail Access Protocol (IMAP) to distribute e-mail messages to users.

Simple Network Management Protocol (SNMP)-Simple Network Management Protocol (SNMP) is a standard protocol used to communicate management information between the network management stations (NMS) and the agents (ex. routers, switches, network devices) in the network elements. By conforming to this protocol, equipment assemblies that are produced by different manufacturers can be managed by a single program. SNMP protocol is widely used via Internet protocol (IP) and operates over UDP well-known ports of 161 and 162. SNMP was originally defined in RFC1098 and is now obsolete and updated by RFC1157.

This figure shows how communication equipment can be configured using SNMP. This diagram shows that one ore more network management systems (NMS) can be connected through a data communication network to devices that can be configured using SNMP messages. Commands (queries) are sent from the NMS to SNMP capable devices (such as a communications router). A portion of the processing section of the communication device contains agent software. An SNMP agent can receive and respond to SNMP commands. The

agent can store and retrieve configuration settings and parameters from the management information base (MIB) that is stored within the device. SNMP capable device can send information to the NMS using event messages.

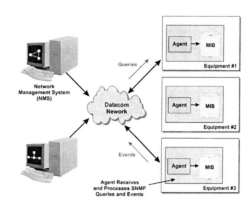

SNMP System Operation

Simplified Network Time Protocol (SNTP)-Simplified network time protocol is a set of commands and processes that allow the requesting and delivery of time information through Internet protocol (IP) networks. SNTP is defined by RFC1769 and RFC2030.

Simulated Annealing -The process of integrating "noise," conflict, and discord into the operational business model in order to encourage change and flexibility within the organization.

Simulation-A mathematical model that employs physical and mathematical quantities to portray a real-life situation and movement conforming natural phenomena.

Simulation Tool-A simulation tool is a software application or service that analyzes and presents information that estimates the results that occur in a system or device when certain conditions occur. Simulation tools may be used to help planners analyze system performance without the need to use the network or system to obtain the results

Simulcast Transmission-Simulcast transmission is a process of transmitting a radio signal on the same frequency (or frequency this is very close) from multiple locations to allow the radio coverage area from adjacent radio transmitters to overlap. This overlap of radio coverage helps to ensure the radio signal more evenly covers a geographic area (preventing dead spots). Simulcast is extensively used in radio paging systems.

Because the transmission time from each of the transmitted signals may not be the same (one of the transmitter towers may be closer to the mobile radio than the other), the information sent by adjacent transmitters may be synchronized in time so that it arrives at the same approximate amplitude and phase as the other transmitter. Otherwise, this may cause radio signal distortion that could cause transmission errors. This is especially important in radio transmission systems that operate at high data rates (e.g. 9600 bps compared to 1200 bps).

This figure shows a paging system that uses simulcast transmission. In this example, the same paging message is sent to two paging transmitters. As can be seen in this example, the challenge with simulcast paging is as the pager is closer to one tower then the other, the transmit delay time can cause the signals to not directly overlap. Because radio signals travel so quickly, this delay is minor. However, it can result in some dead spots due to signal adding or subtracting. This diagram also shows that the ability to simulcast also depends on the distance and data rate of each of the transmitted signals to the radio (paging) receiver.

Simulcasting-Simulcasting is the process of transmitting multiple media channels at the same time.

Single Award Schedule (SAS)-A single award schedule is listing of contracts or projects that are provided to a single supplier.

Single Data Message Format (SDMF)-The caller identification format that transfers only the calling telephone number information.

Single Point of Failure-A single point of failure is potential of a failed device or functional part of a system to disable the overall operation of the system.

Single Source-Single source is the using of one vendor or supplier for all the component, service or product needs.

Site License-A site license is an authorization to use products or software at a specific location or locations.

Site Review-Site review is an inspection of a location to determine existing and/or upgradable capabilities of devices, systems or facilities.

Skill Sets-Skill sets are groups of abilities that are developed through learning. Skill sets can be developed through training, demonstrations, practice drills and/or reviewing safety reference materials.

Skills Based Call Routing-The routing of an incoming call via an automatic call distribution (ACD) system to an extension that has a specific skill set (e.g. salesperson) based on information gathered by the caller (e.g. telephone DTMF key presses).

Skin-Skin is the background or surrounding area of a device or display area.

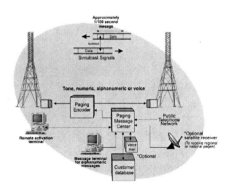

Simulcast Transmission

SKU-Stock Keeping Unit

SLA-Service Level Agreement

Slamming-Slamming is the unauthorized transfer of customer's preferred service provider to a different service provider.

Slander-Behavioral or verbal acts as to misrepresent one in a derogatory or negative way.

SLC-Subscriber Loop Carrier

SLIC-Subscriber Loop Interface Circuit

Slotting Allowances-A slotting allowance is a fee that is paid by a supplier or manufacturer to a retailer for the allocation (slotting) of shelf space for the product.

SM-Security Manager

Small Exchange Carrier Access Billing (SECABS)-Small exchange carrier access billing is a billing standard that was developed by Alliance for Telecommunications Industry Solutions (ATIS) to enable billing records to be exchanged between small service providers.

Small Footprint-A small footprint is a product or software program that only requires a relatively small area or resource to be installed or operate.

Smart Aggregation-Smart aggregation is the use of detailed descriptive information about content that allows for the searching, categorizing and selecting content that helps predict its utility or value to a consumer or group of consumers. For example, smart aggregation allows for the recommendation of programs that directly relate to the desires or viewing habits of specific customers.

Smart Card (SC)-A smart card is a portable credit card size device that can store and process information that is unique to the owner or manager of the smart card. When the card is inserted into a smart card socket, electrical pads on the card connect it to transfer information between the electronic device and the card. Smart cards are used with devices such as mobile phones, television set top boxes or bank card machines. Smart cards can be used to identify and validate the user or a service. They can also be used as storage devices to hold media such as personal messages and pictures.

Smart Switch-Smart switches are interconnection switching systems that can dynamically change its connection points, data rates and services that they provide. Smart switches are used to provide for voice, data, video, and advanced call processing (call feature) services.

This diagram shows how a smart switch can efficiently route information more directly to computers connected within its network. A smart switch builds a routing table based on a device's medium access control (MAC) address. As the data is addressed to a destination computer, the smart switch dynamically maintains a list of the computer addresses that are connected to each of its port's. After it has updated its table, packets are routed directly towards the port that the MAC address table has identified for that address.

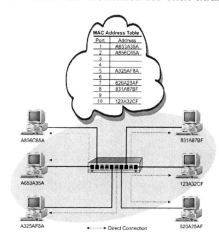

Smart Switch Operation

Smart Telephone (SmartPhone)-A SmartPhone is a telephone device that includes intelligent call processing to provide enhanced audio and display capability. Some of the advanced features offered by SmartPhones are communication list management, enhanced caller information displays, and information storage services.

Smart Terminal-An interface device (terminal) that has both independent (local) computing capability and the ability to communicate with other devices or systems.

SmartPhone-Smart Telephone

SMDR-Station Message Detail Recording

SME-Subject Matter Expert

SMS-Screen Management System

SMS-Service Management System

SMS-Short Message Service

SMS-Subscriber Management System

SMS Service Centre (SC)-The SMS service center (SC) receives, stores, delivers, and confirms receipt of short messages.

SMSC-Short Message Service Center

SMTP-Simple Mail Transfer Protocol

Snail Mail-Snail mail is an Internet term for the process of sending messages or products through the postal service. It is called snail mail because email is near instant and the rate of delivery for the postal service seems as slow as a snail when compared to instant delivery.

Snarfing-Snarfing is the acquiring of documents, media or files without a specific purpose for the use of the documents without the permission or authorization of the owner. Snarfing may be performed using USB memory sticks or devices that have storage capability (e.g. personal digital assistant) that also have connection capability (e.g. Bluetooth).

Snark-A 3 headed beast

Sneakernet-The process of manually delivering information (such as letters) through a network of information users by walking between the distribution points.

SNI-Standard Network Interface

Sniffer-A sniffer is a device or program that receives and analyzes communication activity so that it can display the information to a person or communication system. While sniffers may be used for the analysis of communication systems, they are often associated with the capturing and displaying of information to unauthorized recipients. For example, a sniffer may be able to be setup to look for the first part of any remote login session that includes the user name, password, and host name of a person logging in to another machine. Once this information is captured and viewed by the unauthorized recipient (an intruder or hacker), he or she can log on to that system at will.

SNMP-Simple Network Management Protocol

SNP-Service Number Portability

SNTP-Simplified Network Time Protocol

SO-Service Option

SO-Service Order

SOA-Service Oriented Architecture

SOC-Service Order Code

Social/Psychological Churn -Type of voluntary churn that occurs when customers change carriers in order to enhance their image or self esteem. Often teenagers will churn for these reasons.

Soft Buttons-Soft buttons or definable keys are areas on a display that have the ability to redefine their functions. Soft buttons are typically located adjacent to a display that provides a description of the key function. This allows an electronic assembly to reduce the number of keys, which is especially important for portable handheld telephones.

Soft Console-An access point to an information system that is defined by software that allows users to monitor and/or interact with a system. The use of a soft console allows the controllable features of a console to change based on the authorization and console program software.

Soft Currency-Soft currency is a monetary exchange value that has limited exchange options with other currencies throughout the world.

Soft Goods-Soft goods are products that are in electronic formats. Examples of soft goods include eBooks, music, documents and program application files.

S

Soft Issues-A soft issue is a problem or identifiable attribute that can be analyzed and investigated as a cause of a condition that influences a performance area where the issue is hard to define or measure.

Soft Keys (Softkeys)-Soft keys are buttons on the keypad of an electronic device that have the ability to redefine their functions. Soft keys are typically located adjacent to a display that provides a description of the key function. This allows an electronic assembly to reduce the number of keys which is especially important for portable handheld telephones.

Soft Launch-Soft launch is the process of making a product available for purchase and distribution without significant promotional efforts. Companies may use a soft launch strategy for new products that may have limited available quantities or to test product acceptance and distribution channels prior to a hard launch.

Soft Phone-A soft phone is a software program that operates on a multimedia computing device (such as a personal computer or digital television) to operate as an Internet telephone.

Soft Sell-Soft Selling

Soft Selling (Soft Sell)-Soft selling is the process of using limited conflict communication techniques that help the buyer to identify a need that develops into a desire to purchase products or services.

Soft Switch (Softswitch)-Softswitches are call control processing devices that can receive call requests for users and assign connections directly between communication devices. Soft switches only setup the connections, they do not actually transfer the call data.

Softswitches were developed to replace existing end office (EO) switches that have limited interconnection capabilities and to transfer the communication path connections from dedicated high-capacity lines to other more efficient packet networks (such as packet data on the Internet). This allows a single softswitch to operate anywhere without the need to be connected to high-capacity trunk connections.

Softkeys-Soft Keys

Softswitch-Soft Switch

Software-Software is computer processing data or information.

Software Application-A software application is a software program that performs specific operations to enable a user to apply the software to their specific needs or problems. Software applications in devices may be in the form of embedded applications, downloaded applications or virtual applications.

Software Architecture-Software architecture is the organization of the elements or processes and the relationships between them within a software system.

Software Components-Software components are blocks or sets of program instructions or data that perform functions within a system or computer application.

Software Module-A software module is a program that is a self-contained process. Software modules may be combined with other programs to provide more advanced processes.

Software Specification-A software specification is a document that describes the requirements and features of a software product or application.

Software Tools-Software tools are services or application programs that help developers and programmers to create new services, programs or products.

Software Updating-Software updating is the process of transferring portions of software (patches) or newer versions of software programs (upgrades). Software updates may be performed to correct for operational deficiencies (software bugs) or to add a new feature and/or performance capabilities. Software updates may be performed with the assistance of the user (in the foreground) or it can be performed without the assistance or knowledge of the user (in the background.)

Sole Proprietorship-A sole proprietorship is a form of business where a single person is responsible and liable for the performance and obligations of a business.

Solicitation-Solicitation is the process used to inquire to people or companies if they have an interest in a proposal.

SOP-Standard Operating Procedure

Source Code-Source code is software program instructions that are written in a language that is translated (compiled) into the operating code that a computer can understand.

Source Document-A source document is an original source of information or media formatting.

Source File-A tile on disk that has programming language statements for a specific assembler or compiler. From this data, the assembler or compiler creates an object file which is in machine code or language.

Sourcing-Sourcing is the use of firms or suppliers to obtain materials or services that enable the production of products and assemblies or to perform specific business functions.

South Asian Association for Regional Cooperation (SAARC)-The south Asian association for regional cooperation is an economic and political group of countries created in 1985 covering southeast Asia. In 2008 it consisted of India, Pakistan, Bangladesh, Sri Lanka, Nepal, Maldives, Bhutan, and Afghanistan.

SOW-Scope of Work

SOX-Sarbanes Oxley Act

SPAM-SPAM is unwanted and unsolicited email or messages that are commonly sent by a seller to promote products or services.

SPAM Act-The SPAM Act is a US regulation that was authorized in 2003 to better control the sending of undesired email messages. The CAN-SPAM act defines the necessary relationship requirements before commercial emails can be sent to recipients.

Spam Filter-A spam filter is a process that is designed to redirect, block or remove unwanted mail.

Speakerphone-An audio terminal, consisting of a transmitter and loudspeaker unit, that is part of a telephone for hands-free conversations.

Spec Sheets-Specification Sheets

Special Interest Group (SIG)-A group that works to help develop and promote information about a specific technology, product, or service. A SIG is usually part of or related to an industry association.

Specialty Distribution-Specialty distribution is a route that a product service uses to get from the original manufacturer or supplier that has well defined characteristics that are related to the industry, product or buyer of the product or service.

Specification Sheets (Spec Sheets)-Specification sheets are documents or media items that contain product information such as features, modes of operation and other characteristics.

Specifications-Specifications are documents that describe the essential technical requirements for items, materials or services including the procedures by which it will be determined that the requirements have been met.

Speculative Sample-A speculative sample is a product or service that is provided to a potential distributor that may contain unique characteristics that may interest the prospective buyer (such as their logo or preferred company colors).

Speech Coder-Speech coding (also called voice coding) is a data compression device that characterizes and compresses digital speech information. A speech coder is also called a Coder/Decoder (CoDec).

Speech Coding-Digital speech compression (speech coding) is a process of analyzing and compressing a digitized audio signal, transmitting that compressed digital signal to another point, and decoding the compressed signal to recreate the original (or approximate of the original) signal.

This diagram shows the basic digital speech compression process. The first step is to periodically sample the analog voice signal (5 - 20 msec) into pulse code modulated (PCM) digital form (usually 64 kbps). This digital signal is analyzed and characterized (e.g. volume, pitch) using a speech coder. The speech compression analysis usually removes redundancy in the digital signal (such as silence periods) and attempts to ignore patterns that are not characteristic of the human voice. In this example, this speech compression processes use pre-stored code book tables that allow the speech coder to transmit abbreviated codes that represent larger (probable) digital speech patterns. The result is a digital signal that represents the voice content, not a waveform. The end result is a compressed digital audio signal that is 8-13 kbps instead of the 64 kbps PCM digitized voice.

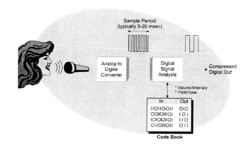

Speech Coding Process

Speech Compression-Speech compression is a technique for converting or encoding audio (sound) information so that a smaller amount of information elements or reduced bandwidth is required to represent, store or transfer audio signals.

Speech Recognition-Voice recognition is the ability of a machine to recognize your particular voice. This contrasts with speech recognition, which is different. It is the ability of a machine to understand human speech — yours and everyone else's. Voice recognition needs training. Speaker independent recognition does not require training.

Speed Dialing-A system or device that allows for automatic dialing of telephone numbers.

Spending Limits-Spending limits are established thresholds that trigger specific actions such as restricted access (service denied) or the creation of alert messages (informing the account guardian).

SPID-Service Profile Identifier

Spiff-The providing of marketing incentives from one company directly to sales representatives of another company (usually a retail sales company). Spiffs are often used to focus sales representatives on demonstrating and giving preference to products or services from a specific manufacturer.

Spin-Offs-(1-television) Spin-offs are media programs (such as a television series) that evolve from a previous media program (such as a movie or other television series). (2-business) Spin-offs are companies or business units that were created (spun off) from another business entity.

Splash Page-A splash page is a web page that has rich media (e.g. animation or video) that is used to make a significant impression (get the attention) of the viewer.

Split Security-Split security is the dividing of a system into components or segment that use different security processes or different keys. An example of split security is separation of the conditional access system that is used to distribute television signals from a content owner to cable television head end from the conditional access system that is used to transport television signals from a headend to an end user.

Sponsor-A sponsor is a company, person or group who provides funding, support or resources for a media program or service.

Sponsored Ringtones-Sponsored ringtones is a segment of media that can downloaded into a communications device so that it can be played when an alerting condition occurs which are paid for by a company or person other than the recipient.

Spoofing-Spoofing is the use of another persons name or address to hide their true identity.

Spooler-A computer program that queues input for later output. For example, a print spooler can accept files at a high transfer rate, then send them to a printer at whatever rate that printer can handle.

Spot Sales-Spot sales are services (e.g. advertising ad insertions) that are sold in a particular region (spot).

Spreadsheet-A spreadsheet is a data presentation format where information or processes are split up into a grid of cells.

Spyware-Spyware that is software that resides on a users' computer that provides information about the user's actions on the computer to a remote location. Spyware program operations may range from simply tracking web browsing habits to the providing of private information such as user account identification and password codes. Spyware may be installed on a users' computer without the user's knowledge or consent.

SQL-Structured Query Language

SQL Server-Structured Query Language Server

SRM-System Renewability Messages

SS-Subscriber Station

SS-Supplementary Services

SSH-Secure Shell

SSL-Secure Sockets Layer

SSL Encryption-A SSL encryption is modifying data for transmission using a secure socket layer (SSL).

Stakeholder-A stakeholder is a person or company that has an interest in the operations or status of a process or business.

Stand Alone Software-Stand alone software are computer programs that can operate without the need for additional services.

Standard-A standard is the operational descriptions, procedures or tests that are part of an industry specification document or series of documents that are recognized by people or companies as having validity or acceptance in a particular industry or technology.

Standard Bearers-Standards bearers are groups of people who work to create or oversee the creation of industry standards.

Standard Generalized Markup Language (SGML)-Standard generalized markup language is an international standard for annotating text with marginal notes or comments. SGML facilitates the exchange of text among different applications, especially computer-based text processing. SGML was standardized in 1986 standard as ISO 8879. SGML defines the formatting a digital document so that it can be modified, viewed, or output on any computer system. SGML documents consist of two parts, a document type definition (DTD) which defines the structure of the document and a document instance (DI) which describes the text or data in the document. SGML was the basis for HTML.

Standard Industry Classification (SIC)-A standard industry classification is a label or code that is assigned to a category of companies or products. Governments may assign or user SICs to help them process items or statistics related company or product types.

Standard Network Interface (SNI)-A device that provides a connection between customer premises equipment and a public switched telephone network. This interface isolates the two systems to the systems from each other (example: protection from lightning strikes).

Standard Operating Procedure (SOP)-A procedure (or set of procedures) that are used (referred to) to ensure the same repeated step by step actions are followed each time an action is required. SOPs are often used to ensure the same quality of service or to ensure the reliable configuration and testing of equipment.

Standardization Body (Standards Body)-A standards body is a group, associated companies or people that produce or oversee the production of industry standards.

S

Standards-Documents that describe an agreed-upon way of doing things such that independent groups or companies can design and build hardware, firmware, software, or combinations there of and have them inter-work with similar products designed and built by others.

Standards Body-Standardization Body

Standards Committee-(1-general) Committees that are commissioned to devel-oped industry standards or operating proce-dures. Standards committees allow technical experts from several manufacturers (often from competing manufacturers) with a defined objectives and generally have a commitment to unbiased development activities. (2-humor) An excuse to travel to interesting places at nice times of the year to drink beer and discuss arcane technical issues with 400 geeks.

Standards Development Organizations (SDO)-Standards development organizations are groups, associations or companies that assist and/or oversee of industry specification documents (industry standards).

Standards Group-A standards group is orga-nization or firm that is charted to create or assist in the creation of product or service operational, performance and other types of specifications. Standards groups usually rep-resent the interests of a group of companies that want to create products or services that are interoperable with each other.

Startup Cost-Startup costs are the fees or charges associated with establishing the ini-tial operations of a business or portion (e.g. division) of a business.

Start-up Fee-A start-up fee is a one-time fee that is charged for the initial setup of a com-munication service. Start-up fees are also called "activation fees."

Startup Venture-A startup venture is a busi-ness or company that has recently been setup.

State-(1-general) The condition or process or system at specific point in time. (2-network element) The operational condition of a net-work part or assembly.

State Information-State information is a set of parameters and/or configurations that rep-resent a specific state of a device, service or system.

Stateful-Stateful is a system or process that remembers and adapts processes for informa-tion or events that have occurred at previous times.

Statement-(1-programming language) A statement is a single unit of program com-mand in a high-level language that is used to define a process or set parameters for objects or variables. (2-billing) A list of invoices and payments for a specific user or account.

Static-(1-general) Static is unchanging over time. (2-electric charge) Static is an unchang-ing value (such as an electrical charge) that exists across capacitor plates. (3-noise) Static noise is audio sounds that are created in a sys-tem (such as a radio receiver) that result from electrical signals that were not created or transmitted (noise).

Static Address-A static address is a unique identifier that remains fixed over time.

Static IP Addressing-Static IP addressing is the process of assigning a fixed Internet Protocol (IP) address to a computer or data network device. Use of a static IP address allows other computers to initiate data trans-mission (such as a video conference call) to a specific recipient.

Static Positioning-Static positioning is the process of gathering location information while the user or location-receiving device is at the same relative position as compared to their reference points. Static positioning may take advantage of recording position informa-tion over extended time periods allowing for the averaging of biases and errors.

Station-(1-general) A device that connects to a communication system that provides input and/or output functions. (2-telecommunica-tions) An installed telephone, computer, or other communications instrument that com-munication service is used or operated.

Station Address-A grouping of numbers that uniquely identifies a station in a local area network. Data transmitted from the station will bear this grouping of numbers as a source address, and data destined for this station will bear this grouping of numbers as a destination address.

Station Hunting-A telephone call-handling feature that causes a transferred call to "hunt" through a predetermined group of telephones until it finds an idle station.

Station Message Detail Recording (SMDR)-A station message detail recording system is the hardware and software services unit that can be integrated with telephone systems (such as a PBX system) or it can be an independent system (such as a computer). The SMDR captures records of all the calls that are made from stations (telephone devices) within a system. A SMDR system is sometimes called a call accounting system.

Statistics-Statistics is the mathematical analysis and representation of information that can be qualified and quantified. The use of statistics allows for the expression of characteristics of information or data in a form that can be used to estimate or predict how additional quantities or future expectations of the information or data will change.

Status-The present operational condition of a device or system.

Status Code-A code that indicates the status of a message or communication session. An example of a status code is a message that indicates that a communication session is active.

Statutory License-A statutory license is a legal authorization that grants specific rights to use of intellectual property based on a legal statute.

STB-Set Top Box

Step Discount-Step Wise Discount

Step Wise Discount (Step Discount)-A step wise discount is a reduction in price that occurs when a particular level or tier of purchases or service levels have been achieved or exceeded.

Stepped Events-Stepped events are assignments of rates or rules for services that vary based on a number of events that have occurred relative to the service. An example of stepped events is a rate of $3 each for the first 5 items purchased and a rate of $2 each for the items above 5.

Stickiness-Stickiness is a measure of the effectiveness of keeping a customer or how long a customer stays on a web site or continues to do business with a company.

Sticky-Sticky is the keeping of a customer or maintaining of a web visitor for an extended period of time.

Sticky Customer-A sticky customer is a person or company who regularly purchases products or subscribes to a service and tends to stay with (sticks to) or use the same company or person who provided the products or services.

Stock Keeping Unit (SKU)-A stock keeping unit is an item and the package size that is used to store or dispense it. Each different package size has a distinct identification number, often represented by a bar coded number on a label. For example, one SKU for electrical resistors is a package of 10 resistors. Another SKU is a roll comprising 1000 resistors held by disposable tapes so that the roll can be used in an automatic component inserting assembly machine.

Stock Options-Stock options are the rights to purchase stock or equity in a company at a pre-defined price.

Stockpile-Stockpile is the storing of inventory or materials in relatively large quantities.

Stop Date-A predetermined or estimated date where an individual is retired from his or her obligated task, putting a stop to further services, can occur prior to completion of correlating project.

Stop Lock-A stop lock is a feature on a system or device that prohibits unauthorized

S

and/or unqualified users from stopping a device from operating. A stop lock feature may be used on a video server or media converter to prevent users or operators from accidentally stopping the device from operating.

Storage-(1-non-volatile) A storage medium in which information can be retained despite the absence of power, and which becomes available again as soon as power is restored. (2-parallel) A storage medium in which all bits, characters, or words are equally available, access time not being dependent upon the order in which they were stored. (3-serial) A storage medium in which access time varies according to the order in which information was stored. Storage can be serial by word, by character or by bit. (4-volatile) A storage medium in which information is lost whenever power is removed from the store.

Storage Area Network (SAN)-A storage area network is a set of data storage devices. A cluster of servers and storage devices that share common resources and users defines the "storage area". Occasionally, the term may be used for a wide area or metro area network that is used for data center redundancy and disaster protection.

SANs typically have very low latency, high throughput and offer assured delivery of block I/O. The most common SAN implementation is Fibre Channel, however this is not the only alternative. Recently there have been efforts, such as iSCSI to implement storage networks with Ethernet and IP (Internet Protocol) infrastructure.

Storage Capacity-The quantity of information that can be retained in a memory system, usually measured in MegaBytes, Gigabytes, or TeraBytes.

Storage Lifetime-Storage lifetime is the duration information can be reliably accessed on a storage medium.

Storage Management-Storage management is the process used in a computing system that manages the assignment and use of media storage in a device or system. Storage management may include short-term storage (RAM or ROM memory) or long term storage (hard disk drive).

Store Editor-A store editor is a software application or service that can be used to modify online store web pages and their associated data links. Store editors commonly include the ability to setup, edit, and delete product pages.

Stored Media-Stored media is a medium in which information can be retained despite the absence of power, and which becomes available again as soon as power is restored.

Stored Value Card-A stored value card is an instrument or identification code (such as a prepaid card) that holds a value or that can reference a value which indicates an amount of services or products that are available to exchange for the value on the card.

Storefront-The location of a store, person, or companies that sells products and/or services. The word "storefront" is sometimes used as a short form of "Online storefront." An online storefront is the web page of a store, person, or companies that sells products and/or services online.

Straight Rebuy-A straight rebuy is the ordering of products, items or model numbers that have been previously purchased.

Streaming Services-Streaming services are audio and video services delivered to customers where the content stays on the server. The delivery of information in streaming form has two main advantages: 1-The customer doesn't have to wait until the entire file downloads to view or listen to the content, 2- The content is kept on the server helps to reduce the copyright fears among content owners.

Streaming Video-Streaming video is the continuous transfer of motion picture information that is sent through a communications network. Streaming video commonly refers to audio and/or video images that are sent through a data network (such as the Internet).

Strengths, Weakness, Opportunities, and Threats (SWOT)-The analysis of strengths, weaknesses, opportunities and threats that compares the strategic position of a company in a given industry or situation.

Structure-Structure is the element types and their relationships with an object or service.

Structured Query Language (SQL)-Structured query language is a set of commands and processes that can organize and process database information. SQL is both a data descriptive language (DDL) and a data manipulation language (DML).

Structured Query Language Server (SQL Server)-An SQL server is a computing device that provides access to database files that uses the SQL database language to find, select and process data records.

STS-System Time Stamp

Studio-A studio is a room or facility that is designed for the recording of audio, video or other form of media.

Studio-Movie Studio

Stuffing Bytes-Stuffing bytes are bits of data that are inserted into a payload of a data packet or a portion of a transmitted signal to fill in the remaining area that has been designated for storage or transmission.

STV-Subscription Television

Style Guide-A style guide is a set of rules or recommendations that help to define how documents, media or processes should appear.

Style Sheet (Stylesheet)-A style sheet is a data file that describes the layout (style) of a document or media file. A style sheet may define the spacing, font types, colors and other characteristics within an electronic document.

Stylesheet-Style Sheet

Sub Network (Subnet)-A sub network is a portion of a larger network.

Subassembly-A functional unit of a system.

Subcommittee-A subcommittee is a group that is created by a committee to perform a specific function such as to analyze or solve a specific problem. Subcommittees are typically created on a temporary basis and are usually composed of members of a committee that created it.

Subcontracting-Subcontracting is the assigning of some or all of the obligations of an agreement or contract to another person or company.

Subcontractor-Substitute Contractor

Subgroup-A subgroup is a subset of a group that is created by a committee or working group to perform a specific function such as to analyze or solve a specific problem. Subgroups are typically created on a temporary basis and are usually composed of members of a committee or working group that created it.

Subject Matter Expert (SME)-A subject matter expert is a person who has a detailed understanding and experience in a particular area of information.

Subnet-Sub Network

Subscribe-The process of indicating to a communication server or other network service provider that the user is requesting services to be provided in the future and where those services can be delivered.

Subscribe Event-A subscribe event is a message or logical indication of the occurrence of a service subscription submittal. A subscribe event may be used to trigger an auto-response message that thanks the person for their submission and provides them with additional information and/or promotional offers.

Subscriber-A subscriber is an end user of a service. A subscriber is sometimes called a "user" or "customer."

Subscriber Acquisition Cost (SAC)-Subscriber acquisition cost is the combined costs that are associated with marketing and adding of a customer to a system or service.

Subscriber Authorization System (SAS)-A subscriber authorization system is the equipment and/or software that can set up and coordinate the processes that enable subscribers to request, access and use media.

Subscriber Data-Subscriber data is information that is associated with a particular user or the devices and services that a subscriber has authorized for use.

Subscriber Database-A subscriber database is an informational and relational database which includes subscriber information, including usage patterns, billing records, personal information and other related data. This base is often "mined" for information which helps identify potential churn candidates as well as useful marketing information.

Subscriber Exporting-Subscriber exporting is the process of transferring a list of subscribers from a data table into another file or data format.

Subscriber Forecast-Subscriber forecast is the expected number of customers, types of access devices and usage characteristics that are expected over a future time period.

Subscriber Fraud-Fraud perpetrated by the end-user in which false user ID information was used to obtain service.

Subscriber Identity Confidentiality-Subscriber identity confidentiality is the protection or processing of identifying information that is associated with a subscriber so the information cannot be obtained or used by and unauthorized person or company.

Subscriber Identity Module Applet (SIM Applet)-A subscriber identity module applet is a small software program stored on a SIM card that uses the Java programming language to request, transfer and process information in a device (such as a mobile telephone).

Subscriber Line-A telephone line that connects from a switching office to a customer's wired phone.

Subscriber Line Charge-A monthly flat-rate charge that recovers a portion of local loop costs paid by an end user. The charge is the result of a Federal Communications Commission (FCC) effort to eliminate unreasonable discrimination and undue preferences among rates for interstate service, to promote efficient use of a local network, to prevent uneconomic bypass, and preserve universal service.

Subscriber List-A subscriber list is a group of people or devices that have requested to receive information or services.

Subscriber Loop Carrier (SLC)-Subscriber loop carrier works in conjunction with digital carrier (e.g. T-carrier or E-carrier) systems to increase the circuit capacity of a distribution plant without adding additional lines. SLC involves adding equipment to end offices (EOs) and remote plant locations that are connected by digital cable pairs, fiber optic cable, or digital radio media. The equipment at the end office (EO) and remote terminals allows more efficient use of cable pairs. The systems that use PAM and PCM technology can generally serve up to 96 subscribers from 10 cable pairs.

Subscriber Loop Interface Circuit (SLIC)-Electronic version of the two to four-wire hybrid interface that supplies an analog signal from a line card to a subscriber's phone or network terminal equipment. It provides what is known as the BORSCHT functions in telephony (Battery Feed, Overvoltage Protection, Ringing, Signaling, Coding, Hybrid, and Test).

Subscriber Management-Subscriber management is a process or system that coordinates the additions, changes, and terminations of subscribers of a service.

Subscriber Management System (SMS)-A subscriber management system coordinates the additions, changes, and terminations of subscribers of a service.

Subscriber Number-The phone number that enables a user to reach a subscriber within the same local network or numbering area. This term is a synonym for directory number.

Subscriber Penetration-Subscriber penetration is the percentage of subscribers who subscribe to one or more service offered by a service provider compared to the total number of potential customers that have the capability to subscribe (e.g. have access to or sufficient signal quality) to the services.

Subscriber Provisioning-Subscriber provisioning is the process used by a service provider to setup the system so connections or services can be provided to subscribers.

Subscriber Station (SS)-A subscriber station is a device that is used by the subscriber or customer or end user to access the telecommunication system. The Subscriber Station is also commonly called a CPE, or Customer Premises Equipment.

Subscriber Unit-A subscriber unit is a communication device or portable radio unit that can subscribe to services that are used in a communication system.

Subscriber Viewing Habits-Subscriber viewing habits are the characteristics that define how a subscriber views media. Viewing habits may include viewing times, channel changing patterns, program categories and media purchase patterns.

Subscription Account-A subscription account is a unique identifier that designates a customer or company that has been established (subscribed) that allows them to continue to receive products or services that are billed or posted to this account.

Subscription Services-Subscription services are value-added services that provide or entitle a customer to receive or gain access to services. Subscription services are typically provided with no fixed termination date and subscription users are often billed periodically (e.g. monthly) for the subscription service.

Subscription Television (STV)-Subscription television is the providing of television signals on a pre-registered basis.

Subscription Video on Demand (SVOD)-Subscription video on demand (SVOD) are on demand services that require a user to have a subscription (pre-authorization) prior to using the on demand services. SVOD service allows a customer to select and watch videos on demand from a predetermined list of video programs for an additional on demand fee. SVOD fees are usually composed of a periodic (e.g. monthly) flat rate fee and/or a usage fee (e.g. per movie).

Subsequent Billing Company (SBC)-A subsequent billing company is a service provider that provides additional billing usage details to a billing record that it has received from another company that provide the initial services to start a billing record. An example of an SBC is a interexchange telephone company (IXC) that adds the long distance portion of a telephone call initiated and billed by a local exchange carrier (LEC).

Subsidiary Rights-Subsidiary rights are the authorizations that allow a publisher to use and/or convert content into another form of publication. For example, the republishing of portions of a book in a magazine series is a subsidiary right.

Substitute Contractor (Subcontractor)-A subcontractor is a person or company who performs the services or responsibilities of another person or company.

Sunk Costs-Sunk costs are investments or payments that have been incurred which cannot be recovered.

Super Distribution (Superdistribution)-Superdistribution is a distribution process that occurs over multiple levels. Superdistribution typically involves transferring content or objects multiple times. An example of superdistribution is the providing of a music file to radio stations and the retransmission of the music on broadcast radio channels.

Super Rep-Super Sales Representative

Super Sales Representative (Super Rep)-A super sales representative is a person who performs sales functions by using more than one sales representative.

Superdistribution-Super Distribution

Supervision-(1-control signals) The act of monitoring a line or trunk for on-hook or off-hook supervisory signals. (2-call status) Signaling that indicates the status of a call or the readiness of an item of equipment to respond to an attempt or release a connection.

Supplementary Services (SS)-Supplementary services provide a network user with capabilities beyond those of elementary call control. Supplementary services enrich the basic service functions and are not specific to a telephone or system features. Often, the subscriber (user) can specify some of the operations of supplementary services (such as call forwarding). Supplementary services may be defined or installed in systems before complete testing or industry consensus can be reached.

Supplicant-A supplicant is a user or device that is attempting to gain access to a system or service.

Supplier-A supplier is a provider of materials, products or services.

Supply Chain-A supply chain is the linking of companies and processes that are involved in production and distribution of products and services to customers.

Supply Chain Integration (SCI)-Supply chain integration is the process of linking or merging separate supplier systems or functions into a common unit.

Supply Chain Management (SCM)-Supply chain management is the identifying and managing of companies and processes that are involved in the supply of products and services to customers.

Support Systems-Support systems are the services and/or equipment that can be provided by a supplier, person or company to maintain the operation of a product or service.

Supported Rates-The data transmission rates that are available for use (hardware and software capable) in a communication device or system.

Supporting Vendors-Supporting vendors are companies that may be used to assist a RFP responder to provide products and services.

Surcharge-A surcharge is an additional charge for a service that is in addition to the basic charge. Examples of surcharges include additional charges for using pay telephone or toll free access lines.

Surge Protector-A device or assembly that restricts the transfer of rapidly changing signals (a power surge) to a device or assembly.

This figure shows a surge protector that provides protection for power overage. This surge protector includes an induction coil that restricts rapid current changes and a fuse that will open in the event of a very large and rapid voltage change. Voltage transients can be caused by power surges and lightning pulses. Surge protectors may not protect against direct lightning strikes or extreme voltage changes.

Surge Protector System

Surround Sound-Surround sound is the reproduction of audio that surrounds the listener with sound that is provided from multiple speaker locations. The use of surround sound can allow a listener to hear audio in a way that they can determine the relative position of sound sources around them (such as in front and behind).

Survivability-The capability of a communications network or system to continue to provide service after major damage to any part of the system.

SVG-Scalable Vector Graphics

SVG 1.1-Scalable Vector Graphics 1.1

SVOD-Subscription Video on Demand

Swap File-A swap file is a set of data or media that is temporarily created to assist in the processing of another file or data block. A swap file may be used to hold or combine multiple segments (fragments) so that the contiguous file can be written to storage (hard disk defragmenting).

Switch-A network device (typically a computer) that is capable of connecting communication paths to other communication paths. Early switches used mechanical levers (crossbars) to interconnect lines. Most switches use a time slot interchange memory matrix to dynamically connect different communications paths through software control. See also: Mobile Switching Center (MSC).

Switch Hook (Switchhook)-A switch hook is an electrical connection device that is located in the handset cradle of a telephone that is connected when the telephone handset is lifted from the cradle.

Switch Hook Flash-A signaling technique whereby the signal is originated at the passive end of a subscriber loop by momentarily depressing the switch hook. This causes an interruption in loop current for an interval of typically 1 to 1.5 seconds. A shorter interruption is ignored or may be considered as equivalent to a wink or dialing the digit 1 with a rotary dial. A longer interruption typically causes a disconnection. See also Cradle Switch Flash.

Switch Port-A resource that allows a Group to communicate with another Group. All Groups implicitly posses a Switch Port as a secondary resource, but in order to use it, the application must explicitly connect the Switch Ports of two Groups.

Switchboard-A manually operated switching system, now rarely used, that connects a limited number of telephones within a building or provides operator services.

Switchboard Operator-A telephone system employee who provides information to customers and assists with the connection of telephone calls. The first switchboard operators connected calls by using cables with plugs to connect calls between connection points on their switchboard panel.

Switched 56 Service-Switched 56 service is a data communications service that provides full-duplex, digital data transmission at 56 kilobits per second (Kbps) via digital network facilities.

Switched Access Transport Services-Transmission of switched voice or data telecom traffic along dedicated facilities between LEC central offices and long distance carrier POPs.

Switched Circuit-A circuit that may be temporarily established at the request of one or more of the connected stations.

Switchhook-The switch of a telephone set that closes and opens a customer local loop circuit. It is used to indicate control (supervisory) signals from the user. The switchhook is usually operated by lifting the handset or returning it to its cradle (holder). A switchhook is also called a hookswitch.

Switchhook-Switch Hook

Switching-Switching is the process of connecting two (or more) points together. Switching may involve a single physical connection (such as a light switch) or it may involve the setup of multiple connections within a network through several communication devices.

Switching Center-A location where either toll or local telecommunication traffic is switched or connected from one line or circuit to another. Also called a switching office.

Switching Cost-Switching cost is the proportion of switching facilities costs that are associated with a specific connection or set of services that use the switching facilities.

Switching Office-A switching center within a building (central office)

SWOT-Strengths, Weakness, Opportunities, and Threats

Symmetrical-Symmetrical transmission is two-way communication that has the same data transmission rates in send (forward) and receive (reverse) directions.

Symmetrical Digital Subscriber Line (SDSL)-Symmetrical digital subscriber line is an all-digital transmission technology that is used on a single pair of copper wires that can deliver near T1 or E1 data transmission speeds. SDSL is a symmetrical service that ranges from 160 kbps to 2.3 Mbps and can reach to 18000 feet from the central switching office.

Synchronization-(1-general) The process of adjusting the corresponding significant instants of signals, such as the zero-crossings, to make them synchronous. The term synchronization often is abbreviated as sync. (2-digital) An arrangement for operating digital switching and transmission systems at a common (synchronized) clock rate to prevent the loss of portions of a bit stream during transmission. The required synchronization pulses commonly are referred to as clock pulses or clock signals. (3-video) The pulses and timing signals that lock the electron beam of the picture monitor in step, both horizontally and vertically, with the electron beam of the pickup tube. The color sync signal (in the NTSC system) is known as color burst. (4-data) Data synchronization is a process that allows for the time sensitive updating of data on application used on one device to be exchanged and updated on another device.

Synchronized Data-Synchronized data is information from queries to multiple databases that are related to each other where the information is selected and/or processed to represent information at a specific time.

Synchronous-A system or signal that involves the transfer of information in a predefined serial time sequence.

Synchronous Link-A synchronous link is a communication connection that exists between devices that can continuously send data at predefined time periods.

Synchronous Modem-A modem that is able to transmit timing information in addition to data. The modem must be synchronized with the associated data terminal equipment by timing signals. A synchronous modem sometimes is referred to as an Isochronous modem.

Syndicate-A syndicate is a company or organization that is formed to represent a common interest to multiple parties.

Syndication-Syndication is the combination of materials our use of materials for a common purpose.

Syntax-The relationships among characters or groups of characters, independent of their meanings or the manner of their interpretation and use.

Syntax Check-A syntax check is a process that determines if the structure (relationships) among words or characters conforms to the rules (such as the commands and parameters of a programming language).

System-Systems are the combination of equipment, protocols and transmission lines that are used to provide communication services.

System Access-System access is the ability for a user or device to connect to a system.

System Administration-System administration is the processes and tasks that are performed by a system administrator to add, change, or disconnect devices, features, and equipment. System administration tasks may include monitoring the performance of the network and making adjustments to the network as necessary.

System Administrator-A person who oversees the operational functionality of a computer or related telephone equipment. Also included in this individual's responsibilities are introducing new user Ids and organizing phone numbers, commonly referred to as moves, adds, or changes (MAC).

System Architecture-System architecture is the functional parts of a system and how they relate to each other.

System Capacity-System capacity is the maximum information or service carrying ability of a communications system. The unit of capacity measurement for the facility or system depends on the type of services or information content that are provided by the system.

System Crash-A system crash is a partial or complete failure of a processing system (such as a customer care system) or network.

System Cutover-System cutover is the process or date where new systems or equipment are used to provide services to customers (old systems are cut off).

System Diagnostics-System diagnostics are the monitoring functions and processes that are used to identify the performance, operation and repair of equipment, protocols and/or transmission lines.

System Field-A system field is a value label and/or its characteristics that are pre-assigned by the system and cannot be changed by the user or operator of the program or service. An example of a system field is the record number ID in a database table.

System Integration-System integration is the process of linking or merging separate systems or functions into a common unit.

System Redundancy-System redundancy is a design of a system or network that includes additional equipment for the backup of key systems or components in the event of an equipment or system failure.

System Reliability-System reliability is the ability of a network or equipment to perform within its normal operating parameters to provide a specific quality level of service. Reliability can be measured as a minimum performance rating over a specified interval of time. These parameters include system availability, bit error rate, minimum data transfer capacity or mean time between equipment failures (MTBF). For communication systems, availability ranges from 99.9% to 99.999% (5 nines reliability).

System Renewability Messages (SRM)-System renewability messages are commands that are used or sent on a communication system that renews the authorization keys and other parameters to enable the use of services or media.

System Time Stamp (STS)-System time stamps are the insertion of information that indicates a time that an event or process has occurred within a system.

System Upgrade-A system upgrade is the addition and/or modification of assemblies and programs that change (usually improve or add features to) the capabilities of a system.

Système International (SI)-The SI version of the metric system of units is compatible with practical electric units such as the volt, ampere, and watt. The basic mass unit is the kilogram, the basic length unit is the meter, and the basic time unit is the second. The basic unit of electric current is the ampere, and the basic unit of electric charge is the coulomb or ampere second (see: charge). It is also called the Georgi system, after the name of the Italian physicist who first formulated the appropriate mechanical units for its use. The older version of the metric system, sometimes called the cgs system, is based on the centimeter, gram and second.

Systems Availability-Systems availability is the ability of a system to perform the services or operate within the parameters defined for the services for features.

Systems Integration-Systems integration is the process of defining, selecting, combining and configuring multiple types of systems to operate together to perform specific functions and/or services. Systems integration can range from porting (simple one-way connections) of systems to each other to full integration (two-way interactive processing) operation.

Systems Integrator-A systems integrator is a company or person who assists with the defining, selecting, combining and configuring multiple types of systems to operate together to perform specific functions and/or services.

T

T-Tera

T Carrier-Trunk Carrier

Tab Delimited Format-Tab delimited format uses tab characters to separate the fields within data records.

Table-A table is a group of structured information. The structure usually includes records (rows) and fields (columns).

TAD-Telephone Answering Device

TADIG-Transferred Account Data Interchange Group

Tagged Image File Format (TIFF)-TIFF is a format for storing and presenting digital images. TIFF image types include black-and-white data, halftones or dithered data, grayscale data, and color.

Taguchi Method-The Taguchi method is the use of statistics to determine the cost of quality control.

Take Rates-Take rates are the percentages of customers or potential customers who subscribe to a new product or service.

Talk Group-A talk group is a predefined group of mobile radios that are capable of receiving and decoding the group messages to or from group members.

Talk Time-The amount of time a mobile telephone or transmitting device can continuously transmit between battery recharges or the replacement of its disposable batteries.

Talk-Group Identifier (TGID)-A talk group identifier is a field that is contained in a radio message that identifies the talk-group for which the radio message is intended to be received.

TAM-Telecom Applications Map

Tamper Resistance-Tamper resistance is the ability of a device, system or service to resist the attempts of an authorized user from obtaining, analyzing or modifying the device, system or service.

Tandem Office-A switching system that is used to interconnect end offices with each other.

Tandem Switch-A tandem switch is an intermediate level switch that connects to other switching exchanges.

Tandem Switching System-A switching system in a tandem office that handles trunk-trunk traffic. Local tandems switch calls from one end office to another within the same area. Access tandems switch calls to and from an interexchange carrier.

Tap-(1-circuit) A branch or intermediate circuit in a communications system. (2-optical) A device for extracting a portion of the optical signal from a fiber. (3-telephony) Short for wire tap.(4-In a local area network an electrical connection permitting signals to be transmitted onto or received from a bus.

TAP-Transferred Account Procedures

TAP-Transferred Accounting Process

TAP3-Transferred Account Procedures 3

Tape Archive Format (TAR)-Tape archive format is a data structure that is designed to store data on a tape format. The TAR format defines a header section that allows the tape system to quickly identify and locate data blocks.

Tape Storage-A tape storage system holds data (information) on a tape.

TAPI-Telephony Application Programming Interface

TAR-Tape Archive Format

Target Costing-Target costing is an accounting system process that associates costs with the planned products or activities.

Targeted Ad Insertion-Targeted ad insertion is the process of inserting an advertising message into a media stream such as a television program based on specific characteristics or preferences that are associated with the viewer or device that will receive the ad.

Tariff-A tariff is a rate and its conditions for provision of a regulated service. Tariffs include a schedule of rates, prices, and regulations for services that have been approved by national or international regulatory bodies (e.g. FCC, OFTEL, etc.).

Tariff Plan-A tariff plan is the structure of service fees that a user will pay to use regulated services. Tariff plans are typically divided into monthly fees and usage fees.

Task-A task is a set of steps or processes a transceiver must take to accomplish a function.

Task Group-A group of people (typically professionals with specific skill sets) that are temporarily assigned to perform a particular task or project.

Task Scheduling-Task scheduling is the process of defining and assigning identification codes to tasks along with priorities and the times or events that the tasks should be executed or performed.

Tax-A tax is a fee or assessment that is imposed (levied) by a government or authorized entity on a product, activity or any other asset that can be valued.

Tax Compliance-Tax compliance is the process of following rules or requirements for recognizing, recording and paying taxes that result from the selling of products or services.

Tax Deductible-Tax deductible is cost that can be applied or used to adjust the tax assessment on a person or business.

Tax Exempt Status-Tax exempt status is a product or business classification that indicates that they should not be charged or assessed taxes for the sale of products or services.

Tax Interfaces-A tax interface is a combination of hardware or software that adapts information or processes in one system (such as billing) to work with modules or external networks that can perform tax assessment, assignment or other tax related processes.

Tax Rates-Tax rates are a listing of taxable categories and their associated rates.

Tax Rules-Tax rules are the conditions and associated processes (such as calculations) that are used to compute tax liabilities.

Tax Stamp-A tax stamp is an identifying mark or attachable label that indicates that a tax has been assessed or paid.

Taxable-Taxable is a condition or option for a product or service that indicates taxes should be calculated and applied to its purchase if certain conditions exist.

Taxation-Taxation is the process of assigning fees or levies on the sale, use, or disposition of products or services.

Taxes-Taxes are charges or levies assessed by a government agency or authority for services that are provided or products that are sold that are defined by the government as a taxable commodity. There can be many types of taxes imposed on a service provider and the calculated tax fees depend on the type of service provided, the location of the service and potentially other criteria. Taxes may be assessed by a combination of national, regional and local authorities.

Taxware-Taxware is computer application or services that assist in the identification, management and payment of taxes.

Tailored Services-A tailored service is the offering or provisioning of a service that has characteristics that are unique to the customer or user.

TCAP-Transaction Capabilities Application Part

T-Carrier-A T-carrier is a transmission system operating at one of the standard levels in the North American digital hierarchy. Each digital signaling level supports several 64 kbps (DS0) channels. T-carrier was initially used in North America and now is used throughout several parts of the world. The different digital signaling levels include;
- DS-1, 1.544 Mbps with 24 channels
- DS-1C, 3.152 Mbps with 48 channels
- DS-2, 6.132 Mbps with 96 channels
- DS-3, 44.37 Mbps with 672 channels
- DS-4, 274.176 Mbps with 4032 channels

TCL-Tool Command Language

TCO-Total Cost Of Ownership

T-commerce-Television Commerce

T-Commerce Order Processing-Television commerce order processing is the steps involved in selecting the products or service from a television catalog or advertising and agreeing to the terms that are required for a person or company to obtain products or services.

TDD-Telecommunications Device For The Deaf

TDD-Teleprinter Device for the Deaf

TDD-Teletypewriter Device for the Deaf

TDMA-Time Division Multiple Access

TDT-Time and Date Table

TE-Terminal Equipment

Tear Sheets-Tear sheets are pages from a publication that are sent to an advertiser that proves that the ad was inserted into the publication.

Tech Support-Technical Support

Technical Qualifications-Technical qualifications are the skill sets and experience that a person or people have that enable them to perform certain types of tasks or projects.

Technical Support-Technical support is the communication, system and processes that are used to help customers or users of products understand how to operate, configure or solve problems related to the product they purchased.

Technical Support (Tech Support)-Technical support is the processes and communication that occurs between users of a product or service and companies to enable the user to resolve problems and successfully use the products or services.

Technician-A skilled craftsman who is capable of diagnosing, servicing, and repairing electronic or electrical assemblies (such as a PBX system or telephone device).

Techno-Geek-A geek is a person who is focused on technology, typically computers who does not tend to conform to mainstream habits such as dressing for success and/or regular bathing.

Technological Churn-Type of voluntary, deliberate churn, where the customer changes carriers because the new carrier has newer and better technological options. (i.e. customers switching to digital from analog)

Technology-Technology is the usage of knowledge, tools and skills to enable new capabilities or assist in the performance of other actions.

Technology Plan-A technology plan is a document that defines the technical capability objectives of a business over a period of time (such as within 5 years), what skills and technologies it needs to have and how it expects to acquire or develop these technologies.

Technology Transfer-Technology transfer is the providing and acceptance of knowledge that can be used to develop, produce or use products or services.

TeLANophy-The use of local access network (LAN) systems as a communication line for telephone systems.

Telco-Telephone Company

Telco Product Market Planning (TPMP)-Discipline employed by telco product managers in order to determine the potential market for a new product. TPMP is utilized to help decide which products to invest in.

Telcom Branding -The process of adding to the value of a telecommunications organization, product line or offer through the creation of "psychological" value. Branding causes prospects to perceive special value in the branded item, encouraging loyalty, acquisition and sometimes higher billing rates.

TelcoTV-Telephone Company Television

Telecom Applications Map (TAM)-Telecom applications map (TAM) is an application system structure that allows communication service providers to define, develop, procure and operate applications on their systems.

Telecom Billing-Telecom billing is the process of grouping telecommunication service or product usage information for specific accounts or customers, producing and sending invoices, and recording (posting) payments made to customer accounts.

Telecom Operations Map (TOM)-Telecom operations map (TOM) is a business system structure that allows communication service providers to define, develop, and procure business and operational support systems. The TOM framework was created by the telemanagement forum.

Telecommunication-The transmission and reception of audio, video, data, and other intelligence by wire, radio, light, and other electronic or electromagnetic system.

Telecommunication Closet-A room or space that is dedicated for telecommunication equipment and cable connection points.

Telecommunication Regulation-The regulation of telecommunications systems and services are developed to help or improve the ability of citizens in a country to reliably communicate with each other at reasonable cost. Telecom rules and regulations are usually imposed by a government agency. These rules are designed to maintain the quality and cost of public utility services.

In the United States, the Telecommunications Act of 1996 was created to allow competition into the telecommunications industry. It provides a national framework for the deregulation of the local exchange market, a deregulation that was already taking place on a state-by-state basis through the actions of state regulatory commissions. Its summary impact on the local exchange market is to require current LECs to remove all barriers to the competition (e.g., interconnect, white and/or yellow pages access, co-location, and wholesaling of facilities restrictions) in return for LEC access to the long distance market.

Telephone companies in the United States are regulated by the government, but not owned by the government. For most European countries and many other countries, local telephone service is provided by government owned post telephone and telegraph (PTT) operators. In some European countries, the post (mail) network has been separated from the operation of telephone and telegraph networks. In some countries, the telephone and telegraph systems have become privatized, and are no longer owned by the government.

Telecommunication Resellers Association (TRA)-A telecom trade association, TRA has merged with the National Wireless Resellers Association (NWRA) and promotes the expansion of telecom services through lobbying, legal efforts and other forms of support.

Telecommunication Services-Telecommunications services are the underlying communications processes that provide information for telecommunications applications. It is common to use the word services in place of applications, especially when the service is very similar to the application. Examples of communication applications include voice mail, email, and web browsing. Telephone services include voice, data, and video transmission. Voice services can be categorized into quality of service and voice privacy. Data services use either circuit-switched (continuous connection) data or packet-switched (dynamically routed) data. Video transmission is the transport of video (multiple images) that may be accompanied by other signals (such as audio or closed-caption text).

Telecommunications-The transmission, between or among points specified by the user, of information of the user's choosing (including voice, data, image, graphics, and video), without change in the form or content of the information.

Telecommunications Act Of 1996-The U.S. Telecommunications Act of 1996 provides a national framework for the deregulation of the local exchange market, a deregulation that was already taking place on a state-by-state basis through the actions of state regulatory commissions. Its summary impact on the local exchange market is to require current LECs to remove all barriers to the competition (e.g. interconnect, white and/or yellow pages access, co-location, & wholesaling of facilities restrictions) in return for LEC access to the long distance market.

Telecommunications Device For The Deaf (TDD)-A small communications terminals with a keyboard and visual display that connects to a telephone circuit to relay written messages to and from hearing- and/or speech-impaired persons. An acoustic coupler is used to send audio tones from a TDD through the handset of a conventional telephone instrument.

Telecommunications Industry Association (TIA)-Telecommunications industry association is an industry trade group that represents the manufacturers of telecom equipment.

Telecommunications Operational Model (TOM)-An industry standard template describing the major operational components that make up a telecommunications organization.

Telecommunications Profitability Formula-A modified business profitability formula that reflects the unique nature of the telecommunications business model where Profit = (Revenue-(CapEx + OpEx))

Telecommunications Relay Service for the Deaf (TRS)-Each state and province in North America operates one or more telecommunications relay service centers for the deaf. Human communications assistants (CAs) at the TRS center provide translation between ordinary speaking and hearing telephone users at one end of a conversation, and deaf or speech impaired users at the other end. The deaf or speech impaired persons use a teletypewriter to communicate with the CA. In the year 2000 the US FCC designated 711 as the universal telephone access number for TRS centers.

Telecommunications Value Chain-An operational model that describes the core functions that are required to deliver telecommunications service to the end customer. The links in the telecommunications value chain include Market Acquisition, Market Research, Network Planning, Network Build, Network Maintenance, Service Order Processing, Provisioning, Activation, Sales, Marketing, Advertising, Billing, Customer Service, Credit Processing.

Telecommuting-The process of an employee that is conducting business related activities at a remote location (usually at a home) through the use of telecommunications services and equipment. Telecommuting allows employees to work at home without the need to commute to the office and reduces the need for the business to maintain office space for workers.

Telecomputing-The use of remotely located computers and databases for computing and access to computer-based services, such as home shopping, banking, and electronic mail.

Teleconferencing-A process of conducting a meeting between two or more people through the use of telecommunications circuits and equipment. Teleconferencing usually involves sharing video and/or audio communications.

Telecopier-Another term for a facsimile (fax) machine.

Teledensity-Teledensity is a measure of the number of communication lines used or installed in a geographic area for a quantity of people. A common unit for teledensity is the number of lines per 100 people.

Teledildonics-Teledildonics are devices that mimic a penis that is controlled through a communication system.

Telegraphy-A form of telecommunications, which is concerned with the process of providing transmission and reproduction at a distance of text material or fixed images. The transmission of such information may be physical transmission facilities or over the air using some form of signaling protocol.

Telemanagement-A technique involving the management of the telephone and telecommunication expenses of an organization. This also refers to third-party software packages directed at monitoring inbound and outbound calling on the PBX switch with the capability to format detailed reports by departments and per-minute charges.

Telemarketing Cost-Telemarketing cost is the combined costs of contacting potential customers by telephone and communicating advertising messages. Telemarketing costs can include script creation, list rental, telecommunication fees, order processing and sales commissions.

Telematics-Telematics is the collection and distribution of measured or machine control data through a communications system.

Telemedicine-Processes that assist with health care service that employ communications services and equipment. Examples of telemedicine include delivery of medical images, remote access to medical records, remote monitoring of heath care equipment and distant monitoring of biological functions such as heart rate and blood pressure.

Telemetrics-Telemetrics is the measurement and transmission of operating or performance characteristics of devices, systems or services.

Telemetry-Telemetry is the transfer of measurement information to a monitoring system through the use of wire, optical fiber, or radio transmission. The term telemetry is often used with the gathering of information.

Telepayment-Telepayment is the transfer or authorization of payment for products and services through the use of telecommunication systems.

Telephone Answering Device (TAD)-A telephone answering device is an assembly or software program in a communication device that can automatically answer telephone calls, play a prerecorded greeting message, store audio information, and allow retrieval and deletion of messages.

Telephone Answering Service-A service company that answers telephone calls and takes messages for subscribers who are away from their homes or offices.

Telephone Company-Telephone companies (also known as service providers or carriers) provide communication services to the general public. They are usually regulated by the government and in some countries, may be partly or wholly owned by the government. For most European countries and many other countries, local telephone service is provided by government owned posts, telephone and telegraph (PTT) operators. In some European countries, the post (mail) network has been separated from the operation of telephone and telegraph networks. In some countries, the telephone and telegraph systems have become privatized, and are no longer owned by the government.

Telephone Company (Telco)-A contraction of telephone company, generally signifying an operating telephone company

Telephone Company Television (TelcoTV)-Telephone company TV is the process of providing television (video and/or audio) services through telephone system broadband (e.g. DSL) transmission lines. Sending television over DSL may use proprietary digital video signals or it may use standard IP data communication protocols to initiate, process, and receive voice or multimedia communications. When television signals are sent over DSL lines using standard IP communication, it is commonly called IPTV.

Telephone Eye-A system that combined television broadcasting with telephone service to provide for early interactive television programming.

Telephone Modem-A telephone modem is a communication device that modulates and demodulates (MoDem) data signals into analog signals that can be sent through telephone lines.

Telephone Network For The Deaf (TND)- A network for relaying the written messages of hearing and/or speech impaired persons. (See also: operator relay services, telecommunications device for the deaf.)

Telephone Numbering Plan-A numbering plan is a system that identifies communication points within a communications network through the structured use of numbers. The structure of the numbers is divided to indicate specific regions or groups of users. It is important that all users connected to a telephone network agree on a specific numbering plan to be able to identify and route calls from one point to another. See also Dialing Plan, open numbering plan, closed numbering plan.

This figure shows the complete international structure of telephone numbers and that these numbers typically identify a specific physical location of a telephone connection point. This diagram shows that a telephone number is composed of a country code, city code, switch code, and an extension code. In this example, the country 001 routes the call to the United States. The 919 city code routes the call to the city Raleigh within North Carolina. The exchange switch code 557 directs the call to one of several switching centers located in

Telephone Numbers

Raleigh. The extension code 2260 directs the call to a specific extension port on that switch. This extension port is connected to the telephone by wires in the local area.

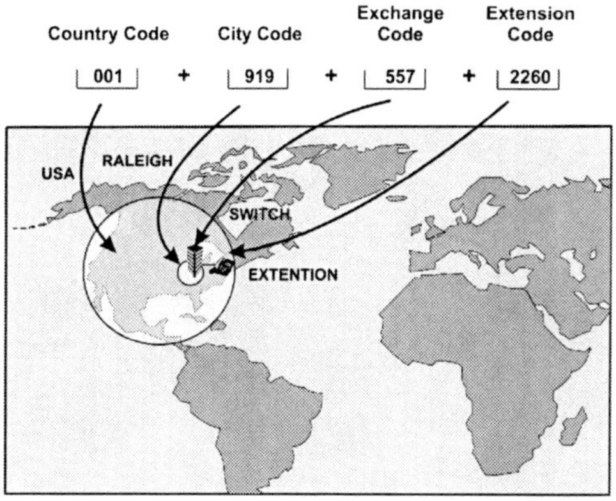

Telephone Routing over Internet Protocol (TRIP)-A protocol that allows for the dynamic assignment of call routes through the advertising of the availability of destination devices (such as telephones) and for providing information relatives the available routes and preferences for these routes to reach the destination device(s).

Telephone Suppression-Telephone suppression is the process of stopping the use of a contact information (e.g. a mailing) or telemarketing to people who have their telephone number on a suppression list.

Telephone Trigger-A telephone trigger number is a unique telephone number that is used to trigger another service such as a callback service. An international user could use a telephone trigger number to initiate a callback service. This would reduce the fees the international traveler pays for international calls as typically it is less expensive to receive incoming calls than it is to initiate outgoing international calls.

Telephony-Telephony is the use of electrical, optical, and/or radio signals to transmit sound to remote locations. Generally, the term telephony means interactive communications over a distance. Traditionally, telephony has related to the telecommunications infrastructure designed and built by private or government-operated telephone companies.

Telephony Application Programming Interface (TAPI)-An industry standard that defines the application interface between computers and telecommunications devices. TAPI was introduced in 1993 as the result of joint development by Microsoft and Intel. The standard supports connections by individual computers as well as LAN connections serving many computers. Within each connection type, TAPI defines standards for simple call control and for manipulating call content.

Telephony Call Dispatcher-A server feature on an AVVID system that allows users to receive and forward calls to other users.

Telephony Number Mapping (ENUM)-A process of relating a standard telephone number (E.164 format) to an Internet address.

Telephony Services Application Programming Interface (TSAPI)-TSAPI is a software communication standard developed primarily by the companies Lucent and Novell to allow PBX or Centrex systems to communicate through the use of NetWare communications software.

Telepresence-Telepresence is the use of communication systems to connect people to devices or systems that are located in other areas where the people can control (interact) the remote devices. This allows people to control communication systems in a way that performs processes or tasks that they would be able to do if they were actually located at the remote location. An example of a telepresence application is the connection of a doctor (e.g. a surgeon) to a medical robot that can perform surgical operations under the commands and control of the doctor. The medical robot would move and provide feedback (e.g. creating resistance or pressure) to movements to the controlling unit as the remote device encounters resistance or pressure during its operation (e.g. such as making a surgical incision).

Teleprinter Device for the Deaf (TDD)-A teleprinter is a teletypewriter device. It can be used for the deaf.

Teleservices-Teleservices are telecommunication services that provide added processing or functionality to the transfer of information between users. Teleservices are categorized by their high level (application) characteristics, the low level attributes of the bearer service(s) that are used as part of the teleservice, and other general attributes. High-level attributes include: application type (for example voice or messaging) and operation of the application. The low-level description includes a list of the bearer services required to allow the teleser-vice to operate with their data transfer rate(s) and types. Other general attributes might specify a minimum quality level for the teleservice or other special condition. The categories of teleservices available include voice (speech), short messaging, facsimile, and group voice.

This diagram shows a typical teleservice in a mobile communication system. In this diagram, a telephone user wishes to send a fax to a recipient who is traveling. The designated recipient has setup a fax forwarding service where the delivery of incoming faxes can be instructed. The sender is given the recipient's fax number. When the sender dials the number, the call is routed through the telephone network to the fax forwarding service provider (step 1). When the incoming call is detected, the fax forwarding service receives the fax into a fax mailbox (step 2). Later that day, the recipient of the fax forwarding service calls in and enters a fax forwarding number (step 3). The fax forwarding service then checks the fax mailbox and automatically sends all the waiting faxes to the new number (possibly a hotel fax number) that has been updated by the recipient (step 4). Because this service involves both the transport and processing of information, it is categorized as a teleservice.

Teletext-Teletext is a service that transfers data information along with a standard television signal to allow the simultaneous display

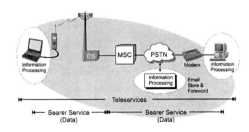

Teleservice Operation

of text and video on the television. Teletext information is usually encoded into the video blanking interval (VBI) and decoded by the receiver in the television or cable converter box.

Teletype-Trade name for teletypewriters made by the Teletype Corporation. Widely used informally as a synonym for teletypewriter.

Teletypewriter (TTY)-A text communications device comprising an alphanumeric keyboard and a display (or paper printer) that connect to an appropriate telecommunications channel. Older teletypewriters, and most teletypewriters used by the deaf, use the Baudot-Murray (ITU alphabet No. 2) 5-bit character code, while newer teletypewriters use ASCII 8-bit character code (similar to ITU alphabet No. 5).

Teletypewriter Device for the Deaf (TDD)-A teletypewriter using a special modem to support text communication for deaf people via the public telephone network.

Television-Television is the providing of visual signals at a distant location through communication systems.

This figure shows a television broadcast system. This television system consists of a television production studio, a high-power transmitter, a communications link between the studio and the transmitter, and network feeds for programming. The production studio controls and mixes the sources of information including videotapes, video studio, computer created images (such as captions), and other video sources. A high-power transmitter broadcasts a single television channel. The television studio is connected to the transmitter by a high bandwidth communications link that can pass video and control signals. This communications link may be a wired (coax) line or a microwave link. Many television stations receive their video source from a television network. This allows a single video source to be relayed to many television transmitters.

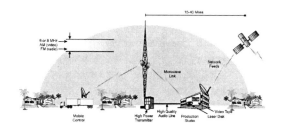

Television Broadcast System

Television Broadcasting-Television broadcasting is the transmission of video and audio to a geographic area that is intended for general reception by the public, funded by commercials, subscription services, or government agencies. The word television originates from the combination of the Greek words "tele" (far off) and "visio" (to see).

Television Channel (TV Channel)-A television channel is a communications channel that transfers audio and visual information (e.g. a television program) from a source to a destination.

Television Commerce (T-commerce)-Television commerce is a shopping medium that uses a television network to present products and process orders.

Television Mail (T-Mail)-Television mail is a process of receiving and sending messages through the use of a television.

Television Networks-Television networks are systems that provide or convert television programming (usually at a headend) through transmission lines that carry television channels (often on a distributed basis) to provide service to video and audio connection points (television outlets). While television network lines may be interconnected and shared by all users in the television network, access to services on the television network may be controlled by a conditional access system.

Television over DSL (TVoDSL)-Television over DSL is the process of providing television (video and/or audio) services through DSL transmission lines. Sending television over DSL may use proprietary digital video signals or it may use standard IP data communication protocols to initiate, process, and receive voice or multimedia communications. When television signals are sent over DSL lines using standard IP communication, it is commonly called IPTV.

Television over Internet Protocol (TVoIP)-Television over Internet protocol is the process of providing television (video and/or audio) services through communication systems that can transfer IP data packets. Sending television over IP transmission networks uses standard IP data communication protocols to initiate, process, and receive voice or multimedia communications. When television signals are sent over broadband lines using standard IP communication, it is commonly called IPTV.

Television Portal (TV Portal)-(1-service) A television portal is an interface between a viewer of television programming and a group of channels or programs that are offered by a television service provider. TV portals are commonly designed to have focus and easily navigate to specific types of content. (2-device) A television portal is an television device that can be used by a user of a product or service to interact with a system that provides the product or services.

Television Services-Television services are the providing of television signals or programs to multimedia viewing devices (e.g. televisions). Television service involves the providing of television signals that can be received, decoded and rendered on a viewing device.

Television Set-A television set is a facility that is used to film programs for television.

Televoting-Televoting is a service that allows for the gathering of votes or opinions via communication systems.

Teleworker-A worker who performs their job at a remote location through the use of a telecommunications connection. An example of a teleworker is a customer service representative who answers customer service inquiry messages from their home.

Template-A template is a base design and/or a set of parameters that can be used as a starting point sample or format structure for information gathering or presentation.

Temporal Method-Temporal method is the process of recording or adjusting events (such as accounting transactions) to reflect the time they occur (such as the currency exchange rate on a specific day).

Temporary Location Directory Number (TLDN)-A temporary location directory number (TLDN) is a temporary identification number that is used to route calls from a home system and a visited communication system. TLDN numbers are commonly used in mobile communication systems where mobile telephones regularly operate in other (visited) systems. The TLDN is usually assigned for each call delivery request received from the home system. When the call is received into the visited system, it is mapped (translated) to the current resources (e.g. cell site and mobile number) that are currently being used in the visited system.

Temporary Mobile Station Identity (TMSI)-A temporary mobile station identity (TMSI) is a number that is used to temporarily identify a mobile device that is operating in a local system. A TMSI is typically assigned to a mobile device by the system during its' initial registration. The TMSI is used instead of the International Mobile Subscriber Identity (IMSI) or the mobile directory number (MDN). TMSIs may be used to provide increased privacy (keeping the telephone number private) and to reduce the number of bits that are sent on the paging channel (the number of bits for a TMSI are much lower than the number of bits that represent an IMSI or MDN).

Tender Document-A tender document is a formal statement of requirements and technical specifications that constitute the needs of a company that is seeking a bid from vendors for the procurement of equipment, services or a system.

Tera (T)-A Tera is a number 10 to the 12th power 1,000,000,000,000. Commonly referred to as one trillion in the American numbering system, and one million million in the British numbering system.

TeraByte-A TeraByte is one trillion bytes (actually 1,099,511,627,776 bytes.)

Term Loan-A term loan is a lending of money or assets for a fixed period of time (term) where the loan is repaid through a single ending payment (term payment).

Term of Agreement-Term of agreement is the time duration the conditions or responsibilities of a contract apply.

Term Plan-A product/service that offers the customer a special discount in return for committing to a certain length of time. Penalties are typically assessed for early termination.

Terminal-(1-circuit) A point at which a circuit element may be directly connected to one or more other elements. (2-communications) Any type of equipment at the end of a communications circuit. User terminals include telephone sets, teletypewriters, and computing equipment. Carrier terminals include modulation, demodulation, and multiplexing equipment used to transmit, combine, and separate communications channels in a transmission system. (3-computer) An input/output device connected to a processor or a computer in order to communicate with it and control processing. An intelligent terminal has some local computing power and an associated data store. (4-loop plant) The hardware that facilitates the connection and removal of drop or service wires to and from cable pairs. Examples include distribution terminals and cross-connection terminals. (See also: central office terminal.)(5-post) A binding past, tag, or lug to which an external circuit may be readily connected.

Terminal Emulation-Terminal emulation is a microcomputer or personal computer that can mimic or pretends to be another type of device (such as a data terminal). It does this with special printed circuit boards inserted into its motherboard and/or special software.

Terminal Equipment (TE)-Equipment that is located at the end of one or more communication lines that send or receive signals for services. Terminal equipment can be wired or wireless devices.

Terminal Multiplexer-A terminal multiplexer is a device that takes typed character streams from several terminals and transfers them through a common transmission medium (sometimes shared with other users) to a distant point where separate character streams appropriate to each terminal are again formed.

Terminals-Devices that typically provide the interface between the telecommunications system and the user. Terminals may be fixed (stationary) or mobile (portable).

Terminating Point Master File (TPMF)-Also known as "V&H file", it contains a correlation between an NPA-NXX and the name of the locale where the serving switch resides. This "Place Name" appears in the detail section for each call. This file also contains the Vertical & Horizontal coordinates of the NPA-NXX; this is used by the rating engine for distance-based rating.

Terminating Zone-A terminating zone is a geographic area, region, or a group of points that shares a common identifier or characteristic where a service or process is received.

Termination-(1-general) Termination is the end of a circuit or connection (2-circuit) A circuit termination is an impedance matching device (termination) that is connected to the end of a circuit under receive and terminate signals without causing a reflected signal. (3-service) Service termination is the ending of authorization or providing of service.

Termination Charges-Fees paid by telephone operators to other access providers for terminating (routing calls to their destination) through the other access networks. An example of this is the payment by wireless carriers to local exchange carriers (local telcos) for the termination of mobile telephone calls into the wired public telephone network.

Termination Gateway-A gateway that is used as a termination point for a communication service. An example of a termination gateway is an audio gateway that is used to connect Internet telephone calls (the termination of the VoIP connection) to a public telephone network.

Termination Rights-Termination rights are the authorizations or terms that remain in effect after the ending of a contract.

Terrestrial-Terrestrial is the operation of systems or services on the ground (surface) of a planet (typically Earth). The term terrestrial as it is applied to radio or optical involves the transmission of radio signals along the surface of the Earth.

Test-A test is the sequential operations performed on a component, circuit or system to determine whether the design objectives have been met, verify correct operation, locate trouble, and determine behavior characteristics. Tests are performed at installation and at intervals throughout the life of a circuit, device, or system.

Test Data Form-A test data form is a record information entry sheet that is used to store test measurement information. Test data forms are often designed for specific types of measurements.

Test Equipment-Test equipment is a device or assembly that can measure or verify that a particular product or system to determine if it meets specific requirements or if it is capable of performing specific functions or actions.

Test Line-A central office test facility, including testing equipment, circuits, and communication channels, used for the maintenance of trunks. A test line can range from a simple passive termination and tone generations to complex electronic equipment for signal and transmission testing. A test line also can be referred to as a test termination.

Test Plan-A test plan is a set of tests that are structured to identify the operational and/or performance characteristics of a device, service or system.

Test Procedures-Test procedures are a set of measurements and processes that are used to obtain the operational status or performance of a product or service.

Test Specification-A test specification is a document that is primarily used for the validation of operational performance of a device or system. A test specification typically includes the device and test equipment configurations (connections) along with required signal inputs and expected signal outputs.

Test Suite-A test suite is a set of tests that can be used to ensure a product or service conforms to a specification or if the product will operate correctly in specific environments or conditions.

Testing-Testing is the performing of measurements or observations of a device, system or service to obtain the characteristics of its operation that indicate its successful operation and/or performance.

Text-Text is a sequence of characters.

Text Advertisements-Text advertisements are promotional messages that are in text form. Text advertisements within a web page may be embedded with other content and it may use hyperlinks to redirect users to a promotional page (a landing page).

Text File-A text file is a data segment where its information elements can be represented in the form of text characters.

Text Link-Text links are hyperlinks that contain text (usually underlinked characters) that allows the link to redirect the source of information to another document or file.

Text Messages (Textual Messages)-A message that contains strings of characters (text) to define meaning of the message. The use of text messages results in easy to read messages at the expense of using more data bits for each message.

Text Messaging-Text messaging is a communication service that typically transfers small amounts of text (several hundred characters) between communication devices. Text messaging services can be broadcast without acknowledgement (e.g. traffic reports) or sent point-to-point (paging or email).

Text Mode-Text mode is a condition of operation where a display or interface is limited to the use of text characters.

Text Paging-Text messages which are sent via operator or computer to a Text Pager. A text pager is commonly called an Alpha pager.

Text To Speech (TTS)-The conversion of text (ASCII) information into synthetic speech output. This technology is used in voice processing applications that require the production of broad, unrelated and unpredictable vocabularies, e.g., products in a catalog, names and addresses, etc.

Textual Messages-Text Messages
TGID-Talk-Group Identifier
Theft Of Service-Theft of service is the usage of transmission, information or other type of service where the user is not authorized to use the service without payment and does not pay for the service.

Third Generation (3G)-A term commonly used to describe the third generation of technology used in a specific application or industry. In cellular telecommunications, third generation systems used wideband digital radio technology as compared to 2nd generation narrowband digital radio. For third generation cordless telephones, products used multiple digital radio channels and new registration processes allowed some 3rd generation cordless phones to roam into other public places.
This diagram shows a 3rd generation broadband wireless system. This system uses two 5

MHz wide radio channels to provide for simultaneous (duplex) transmission between the end-user and other telecommunication networks. There are different channels used for end- user to the system (called the "uplink") and from the system to the end-user (called the "downlink"). This diagram shows that 3G networks interconnect with the public switched telephone network and the Internet. While the radio channel is divided into separate codes, different protocols are used on the radio channels to give high priority for voice information and high-integrity to the transmission of data information.

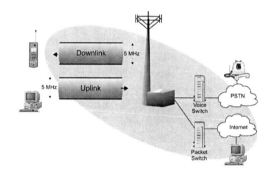

Third Generation (3G) Wireless

Third Line Forcing-Third line forcing is the pressuring of a buyer to purchase products they do not want as a condition of supplying them with the products they do want.

Third Party Billing-The billing of customers by one telephone company ("the Biller") on behalf of another Telco (the "Third Party").

Third Party Developers-Third party developers create programs or applications that are used on a system or product that is supplied by a company that does not produce the system or product.

Third Party Referral-A third party referral is contact information for a prospective customer that is provided by another person or other source.

Third Party Settlements-Third party settlement is the processes that are used to pay the financial difference between revenues generated from charging for services provided by other companies to the revenues generated from charging for services provided internally.

Third Party Verification-Third party verification is a process of using a third entity (a person or company) to review or validate information on the products or services that are used, exchanged or transferred between two other entities. Typically, the third party is trusted by both the other parties. The third party may be used to validate the information (such as quantities of products produced under licensing agreements) through the viewing of information that may be proprietary or private.

Three Way Calling (3WC)-A service that provides a telephone service customer the capability of adding another party (third party) to an established two-party call. During a three-way call, all three parties may communicate at the same time. However, to prevent annoying audio feedback, sophisticated audio volume control is typically offered that reduces the amplified audio signal level for parties that are not talking.

Throughput-(1-telephone system) The number of telephone call attempts successfully completed each second. (2-data communication) The number of bits, characters, or blocks of data that a system working at maximum speed can process during a specified period of time. (3-ISDN/PPSN) In the Integrated Services Digital Network (ISDN) and the Public Packet-Switched Network (PPSN), the average information rate of a particular virtual circuit.

TIA-Telecommunications Industry Association

TIA-41-TIA-41 is an industry standard that defines the communication aspects of intersystem signaling (such as in the SS7 system) that is used for the AMPS (analog), IS-136 TDMA, and IS-95 CDMA mobile communication systems.

Tier-A tier is a grouping of a product or service into groups (tiers) that have similar characteristics or hierarchy levels.

Tier Revenue-Tier revenue is the generation of revenue from the sale of products or services that have specific characteristic ranges (tiers).

Tiered Pricing-Tiered pricing is the assignment of value to products or services that are offered for sale that vary with specific characteristic ranges (tiers).

TIFF-Tagged Image File Format

Tilde-Tilde is the character ~. Because the tilde character is not commonly used, it may be used in data files to separate fields or characters.

Time and Date Table (TDT)-This table provides updated time and date information (changes and corrections).

Time Band-A time band is a range of time between starting and ending times. A time band may be used to define service categories based on its operation in a specific time interval (such as peak or off peak time).

Time Division Multiple Access (TDMA)-Time division multiple access (TDMA) is a process of sharing a single radio channel by dividing the channel into time slots that are shared between simultaneous users of the radio channel. When a mobile radio communicates with a TDMA system, it is assigned a specific time position on the radio channel. By allowing several users to use different time positions (time slots) on a single radio channel, TDMA systems increase their ability to serve multiple users with a limited number of radio channels.

Time Gap Validation-Time gap validation is the process of comparing the differences in time events (such as service start and stop times) to determine or ensure that billing records or other associated information is correct.

Time of Day (TOD)-The time of day is the time an event, service or action occurs within a specific day.

Time Of Day Routing-Time of day routing is a call-processing feature that automatically changes the route used to connect calls or services through a network based on the time of day.

Time Poll-A time poll is the process of sending a data request message (polling message) to a device at a defined time that initiates the collection of data or media from the device.

Time Rate-Time Rating

Time Rating (Time Rate)-Usage rating is the assignment of a value to a usage record that is based on the amount of time a service or function is used.

Time Shift Television (TSTV)-Time shift television is a television distribution service that allows viewers to select and play programs at a time other than that at which it was originally received or scheduled to be received.

Time Shifting-Time shifting is the viewing of a program or information at a time other than that at which it was originally broadcast.

Time Stamp (Timestamp)-Time stamps are the insertion of information that indicates a time that an event occurred into a record. Time stamps are used to indicate the time a file was created or when an error or network failure has occurred in a system.

Time to First Fix (TTFF)-Time to first fix is duration that occurs between when a process is started and the first position location that can be determined.

Time To Live (TTL)-Time to live is a field within a data packet that is used to limit the maximum number of routing or switching points a packet may pass through during transmission in a data network. The TTL counter is decreased as it progresses through

each router or switching point in the network. If the TTL counter reaches 0, the packet can be discarded. The use of TTL ensures packets will not be transmitted in an infinite loop.

This diagram shows a packet data network uses a time to live (TTL) control field to avoid the potential for routing packets through long travel paths or infinite loops. This diagram shows that a data packet enters into a packet network contains its destination address, time to live field, and data payload. The TTL field is initially set by the sending computer and its value may vary dependent on the type of service (e.g. real time voice compared to file transfer.) As the packets are routed through the packet network, each router (packet switching device) forwards the packet towards the destination it believes will send the packet towards its destination. Each time the packet passes through a router, the TTL field value is decreased. This diagram shows that the routers in this diagram accidentally send the packet into an infinite loop due to a broken line between routers that force packets to be rerouted through alternate paths. Eventually the routers will adjust their packet forwarding tables. However, in this case the packet travels through too many routers and the TTL field expires and the packet is discarded.

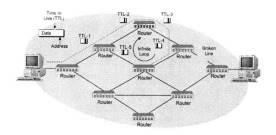

Time to Live (TTL) Operation

Time Varying Media (TVM)-Time varying media is content or information that changes over time (such as video).

Timed License-A time license is a set of authorizations that have time duration limits. A time license may be used to authorize a user or device to have access to content or services for a period of time from when an event occurs (such as 24 hours from when a user purchases the rights to watch a movie).

Timed Release Disconnect-A disconnect process that is a result of exceeding a predetermined maximum time interval for a process or service.

Timers-Timers are counters or timing reference generators that can be used to coordinate or control other processes or actions.

Timestamp-A timestamp is the recording of the time that an event occurs. A timestamp may be created using an internal (unsynchronized) time clock (such as a clock in a personal computer) or it may be recorded using a more accurate time through a network connection such as using network time protocol (NTP).

Timestamp-Time Stamp

TLD-Top Level Domain

TLDN-Temporary Location Directory Number

TLS-Transport Layer Security

TLV-Type Length Value

TM Services-CLASS

T-Mail-Television Mail

TMSI-Temporary Mobile Station Identity

TND-Telephone Network For The Deaf

TOD-Time of Day

Token Ring-Token ring is a LAN system that passes a token to each computer connected to the network. Holding of the token permits the computer to transmit data. The token ring specification is IEEE 802.5 and token ring data transmission speeds include 4 Mbps, 16 Mbps, 100 Mbps and 1 Gbps.

Token ring networks are non-contention based systems, as each device connected via the token ring network must receive and hold a token before it can transmit. This ensures only one device will transmit data at any given time. Token ring systems provide an efficient control system when many devices are interconnected with each other. This is the reason token ring systems will not see data traffic degradation when many new stations are added, compared to collision-based (nonswitched) Ethernet systems which degrade exponentially as new stations are added. Passing tokens does add a small overhead (additional control messages) and can slightly increase a packet's transmission time if the transmitting station must wait for a free token to arrive in efficient control system when many computers are interconnected with each other. This is the reason token ring systems will not see data traffic degradation when many new users are added compared to Ethernet systems. However, passing tokens does add overhead (additional control messages) that reduces the overall data transmission bandwidth of the system.

Token Ring Network-IBM's implementation of the IEEE 802.5 token-ring standard. This protocol implements a token-passing mechanism at rates of 4 or 16Mbps on shielded or unshielded twisted pair cabling.

As many as 250 stations may be connected on a single ring. Multiple rings may be connected with bridges to create virtually unlimited sized networks. Token Ring Networks utilize a star-wired topology with a physical ring created by the wiring hubs, also called Multistation Access Units (MAU). MAUs can be chained together to produce large rings.

On a token-ring network, one station is elected to be the controlling network interface card, known as the Active Monitor. The Active Monitor is responsible for providing the synchronization clock for all stations on the ring, as well an ensuring that a token is always circulating on the ring. If the Active Monitor does not receive a token every 10 milliseconds, it will begin a ring recovery protocol to ensure that there are no faults in the ring and produce a new token.

When a station has data to transmit, it captures the token, sets its Access Control Field to busy, and then adds a MAC header and user data. All other nodes continuously receive data from the ring and search for valid frames and their own destination MAC address. If a frame is destined for a station, it will copy the packet to a buffer while repeating the frame to the next station. This enables the transmission of multicast frames. For unicast frames, the receiver will set an Address Recognized (AR) bit in the Ending Delimeter to inform the sender that the frame was seen by the receiver. In addition to the AR bit, a Frame Copied (FC) bit is set by the receiver if it has buffer space available to receive the entire frame. The sender is responsible for removing any packet it transmits. If, due to a sender error, a frame is seen by the Active Monitor more than once, it will destroy the frame and begin the ring recover protocol.

This diagram shows a typical token ring LAN. This diagram shows that token ring networks are non-contention based systems, as each computer connected via the token ring network must have received and hold a token

General:
Specified by IEE 802.5
Protocol: Singular token passing
Standard Speeds: 4Mbps, 16Mbps, and 100Mbps

Token Ring (T/R)
Ring Topology

Token Packet

Star Token Ring

Token Ring (Ring Topology)
Obsolete
Cable: IBM Type 1, 2, 3
Token Direction: Counterclockwise
Speed: 4/16Mbps

Token Ring (Ring Topology)
Logical Ring
Hubs:
IBM MAU, MSAU, SMAU,
non-IBM T/R Hubs (various)
Cable:
IBM Hubs use IBM Type 1 cable, STP/UTP
non-IBM Hubs use Cateogory 5 UTP/STP
Token Direction: Either direction
Speed: 4/16Mbps

Token Ring System

before it can transmit. This ensures computers will not transmit data at the same time. Token ring systems provide an efficient control system when many computers are interconnected with each other. This is the reason token ring systems will not see data traffic degradation when many new users are added compared to Ethernet systems. However, passing tokens does add overhead (additional control messages) that reduces the overall data transmission bandwidth of the system.

Tolerance-A tolerance is a permissible amount of variation from an established expected or measured value.

Toll-A toll is a charge for telecommunications service that results when a call is routed beyond a local calling area. Due to telecommunications deregulation, toll charge boundaries have changed and in some cases, have been eliminated throughout entire country areas so all calls are charged the same rate even when calling out of their local service area.

Toll Bypass-Toll bypass is the routing of calls or communication sessions around any other networks facilities to avoid toll charges. An example of toll bypass is the use of voice over internet protocol (VoIP) services that allows customers to bypass the public switched telephone network (PSTN) switches in order to utilize the packet network (ex. IP data network) for long-distance (also known as toll) voice calls. This technique is used in conjunction with the H.323 protocol.

Toll Call-A toll call is a communication connection that is made beyond a subscriber's local calling area. The charges for such calls vary, but may take into account the distance, duration, day of week, and time of day.

Toll Fraud-The theft of long-distance service including hacking or using stolen credit cards, computers, and 800 numbers to access a switch and determine a method by which other calls can be placed.

Toll Free-A service that allows callers to dial a telephone number without being charged for the call. The toll free call is billed to the receiver of the call. In the United States, toll free calls are preceded by a 1-800, 1-888 or 1-877 exchange. In Europe and other parts of the world, toll free calls are called Freephone and typically begin with 0-800.

Toll Free Billing-Toll free billing is the process of gathering toll free call usage records to their appropriate accounts, rating these calls, creating invoices and recording (posting) payments made to toll free customer accounts.

Toll Quality-Toll quality is a measure of the quality of service (QoS) that is acceptable for voice communication on a telephone system. Because the measurement of voice quality is subjective, it is measured by a mean opinion score (MOS). The MOS is a number that is determined by a panel of listeners who subjectively rate the quality of audio on various samples. The rating level varies from 1 (bad) to 5 (excellent). Good quality telephone service (called "toll quality") has a MOS level of 4.0.

Toll Service Operator-A toll service operator is a company who controls the system that provides communication services beyond the local calling areas.

TollGate-A milestone step in a product development process that is used as an evaluation point to determine if further product development steps will be taken. There are often tollgates between concept, feasibility, planning, development, and introduction phases. Tollgates may require specific types of documents such as product descriptions, marketing evaluations, and intellectual property reviews.

TOM-Telecom Operations Map

TOM-Telecommunications Operational Model

Tone Paging-Paging service whereby the user is notified of a message via a tone. This tone usually is designated to mean a callback to a single location is requested.

This figure shows basic operating of a tone paging system. This diagram shows how a tone paging system receives incoming calls and creates a message that can be sent to a tone pager to alert the tone pager user that a caller has called and left a message. In this

example, a caller dials the tone pagers assigned telephone number. When the tone paging system receives the incoming call, it plays in interactive voice response message to the caller to leave a message. This message may be stored in a voice mailbox. The paging system ten converts the telephone number to the pagers identifying address and sends the message to the paging transmitters. This diagram shows that the tone pager uses paging groups and this pager is assigned to paging group 4. When paging group 4 starts, the pager wakes up and looks for messages with its address. When it finds this message, it alerts the paging user with a tone ("beep-beep"). When the user hears the tone, the user will go to a telephone and call a predetermined number (possibly the voice mailbox number).

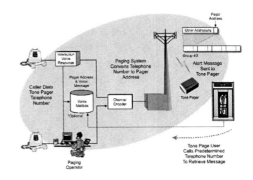

Tone Paging Operation

Tool Command Language (TCL)-A scripting language that is commonly used by communication service providers to customize their interactive voice response (IVR) systems. TCL uses relatively simple (and modifiable) syntax and can be used as a separate program or combined (embedded) in other programs. TCL is open source language so its' use is free.

See http://www.tcl.tk for more information.

Tool Set-A tool set is a combination of software programs and/or hardware devices that are used to perform processes or develop services.

Toolbars-Toolbars are groups of features or processing functions that are added to software applications.

Toolkit-Toolkits are a group of software programs that assist designers or developers to create products or services. These software tools will allow vendors to create products such as end-points (IP telephones and messaging systems), servers (proxy, redirection, and registrar), media gateways, and application servers.

Top Level Domain (TLD)-A top level domain is the uppermost group of items in a hierarchical system. For file systems it is the root directory. For the internet, it is the right-most part of the web address (e.g. .COM, .EDU, .NET).

Torrent-(1-file transfer) A torrent is a rapid file transfer that occurs when multiple providers of information can combine their data transfer into a single stream (a torrent) of file information to the receiving computer. (2-water) A fast moving stream or flow of water. This diagram shows how to transfer files using the torrent process. This example shows that 4 computers contain a large information file

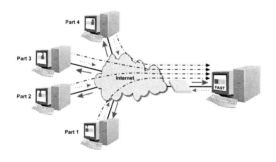

Torrent File Transfer Operation

(such as a movie DVD). Each of the computers is connected to the internet via high-speed connections that have high-speed download capability and medium-speed upload capability. To speed up the transfer speed for the file transfer, the receiver of information can request sections of the media file to be downloaded. Because the receiver of the information has a high-speed download connection, the limited uplink data rates of the section suppliers are combined. This allows the receiver of the information to transfer the entire file much faster.

Tort-A tort is wrongful act such as a breach of a contract term, which may result in the award of compensation by a legal authority.

TOS-Type Of Service

Total Cost Of Ownership (TCO)-Total cost of ownership is the direct costs of hardware and software along with the costs for operation and maintenance.

Total Running Time (TRT)-Total running time is the duration of a media program from the beginning of the indicated media start time to the indicated end time.

TP-Transaction Processing

TPMF-Terminating Point Master File

TPMP-Telco Product Market Planning

TRA-Telecommunication Resellers Association

Track-(1-memory device) The portion of a moving-type storage medium that is accessible to a given reading station. (2-videotape or digital video) The section of a videotape or digital media file where a particular signal is recorded. In an analog recorder, there typically are separate tracks for video, audio, and time code.

Tracking-(1-services) Tracking is the monitoring or following of the flow or usage of information, transmission or services. (2-radio) The locking of tuned stages in a radio receiver so that the system changes as the receiver tuning is changed (tracking the changes).

Tracking Number-A tracking number is an identification code that is used to track the status of an order, process or shipment.

Trade Discount-Trade discounts are cost reductions that are provided to a purchaser for goods or services that operate in the same industry. A trade discount is typically provided in return for the ordering of additional quantities of products or services over a period of time.

Trade Laws-Trade laws are a set of rules and processes that are applied to the interpretation, application and enforcement of trades or exchanges between people or companies

Trade Leads-A trade lead is information about the product needs for a person or company. Trade leads may be displayed or posted on message areas of trade groups or associations as a way to find suppliers of products and services.

Trade Practices-Trade practices are standards or procedures that are used when exchanging products or services of a certain type or region between people or companies.

Trade Restrictions-Trade restrictions are a set of rules or lists items or must receive special authorization or testing to be allowed to be traded with other people or countries. Trade restrictions may include items such as technology, chemicals or weapons.

Trade Secret-Information, data, documents, formulas, or anything which has commercial value and which is kept and maintained as confidential.

Trade Show-Industry trade shows are events that bring industry vendors and qualified customers together to share product, technology, and industry information.

Trade Show Host-A trade show host is a company or association that obtains the facilities, gets exhibitors, provides marketing, and manages attendees to a trade show.

Trademark-A unique symbol, word, name, picture, design, or combination thereof used by firms to identify their own goods and distinguish them from the goods made or sold by others.

Trademark rights are Intellectual Property Rights which give the owner the right to prevent others from using a similar mark which could create confusion among consumers as to the source of the goods.

Trademark protection must be registered in the country, or region, where it is desired. In most countries and regions, patents and trademarks are administered by the same government agency.

Trademark Infringement-Trademark infringement is the unauthorized use of a trademark by a company or person.

Trading Company-A trading company is a business entity that assists companies or people in the selling and buying of goods. Trading companies are commonly associated with international trading.

Trading Partner-(1-general) A trading partner is an entity that exchanges products or services with other entities. (2-EDI) An entity that sends or receives electronic data interchange (EDI) messages.

Traffic-The amount of data transferred over a communications link or number of messages processed by a communication server over a specified period of time. An example of traffic measurement is centum call seconds (CCS). A single CCS equals 100 seconds of a call on the same communication circuit.

Traffic Capacity-The maximum amount of communication traffic (users or data transfer rate) per unit time that can be carried by a specified telecommunication system, sub system, or device under specified conditions.

Traffic Classification-Traffic class is the categorization of the connection or media stream types. Traffic classification can be used to select alternate routes or how media streams or data packets may be processed (e.g. prioritization).

Traffic Load-Traffic loading is the volume of traffic that equals the sum of the holding times of several calls or attempts. Loads normally are expressed in either hundred call seconds (CCS) or Erlangs. A statement of load is inherently an average of all the instantaneous loads over a basic interval, such as an hour. This term often is called simply traffic load.

Traffic Priority-Traffic priority a code or value that is assigned to a data connection or communication session to enable preferences and priorities on how the routing and/or processing of connections or sessions will occur.

Traffic Routing-Traffic routing is the selection of communication routes, paths, or the choice of preferred communication carriers (such as long distance) based on dialed digits, cost of service, traffic congestion, line failure, or other criteria that may affect choice of routing path.

Traffic Use Code-A traffic use code is a unique identifier that is assigned for specific types of traffic offered on a trunk group.

Train the Trainer-Train the trainer is the providing of information materials in a format that helps a person or group of people to gain knowledge or skills that enable them to provide the training in the future. Train the trainer may involve providing more detailed information on the topic to allow the future trainer to answer questions that are beyond the course contents.

Training-Training is the providing of information materials in a format that helps a person to gain knowledge or skills. Training may be provided in various forms including instructor led training, online training, web seminars, and self-study courses.

Transaction-A transaction is an activity within system or domain that transfers the ownership or rights to assets or other items that have value or that can be characterized.

Transaction Audit Trail-A transaction audit trail is the availability of information elements (such as financial transactions and usage events) that can be linked to determine the origin and history of the usage of a product, service or information (content).

Transaction Authority Markup Language (XAML)-Transaction authority markup language is an industry standard that can be used to coordinate and process online business transactions. XAML was initially developed jointly by Bowstreet, Hewlett-Packard, IBM, Oracle and Sun.

Transaction Capabilities Application Part (TCAP)-The portion of the SS7 protocol which is responsible for information transfer between two or more nodes in the signaling network. The application layer of the Transaction Capabilities protocol consists of transaction capabilities that manage remote operations.

Transaction Costs-Transaction cost is the combined expenses that are incurred when orders are processed. Examples of transaction cost expenses include order-processing labor, telephone usage and data entry.

Transaction Fee-A transaction fee is a value that is assessed for each event (transaction) that occurs.

Transaction Gateway-A transaction gateway is a message processing device that can receive, process and forward transaction messages between systems. A transaction gateway may be used to adapt transaction messages from one format into another format.

Transaction History-Transaction history is a set of records or event information that identifies or provides information on each transaction that is associated with a user or account.

Transaction Processing (TP)-(1-general) Transaction processing are the steps and processes taken to complete a transaction (such as an order). (2-credit card) The processing of credit card payments by a merchant credit card processor.

Transaction Sales-Transaction sales are revenue that is generated from a single event (a transaction). An example of transaction sales is the purchasing of a movie viewing right (pay per view).

Transaction Set-A transaction set is a group of activities such as orders or purchases. An example of a transaction set is a group of purchases that are sent in an electronic data interchange (EDI) file that is sent from a distributor and a vendor.

Transaction Volume-Transaction volume is the number of transactions that occur or can occur over a period of time.

Transactional Billing-Transactional billing is the charging for services on a per event (per transaction) basis.

Transactional Data-Transactional data is information that is part of a purchase or transaction. Transactional may include order ID, product IDs, billing address, shipping address, shipping cost, sales tax, total cost, and other associated items and costs.

Transactional Metadata-Transactional metadata is information that describes the requirements for the business requirements for the transfer or use of media. Examples of transactional metadata include the cost, usage time, authorized types of users and other usage rights restrictions.

Transactional Rights-Transactional rights are actions or procedures that are authorized to be performed by individuals or companies that granted as the result of a transaction or event. An example of a transactional right is the authorization to read and use a book after it is purchased in a bookstore.

Transactive Pricing-Transactive pricing is the assigning of a cost or value to products on each transaction or specific transactions

Transceiver-A transceiver is combination of a radio transmitter and receiver into one radio device or assembly. A portable cellular phone is a transceiver.

This diagram shows a block diagram of a mobile radio transceiver. In this diagram, sound is converted to an electrical signal by a microphone. The audio signal is processed (filtered and adjusted) and is sent to a modulator. The modulator creates a modulated RF signal using the audio signal. The modulated signal

is supplied to an RF amplifier that increases the level of the RF signal and supplies it to the antenna for radio transmission. This mobile radio simultaneously receives another RF signal on a different frequency to allow the listening of the other person while talking. The received RF signal is then boosted by the receiver to a level acceptable for the demodulator assembly. The demodulator extracts the audio signal and the audio signal is amplified so it can create sound from the speaker.

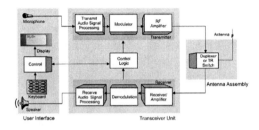

Mobile Radio Functional

Transceiver Cost-Transceiver cost is the purchase cost of a transceiver. The transceiver cost may or may not include a licensing fee that is paid for the use of technology (IPR) or software in the transceiver.

Transfer Pricing-Transfer pricing is the assigning of a cost or value to products that are transferred between departments within a company.

Transfer Protocol-A transfer protocol is the commands, procedures and processes that are used to coordinate access to transmission mediums and to transfer data between two or more communication devices.

Transfer Scheduler-A transfer scheduler is a device or application that is used to setup and manage the transfer of media at specific times.

Transferred Account Data Interchange Group (TADIG)-Transferred account data interchange group is a core working group within the GSM association that develops and maintains billing data interchange and testing specifications and standards for GSM carriers. Examples include the Transferred Account Procedure (TAP), TAP Testing Toolkit, Near Real Time Roaming Data Exchange, and Electronic Invoicing Data.

Transferred Account Procedures (TAP)-Transferred accounting process (TAP) is a standard billing format that is primarily used for global system for mobile (GSM) cellular and personal communications systems (PCS). As of 2001, the versions of TAP, TAP II, TAP II+, NAIG TAP II, and TAP 3. Each successive version of TAP provided for enhanced features.

Due to the global nature of 3G wireless and GSM, the TAP billing standard provides solutions for multi-lingual and multiple exchange rate issues. TAP3 was released in 2000 as a significant revision of TAP2. TAP3 has changed from the fixed record size used in TAP2 to variable record size and TAP3 offers billing information for many new types of services such as billing for short messaging and other information services. The TAP standard is managed by the GSM association at www.GSMmobile.com.

Transferred Account Procedures 3 (TAP3)-Transferred account procedure 3 (TAP3) is a standard billing format that is primarily used for GSM and GPRS systems. TAP3 was released in 2000 as a significant revision of its predecessor TAP2. TAP3 uses a variable record size and it offers billing information for many new types of services such as billing for short messaging and other information services. The GSM Association at www.GSMWorld.com manages the TAP standard.

Transferred Accounting Process (TAP)-Transferred accounting process (TAP) is a standard billing format that is primarily used for global system for mobile (GSM) cellular and personal communications systems (PCS). As of 2001, the versions of TAP, TAP II, TAP II+, NAIG TAP II, and TAP 3. Each successive version of TAP provided for enhanced features.

Due to the global nature of 3G wireless and GSM, the TAP billing standard provides solutions for multi-lingual and multiple exchange rate issues. TAP3 was released in 2000 as a significant revision of TAP2. TAP3 has changed from the fixed record size used in TAP2 to variable record size and TAP3 offers billing information for many new types of services such as billing for short messaging and other information services. The GSM association at www.GSMmobile.com manages the TAP standard.

Translate-(1-radio frequency) The process of converting a signal or band of signals to a different frequency spectrum. (2-data) The process of converting information from one language or code to another.

Translated Mirror Website-A translated mirror web site is a web server that contains the same content as another web site that is translated into another language. Translated mirror websites may be located on the same web server or they may be located on web sites in other locations (such as having a translated mirror web site in different countries).

Translation-The conversion of information from one form to another or the conversion of all or part of a telephone address destination code to routing instructions or routing digits.

Transmission-(1-general) The transfer of electric power, signals, or intelligence from one location to another by wire, fiber optic, or radio means. (2-beam) The use of a directional radio antenna to concentrate radio power into a small angle. (3-parallel) The simultaneous transmission of a number of signals through one or more systems. (4-radio) The transmission of electromagnetic radiation at radio frequencies. (5-serial) The transmission of sequential signals over a given system. (6-

stereophonic) The use of various methods to transmit two separate, but related, channels of audio information in order to convey the stereo effect

Transmission Certificate-A transmission certificate is a document or electronic message that confirms that a program or media item (such as an advertising message) was broadcasted or transmitted. Transmission certificates are sent to advertisers to prove that their promotion was performed as agreed so that they will pay their invoice for the advertising.

Transmission Elements-Transmission elements are the parts of a network that transfer information or signals. Transmission elements include communication lines, routers, switches, hubs and other signal switching and transmission functions.

Transmission Facility-A transmission facility is the transmission line and all the associated equipment required to transmit information.

Transmission Line-(1-general) Any transmission medium, including free space. (2-transmitter) A circuit that connects a transmitter to a load over a distance. (3-wave) Any circuit whose dimension is large compared with the wavelength of the signals passing through it.

Transmission Queue-A transmission queue is a list of data, tasks or information that are waiting to be transmitted.

Transmission Rate-Transmission rate is the amount of information that is transferred over a transmission medium over a specific period of time.

Transmission System-Transmission systems interconnect communication devices (end nodes) by guiding signal energy in a particular direction or directions through a transmission medium such as copper, air, or glass. A transmission system will have at least one transmitting device, a transmission medium, and a receiving device. The transmitting communication devices is capable of converting information to form electrical, electromagnetic wave (radio), or optical signals that allows the information to be transferred through the medium. The receiving communication device converts the transmitted signal into another form that can be used by the device or other devices that are connected to it. Transmission systems can be unidirectional (one direction) or they can be bi-directional (two directions).

Transmission Testing-Transmission testing is the measurement of specific performance characteristics of a transmission line that may include signal loss, frame slips or error seconds.

Transmit Delay-The time delay between when a signal is first originated to when it is first received at its destination. Also called transmission time or propagation delay for radio signals.

Transmitter-(1-telephone) Older term for the microphone in a telephone handset, and used as an adjective to identify signals associated with that microphone. (2-radio) A device that converts an electrical or acoustic signal into a transmitted carrier signal.

Transparency-A property of a communications medium (such as a coaxial cable or optical fiber) to carry a signal without altering or other-wise affecting the photonic or electrical characteristics of the signal.

Transparent-Transparent processes and operations are processed automatically in computer software or hardware systems but are not visible to the user or operator

Transparent Content Distribution-Transparent content distribution is the ability to transfer media or information into a device or system (such as a television set top box) where the transfer is not visible to the recipient or user.

Transparent Interface-A transparent interface is a connection point between two devices or systems that enables their interaction without the need to modify system operations on either side of the interface.

Transport Access-Transport access is the transmission medium that is used to send or receive data, media or information services.

Transport Aggregator-Transport aggregator is a company or system that combines multiple transmission channels for distribution through other communication channels.

Transport Layer Security (TLS)-Transport layer security is a set of commands and processes that are used to ensure data transferred across a connection is private. TLS is composed of two protocol layers; TLS record protocol and TLS handshake protocol. TLS is the evolution of secure socket layer (SSL) protocol that was developed by Netscape. TLS is defined in RFC 4346.

Transportable-(1-General) A transportable device is a product or assembly that performs processing functions at more than one location. (2-Mobile Radio) A mobile telephone that is capable of operation at multiple locations which usually has higher transmission power and more battery life capacity than portable mobile telephones.

Transraters-Transraters are transmission line processing devices or assemblies that adapt or convert the transmission rate of one communication line and/or circuits to the transmission rate of another communication line and/or circuit.

Transrating-Transrating is the process of converting information from one transmission rate to another transmission rate. An example of transrating is the conversion of a high-speed digital video signal that is received from a satellite into a medium-speed digital video signal that is transferred through the Internet. Transrating products are also called digital turnaround devices.

Travel Advances-Travel advances are the providing of funds or assets for travel costs that will be used in the future.

Travel Arrangements-Travel arrangements are the reserving or booking of travel related activities such as buying airline tickets, reserving hotels or booking car rentals.

Travel Authorization-Travel authorization is the process of submitting and getting approval for travel plans. The travel authorization process can range from a blanket agreement authorizing all travel plans that meet certain conditions to submission of detailed travel plans, costs and reasons for the travel.

Travel Class-Travel class is the level of service or quality of travel experience. Examples of travel class include economy, business and first class.

Travel Expenses-Travel expenses are costs incurred during travel between locations. Travel expenses may include transportation ticket costs (plane, train, taxi), leased vehicle cost, automobile usage (mileage), and meals that occur during travel.

Travel Policies-Travel policies are a set of rules or guidelines that define acceptable types of travel costs and how the travel should be booked, paid for and processed.

Treatment-The handling of customers or vendors for a specific purpose such as recovering overdue balances. When special treatment involves collection activities, it is called "dunning".

Trickledown Video Delivery-Trickle down video delivery is the process of transferring media programs by streaming the program at a low bit rate to a storage device (such as a set-top box that has disk storage) so the program can be directly viewed from the storage after the transfer is complete. Trickledown video delivery can simulate a video on demand experience where the user selects to watch a movie on demand that has already been transferred to a set top box without requiring the use of significant network resources.

Trigger-The process of initiating an action as a result of a specific event occurring or a parameter associated with an event (such as exceeding a voltage level).

TRIP-Telephone Routing over Internet Protocol

Triple Data Encryption Standard (3DES)- Triple data encryption standard is a variation of the data encryption standard that adds complexity to the encryption process by increasing the difficulty to break the encryption process.

Triple Play-Triple play refers to providing of three main services such as data, voice, and video on one network. For cable MSOs, this usually means building out the next generation network to DOCSIS 2.0 specifications, for Carriers this often means building out fibre or VDSL (very fast DSL) networks. Usually it is the larger MSOs and Telecom Carriers that roll out triple play services, and the advantage is that they can sign customers to a bundle of three services, thereby increasing revenue and customer loyalty.

Trojan Horse-A Trojan horse is a type of computer virus that is disguised to be a known or useful program but contains unexpected or harmful program codes that are activated when it is used. Trojan viruses are often provided in games, utilities, or other executable programs.

Trouble-A failure or fault affecting the service provided by a system.

Trouble Ticket-A trouble ticket is a maintenance work record that identifies a particular problem. The unique trouble ticket identification number allows for the gathering of actions and resources assigned to solve the problem.

Troubleshooting-Troubleshooting is the process of investigating, localizing and correcting (if possible) a fault or out of tolerance condition.

Troubleshooting Tools-Troubleshooting tools are programs, systems or processes that can be used to assist in the investigating, localizing and correcting (if possible) a fault or out of tolerance conditions.

TRS-Telecommunications Relay Service for the Deaf

TRT-Total Running Time

Truck Roll-A truck roll is the dispatching of an installer or technician to a customer's location.

Truncate-Truncating is the shortening of data or information.

Trunk-A communication path that connects two network elements (such as switching systems, networks or data devices). Trunks are usually shared by many users. Trunks may be classified by the type of equipment they connect. For example, a PBX trunk connects a PBX system to the public switched telephone network (PSTN). Trunks carry conversations for different subscribers at different times. The name comes from a trunk of a tree, via the prior usage for railroad trunk lines.

Trunk Cable-(1-telephone) A trunk cable is a communication line that transports multiple communication channels. (2-cable television) A cable television trunk is a main transmission line that supplies multiple branch connections.

Trunk Carrier (T Carrier)-Trunk carriers (T Carrier) are a hierarchy of digital communication lines that are used for multiple channel transmission lines that range from 1.544 Mbps to 565 Mbps. T carrier is the actual transmission channel as opposed to the digital signal (DS) channel that is the digital format into the channel transmitter. Tx has been used to represent the digital transmission standards where the "x" denotes which service is under discussion.

Trunk Circuit-A trunk circuit is a communication channel that is transported over a multi-channel communication line (trunk line).

Trunk Code-A trunk code is a unique character identifier that is associated with a specific trunk to allow for the routing of calls or communication circuits to or through specific trunks or trunk groups.

Trunk Forecast-A trunk forecast is an estimate of the number of communication circuits that is likely to be required to provide a level of service for a system, an area or particular user.

Trunk Group Alternate Route-Trunk group alternate route is an alternate path that may be used in the event of a trunk group failure or congested condition.

Trunk Group Data-Trunk group data is the measurements of the amount of usage (traffic) on a trunk group.

Trunk Occupancy-The percentage of some time period, usually an hour, during which a trunk is in use. Trunk occupancy also may be expressed as the carried hundred call seconds per hour per trunk.

Trunks-Groups of wires or fiber optic communication lines that are used to interconnect communications devices. Trunks usually have many physical and/or logical communication channels.

Trusted Authority-A trusted authority is an information source or company that can issue or validate certificates. A trusted authority is sometimes called a "certificate authority."

Trusted Boundary-A trusted boundary is an interface point for a system or domain that has been recognized to communicate with some level of authenticity or validity.

Trusted Certificate-A trusted certificate is an identification document that is signed, typically by a trusted authority (such as a certificate authority).

Trusted Device-(1-general) A trusted device is an equipment or assembly that is previously known or suspected to only communicate information that will not alter or damage equipment of stored data. Trusted devices are usually allowed privilege levels that could allow data manipulation and or deletion. (2-Bluetooth) A device using Bluetooth wireless technology that has been previously authenticated and allowed to access another Bluetooth device based on its link-level key.

Trusted Third Party (TTP)-A trusted third party is a person or company that is recognized by two (or more) parties to a transaction (such as an online) as a credible or reliable

entity who will ensure a transaction or process is performed as both parties have agreed.

Trustmark-A trustmark is a symbol or brand that provides users or consumers with confidence about the operation, reliability or other characteristics of a product or service.

TSAPI-Telephony Services Application Programming Interface

TSTV-Time Shift Television

TTFF-Time to First Fix

TTL-Time To Live

TTP-Trusted Third Party

TTS-Text To Speech

TTY-Teletypewriter

Turn Key (Turnkey)-Turn key is the process of defining, developing or providing a system or service that is completely ready for use without additional actions or tasks to be performed by the purchaser or user.

Turn Up-Turn up is process that enables communication equipment or a system to be put into service.

Turnkey-Turn Key

TV Bypass-TV bypass is the obtaining of television programs without the use of broadcasting services from other companies such as cable TV, satellite or terrestrial television. TV bypass service may allow program access over the data network portion of broadcast networks.

TV Channel-Television Channel

TV Cookie-A TV cookie is a small amount of information that is stored on a set top box (a TV client) that is used by an IPTV service provider to help control the content and format of information to the user during future visits to IPTV channels.

TV Portal-Television Portal

TVM-Time Varying Media

TVoDSL-Television over DSL

TVoIP-Television over Internet Protocol

Tweaking-The process of slightly adjusting an electronic assembly circuit to optimize its performance or to bring the system within the required specifications.

Type Length Value (TLV)-Type length value is a method that can be used to identify and extend the parameters or data that is sent within a message.

Type Of Service (TOS)-(1-data packet) The field within the IP datagram header that indicates the type of packet so the system can vary the type of service (typically the priority of routing) that is performed on the IP data packet. The TOS field as specified in RFC 760 and RFC 2475 defines an application of TOS as used in DiffServ networks. (2-billing) A type of service is a classification or category of processes (such as voice communication) or function (such as data processing) that is authorized or provided.

Type of Use Fraud-Type of use fraud is the usage of media, products or services in a way that is not permitted by its licensing agreement.

Typeface-Typeface is the character images that are associated with the underlying text (type).

U

UA-User Agent

UAT-User Acceptance Testing

Ubicomp-Ubiquitous Computing

Ubiquitous Computing (Ubicomp)- Ubiquitous computing (or "ubicomp") is a post-desktop model of human-computer interaction in which information processing has been thoroughly integrated into everyday objects and activities. As opposed to the desktop paradigm, in which a single user consciously engages a single device for a specialized purpose, someone "using" ubiquitous computing engages many computational devices and systems simultaneously in the course of ordinary activities and may not necessarily even be aware that they are doing so.

UBR-Unspecified Bit Rate

UC-Usage Control

UCC-Uniform Code Council

UCC-Uniform Commercial Code

UCD-Uniform Call Distribution

UCITA-Uniform Computer Information Transactions Act

UDP-Uniform Dial Plan

UDR-Usage Data Record

UE-User Equipment

UI-User Interface

Ultra Broadband-Ultra broadband is a term that is commonly associated with very high-speed data transfer connections. When applied to consumer access networks, ultra broadband often refers to data transmission rates of 10 Mbps or higher, and is generally associated with the delivery or ability to deliver triple play services. Ultra broadband is increasingly associated with IP-based networks and communications.

UM-Unified Messaging

Umbrella-Umbrella is a boundary or coverage area that one or more systems or services may be offered.

Umbrella Principle-The umbrella principle is the application of the same charging rates (tariffs) to all visiting customers in a mobile communication system irregardless of which home system they are registered with.

Umbrella Statute-An umbrella statute is a law or rule created by a government or company that defines how the regulation applies to a group or category of products.

Unacknowledged Mode-Unacknowledged mode is a communication process that does not require the receiver of information to send indications back to the sender that it has successfully received the information that was sent.

Unattended Call Screening-A process that automatically redirects incoming telephone calls to other places dependent on preprogrammed criteria. These criteria may include forwarding specific telephone numbers or rerouting all calls during a specific time period.

Unattended Operation-Unattended operation is the use of a system without using or needing the assistance of an attendant or operator. An example of unattended operation is an interactive voice response (IVR) system that allows users to select extensions within a telephone system without the use of a person.

Unauthorized Access-Unauthorized access is the use of services or systems without obtaining the proper authorization.

Unavailability-A measure of the degree to which a system, subsystem, or piece of equipment is not operable and not in a committable state at the start of a mission when the mission is called for at a random point in time.

Un-billable Time-Un-billable time is the duration of services that are provided that are not billable to a user. An example of un-billable time is the duration of a toll free or free phone conversation.

Unbundled-A term describing services and programs that are sold separately by the manufacturers of telephone and computer-related equipment.

Unbundled Loop-The portions of a telephone system local loop that can be separated and leased or provided to other companies for their own use.

Unbundled Network Element (UNE)-Unbundled network elements are identifiable portions of a network that may be leased or assigned for use by other entities. Deregulation and privatization of communication systems (such as the Telecommunications Act of 1996) has required some communication system owners to allow other companies to lease UNE portions of their network to rapidly deploy new services and to allow for more competitions.

Unbundling Services-Unbundling is the process of separating portions of a telecommunication network that are owned or operated by a service provider. Unbundling is a common term used to describe the separation of standard telephone equipment and services to allow competing telephone service providers to gain fair access to parts of incumbent telephone company systems. An example of an unbundled service is for the incumbent phone company to lease access to the copper wire line that connects an end user to the local telephone company. The competing company may install high-speed data modems (such as ADSL) on the copper line to enhancing the value of the telecommunications service.

Uncertainty-An expression of the magnitude of a possible deviation of a measured value from the true value. Frequently, it is possible to distinguish two components: the systematic uncertainty and the random uncertainty. The random uncertainty is expressed by the standard deviation or by a multiple of the standard deviation. The systematic uncertainty generally is estimated on the basis of the parameter characteristics.

Unchannelized Carrier-An unchannelized carrier is a communication line (carrier) that allows the user to have unrestricted access to the entire data transmission capacity (after the line control overhead is removed) of the communication bearer circuit. Unchannelized carriers are sometimes called unstructured carriers.

Uncompress-Uncompressing is the process of converting a compressed file into its original form.

Unconditional-Unconditional is the processing of a command or request without regard to pre-existing conditions or the state of an equipment or assembly.

Uncontrollable Costs-Uncontrollable costs are fees or assessments that will be incurred which cannot be changed by the person or company who receives them. An example of uncontrollable costs is business license fees.

Undelete-Undelete is a command or process that is used to restore a file that has been deleted or that has been marked for deletion. Undelete commands only work if the data area for the file storage have not been modified (e.g. over written by other data) since the file was deleted.

Underlying Carrier-A common carrier that provides services or facilities to other common carriers.

Underwriting-Underwriting is obtaining of finances for a program or project in return for an acknowledgement of support. For television programming, underwriting typically restricts the acknowledgement of support through a short acknowledgement message that occurs during the playing of the program.

Undeveloped Areas-Undeveloped areas are geographic regions that do not have communication lines or services because they are not developed.

Undue Enrichment-Undue enrichment is the obtaining of value from others by using assets or resources of others without their authorization or willingness.

UNE-Unbundled Network Element

UNI-User Network Interface

Unicode-Universal Code

Unified Messaging (UM)-Unified messaging allows you to store, manage, and transfer different forms of messages from a variety of access devices. Unified messages include audio (voice messages), electronic mail (email), data messages (such as fax or files), and video (video mail). Unified messaging provides you with access to these multiple types of messages using standard telephones, (text to audio), Internet web pages (playing back voice messages), and other devices such as fax machines and mobile telephones.

Uniform Call Distribution (UCD)-Uniform call distribution is the process of routing calls to a group of lines so that they are assigned to stations on a uniform (evenly distributed) basis.

Uniform Code Council (UCC)-The uniform code council was a group that defined several commerce standards such as barcodes and EDI. The UCC has now become GS1.

Uniform Commercial Code (UCC)-Uniform commercial code is set of models and rules that define how the sale of goods, credits and bank transactions should occur in the United States.

Uniform Computer Information Transactions Act (UCITA)-The uniform computer information transactions act is a regulation (statute) that covers transactions that involve digital information. UCITA is similar to the uniform commercial code (UCC).

Uniform Dial Plan (UDP)-The dialing plan (digit sequences) that are used by all standard users within a telephone system. The use of UDP in private telephone systems (such as a PBX) allows callers to dial telephone extensions (other users within the private system) using predefined dialing codes (such a 4 digit or 5 digit extension codes).

Uniform Numbering Plan-A uniform numbering plan is a identifying number assignment system that allows for the initiation of calls or communication sessions using a uniform addressing (dialing) process. An example of a uniform numbering plan is the assignment of numbers to a private branch exchange (PBX) system that allows PBX telephones to place telephone calls over shared communication lines.

Uniform Resource Identifier (URI)-A uniform resource identifier is set of characters that identify a resource or the location of a resource. A URI can be contained in a hypertext transfer protocol (HTTP) message header (such as an image or media file) to identify where the resource exists on a network (such as on the Internet).

Uniform Resource Indicator (URI)-A uniform resource locator is a label (a string of characters) that is used to identify the location of a resource.

Uniform Resource Name (URN)-Uniform resource name (URN) is a naming process defined by the Internet Engineering Task Force (IETF) that is used as an identifying name for Internet resources. The URN has no relation to where the resources are located within the Internet.

Uniform Standard Tape Archive Format (USTAR)-Uniform standard tape archive format is a data structure that is designed to store data on a tape format. The USTAR format defines a header section that allows the tape system to quickly identify and locate data blocks.

Uninterruptible Power Supply (UPS)-A battery backup system designed to provide continuous power in the event of a commercial power failure or fluctuation. A UPS system is particularly important for network servers, bridges, and gateways.

Unit-(1-equipment) A unit is a device or an assembly of equipment that along with other devices or units form a complete system or subsystem. (2-quantity) A specified quantity in terms of which other quantities can be measured.

Unit Billing-Unit billing is a value or characteristic that is used in the measurement of service where the measurement is expressed in units. Examples of a unit billing include 30 seconds per unit.

Unit Configuration-Unit configuration is the operational parameters and feature options that are used to adjust or modify operation of a device or system. Unit configuration is typically performed to adapt a device to the specific requirements of the system is communicating or operating with.

Unit Conversion Table-A unit conversion table is a group of structured information that can be used to convert or filter data from one unit source into another unit form. An example of a unit conversion table is a set of currency conversion rates that are used to convert prices or sales amounts from multiple currencies into a form that uses a single currency.

Unit Price-A unit price is an offer value that is assigned to a defined amount of a product or service (such as $10 per 1 MB unit).

Universal Code (Unicode)-A 16-bit, fixed-width character encoding standard that encompasses virtually all of the characters commonly used on computers today-this includes most written languages, plus publishing characters, mathematical and technical symbols, and punctuation marks. Unicode support makes it easier and faster for developers to create products and localize them different languages.

Universal Plug and Play (UPnP)-Universal plug and play is an industry standard that simplifies the installation, setup, operation removal of consumer electronic devices. UPnP includes the automatic recognition of device type, communication capability, service capabilities, activation and deactivation of software drivers and system management functions. More information about UPnP can be found at www.UPnP.org.

Universal Ports-Universal ports are connection points on communication systems that can be configured as different types of connections

Universal Resource Locator (URL)-A standardized addressing process used to identify resources that are connected to the Internet. The URL is a text string that defines the location of a resource (such as an address of a web site the Internet), as well as the protocol to be used to access the resource.

Universal Serial Bus (USB)-Universal serial bus is an industry standard data communication interface that is installed on personal computers. The USB was designed to replace the older UART data communications port. There are two standards for USB. Version 1.1 that permits data transmission speeds up to 12 Mbps and up to 127 devices can share a single USB port. In 2001, USB version 2.0 was released that increases the data transmission rate to 480 Mbps.

This diagram shows how a universal serial bus (USB) system interconnects devices in a personal distribution network (PDN). This example shows that a USB system uses a host controller interface (HCI) to coordinate the access to all other devices that it is attached to. As each device is added, the host controller registers the device (called device enumeration) and coordinates all communication to and from the devices. This diagram also shows that there are two types of connectors used in the system to ensure that a host device is not accidentally connected to another host device.

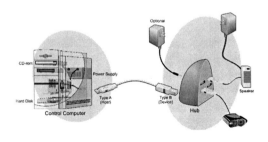

Universal Serial Bus (USB) System

A hub is used to allow the connection of additional devices (up to 127 can be attached to one host system). The USB system allows for the supply of power through the USB cable (5 volts) or an external power supply can be used.

Universal Service-The objective set by many state regulatory agencies and the Federal Communications Commission to keep telephone services affordable for as many customers as possible.

Universal Service Charges-Universal service charges are fees that are collected from service users in a communication system to subsidize the construction of communication systems in rural areas to allow people in all areas within a country to have affordable access to communication services.

Universal Service Fund (USF)-The universal service fund is a financial account that receives money from users of a services from established systems to subsidize the development or providing of services in developing or rural areas so that a majority of the population can have fair access to similar types of services.

Universal Standard Products and Services Classification (UNSPSC)-Universal standard products and services classification is a global coding system that is used to classify products and systems. UNSPSC codes are used for product catalogs and accounting systems.

Universal Time 1 (UT1)-Universal time 1 is a reference time scale that adds leap seconds to the time UTC value maintained by the Bureau International de l'Heure (BIR). UT1 adjusts time for small changes in the Earth's rotation.

Universally Unique Identifier (UUID)-Universally unique identifier is a 128 bit (16 Byte) identification code that is used to identify devices and software. UUIDs are globally unique. The standard for UUIDs is defined by the open software foundation (www.opengroup.org). UUIDs can be created on a tempo-rary basis (such as when a new service capability is added to a software program) or it can be permanently assigned.

Universe of Possible Buyers-The total number of potential possible buyers of a product or service.

UNIX-A computer operating system originally developed and deployed by the Bell Telephone laboratories and now an industry standard. UNIX is a registered trademark mark of UNIX System Laboratories.

Unlicensed Frequency Band-Unlicensed frequency bands are a range of frequencies that can be used by any product or person provided the transmission conforms to transmission characteristics defined by the appropriate regulatory agency.

This figure shows typical types of unlicensed radio transmission systems. This example shows that there are several different communication sessions that are simultaneously operating in the same frequency band and that the transmission of these devices are not controlled by any single operator. These devices do cause some interference with each other and the types of interference can be continuous, short-term intermittent, or even short bursts. For the video camera (such as a wireless security system), the transmission is continuous. For the wireless headset, the

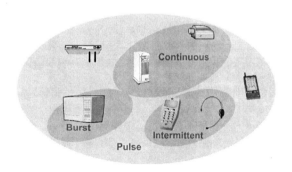

Unlicensed Radio Systems

transmission is on for several minutes at a time. For the microwave oven, the radio signals (undesired) occur for very short bursts only when the microwave is operating.

Unlisted-Unlisted is the keeping of identification and/or other information related to a user or account from information listings (such as telephone directories).

Unsold Inventory-Unsold inventory is products or services (such as advertising time slots) that have not been sold.

Unsolicited Grant-An unsolicited grant is a command authorization from a system or server that grants access or services (such as bandwidth allocation) which is not a result of a service request.

Unspecified Bit Rate (UBR)-Unspecified bit rate (UBR) is a category of telecommunications service that provide an unspecified data transmission rate of service to end user applications. Applications that use UBR services do not require real-time interactivity nor do they require a minimum data transfer rate. UBR applications may not require the pre-establishment of connections. An example of a UBR application is Internet web browsing.

UNSPSC-Universal Standard Products and Services Classification

Unsubscribe-Unsubscribing is the process of removing a name, account or device from a list of users or service subscribers.

Untethered-A system that is not bound by wires or cables.

Unzipping-Unzipping is a process that identifies components within a compressed data file and expands these files into their original form.

Up Selling (Up-Sell)-Up selling is a marketing activity that is designed to encourage customers to buy different products from the same company that have a higher value than the initial product that is requested by the customer. In telecommunications, up selling could include selling a bundled package of services or services that have higher or improved performance characteristics.

Updating-Updating is the process of providing corrective data or new programs or products that correct deficiencies or improve the functionality of a product or service.

Upfront Advertising (Upfronts)-Upfronts are advertising messages (e.g. commercials) for programs are provided or distributed. Upfronts are commonly associated with advertising that is prepaid or sold for a season of television programs.

Up-Front Buys-Up-front buys are the purchases of products or services in advance. An example of an up-front buy is the purchasing of broadcast media time before the ads are planned or scheduled.

Upfronts-Upfront Advertising

Upgradeable Computer-An upgradeable computer is a computing system that is designed to allow devices, cards or internal systems to be removed, changed or replaced with newer components and/or as technology innovations become available.

Upgrade-An upgrade is the process and/or the equipment and programs that are used to modify, alter or improve the features or services of a product or system.

Upgrade Kit-An upgrade kit is a package that contains components, wires, software or other items that enable a device or system to be modified to provide new features and/or services.

Uploading-Uploading is a process of transferring information or programs from a device or computer to a network or a computer server on that network. An example of upload is the transferring of web page files from a computer to a web site server.

UPnP-Universal Plug and Play

Upper Application Layers-A protocol interface layer that represents the application that is currently being used, such as voice and SMS transactions, media player operation or over the air programming.

UPR-User Performance Requirements

UPS-Uninterruptible Power Supply

Up-Sell-Up Selling

Upset Price-An upset price is xxx Defined by the WGA, the lowest amount of money due to be paid to the writer of a product, which allows a buyer to acquire abstracted rights to that product.

Upstream-(1-general) A device or system placed ahead of other devices or systems in a signal path. (2-network) The direction opposite the direction of distribution of network timing signals. (3-video keyer) A term that describes the location of keyers in a mix/effects level or in the overall switcher architecture. (4- video switcher) A term relating the priority of the video signals as they are combined through a production switcher.

Uptime-The uninterrupted period of time that network or computer resources are accessible and available to a user.

Upward Compatible-Upward compatible is the ability of a computer or system to perform operations (such as running programs) from older (e.g. legacy) computers or systems.

Urban Service-Urban services are communication services provided to users located in urban (developed) areas. For 3rd generation systems, rural services may obtain data transmission rates of 2 Mbps.

URI-Uniform Resource Identifier
URI-Uniform Resource Indicator
URL-Universal Resource Locator
URN-Uniform Resource Name

Usability-Usability is the processes and features that enable users to access and interact with applications.

Usability Testing-Usability testing is the processes that are used to evaluate the ability of people to use products, features or services.

Usage Accounting-Usage accounting is the tracking of data obtained from network transactions for applications in billing, traffic analysis, settlement, fraud, and customer analysis.

Usage Based Charging-Usage based charging is the rating of billing cost that is determined by the amount of data or service used

regardless of the duration (start to end) time of the service.

Usage Collection-Usage collection is the gathering of information that is used to determine the amount of usage of a product or service.

Usage Control (UC)-Usage control is the process of monitoring and coordinating the usage of media or services.

Usage Data-Usage data is information related to a communication session or the type, quantity and characteristics of service usage.

Usage Data Record (UDR)-A usage data record holds information related to a communication session or service usage data. This information usually contains the origination and destination address of the session, time of day the session was connected, added charges through other networks or systems, and the duration of the service.

Usage Fees-Usage fees are the billing charges to a customer for the usage of a product or service.

Usage Investigation-Usage investigation is the process of identifying, gathering and analyzing events or processes related to the usage of products or services.

Usage Level-Usage level is a measure of an amount of resource that is used (such as minutes used or amount of data that is transferred) over a given time period or for a particular device or service.

Usage Management-Usage management is the analysis and application of data obtained from network transactions that are used for authorization, mediation, charging, accounting, and various types of business intelligence.

Usage Metering-Usage metering is the process of tracking a quantity of service or material over a period of time or event period. This figure shows some of the common types of service usage metering. This diagram shows that the types of usage metrics may include the amount of time a service has been used, how much of a service may be used, the num-

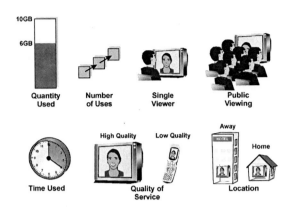

Billing Usage Metering

ber of times a service has been used or activated, the type of use (e.g. single viewer or public viewers), quality of service (e.g. high resolution or low resolution) or the location of the service access point (e.g. home or at a visited/away location).

Usage Model-A usage model is a representation of the typical usage of a product or service. A usage model is created to help designers define the required operations and communication that is required to satisfy the needs of the usage model.

Usage Monitoring-Usage monitoring is the measurement of user actions that involve content or services.

Usage Profiling-User profiling is the process of monitoring, measuring and analyzing usage characteristics for a user of a product or service.

Usage Rating-Usage rating is the assignment of a value to a usage record.

Usage Reporting-Usage reporting is the providing of usage activity data in a form that can be used to evaluate the performance or effectiveness of a media promotion or service.

Usage Royalties-Usage royalties are fees that are paid to an author or composer for the right to use each copy of a work that is sold, performed, or produced under license of an exclusive right (such as patent rights).

Usage Sensitive Pricing-Usage sensitive pricing is the selection of usage charging rates based on actual usage of services or resources. Usage sensitive pricing may vary based on the time of day, the duration of data transfer or call, or the type of media transferred.

Usage Statistics (Usage Stats)-Usage statistics is the mathematical analysis and representation of usage information that can be qualified and quantified. Usage statistics allows for the expression of characteristics of information or data in a form that can be used to help understand specific aspects of information (such as average usage rates for specific types of users). Usage statistics can also be used to predict or estimate the usage of a product or service based on the related factors (such as changes of price or types of customers).

Usage Stats-Usage Statistics

Usage Tracking-Usage tracking is the recording of a quantity of service or material that is transferred over a period of time or between events.

USB-Universal Serial Bus

USC-User Selectable Content

usec-Microsecond

Useful Bandwidth-Useful bandwidth is the useable amount of transfer that may occur in a communication channel. The useful bandwidth is usually the portion of a channel frequency band that can carry signals (such as between guard bands).

User-A user is a person, company, or group that receives processes or takes some form of action on services or products. A communication user transfers or processes voice, data, video or other information..

User Acceptance Testing (UAT)-A set of tests that are performed by or for a potential user (often a buyer) of a piece of equipment or system that is supposed to ensure the equipment or system will meet the functional requirements of the user.

User Accessible Tables-User accessible tables or lists (databases) that are stored or accessible by users. User accessible tables may be stored inside a communication system (such as a PBX system) to hold feature options and their associated privileges.

User Account-A user account is a security system that controls or monitors access to a network by assigning users accounts. A user account may include a user identification code, password, and usage rights.

User Agent (UA)-End user devices in a SIP system are called user agents (UA). The UA is a conversion device that adapts signals from a data network into a format that is suitable for users. Examples of user agents include dedicated IP telephones (hardphones), analog telephone adapters (ATAs), or software (softphones) that operate on a computer that has multimedia (audio) capabilities.

User Behavior-User behavior is the trends or characteristics of user actions during the usage of services by customers in order to obtain useful data needed to allocate resources for the delivery of those and future services.

User Codes-User code is a set of instructions that are created by a user to perform processes or tasks in a computing device (such as a software script that is used to automatically enter data into a spreadsheet program).

User Definable Fields-User definable fields are a portion of a record (a field) that can be changed or adapted by the user to meet their specific needs or interests.

User Defined-User defined is an element that can be changed or adapted by the user to meet their specific needs or interests.

User Defined Rules-User defined rules are requirements or conditions for a process that are created by a user. An example of user defined rules is the ability of a company or user to define new data entry requirements for

a billing system without the need to make changes to the underlying software program or service.

User Domain-A user domain is a set of devices, access points or physical areas where a user can obtain access or control devices, applications or services.

User Equipment (UE)-User equipment is a device (such as a mobile radio telephone) that can connect to a communication system such as a universal mobile telephone system (UMTS). UE devices can be multimedia mobile telephones, personal computers, transceivers that are installed in vehicles or fixed wireless units.

User Experience-User experience is the perceived interactions that a person has with a product or service.

User Forums-User forums are groups of people that share common needs and interests for products or services and are willing to share questions and solutions to be part of the group.

User Friendly-User friendly is a product or service that is designed to be easily understood and operated by a user.

User Grade-A quality of service (QoS) that is acceptable for users or a service. User grade QoS usually higher levels of error rates and lower reliability than carrier grade services.

User Group-A user group is a number of users of a specific product or software system that share information. User groups may have newsletters and chat rooms to help gather and distribute information relative to a product or service.

User ID-User Identification

User Identification (User ID)-A user identification number is a unique identifier that is assigned to a user. The User ID may be a temporary or permanent number.

User Interface (UI)-A user interface is a portion of equipment or operating system that allows the equipment to interface with the user. A user interface is also called the man machine interface.

User Interface Developer-User interface developers are companies or people that develop software and/hardware that allows equipment (such as television set top boxes) to interface with the user.

User Interface Layer-A user interface layer coordinates the presentation and interaction of information between a device and the user of the device. The user interface layer receives data from the underlying protocols and processes this information into a form required or requested by the user or endpoint device.

User Interface Specification-A user interface specification describes the requirements and operation of the interaction (interface) with a device or system.

User Key-User keys are unique codes that are specific to each user or device.

User Lock Out (Lockout)-User lock out is the inhibiting of a connection or future connection to a system or service.

User Manual-A user's manual describes the typical operation and requirements of devices or equipment.

User Network Interface (UNI)-The interface between an end user and a telecommunications network. A UNI could be a industry standard set of protocol rules and data transmission specifications or may be a proprietary protocol.

User Performance Requirements (UPR)-A set of requirements that are necessary to meet the needs or desires of typical users of a system and/or service.

User Priority-User priority is the assignment of levels or categories that are used to determine how the assignment of resources and services will occur during periods of limited resources.

User Profile-A user profile is the characteristics that are associated with a specific user, company or account.

User Profiling-User profiling is the process of monitoring, measuring and analyzing usage characteristics of a user of a product or service. IPTV service offers the possibility for recording and using viewer information to better target services to users. An example of user profiling is the offering of movie service packages (such as 5 films for children at a discount price) that is based on the previous viewing habits (such as watching 5 children's movies in the past 2 weeks).

User Rights Policies-User rights policies are the rules that are used to determine and assign rights to users, groups or companies to access and use products or services.

User Selectable Content (USC)-User selectable content is media on a display or interface device that a user can use to interact with or choose options.

User Session-(1-web) A user session is a single visitor's access of a web site from the initial request to the termination of activity on that web site. Because a single visitor usually requests more than one web page from a web site, a session can be tracked by the IP address of the visitor or through the review of a cookie that is stored on the visitor's computer.

User Template-A user template is a document, data file, form or sample that is used by a person or company as a guideline or process to assist in the creation of content or data.

User Tracking-User tracking is the process of sensing and recording the activities of a user of a program, product or service.

USF-Universal Service Fund

USTAR-Uniform Standard Tape Archive Format

Usury-Usury is the charging of interest or fees that are in excess of acceptable levels. Governments or their agencies may define acceptable interest rates.**UT1**-Universal Time 1

UTC-Coordinated Universal Time

Utility-(1-general) A company that provides services to the public. (2-datacom) A part of a protocol part that provides service information.

Utility Billing-Utility billing is the process of grouping utility service usage information for specific accounts or customers, producing and sending invoices, recording (posting) payments made to customer accounts.

Utilization-(1-Facilities) The use of telecommunications facilities and equipment, expressed as a percentage of working to working plus spare facilities. (2-Performance Management) Utilization as used in performance management for network management measures the use of a particular resource over time. The measure is usually expressed in the form of a percentage in which the usage of a resource is compared with its maximum operational capacity, like bandwidth utilization on a network link.

UUID-Universally Unique Identifier

V

VAC-Visitor Acquisition Cost

Vacant Code-A vacant code is a code (number) in a numbering plan that is not yet assigned.

Validation-Validation is the determination of conformance or acceptability of information or systems.

Validation Rules-Validation rules are a set of requirements or conditions for the determination of conformance or acceptability of information or systems.

Validity Check-A validity check is the transfer of information that is used to ensure that the service or quality of transmission is within established limits.

Validity Period-A validity period is a time interval that an action, service or authorization will remain active.

Valorization-Valorization is the process of influencing or controlling the value of a commodity or currency through external forces (such as a government controlling the value of its currency by controlling the supply of currency or loans).

Value Added Network (VAN)-A value-added network is a communication system that provides products or services that go beyond the basic transfer or management of information. Examples of features contained in value added networks include database management systems, protocol conversion, end-user support and media adaptation.

Value Added Reseller (VAR)-A company or organization that adds assemblies, software, or documentation to products produced by another manufacturer or service provider so they may be sold in their sales and distribution system. VARs may modify a standard product (such as a laptop computer) and modify for use in a specific industry (called a vertical application.)

Value Added Services (VAS)-Services that provides benefits to a customer that are not part of the standard telecommunications services associated with a basic communication service. VAS services include voice mail, information services and content delivery.
Services offered by prepaid provider (e.g., voice mail, fax store and forward, interactive voice response, and information services) in addition to calling time.

Value Added Tax (VAT)-A tax that is added on to the value of the product or service.

Value Chain-An operational model that describes the core functions that are required to deliver products or services to the end customer. The blocks in a typical retail value chain include marketing, sales, order management, and customer support.

Value Chain Pirate-A value chain pirate is a company or person who targets and acquires parts of another company's value chain. An example of a value chain pirate is an online bookstore. The online bookstore can more cost effectively display book information than a bricks and mortar retail bookstore.

Value Engineering-Value engineering is the process of selecting, designing and changing products or services to help improve their market value.

Value Proposition-A value proposition is statement, made by a business person, which describes how the business will make use of the information delivered to improve operational efficiency, profitability, or market strength.

VAN-Value Added Network

Vaporware-Vaporware is a sarcastic name for a product that has been announced but is not available and has not been produced. Vaporware products often do not get released.

VAR-Value Added Reseller

VAR Account-Value added reseller accounts are companies or organizations that sell products or services that the produce by enhancing (adding value) to products or services from other companies. VAR account sales representative support needs may include the ability to access and communicate with technical staff and providing design details of products and services.

Variable-A variable is a symbol or label that contains a value that can change. A variable is commonly used in software programs to allow the temporary storage and manipulation of data or information.

Variable Bandwidth-A communication system that allows for a variable data transmission rate or changes in communication channel frequency bandwidth dependent on the need of the end user applications and/or the ability of the system to provide the desired data transmission or frequency bandwidth. Because variable bandwidth systems help match the system resources used to the actual data transmission needs of the end customer (e.g. reduce the bandwidth when the user has nothing to send or say), variable bandwidth systems are more efficient that constant bandwidth systems.

Variable Bit Rate (VBR)-A category of telecommunications service that provides a variable data transmission rate of service to end user applications. Applications that use VBR services usually require some real-time interactivity with bursts of data transmission. An example of a VBR application is videoconferencing.

Variable Costs-Variable costs are expenses that occur and increase as products or services are produced or provided.

Variable Length Records-Variable length records are sets of data that can have different field structures and data lengths. Variable length records are more flexible to work with than fixed length records.

Variant-A variant is a product or service that is related to other products or services but has slightly different features or characteristics.

VAS-Value Added Services
VAT-Value Added Tax
VBR-Variable Bit Rate
VBS-Voice Broadcast Service
VC-Venture Capitalist
vCalendar-A format for calendar and scheduling information. The vCalendar specification was created by the Versit consortium and is now managed by the Internet Mail Consortium (IMC).

vCard-A format for personal information such as would appear on a business card. The vCard specification was created by Versit consortium and is now managed by the Internet Mail Consortium (IMC).

V-Chip-Video Chip
VCO-Video Central Office
VCR-Video Cassette Recorder
VCS-Virtual Circuit Switch
VDN-Video Distribution Network
VDT-Video Display Terminal
Vendor-A vendor is a person or company that produces, manufactures or sells products or services.

Vendor Code-A vendor code is a number or label that identifies a company or person who produces products or services.

Vendor Managed Inventory (VMI)-Vendor managed inventory is the process of allowing vendors to identify the quantity of inventory assets (stock) that are available, determining how many will be needed in the future and coordinating the ordering of the stock items.

Vendor Management-Vendor management is the process that identifies, defines and tracks vendors who supply products and services.

Vendor Selection-Vendor selection is the identified company or process used to identify one or more companies to supply products or services. When vendor selection is performed as a result of a request for proposal (RFP), vendor selection may be performed through the use of a criteria evaluation. The criteria evaluation may be itemized and weighted as to importance of each evaluated item.

Venture Capital-Venture capital is money that is invested in companies during their early growth mode.

Venture Capitalist (VC)-Venture capital is a person who typically invests money in companies in the early growth stages of a business with the anticipation of making relatively high rates of return on their investment.

Versatility-Versatility is the ability of a product, system, or service to be expanded or adapted to support or provide other services and applications.

Version-A version is product or service that has a specific configuration or set of features that varies from a related item.

Version Control-Version control is the identification, assignment of version numbers and management of the products or services that have multiple configurations or feature sets.

Version Number-A number that identifies a particular software or hardware product that uses the same name as other products. These products usually undergo revisions or updates and the version number often relates to the date of release. The version number is usually assigned by the manufacturer or developer of the product. That often includes numbers before and after a decimal point; the higher the number, the more recent the release. The version number is important as features and operation of a product or software program may vary between different versions. The ability to determine the specific version number of a product may allow more reliable interaction between programs and products. Version number 1.0 often indicates an initial version of a product or software.

Versioning System-A versioning system is a management tool that allows an application to install, maintain or revert to specific versions of a program or service.

Vertical and Horizontal Coordinates (VH Coordinates)-Vertical and horizontal coordinates are a pair of geographical identifying positions that are used to calculate the rates for communication services between locations. VH coordinates are created using latitude and longitude locations.

Vertical Application-A vertical application is a program or software that is designed or used for a specialized industry application or profession. An example of a vertical application is a software programmed to provide wireless meter reading services to utility companies.

Vertical Discount-A vertical discount is a price reduction in the cost of media (such as broadcast time (or space in a print publication) that is defined in an agreement to advertise over a short period of time.

Vertical Market-A market that is defined for a specific category of product or service.

Vertical Services-Products or services that customers can add, at an additional charge, to enhance their basic exchange service. (See also: custom calling services, enhanced services.)

Vesting Schedule-A vesting schedule is a wage payment timetable that indicates which days people receive their pay.

VGCS-Voice Group Call Service

VH Coordinates-Vertical and Horizontal Coordinates

Vidcast-Video Podcast

Video-An electrical or optical signal that carries moving picture information.

This figure demonstrates the operation of the basic NTSC analog television system. The video source is broken into 30 frames per second and converted into multiple lines per frame. Each video line transmission begins with a burst pulse (called a sync pulse) that is followed by a signal that represents color and intensity. The time relative to the starting sync is the position on the line from left to right. Each line is sent until a frame is complete and the next frame can begin. The television receiver decodes the video signal to position and control the intensity of an electronic beam that scans the phosphorus tube ("picture tube") to recreate the display.

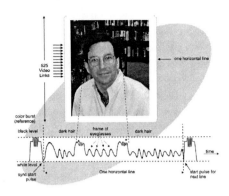

Analog Video

Video Camera-A device that converts images (light signals) into electrical video (multiple frame) signals.

This diagram shows how a video camera uses a cathode ray tube (CRT) to convert light energy into a video signal. This diagram shows that the CRT includes a photosensitive plate that receives an optical signal through a lens and also receives energy from an electron beam signal. When there is light on the plate and the electron beam hits the plate in a specific spot, a small amount of current flows from the CRT tube. The video signal generator controls the horizontal and vertical position of the electron beam through the deflection coils. Because the video generator knows the exact

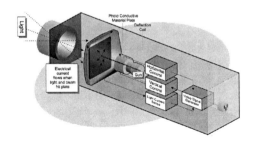

Video Camera Operation

position of the beam, it can create a composite video signal that represents the intensity (amplitude) and position (timing) of the image (light).

Video Capture-Video capturing is the process of receiving and storing video images. Video capture typically refers to capture of video images into digital form.

Video Capture Card-A video capture card is a printed circuit board or electronics assembly that is designed to be inserted (plugged-in) into a computer or electronic device and capture video information. A capture card may have several types of inputs including several formats of analog video, RF television channels, and digital video.

Video Cassette Recorder (VCR)-A video-cassette recorder is a video media storage device that is used to record and play back audio-visual programs on magnetic cassette tapes. VCRs were developed in the 1960s and become commercially available by the 1970s.

Video Casting (Videocast)-Videocasting is the recording and making available of video programs related to a subject topic on a per download or subscription basis.

Video Catalog-A video catalog is the presenting of items available for selecting or ordering in a video format. Video catalog formats can range from a linear progression of products (such as a television shopping channel) to an interactive video shopping cart that allows users to search and find items.

Video Central Office (VCO)-A video central office is an edge distribution facility for video signals in a IPTV network.

Video Chip (V-Chip)-A video chip is an integrated circuit (chip) that is designed to identify and block the viewing of certain video or television programs.

Video Communication-Video communication is the transmission and reception of video (multiple images) and other signals that can be represented by the frequency band used for video signal transmission.

Telecommunications systems can transfer video signals in analog or digital form.

This figure shows the basic process used for video signal transmission. In this example, a television camera converts an image and audio sounds to electrical signals. The video signal is created by a camera scanning the viewing area line by line. At the beginning of each line scan, the camera creates a synchronization pulse and the image (light level) is created by varying the electrical signal level after the synchronization pulse. The audio signal is created by using a microphone. These video and audio electrical signals are combined to form a composite video electrical signal. The composite video signal (baseband) modulates the radio transmitter frequency (broadband) signal. This low level radio signal is amplified to a very high power level for transmission. A video receiver (typically a television) receives the radio signal and many others from its antenna. It's receiver selects the correct radio signal by using a variable frequency filter (television channel selector) that demodulates the incoming radio signal to create the original video and audio electrical signals. The video signal is connected to a display device (typically a picture tube) and the audio signal is connected to the speaker.

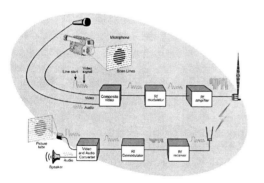

Video Transmission Operation

Video Dial Tone-An access and transport service for carrying full-motion video in much

the same way as a dial-up call is carried on a conventional voice network.

Video Display-A computer output device that presents data to the user in the form of an image, including text and/or graphics.

Video Display Terminal (VDT)-A computer terminal equipped with a keyboard and an electronic readout, such as a cathode ray tube or liquid crystal display. Video display terminals often are used to connect remote locations to a distant host computer.

Video Distribution Network (VDN)-A video distribution network is a system of communication lines and signal routing equipment that links video sources to viewing devices.

Video Gateway-A video gateway is a device or assembly that transforms video that is received from a device or system into a format that can be used by another network. A video gateway can convert video formats (such as analog video) into other formats (such as IP video).

Video Hub-A video hub is a distribution device or assembly that receives copies and redistributes the same information on other ports of the hub.

Video Library-A video library is a collection of video or television programs that can be accessed or used.

Video Mail (VMail)-Video mail is a system that can receive, manage and distribute short video clips (typically 1-2 minute video clips) via a communication system. Optionally, the user may be able to edit and process video clips that are available in their video mailbox. Video mail messages may be sent using integrated systems (multimedia messaging) or as an attachment to standard Email addresses.

Video Mailbox-A video mailbox is a portion of memory, usually located on a computer hard disk, which stores and plays video messages. The video messages are often in compressed digital video format.

Video Mixer-A video mixer (also called a video switcher) is a signal connection device or assembly that enables an operator to select the input source that will be connected to other lines or devices. A video mixer may include some media processing capabilities including fades, wipes and mixing.

Video Network-A video network is a system that contains a series of points that are interconnected by communications channels, often on a distributed (tree structure) basis.

Video On Demand (VOD)-Video on demand is a service that provides end users to interactively request and receive video services. These video services are from previously stored media (entertainment movies or education videos) or have a live connection (news events in real time).

This figure shows a video on demand (VOD) system. This diagram shows that multiple video players are available and these video players can be access by the end customer through the set-top box. When the customer browses through the available selection list, they can select the media to play.

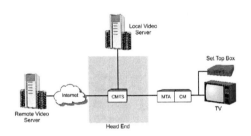

Video On Demand (VOD) Operation

Video on Demand Server (VOD Server)-The video on demand server is an application server that receives requests for video or multimedia programs and provides access to the requested media for authorized users.

Video on Demand Sponsorship (VOD Sponsorship)-Video on demand sponsorship is the providing of video on demand services to users that are paid for by another person or company (a sponsor).

Video Pass Through-Video pass thorough is the transferring of video signals from the input to an output of an assembly without any processing or changes.

Video Podcast (Vidcast)-Video podcast (sometimes shortened to vidcast or vodcast) is a term used for the online delivery of video on demand video clip content via Atom or RSS enclosures. The term is an evolution specialized for video, coming from the generally audio-based podcast and referring to the distribution of video where the RSS feed is used as a non-linear TV channel to which consumers can subscribe using a PC, TV, set-top box, media center, or mobile multimedia device.

Video Quality-Video quality is the ability of a display or video transfer system to recreate the key characteristics of an original video signal. Traditional video quality impairment measurements include blurriness and edge noise. Digital video and transmission system impairments include tiling, error blocks, smearing, jerkiness, edge business and object retention.

Video Quality Measurement (VQM)-Video quality measurement is a video quality rating score that is calculated using a combination of packet loss rate, packet loss distribution (bursty or distributed), types of packets lost and codec bit rate. The factor score ranges from 1 to 5.

Video Quality of Service (VQOS)-Video quality of service (QoS) is one or more measurement of desired performance and priorities of a video communications system. VQoS measures may include blocking, freeze frames or error free seconds.

Video Repository-A video repository is a storage system that is used to hold video media.

Video Ring Tone (Video Ringtone)-A video ringtone is a video and audio sequence that is used to announce an incoming telecommunications call. There may be several different types of ring tones and some telecommuni cations devices allow the user to program in their own unique selection for a ring tone.

Video Ringtone-Video Ring Tone

Video Scaling-Video scaling is the process of adjusting a given set of video image attributes (such as pixel location) so they can be used in another format (such as a smaller display area). Video scaling can be used to reduce the size of a display to an area that allows for placing it or surrounding it within a video picture or within in a graphics image. When a video program is placed within a graphic image, it is called picture in graphics. Picture in graphics may be used to display other media items (such as a channel menu).

Video Search Engine-A video search engine is a software program that searches through data records of video files or programs to find matches to specific words or items.

Video Server (VS)-The video server is an application server that provides video and/or specialized television capabilities. Video servers receive requests for video and/or media delivery, find the matching media, and deliver the video program as requested.

Video Serving Office (VSO)-A video serving office is an edge distribution facility for video signals in an IPTV network.

Video Sharing-Video sharing is the providing of access rights and the ability to identify, access and potentially modify video files between computing equipment which are usually connected to a network.

Video Streaming-Video streaming is the process of delivering video, usually along with synchronized accompanying audio in real time (no delays) or near real time (very short delays). Upon request, a video media server system will deliver a stream of video and audio (both can be compressed) to a client. The client will receive the data stream and (after a short buffering delay) decode the video and audio and play them in synchronization to a user.

Video Surveillance-Video surveillance is the capturing of video for the observation of an area or location at another location.

This figure shows how wireless video devices can be used to provide security monitoring services. This example shows how a police station can monitor multiple locations (e.g. several banks) through the addition of wireless digital video connections. In this example, a trigger alarm occurs at a bank (such as when a bank teller presses a silent alarm button) which alerts the police that a robbery is in progress. Using the wireless video connections, the police can immediately see what is occurring at the bank in real-time. Because the images are already in digital format, it may be possible to send these pictures to police cars in the local area to help identify the bank robbers.

Video Surveillance

Videocast-Video Casting

Videophone-A videophone is a communication device that can capture and display video information in addition to audio information. A videophone converts multiple forms of media; audio and video into a single transmission format (such as Internet Protocol). The use of videophones with an Internet telephone service allows the video portion of the communications session to share the data connection.

Viewer-A viewer is a software program and/or display device that allows a user to view media content. An example of a viewer is a web browser that allows a user to view media files that are transferred through the Internet.

Viewer Consumption-Viewer consumption is the programs viewed or accessed from a video distribution system (such as from a television broadcast network).

VIP Alert-VIP alert is a telephone CLASS service feature that provides a distinctive ring for a list of callers.

Virtual-(1-experience) Virtual is a sensory experience that is emulated or occurs as a result of an external event. (2-system) A virtual system is a facility or arrangement that gives the effect of being in a dedicated network or facility but is a shared resource (e.g. virtual private network).

Virtual Call Center-A call center where calls are answered and originated, typically between a company and a customer that uses customer service representatives (CSRs) that can be located at different places via virtual connections. Virtual call centers can assist customers with requests for service activation and help with product features and services as if they were located in a company office. Virtual call centers may use many virtual connections to connect CSRs with customers.

Virtual Call Service-A virtual call service is the providing of telephone calls through networks or systems that are not directly controlled or managed by the person or company that initiates, uses or manages the telephone calls.

Virtual Circuit Switch (VCS)-A virtual circuit switch is a device or assembly that receives signals on input ports and provides (transfers) a continuous logical (virtual) connection of the signal to output ports.

Virtual Desktop-A desktop workplace for the employee that consists primarily of computing devices. The virtual desktop devices usually include a computer, printer, and a telephone.

Virtual Keyboard-A virtual keyboard is a software program that operates on a set top box or other interactive display device to provide a user with the ability to enter keyboard (e.g. alphanumeric) information.

Virtual Keypad-A virtual keypad is a software program that operates on a computer or other interactive display device to provide a user with the ability to enter keypad or keyboard information.

Virtual LAN-A number of devices (a subset) that are linked to each other within a larger network by logical channels to allow each device to communicate with other devices in the virtual network using these logical channels. Virtual networks often appear as a separate network to the users of the network. An example of a virtual network is the connection of computers in a city to computers in another city via logical channels (and encrypted channels for security) through the Internet. This allows the computers in one city to access the computers in the other city as if they were connected as a separate network.

Virtual Local Area Network (VLAN)-A virtual local area network is a data communication network that interconnects computers and related equipment in a limited geographic area where multiple logically setup networks (virtual networks) can be setup and managed through the underlying communication network. VLAN connections are setup to allow data to safely and privately pass over other types of data networks (such as the Internet).

Virtual Machine (VM)-A virtual machine is a data processing device (such as a computer or television set top box) which is designed to allow software programs to operate as if they were on processing devices of another type (a different computer type). A virtual machine allows software that was designed for another type of computing device to operate within its system without changes to the software.

Virtual Mall-A shopping medium that uses electronic networks (such as the Internet or telecommunications) to present products and process orders.

Virtual Money-Virtual money is a financial unit measurement that is defined by a company or other authorized entity. Virtual money may be stored in accounts that can be used to purchase items or enable a person or the device they control to perform actions or access services.

Virtual Monopoly-A virtual monopoly is a company or organization that is provided authorization to operate or has some unique quality that restricts other companies from effectively competing against it.

Virtual PBX (vPBX)-A virtual PBX offers business users the ability to make and receive calls through the company's PBX system using telephones that can be connected to any of the company's PBX systems at locations that have the ability to connect to a PBX access port (such as an Internet connection).

Virtual Phone-A software program that operates on a computer to provide telephone service.

Virtual Private Network (VPN)-Virtual private networks are private communication path(s) that transfer data or information through one or more data network that is dedicated between two or more points. VPN connections allow data to safely and privately pass over public networks (such as the Internet). The data traveling between two points is usually encrypted for privacy.

This figure shows the operation of a virtual private network (VPN). This diagram shows that the virtual private network is constructed of network access points that are under the control of the network operator. These network access points usually encrypt the data entering into the network to provide secure private communication path(s) through the network. These secure VPN connections allow a company to safely and privately pass over

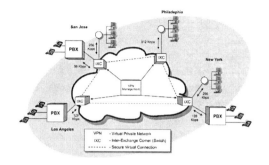

Virtual Private Network (VPN)

public networks (such as the Internet). A VPN management system is used to program the access points (e.g. IXC switch) for key parameters (e.g. data rates and QoS.) While this diagram shows virtual paths, the connections may actually pass through one or more switches have been set so a reserved amount of bandwidth is assigned so the end user can reliably receive a Quality of Service (QoS) characteristics that allows the connections to appear as dedicated lines.

Virtual Switch-A virtual switch is a connection control process that allows devices to setup connections through communication networks (such as the Internet) where the connection can be dynamically setup and/or changed. A virtual switch only sets up the connections. The data transfer process is independent of the virtual switch.

Virtualization-Virtualization is the creation of a view of information that has been adapted or customized for the user or the needs of a user. Virtualized objects are created using underlying data that is combined and processed with the user's characteristics.

Virus-A software program spread by automatic copying from disks or computer networks and intended to interrupt or destroy the functioning of a computer.

Visitor-(1-Web Site) A person who visits a web site. (2-Mobile Telephone) A mobile telephone that is operating in a system other than its system of home registration.

Visitor Acquisition Cost (VAC)-Visitor acquisition cost is the combined costs that are associated with marketing operations that motivate people or devices to visit a web site.

Visitor Identification Code (VisitorID)-A visitor identification code is a unique number or label that is assigned to a web page visitor that can be used to track their progress (click trail) through a web site.

Visitor Location Register (VLR)-A visitor location register is a database part of a wireless network (typically cellular or UMTS) that holds the subscription and other information about local or visiting subscribers that are authorized to use the wireless network.

VisitorID-Visitor Identification Code

Visual Call Control-Visual call control is the process of using images or graphics to setup, view and manage call-processing systems.

Visual Model-A visual model is a graphic representation of a system or process that can be used to display how the system or process may operate or perform when certain conditions or events occur.

Visual Voice Mail-An application displaying and controlling voice messages on a desktop computer. Visual voice mail may be associated with unified messaging.

Virtual Phone Network (VPN)-A virtual phone network is a telephone system that uses other networks to provide part or all of the transfer and call processing functions.

VLAN-Virtual Local Area Network

VLR-Visitor Location Register

VM-Virtual Machine

VM-Voice Mail

VM/UM-Voicemail/Unified Messaging Server

VMail-Video Mail

VMI-Vendor Managed Inventory

VMS-Voice Mail System

VoB-Voice over Broadband

VoCable-Voice Over Cable

VoCoder-Voice Coder

VOD-Video On Demand

VOD Server-Video on Demand Server

VOD Sponsorship-Video on Demand Sponsorship

VoDSL-Voice Over DSL

VoFR-Voice Over Frame Relay

Voice Broadcast Service (VBS)-A voice communications service that allows a single voice conversation or message to be transmitted to a geographic coverage area to be received by subscribers that are capable of identifying and receiving the voice communications.

This figure shows the basic operation of voice broadcast service. This example shows how an urgent news message (traffic alert) can be sent to all mobile devices that are operating within the same radio coverage area.

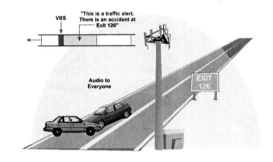

Voice Broadcast Service

Voice Card-A communication card that is inserted into a computer that can process calls.

Voice Channel-A channel in a communication system on which voice or data communication occurs. During voice communication, brief digital control messages may be sent (such as from a mobile telephone to a base station).

Voice Circuit-A circuit for the interchange of human speech. Normally, the standard band provided is 300 Hz to 3400 Hz, but narrower bands also provide commercially acceptable circuits in some circumstances.

Voice Coder (VoCoder)-A voice coder is a digital compression device that consists of a speech analyzer that converts analog speech into its component speech parts. A speech decoder recreates the speech parts back into their original speech form. Voice coders are only capable of compressing and decompressing voice audio signals.

Voice Communication-Voice communication is the transmission and reception of audio and other signals that can be represented by the frequency band used for voice signal transmission. Telephone systems transfer voice signals in a variety of forms, by wire, radio, light, and other electronic or electromagnetic systems. These forms include analog and digital voice signals. Options for voice communications include different voice quality of service levels and voice privacy options.

Voice Compression-Refers to the process of electronically modifying a 64 Kbps PCM voice channel to obtain a channel of 32 Kbps or less for the purpose of increased efficiency in transmission.

Voice Dialing-A process that uses the caller's voice to dial a call. Voice dialing involves the activation of the voice dialing feature (either by pressing a key or by saying a key word), saying words in the vocabulary of the voice dialing processor, and providing feedback to the user (usually by audio messages) of the status of the voice dialing process. Voice dialing can be a system (network provided) or device (stored in the telephone device) feature. There are two basic forms of voice dialing; speaker independent and speaker dependent. Speaker independent voice dialing allows any user to initiate voice commands from a predefined menu of commands. Speaker dependent voice dialing requires the user to store voice commands so these voice commands can be

activated by the user and others are unlikely to match the speaker dependent voice commands. Speaker dependent voice recognition allows a user to program specific names into the telephone or network voice recognition system.

This diagram shows different types of dialing using voice commands. In this example, both the telephone set and telephone network have voice dialing control capability. When the telephone is used for voice control, the voice from the user is converted to digital form by and analog to digital converter. After the audio is converted to digital form, it is analyzed for patterns and matched to previously stored voice control digital sound patterns. This example shows that the telephone set has some speaker independent patterns (such as start and digits) that have been previously stored. It also shows that this telephone also has a speaker dependent memory storage area that allows the user to store specific names. When these specific names are spoken, the telephone set will retrieve the pre-stored telephone numbers or extensions.

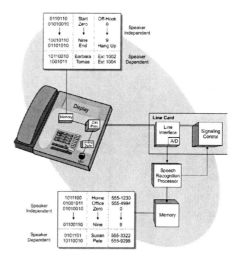

Voice Dialing Operation

This diagram shows similar voice dialing capabilities that are located in a telephone network. This network voice control system has more accurate voice processing capability than the telephone set and each voice control module can service many line cards as users only use voice control for brief periods.

Voice Digitization-This figure shows how an analog signal is converted to a digital signal. An acoustic (sound) signal is first converted to an audio electrical signal (continuously varying signal) by a microphone. This signal is sent through an audio band-pass filter that only allows frequency ranges within the desired audio band (removes unwanted noise and other non-audio frequency components). The audio signal is then sampled every 125 microseconds (8,000 times per second) and converted into 8 digital bits. The digital bits represent the amplitude of the input analog signal.

This figure shows how an analog signal is converted to a digital signal. An acoustic (sound) signal is first converted to an audio electrical signal (continuously varying signal) by a microphone. This signal is sent through an audio band-pass filter that only allows frequency ranges within the desired audio band (removes unwanted noise and other non-audio frequency components). The audio signal is then sampled every 125 microseconds (8,000

times per second) and converted into 8 digital bits. The digital bits represent the amplitude of the input analog signal.

Voice Gateway-A voice gateway is a communications device or assembly that transforms audio that is received from a telephone device or telecommunications system (e.g. PBX) into a format that can be used by a different network. A voice gateway usually has more intelligence (processing function) than a bridge as it can select the voice compression coder and adjust the protocols and timing between two dissimilar computer systems or voice over data networks.

This diagram shows the functional structure of a voice gateway device. This diagram shows that this voice gateway interfaces between a public telephone network to a packet data network. Input signals from the public telephone network pass through a line card to adapt the information for use within the voice gateway. This line card separates (extracts) and combines (inserts) control signals from the input line from the audio signal. If the audio signal is in analog form, the voice gateway converts the audio signal to digital form using an analog to digital converter. The digital audio signal is then passed through a data compression (speech coding) device so the data rate is reduced for more efficient communication. This diagram shows that there are several speech coder options to select from. The selection of the speech coder is negotiated on call setup based on preferences and communication capability of both voice gateways. After the speech signal is compressed, the digital signal is formatted for the protocol that is used for data communication (e.g. IP packet or Ethernet packet). This call processing section of the voice gateway may insert control commands (in-band signaling) to allow this gateway to directly communicate with the remote gateway. These digital signals are sent through a data access device (e.g. router shown here) so it can travel through the data communication network. The overall opera-

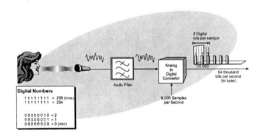

Voice Digitization

tion of the voice gateway is controlled by the call processing section. The call processing section receives and inserts signaling control messages from the input (telephone line) and output (data port). The call processing section may use separate communication channels (out-of-band) to coordinate call setup and disconnection.

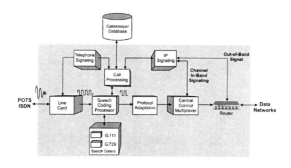

Voice Gateway Operation

Voice Grade-Voice grade is a quality of service that provides audio services that are acceptable for telephony applications. Voice grade services transfer audio frequencies within the range of 300 Hz to 3400 Hz.

Voice Grade Facility-A voice grade facility is the equipment, channel, circuit or service that can transfer signals in a frequency range of 300Hz to 3200Hz.

Voice Group Call Service (VGCS)-Voice group call service (VGS) is the process of transmitting a single voice conversation on a channel or group of channels so it can be simultaneously received by a predefined group of service subscribers.

Voice Mail (VM)-A service that provides a telephone customer with an electronic storage mailbox that can answer and store incoming voice messages. Voice mail systems use interactive voice response (IVR) technology to prompt callers and customers through the options available from voice mailbox systems. Voice mail systems offer advanced features not available from standard answering machines including message forwarding to other mailboxes, time of day recording and routing, special announcements and other features.

This diagram shows how a voice mail system provides electronic storage mailboxes to users within the telephone system. In this example, the voice mailbox system connects to a switching system through 2 extensions (ports) on the switching system (other voice mail systems may have many more ports). To access the voice mail system, users may select the voice mailbox system extension (usually programmed into a button on a telephone set that says voice mail). In this example, when a user dials into the telephone system to reach extension 1001, the line is busy. The system has been setup to forward calls to extension 1015 (the voice mail system) when extension 1001 is busy. To help ensure the voice mail system is accessible, if extension 1015 is busy, the call will be forwarded to extension 1016. When the call has entered the voice mail system, the interactive voice response (IVR) system will prompt the caller or user to enter information using touchtone or voice commands. This will

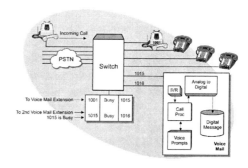

Voice Mail Operation

allow callers or users to either store or retrieve messages from the digital message storage area (e.g. a computer hard disk drive).

Voice Mail System (VMS)-The voice mail system is a telecommunications system that allows a subscriber to receive and play back messages from a remote location (such as a PBX telephone or mobile phone). The VMS consists primarily of memory storage (for messages), telephone interfaces (to connect to the communication system), and message recording, playback, and control features (typically via DTMF tones).

Voice Mailbox-A voice mailbox is a portion of memory, usually located on a computer hard disk, that stores and plays audio messages. The audio messages are often in compressed digital audio format.

Voice Messaging-A storage and retrieval system for voice messages. Commonly called "voice mail."

Voice On the Net (VON)-The process of sending voice over a data network (such as sending voice over the Internet).

Voice over Broadband (VoB)-The process of sending digitized voice over a high-speed broadband (typically 1 Mbps or more) connection.

Voice Over Cable (VoCable)-Voice over Cable solution is a complete Voice over Internet Protocol (VoIP) packet based broadband solution that supports DOCSIS and the PacketCable 1.0 specification.

This diagram shows how a cable television can offer telephony services. In this example, the cable television system has been modified to offer telephone service by adding voice gateways to the cable network's head-end cable modem termination system (CMTS) system and multimedia terminal adapters (MTAs) at the residence or business. The voice gateway connects and converts signals from the public telephone network into data signals that can be transported on the cable modem system. The CMTS system uses a portion of the cable

modem signal (data channel) to communicate with the MTA. The MTA converts the telephony data signal to its analog audio component for connection to standard telephones. MTAs are sometimes called integrated access devices

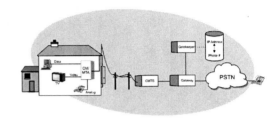

Voice over Cable Television Operation

(IADs).

Voice Over Data Networks-A process of sending digitized voice signals over data networks (such as the Internet).

Voice Over DSL (VoDSL)-Sending voice over a digital subscriber line system (VoDSL) is a process that sends audio band (also called "voice band") signals (e.g. voice, fax or voice band modem) via a digital channel on a digital subscriber line (DSL) system. VoDSL requires conversion from analog signals to a digital format and involves the formatting of digital audio signals into frames and time slots so they can be combined onto a digital (DSL) channel.

To communicate to other users, VoDSL requires one or more communication device that are capable of sending and receiving with the DSL network and conversion of a digital channel back into its analog voice band signal. This can be as simple as a computer with a sound card, a DSL modem and VoDSL software or as complex as a companies telephone network with an integrated access device (IAD). Optionally, some DSL systems have a

PSTN gateway that can convert digital audio on a DSL system into telephone signals that can be sent through the public switched telephone network.

Voice Over Frame Relay (VoFR)-A process of sending digitized voice signals over frame relay data networks.

Voice Over Internet Protocol (VoIP)-A process of sending voice telephone signals over the Internet or other data network. If the telephone signal is in analog form (voice or fax) the signal is first converted to a digital form. Packet routing information is then added to the digital voice signal so it can be routed through the Internet or data network.

This diagram shows how an Internet network (public or private) can be used to provide telephone service. In this example, a calling telephone or multimedia capable computer dials a telephone number. This telephone number is provided to a voice gateway. The voice gateway decodes the dialed digits and determines the destination address (IP address) of the gateway that can service the dialed telephone number. The remote gateway signals the caller of an incoming call (rings the phone or alerts a multimedia computer). When the user answers the call, a message is sent between the gateways and a virtual path can be created between the gateways. This virtual path takes the audio, converts it to digital form, compresses and packetizes the information, adds the destination gateway address to each

packet, routes the packets through the Internet to the destination gateway, and converts the digital audio back to its original analog form.

Voice Paging-Refers to paging service whereby the transmission of information is in the form of actual voice data. Messages can be stored and different volume controls are selected when receiving pages, including an ear piece which allows privacy.

This diagram shows how a voice paging system receives voice messages from callers and forwards these messages on to a voice pager. In this example, a caller dials a paging access number. This number either connects the caller to an interactive voice response unit or an operator that can direct the caller to a voice mailbox associated with the voice pager. After the caller's message is stored in the voice mailbox, it will be placed in the queue for the voice mail system. When the message reaches the top of the queue (available time to send), it will be encoded (formatted) to a form suitable for transmission on a radio channel. In this example, the message is sent as part of group 4. Sending the messages in groups allows the pager to sleep during transmission of pages from other groups that are not intended to

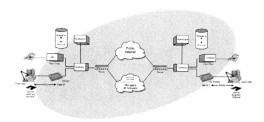

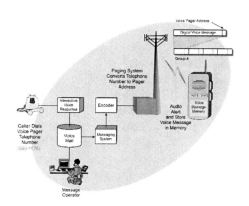

Voice over the Internet (VoIP) System

Voice Paging Operation

reach the voice pager. The voice message includes the voice pager address along with the voice message in digital form. During the reception of the message, it is stored into the voice message memory area so the voice pager can play the message one or many times after it is received.

Voice Privacy-Voice privacy is a process that is used to prevent the unauthorized listening of communications by other people. Voice privacy involves coding or encrypting of the voice signal with a key so only authorized users with the correct key and decryption program can listen to the communication information.

Voice Quality-Voice quality is a measurement of the level of audio quality, often expressed in mean opinion score (MOS). The MOS is a number that is determined by a panel of listeners who subjectively rate the quality of audio on various samples. The rating level varies from 1 (bad) to 5 (excellent). Good quality telephone service (called toll quality) has a MOS level of 4.0.

Voice Recognition-A computer-based technology that analyzes audio signals (typically spoken words) converts them into digital signals for other processing (e.g. voice dialing).

Voice Response Unit (VRU)-An equipment that provides a caller with audio messages in response to their touch-tone(tm) key presses or voice commands . VRU are part of interactive voice response (IVR) systems.

Voice Service-Voice service is a type of communication service where two or more people can transfer information in the voice frequency band (not necessarily voice signals) through a communication network. Voice service involves the setup of communication sessions between two (or more) users that allows for the real time (or near real time) transfer of voice type signals between users.

Voice Store And Forward (VS&F)-Voice store and forward is a communication service that allows audio messages to be stored and retrieved remotely.

Voice Trigger-A voice trigger is a signal or event that begins the creation of an audio (voice) signal.

Voicemail/Unified Messaging Server (VM/UM)-Voicemail/Unified Messaging Server provides call management services that allow the storage, retrieval and redirecting of voice messages through voice and email communications.

Voiding Payments-Voiding payments is the process of ending and possibly reversing the tasks that were associated with collecting payments or the collection of assets. Payment voiding may be performed before a transaction is posted or batched to financial centers to avoid processing fees.

VoIP-Voice Over Internet Protocol

Volume-(1-sound) The loudness (intensity) of a sound. (2-data) A certain portion of data, together with its data carrier, that can be handled conveniently as a unit. (3-graphics) A three-dimensional array of raster data. (4-general) The amount of a cubic space measure (cubic meters, cubic inches, etc.) contained in a given three-dimensional region of space.

Volume Based Charging-Volume based charging is the measurement of a quantity of service provided by a server or system. For voice communication, this is commonly expressed in hundred call seconds (CCS) per hour or Erlangs. For data communication, this is usually expressed in bytes of information used or transferred.

Volume Discounting-Volume discounting is the applying of different discount rates to a fee or charge that is determine by the quantity of a service that is provided or consumed at an event (such as the number of items purchased) or over a specific time duration.

Volume Label-A volume label is a name assigned to a storage device (such as a disk).

Volume Spanning-Volume spanning is the process that allows storage systems to use or access multiple physical hard disks (multiple drive volumes).

Voluntary Churn -A disconnection of service

that occurs when a customer decides to drop a service from an existing carrier and initiate the service with another carrier.

VON-Voice On the Net

Voucher-A voucher is a card, printed brochure or electronic record that authorizes a customer to pay for products or services.

vPBX-Virtual PBX

VPN-Virtual Private Network

VPN-Vitual Phone Network

VQM-Video Quality Measurement

VQOS-Video Quality of Service

VRU-Voice Response Unit

VS-Video Server

VS&F-Voice Store And Forward

VSO-Video Serving Office

Vulnerabilities-Vulnerabilities are processes that can be used on systems or services that can produce events or outcomes that are undesirable.

V

W

W3C-World Wide Web Consortium
W911-Wireless 911
WAE-Wireless Application Environment
Wait Time Announcement-Wait time announcement is a message that provides a caller with an estimated amount of time the call will hold before an operator will answer or a call transfer will occur.
Waiver-A waiver is an authorization that permits a person or company to take actions or receive benefits that they would not have been allowed to perform under existing agreements or terms.
WAN-Wide Area Network
WAP-Wireless Access Point
WAP-Wireless Access Protocol
WAP-Wireless Application Protocol
Warehouse-A warehouse is a building or facility where goods can be received, stored and shipped.
Watermark-An imperceptible signal hidden in another signal, such as audio or an image, which carries information. Watermarking is related to the general field of stenography, or information hiding. Ideally a watermark would not be destroyed (that is, the signal altered so that the hidden information could no longer be determined) by any imperceptible processing of the overall signal, for example high-quality lossy compression, slight equalization, or digital-to-analog-to-digital conversion. Sophisticated techniques for successfully destroying watermarks make that ideal difficult to achieve.
Watermarking-Watermarking is a process of adding (data embedding) or changing information in a file or other form of media that can be used to identify that the media is authentic or to provide other information about the media such as its creator or authorized usage.
WATS-Wide Area Telecommunications Service

WAV-Wave Audio
Wave Audio (WAV)-Wave audio is a coding form for digital audio used by Win32. Wave audio files commonly have a. WAV extension to allow programs to know it is a digital audio file in Wave coding format.
WBB-Wireless Broadband
Weak Password-A weak password is a user access code that is easy to guess or obtain through the use of word lists or sequential password access attacks. Weak passwords are usually short (9 characters or less) and usually contain well used words, names or birthdates that are related to the user.

Web-A web is an interconnection of devices or systems. The web allows devices to connect to other devices that are connected to the web.
Web 2.0-Web 2.0 refers to a perceived second generation of web-based communities and hosted services such as social-networking sites, wikis, and folksonomies, which aim to facilitate creativity, collaboration and sharing between users. The term gained currency following the first O'Reilly Media Web 2.0 conference in 2004. Although the term suggests a new version of the World Wide Web, it does not refer to an update to any technical specifications, but to changes in the ways software developers and end-users use webs.
Web Accessible-Web accessible is the ability of a device, system or service to be controlled via a web page or Internet access device.
Web Address-A web address is text or numeric identifier that can be used to access a web site on the Internet.
Web Advertising-Web advertising is the communication of messages or media content to one or more potential customers through the use of the Internet.

Web Analytics-Web analytics are the processes that are used to evaluate the operations and performance of programs or services that operate on the Internet.

Web Based Setup Routine-A software program that is accessed through the Internet web that coordinates the installation of a program or system.

Web Billing-Web billing is the process of grouping service or product usage information for specific accounts or customers, producing and sending invoices, recording (posting) payments made to customer accounts and providing access or control of this information through the Internet.

Web Browser-A web browser is software that is used to graphically view information retrieved from Web servers. Web browsers request, receive, and reformat information receives from web servers.

Web Channel-A web channel is a video program (such as a TV program) that is viewable or accessible through the Internet.

Web Communities-Web communities are people who have common interests and share information on specific subjects through the Internet (web).

Web Component-A web component is a portion or functional part of a web site.

Web Conferencing-Web conferencing is the conducting of meetings or functions that use and/or provide media (such as audio and video) via the Internet.

Web Content-Web content is information and/or media contained within a web page or web site.

Web Content Accessibility-Web content accessibility is the ability of devices and their users to access and interact with content on the Internet.

Web Editor-Web Page Editor

Web Engine-A web engine is a software program that can receive, process and respond to an end user's (client's) request to send, receive and process information from web sites.

Web Feeds-A web feed is an logical link that can transfer a sequence of data (usually in XML format) from a source (such as a news service) where each item can be identified by codes and categories to allow the recipient to extract information (such as news stories with a particular category) that can be displayed or accessed on a web site.

Web Host-A web host is a company or person that performs web hosting services.

Web Hosting Support-Web hosting support is supplying the resources and services that are requested or required by a person or company that is using web hosting services. An example of web hosting support is the providing of information on how to setup and operate scripts or programs that link web pages to databases.

Web Interface-A web interface is the software program that enables a device, system or service to be controlled via a web page or Internet access device.

Web Link-Web links (Hyperlinks) are tags, icons, or images that contain a crossed reference address that allows the link to redirect the source of information to another document or file. These documents or files may be located anywhere the link address can be connected to.

Web Ontology Language (OWL)-Web ontology language is a set of commands and process that can be used to define and instantiate relationships between data or media. OWL ontology may include descriptions of classes, along with their related properties and instances. OWL is designed for use by applications that need to process the content of information instead of just presenting information to humans. It facilitates greater machine interpretability of Web content than that supported by XML, RDF, and RDF Schema (RDF-S) by providing additional vocabulary along with a formal semantics. OWL is based on earlier languages OIL and DAML+OIL, and is now a W3C recommendation.

Web Page-A web page is a file located on a computer that is connected to the Internet that has a format that allows the file to display (format) information on a user's display (web browser).

Web Page Editor (Web Editor)-A web page editor is a software application or service that can be used to modify web page files. Web page editors commonly include added editor features such as converting layout and styles into their HTML codes.

Web Portal-Web portals are Internet web sites that act as an interface between a user and an information service. This interface is typically specialized for specific functions such as searching or displaying certain types of information.

Web Presence-Web presence is the activity or perceived activity and availability on the Internet web.

Web Seminar (Webinar)-A web seminar (webinar) is an online instruction session that uses the Internet Web as a real time presentation format along with audio channels (via web or telephone) that allow participants to listen and possibly interact with the session. Webinars allow people to participate in information or training sessions from anywhere that has Internet and audio access.

Web Server-Web servers are computer systems that are used provide access to data that is stored and retrieved by commands in Hypertext Transfer Protocol (HTTP). HTTP is a protocol that is used to request and coordinate that transfer of documents between a web server and a web client (user of information). The typical use of web servers is to allow web browsers (graphic interfaces for users) to request and process information through the Internet.

Web Services-Web services are the providing of information transfer, processes or authorizations that enable users, devices or systems to perform actions that they desire via web hosts. Web services may use well-known addresses along with extensible Markup Language (XML) data tables or structured messages.

Web Services Description Language (WSDL)-Web services description language is a set of commands and processes that can be used to interact with XML documents to find and describe Web services.

Web Site-A file or group of files located on a computer that is connected to the Internet. These files are generally accessible by other users that are connected to the Internet through the use of Internet protocols.

Web Site Address-A web site address is the identifier that is used to connect to a web page. The identifier is usually a numeric Internet protocol (IP) address.

Web Site Design-Web site design is the setup and positioning of media elements for a web page.

Web Site Functionality-Web site functionality is the capabilities of web hosting systems to perform supporting services or functions. An example of web site functionality is the ability of a web server to run certain types of files such as Javascript, Unix or Windows based programs.

Web Surfing-Web surfing is the process of viewing web pages on the Internet and following links from these web pages to new web pages.

Web Television (WebTV)-A set-top box (cable converter) that provides the user with the ability to use and display Internet services on a standard television.

Web Traffic-Web traffic is the amount number of people who visit web sties.

Web Traffic Analysis-Web traffic analysis is the processes that are used to capture data that can be used to evaluate the characteristics (such as the types of interest) of visitors to web sites.

Web Widget-A web widget is a portable chunk of code that can be installed and executed within any separate HTML-based web page by an end user without requiring additional compilation. They are derived from the idea of reusable code that has existed for years. Nowadays other terms used to describe web widgets including gadget, badge, module, capsule, snippet, mini, and flake. Web widgets often but not always use DHTML, Adobe Flash or JavaScript programming languages.

WebCam-A webcam is a PC video camera that captures and posts live images to a website. These images are refreshed every few seconds. Webcams can be used for video email and video conferencing and video instant messaging.

Webcast-The live presentation of information in a continuous (streaming) format delivered through the Internet web. A webcast might be associated with other web pages or other web-browser-based content in addition to the live stream.

Webinar-Web Seminar

Webmaster-A webmaster is the person who is responsible for maintaining and administering a Web site.

WebTV-Web Television

Weighting-Weighting is the adjustment or modification of a measured value or series of measured values to adjust for conditions or system performance requirements.

WFM-Workflow Management

What you See is What You Get (WYSI-WYG)-An expression that is used for a computer system that displays information in the same style in which it will be printed. It is pronounced "wizzy-wig."

White Goods-White goods are a type of consumer durable products such as washing machines, refrigerators and dishwashers. The name white goods was initially created because these items commonly had a white finish.

White Label-White label is a product or service that has blank or generic areas that are available for branding or promotional use by other distribution companies.

White Pages-White pages are a directory of telephone customer listings that commonly includes the name of a person or business, their address and telephone number.

White Paper-A white paper is a document or reference material that provides basic viewpoint or overview information on a subject. A white paper may provide summary results of testing or implementation of a product or service.

Whiteboard-A device that can capture images or hand drawn text so they can be transferred to a video conferencing system. Whiteboards allow video conferencing users to place share documents, images and/or hand written diagrams with one (or more) video conference attendees.

This figure shows how a whiteboard can be used during an Internet telephone call to transfer hand drawn images. In this example, an instructor is drawing a diagram on a white pad. While the instructor is drawing, the image is being displayed on both the instructors monitor and the students monitor.

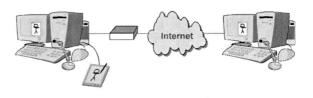

Internet Telephone Whiteboard Operation

Wholesale Boundaries-Wholesale boundaries are the limits of a system or the processes at a company to allow or work with other companies to provide services on a wholesale basis.

Wholesale Line Rental (WLR)-Wholesale Line Rental is a service that enables a communication service provider to offer their own branded telephony service to their customers using the incumbent network.

Wholesale Price-A wholesale price is an offer value that is assigned to a product or service for sale of products to companies that resells the products to distributors.

Wholesale Rating-Wholesale rating is the assigning of wholesale prices to usage records.

Wholesaler-A wholesaler is a company or individual that buys products in quantity and sells tem to distributing companies.

Wide Area Network (WAN)-A communications network serving geographically separate areas. A WAN can be established by linking together two or more metropolitan area networks, which enables data terminals in one city to access data resources in another city or country.

This figure shows that a WAN is usually composed of several different data networks. Different types of communication lines such as leased lines, packet data systems, or fiber transmission lines can interconnect these networks.

Wide Area Telecommunications Service

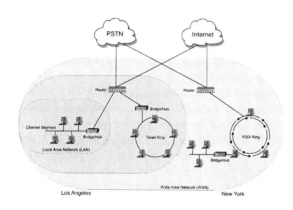

Wide Area Network (WAN) Systems

(WATS)-A service that allows companies to be billed using different rate (such as flat rate) structures for calls within their wide area telephone system (WATS) area.

Wide Screen-Wide screen is the display of a video or motion picture on a screen that has a wider aspect ratio than standard 4:3 (width to height ratio). For example, a motion picture has a 16:9 aspect ratio.

Widescreen-Widescreen is a video or image display format that has a larger width to height ratio than a standard display. For widescreen televisions, ratios that are above 1.7 (16:9) are considered widescreen.

Widget-A widget is a hypothetical device.

Wiki-A Wiki is an application that allows multiple people to easily and quickly contribute (wiki means fast in Hawaiian) to a common content base. Wikis are often used to create collaborative websites, power community websites and are increasingly being installed by businesses to provide affordable and effective Intranets or for use in knowledge management.

Wildcards-Wildcards are special characters that can be used to represent one or more characters in a filename. A typical wildcard is an asterisk (*) which commonly represents any characters in a string and a question mark (?) that may represent a single character in a string of characters.

Window-(1-transmission buffer) An indication of the amount of time or data that should not be exceeded waiting for successful reception of information. If this amount is exceeded, retransmission may occur. (2-video) Video containing information or allowing information entry from a keyboard, time code generator, or other device. A window dub is a copy of a videotape with time code numbers keyed into the picture. (3-display) A window is an area of a screen display that is used to display information associated with a specific program, application or function.

Winning Responder-A winning responder is a person or company that has been selected as a vendor or provider of services as the result of their proposals or responses that they sent from a request for proposal.

WIPO-World Intellectual Property Organization

Wire Transfer-A wire transfer is an exchange of value from one financial institution (such as a bank) to another financial institution (such as another bank) where the transfer is performed by electronic means. While the transfer is performed by electronic means, it may not occur in real time.

Wireless-Communication without the use of cables or devices that transmit over wireless networks rather than over telephone lines. Historically, at various times during the 20th century, this had specialized meanings that have come and gone. For many years, British English used the word "wireless" while North American English used "Radio" instead. In the past, the word "wireless" was occasionally used to describe the transmission via radio of Morse code, but not voice. Today the term wireless is used primarily for cellular systems, and secondarily also for other short range radio systems used directly by end users, such as for example 802.11b short range data transmission.

Wireless 911 (W911)-Wireless 911 is a feature of the telephone calling system that provides emergency dispatchers with additional information about a mobile subscriber which may include name, registered address and actual location.

This figure shows how mobile communication systems can be enhanced through the use of GPS technology to provide for emergency location services. This example shows that a mobile telephone has both mobile communication and GPS reception capability. When the user needs emergency assistance, the GPS information can be sent to the automatic location identification (ALI) database which determines which public safety access point (PSAP) to route the call to. The mobile system sends the call with the location information to the PSAP which can display on a map.

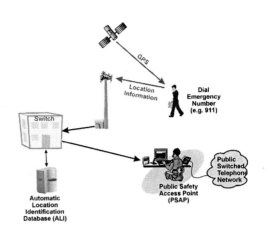

GPS Mobile Location ALI Assisted

Wireless Access Point (WAP)-A wireless access point contains radio transceivers that convert digital information to and from radio signals that can be exchanged with other wireless communication devices. The most basic forms of wireless access points simply for wireless connections. A wireless access point that includes the ability of DHCP and network address translation (NAT) is typically called a wireless gateway.

Wireless Access Protocol (WAP)-Wireless access protocol is a standard protocol specification that allows advanced messaging and information services to be delivered to wireless devices independent of which wireless technology they use.

Wireless Application-Wireless applications are systems and services that are designed and perform operations using commands or information that are transferred between devices without physical connections.

Wireless Application Environment (WAE)-A software application that utilizes wireless communication. Wireless communication often has limited costly access to bandwidth.

Wireless Application Protocol (WAP)-
WAP is a collection of protocols and standards
that enable communication and information
applications to run efficiently on mobile
devices. WAP is to wireless devices what
hypertext transfer protocol (HTTP) is to Web
browsers.

Wireless Billing-The recording and processing of wireless transmission events for billing
purposes.

Wireless Broadband (WBB)-Wireless
broadband is the transfer of high-speed data
communications via a wireless connection.
Wireless broadband often refers to data transmission rates of 1 Mbps or higher.

Wireless Broadband Television-Wireless
broadband television is the sending of digital
television signals via a wireless broadband
(e.g. 1 Mbps+) data connection.

Wireless Cable-"Wireless Cable" is a term
given to land based (terrestrial) wireless distribution systems that utilize microwave frequencies to deliver video, data and/or voice signals to end-users. There are two basic types of
wireless cable systems, multichannel multipoint distribution service (MMDS) and local
multichannel distribution service (LMDS).

This figure shows a overview of a wireless
broadband communication system. This system uses radio towers (usually called "base
stations") that are located within a few miles
of the customer to transmit relatively low
power RF signals directly to a customer's radio
receiver. The radio receiver uses a directional
high-gain antenna to capture and focus radio
signals for transmission between the radio
tower and the customer's house. The radio
transceivers (transmitter and receiver pair) in
the base station transmit on radio channels up
to several MHz wide each. The radio receiver
converts these radio channels back to its original digital form so it can be provided to the
customer's computer. Many MMDS and LMDS
systems provide a data transmission rate of

approximately 10 Mbps from the radio tower
to the customer's receiver and approximately 1
Mbps from the customer's receiver back to the
radio tower.

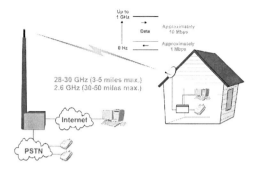

Wireless Cable Overview

Wireless Communication-Wireless communications is the transmission of information
without wires. Wireless communication may
use radio or optical transmission for communication.

Wireless Data-A system or the transmission
of digital information through a wireless network such as wireless packet data systems or
cellular mobile communications. Wireless data
systems are specifically designed to reliably
transfer information (data) between a sender
and receiver. The term wireless data can apply
to mobile or fixed devices and the transmission
may be in the form of radio or optical (e.g.
infrared systems) communication systems.
Wireless data transmission can be sent over
dedicated wireless data communication systems (such as Reflexion, Ardis, or Mobitex) or
the transmission may share a common channel for voice and data (such as on GSM or 3G
cellular systems).

This figure shows the three key types of wireless data networks. This diagram shows a
wireless LAN system that has multiple access
nodes. These access nodes operate as gateways

between the data communication devices (e.g., mobile computer) and the data network hub. Building 1 uses an older 801.11 wireless LAN system that operates from 902-928 MHz at 2 Mbps. Building 2 uses a newer 802.11 wireless LAN system that operates at 2.4 GHz providing up to 11 Mbps data transfer rate. This diagram also shows a microwave data link that provides a 45 Mbps interconnection between campus buildings. Finally, a user who is operating in a remote area outside the core campus is using the wide area mobile system to transfer data files (at a data transfer rate below 28 kbps).

ming. This diagram shows a wireless local loop system. In this diagram, a central office switch is connected via a fiberoptic cable to radio transmitters located in residential neighborhoods. Each house that desires to have dial tone service from the WLL service provider has a radio receiver mounted outside with a dial tone converter box. The dial tone converter box changes the radio signal into the dial tone that can be used in standard telephone devices such as answering machines and fax machines. It is also possible for the customer to have one or more wireless (cordless) telephones to use in the house and to use around the residential area where the WLL transmitters are located.

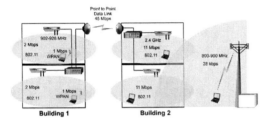

Wireless Data Networks

Wireless Device-A wireless device is a product, assembly, network unit that can receive and or transmit information with one or more communication devices without the use of a physical transmission line. Wireless devices may use radio, optical or acoustic transmission.

Wireless Local Loop (WLL)-Wireless local loop (WLL) is the providing of local telephone service via radio transmission. Wireless local loop systems often use a radio conversion device located at the home or business to allow the use of standard telephones. Although WLL systems may provide for traditional dial tone service, WLL systems commonly provide for multiple types of services such as telephone service, Internet access, and video program-

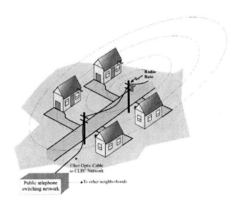

Wireless Local Loop (WLL) System

Wireless Markup Language (WML)-Wireless markup language is an efficient version of hypertext markup language (HTML) that is used between wireless communication devices (such as a PDA) and a server to allow world wide web (WWW) to efficiently operate over low speed data connections. Wireless markup language (WML) is part of wireless access protocol (WAP).

JJJ Wireless Markup Language" is a tag-based language used for describing the structure of documents to be delivered to wireless devices. It is to wireless devices what HTML (Hyper Text Markup Language) is to Web browsers. WML is used to layout "pages" (or cards as they are called in WML) to be viewed in wireless devices. WAP is the protocol for WML. WML is less forgiving and stricter on syntax than HTML. It is more like a programming language this way than HTML is.

Wireless Mesh Network-A wireless mesh network is a radio communication system where each communication device (typically radio access points) can be interconnected to multiple nodes so data packets can travel through alternate paths to reach their destination. For some wireless mesh networks, the relaying of packets through multiple access points forms the backbone connection (interconnection) of the network.

Wireless Metropolitan-Area Network (WMAN)-WMANs are usually private wireless packet radio networks often that cover an urban or city geographic area. They are commonly used for law-enforcement, utility or public safety applications.

Wireless Network-Wireless networks are primarily designed to transfer voice and or data from one point to one or more other points, (multipoint). Many networks make use of some wireless technologies as a transport medium even though we do not consider them to be wireless networks. Examples of wireless networks include cellular, personal communication service, (PCS), paging, wireless data, satellite, and broadcast radio and television. Wireless network is a term commonly used for wireless local area network (WLAN).

This figure shows the different types of wireless networks. This diagram shows a private land mobile radio system, television broadcast system, paging system, mobile telephone system, a wireless broadband system and a satellite communication system. Although all wireless networks can transmit information from one point to another, different types of net-

works better suited to provide specific types of services (e.g., paging compared to television broadcasting).

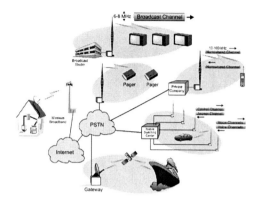

Wireless Networks

Wireless Personal Area Network (WPAN)-Wireless personal area networks (WPANs) are temporary (ad-hoc) short-range wireless communication systems that typically connect personal accessories such as headsets, keyboards, and portable devices to communications equipment and networks.

Wireless Portal-A wireless portal is an Internet Web site that acts as an interface between a mobile device and an Information service.

Wireless Private Branch Exchange (WPBX)-A WPBX offers business users the ability to make and receive calls through the company's PBX system using cordless telephones anywhere on a company's premises that has a radio port (wireless access node).

This diagram shows a sample WPBX radio system. A WPBX system typically has a switching system that is located at the company. The WPBX switch interfaces a PSTN communication line and multiple radio base stations. Radio base stations communicate with wireless office telephones that can move

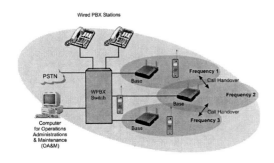

Wired PBX Stations

Frequency 1
Call Handover
Frequency 2
Call Handover
Frequency 3

PSTN

WPBX
Switch

Base

Base

Base

Computer
for Operations
Administrations
& Maintenance
(OA&M)

Wireless Private Branch Exchange (WPBX)
System

throughout the system. A control terminal is used to configure and update the WPBX with information about the wireless office telephones and how they can be connected to the PSTN.

Wireless Security-Wireless security is the ability of a wireless system or service to maintain its desired operation without damage, theft or compromise of its resources from unwanted people or events. Wireless security may use access security, authentication and encryption systems to maintain the security of the system.

Wireless Session Layer (WSL)-The wireless session layer (WSL) is the management layer of a communication session that is linked by a wireless channel.

Wireline-Telecommunications services provided by wireline common carriers, such as telephone companies. This term also refers to the use of copper wire for transmission of signals rather than radio links.

Wiretap-A wiretap is a connection of a wire or wire pair to a communication line.

Without Prejudice-Without prejudice is a disclaimer that may be used on a document to reduce or eliminate the ability to use the information in the document to influence other legal rights or obligations.

Wizzard-A wizzard is a software tool or service that assists a user in adding, setting up or

configuring a file or object they are working on.

WLL-Wireless Local Loop

WLR-Wholesale Line Rental

WMAN-Wireless Metropolitan-Area Network

WML-Wireless Markup Language

WMS-Workforce Management System

Word-In data communications, a character string, binary element string, or bit string that is considered as an entity.

Word Processing-Word processing is the use of computing system to assist in editing, formatting and printing of media materials.

Work Order-(1-communication system) A detailed drawing or print that indicates the addition, removal, or rearrangement of an outside plant, also called a work print. (2-production) A work order is a record that contains information that defines and quantifies a process that is used in the production of media (e.g. television programs) or services.

Work Package-A work package is a document or report that describes the work to be performed by installers, technicians or workers.

Work Station (Workstation)-A computer that is attached to the network. A workstation has the capability of processing information in addition to requesting and sending information through a network.

Work Unit-A work unit is a standard measurement amount of time or resources that is associated with a particular function or task.

Workflow-Workflow is the tasks and processes that are necessary to perform projects or assignments.

Workflow Automation-Workflow automation is the process of using a system that has established rules or procedures that allows for the acquisition, creation, scheduling or transmission of content assets.

Workflow Islands-Workflow islands are program and project management systems that have information processes that are stored or processed in separate areas.

Workflow Management (WFM)-Workflow management is the tasks and processes that are necessary to coordinate the flow of work between participants (employees, contractors and vendors) to perform projects or assignments. It manages the flow of information from system to system, essentially checking off the tasks associated with any process as it goes. They are sometimes industry specific (such as telecom specific), but just as often they are general IT products that can function effectively in any environment from telephony to financial services to manufacturing. For the television industry, workflow management commonly uses software is the programs or services that are used to identify, setup, manage and track tasks associated with projects.

Workflow Software-Workflow software is the programs or applications that are used to setup, manage and complete projects.

Workforce Management System (WMS)-A workforce management system is a combination of hardware (e.g. computers) and software (scheduling programs) that can identify, assign, track, and manage work related tasks and assignments.

Workgroup-A workgroup is a list of devices or users who share information and access to files and databases over a communication network (such as a local area network).

Workgroup Switch-A workgroup switch is a device that can interconnect users or devices within a department or workgroup.

Working Capital-Working capital is funds or resources that are used for short-term operations.

Working Group-A working group is a team of experts that are responsible for the development of documents or standards that relate to a particular technology or industry.

Working Key-A working key is a word, algorithm, or program that used to encrypt or decrypt a message or block of data. A working key is typically a temporary key that is used to decode media during a communication session.

Working Lines-Working lines are wires or pairs that are providing service in a as part of a communication system (plant facilities).

Working Model-A working model is a specification or system that is an early version of an industry specification that allows people or companies to begin development and/or testing while the approved or final version is completed.

Worksheet-A worksheet is a template or form that contains information entry areas for a student or participant to enter information that will be used by an instructor or process to solve a problem or to analyze the amount of understanding the worksheet user has.

Workstation-Work Station

World Intellectual Property Organization (WIPO)-World Intellectual Property Organization is an agency of the United Nations that focuses on the defining and enforcement of rules and regulations for the protection of intellectual property.

World Numbering Plan-The numbering plan that assigns each telephone customer in the world a unique telephone number that consists of a country code followed by a national number.

World Wide Web (WWW)-A service that resides on computers that are connected to the Internet that allows end users to access data that is stored on the computers using standard interface software (browsers). The WWW (commonly called the "web") is associated with customers that use web browsers (graphic display software) to public users to find, acquire and transfer information.

World Wide Web Consortium (W3C)-The World Wide Web consortium (W3C), formed in 1994, is a group of companies in the wireless market whose goal is to provide a common markup language for wireless devices and other small devices with limited memory.

WORM-Write Once Read Many

Worm Virus-A Worm is a type of computer virus that replicates itself onto other programs. Worms are commonly used with email messages allowing it to move from file to file and computer to computer.

WPAN-Wireless Personal Area Network

WPBX-Wireless Private Branch Exchange

Write-Writing is the moving or storing of information into memory or onto a storage medium.

Write Once Read Many (WORM)-A write once read many device is a storage system that allows information to be written (stored) to media one time and allows the information to be read (retrieved) many times.

Write-Off-A financial transaction which records as a loss to a company billed services or fees that cannot be collected, either due to customer non-payment, or due to an inability to bill for service or usage.

Written Consent-Written consent is the providing of authorization in marking (e.g. signature) or document form. Agreements may contain a written consent clause to require changes to be made in writing to avoid ambiguity.

WSDL-Web Services Description Language

WSL-Wireless Session Layer

WWW-World Wide Web

WYSIWYG-What you See is What You Get

X

X Axis-An X-axis is the horizontal axis in a coordinate system.

X.400-X.400 is an ITU recommendation that defines how mail messages are encoded and transferred between computers, terminals and computer networks.

X.500-X.500 is an ITU industry standard that defines common naming and addressing for directory services such as e-mail systems.

X.509-X.509 is an industry specification that defines the format of public key certificates. It is part of the ITU-T X.500 standard.

X12-X12 is a standard electronic data interchange (EDI) format that developed by the accredited standards committee (ASC).

XAML-Transaction Authority Markup Language

xBASE-xBase is a universal name that identifies database management systems that have dBASE-compatible (.DBF format) data files.

XBRL-Extensible Business Reporting Language

XCBL-XML Common Business Library

XDR-External Data Representation

xDSL-A set of large-scale high bandwidth data technologies that can use standard twisted-pair copper wire to deliver high speed digital services (up to 52 Mbps).

xHTML-Extensible Hypertext Markup Language

XML-Extensible Markup Language

XML Common Business Library (XCBL)-Extensible markup language common business library is an extensible markup language e-business platform that can be used by companies to exchange electronic business documents or data over electronic communication networks. More information about XCBL can be found at www.XCBL.org.

XML Stylesheet Language Transformation (XSLT)-XML stylesheet language transformations is a software standard that is used to format or transform XML documents or reformat elements of a web (HTML) page into a standardized display format. XSLT was developed in 199 by the World Wide Web Consortium (W3C).

XQL-Extensible Query Language

XrML-Extensible Rights Markup Language

XSLT-XML Stylesheet Language Transformation

Y

Y Axis-The vertical axis in a coordinate system.

Yearly rate-Yearly rate is the fee structure for a service that is purchased on a year to year basis.

Yellow Pages (YP)-Yellow pages are a portion of a telephone directory or a separate book that provides business listings.

YP-Yellow Pages

Yuppie-A yuppie is an ambitious person.

Z

Zero Touch Provisioning-Zero touch provisioning is an automatic operation of all configuration and system processing tasks to achieve service activation. This feature is possible only when the service delivery (fulfillment) systems are completely integrated.

Zip Code-A zip code is a sequence of numbers that are used by the United States to assist in the routing of letters or packages through the postal distribution system.

Zipping-Zipping is a process of selecting files to be compressed and combined into another file.

Zone-A zone is a geographic area, region or a group of points that share a common identifier or characteristic. A zone may be defined as an area where services may originate or terminate from.

Zone Based-Zone based is a type of service that is determined by a distance between service zones.

Zone Pricing-Zone pricing is the assigning of a value to a product or service that is dependent on what region (zone) it will be sold or used in.

Z

Associations

160 Characters
10 Upper Close
Forest Row
East Sussex
RH18 5DX
UK
Tel: +44 (0)1342 825169
www.160characters.org/

3G Americas
112th Avenue NE Suite B220 Bellevue, WA
98004 USA
Phone: 425-372-8922
www.3gamericas.org

ACCU
1330 Trinity Dr.
Menlo Park, CA 94025
Tel: 650-233-9082
www.accu.org

Alliance for Telecommunications Industry
Solutions (ATIS)
1200 G St. NW, Suite 500
Washington, DC 20005
Tel: 202-628-6380
Fax: 202-393-5453
www.atis.org

American National Standards Institute
(ANSI)
1819 L St. NW 6th floor
Washington, DC 20036
Tel: 202-293-8020
Fax: 202-293-9287
www.ansi.org

American Registry for Internet Numbers
(ARIN)
3635 Concorde Pkwy., ste. 200
Chantilly, VA 20151-1130
Tel: 703-227-9840
Fax: 703-227-0676
www.arin.net

American Teleservices Association
3815 River Crossing Parkway, Suite 20
Indianapolis, IN 46240
Tel: 317.816.9336
www.ataconnect.org

Association Management Solutions (AMS)
48377 Fremont Blvd., Suite 117
Fremont, CA 94538
Phone (510) 492-4000
www.amsl.com

Association of TeleServices International
(ATSI)
12 Academy Avenue
Atkinson, NH 03811
Tel: 603-362-9489
Fax: 603-362-9486
www.atsi.org

Association of Wireless Technology (AOWT)
1 Grove Road
Maidenhead
Berkshire
SL6 1LW
UK
Tel: 0044 1628 666399
www.aowt.org

Bluetooth Special Interest Group
500 108th Avenue NE
Suite 250
Bellevue, WA 98004
Phone: +1 425.691.3535
www.bluetooth.com

British Standards Institution
BSI British Standards
389 Chiswick High Road
London
W4 4AL
United Kingdom
Tel: +44 (0)20 8996 9001
Fax: +44 (0)20 8996 7001
www.bsi-global.com

Broadband Wireless Association
54 Mancetter Road
Atherstone
Warwickshire
CV9 1NY
UK
Tel: 44 7968 845016
Fax: 44 1827 716299
www.broadband-wireless.org

Building Industry Consulting Service
International (BICSI)
8610 Hidden River Parkway
Tampa, FL, 33637
Tel.: 813.979.1991
Fax: 813.971.4311
www.bicsi.org

Business Technology Association
12411 Wornall Road, Suite 200
Kansas City, MO 64145
Tel.: 816.941.3100
Fax: 816.941.2829
www.bta.org

Cable Television Laboratories, Inc
(CableLabs)
858 Coal Creek Circle
Louisville, CO 80027-9750
Tel: 303-661-9100
Fax: 303-661-9199
www.cablelabs.com

California Cable & Telecommunications
Association (CCTA)
4341 Piedmont Ave. (P.O. Box 11080)
Oakland, CA 94611
Tel: 510- 428-2225
Fax: 510-428-0151
www.calcable.org

Canadian Wireless Telecommunications
Association (CWTA)
130 Albert Street, Suite 1110
Ottawa, ON, K1P 5G4
Canada
Tel: 613-233-4888
Fax: 613-233-2032
www.cwta.ca

CDMA Development Group (CDG)
575 Anton Blvd., Ste. 560
Costa Mesa, CA, 92626
Tel.:1-888-800-CDMA
Fax: 714-545-4601
www.cdg.org

Cellular Telecommunications Internet
Association (CTIA)
1250 Connecticut Ave NW, Suite 800
Washington, DC 20036
Tel: 202-785-0081
Fax: 202-785-0721
www.ctia.org

Communications Fraud Control Association
(CFCA)
3030 North Central Avenue, Suite 707
Phoenix, AZ 85012
Tel: 602-265-2322
Fax: 602-265-1015
www.cfca.org

COMPTEL
888 17th Street, NW, 12th Floor
Washington, DC 20006
Tel: 202-296-6650
www.comptel.org

Computer and Communications Industry
Association (CCIA)
666 11th St. NW
Washington, DC 20001
Tel: 202-783-0070
Fax: 202-783-0534
www.ccianet.org

Defense Advanced Research Projects Agency
(DARPA)
3701 Fairfax Drive
Arlington, VA 22203-1714
Tel.: 703-526-6630
Fax: 703-528-1943
www.darpa.mil

DSL Forum
39355 California Street, Suite 307
Fremont, CA 94538
Tel: 510-608-5905
Fax: 510-608-5917
www.dslforum.org/

Electronic Industries Association (EIA)
2500 Wilson Blvd.
Arlington, VA 22201
Tel: 703-907-7500
Fax: 703-907-7501
www.eia.org

Enterprise Wireless Alliance
8484 Westpark Dr.
Suite 630
McLean, VA 22102
Tel: 703-528-5115
Fax: 703-524-1074
www.ita-relay.com

European Association for Standardizing
Information and Communication Systems
(ECMA)
114 Rue du Rhône
CH-1204 Geneva, Switzerland
Tel: 41228496000
Fax: 41228496001
www.ecma-international.org

European Telecommunications Standards
Institute (ETSI)
650, route des Lucioles
Sophia-Antipolis, Cedex, 06921
France
Tel: 33 (0)4 92 94 42 00
Fax: 33 (0)4 93 65 47 16
www.etsi.org

Federal Communications Commision (FCC)
445 12 St. SW
Washington, DC 20554
Tel: 888-CALL-FCC
Fax: 202-418-0232
www.fcc.gov

Global Mobile Suppliers Association (GSA)
PO Box 5817
Sawbridgeworth
CM21 0BH
UK
Tel: 44 1279 439 667
Fax: 44 1279 435 443
www.gsacom.com

Global VSAT Forum
Fountain Court
2 Victoria Square
Victoria Street
St Albans
Hertfordshire
AL1 3TF
UK
Tel: 44 1727 884739
www.gvf.org

GSM World Association
1st Floor Mid City Place
71 High Holborn
London
WC1V 6EA
UK
Tel: 353 1 289 1800
www.gsmworld.com/

HomePNA
Bishop Ranch 2, 2694 Bishop Drive, Suite 105
San Ramon, CA 94583
Tel: 925-275-6620
Fax: 925-886-3613
www.homepna.org

IMS Forum
211 Summit Place #292, (Box 10,000)
Silverthorne, CO 80498
Tel: 970-262-6100
Fax: 407-641-9595
www.IMSForum.org

IMSA FREQUENCY COORDINATION
Suite 6, 200 Metro Center Blvd.
Warwick, RI 02886
Tel: 401-738-2220
Fax: 401-738-7336
www.imsasafety.org

Indiana Telecommunications Association
(ITA)
54 Monument Circle, Suite 200
Indianapolis, IN 46204
Tel: 317-635-1272
Fax: 317-635-0285
www.itainfo.org

Infared Data Association (IrDA)
P.O. Box 3883
Walnut Creek, CA, 94598
www.irda.org

Information Technology Association of
America (ITAA)
1401 Wilson Blvd. Ste. 1100
Arlington, VA 22209
Tel: 703-522-5055
Fax: 703-525-2279
www.itaa.org

Information Technology Industry Council
1250 Eye Street NW, Suite 200
Washington, DC 20005
Tel: 202-737-8888
Fax: 202-638-4922
www.itic.org

Institute of Electrical and Electronics
Engineers, Inc. (IEEE)
1828 L Street, N.W., Suite 1202
Washington, DC 20036-5104
Tel: 202-785-0017
Fax: 202-785-0835
www.ieee.org

Insulated Cable Engineers Association
(ICEA)
P.O. Box 1568
Carrolton, GA 30117
Tel: 508-394-4424
www.icea.net

International Engineering Consortium (IEC)
300 W. Adams Street
Suite 1210
Chicago, IL 60606-5114
Tel: 312-559-4100
Fax: 312-559-4111
www.iec.org

International Internet Marketing Association
(IIMA)
PO Box 4018
Vancouver Main
349 West Georgia Street
V6B 3Z4
Vancouver, BC
Tel: 866-281-4462
www.iimaonline.org

International Multimedia Teleconferencing
Consortium, Inc. (IMTC)
Bishop Ranch 2, 2694 Bishop Drive, Suite 275
San Ramon, CA 94583
Tel: 925-275-6600
Fax: 925-275-6691
www.imtc.org

International Municipal Signal Association
(IMSA)
165 East Union Street (PO Box 539)
Newark, NY 14513-0539
Tel: 315-331-2182 1-800-723-IMSA
Fax: 315-331-8205
www.IMSAsafety.org

International Telecommunications Union
(ITU)
Place des Nations
 Geneva 20, Geneva, CH-1211
Switzerland
Tel: 41 22 730 51 11
Fax: 41 22 733 7256
www.itu.int

Internet Assigned Numbers Authority (IANA)
4676 Admiralty Way, Suite 330
Marina del Rey, CA 90292
Tel: 310-823-9358
Fax: 310-823-8649
www.iana.org

Internet Mail Consortium
127 Segre Place
Santa Cruz, CA, 95060
Tel: 831-426-9827
Fax: 831-426-7301
www.imc.org

Internet SOCiety (ISOC)
1775 Wiehle Ave., Suite 102
Reston, VA 20190
Tel: 703-326-9880
Fax: 703-326-9881
www.isoc.org/isoc

InterNIC
P.O. Box 1656
Herndon, VA, 22070
www.internic.net

Manufacturers Radio Frequency Advisory
Committee, Inc. (MRFAC)
899-A Harrison Drive SE
Leesburg, Virginia 20175
Tel: 800-262-9206
www.mrfac.com

Mid-America Cable Telecommunications
Association (Mid-America)
P.O. Box 3306
Lawrence, KS, 66046
Tel: 785-841-9241
Fax: 785-841-4975
www.midamericacable.com

Minnesota Telephone Association
30 East 7th Street
St. Paul, MN 55101
Tel: 651-291-7311
Fax: 651-291-2795
www.mnta.org

Mobile Satellite Users Associations (MSUA)
1350 Beverly Rd., Suite 115-341,
McLean, Virginia 22101
Tel: 410-827-9268
www.msua.org

Multiservice Switching Forum (MSF)
39355 California Street #307
Fremont, CA 94538
Tel: 510-608-5922
Fax: 510-608-5917
www.msforum.org

National Association of Broadcasters (NAB)
1717 N Street, NW
Washington, DC 20036-2891
Tel: 202-429-5300
Fax: 202-429-4199
www.nab.org

National Association of Radio and
Telecommunications Engineers (NARTE)
P.O. Box 678 (mail)
167 Village Street (surface)
Medway, MA 02053
Tel: 508-533-8333
Fax: 508-533-3815
www.narte.org

National Association of Regulatory Utility
Commissioners (NARUC)
1101 Vermont Avenue NW, Suite 200
Washington, DC 20005
Tel: 202-898-2200
Fax: 202-898-2213
www.naruc.org

National Association of State
Telecommunications Directors (NASTD)
PO Box 11910 or 2760 Research Park Dr.
Lexington, KY 40578-1910
Tel: 859-244-8186
Fax: 859-244-8001
www.nastd.org

National Association of Tower Erectors
8 Second Street SE
Watertown, South Dakota 57201-3624
Tel: 888-882-5865
www.natehome.com

National Cable TV Association (NCTA)
1724 Massachusetts Ave., N.W.
Washington, DC 20036
Tel: 202-775-3669
Fax: 202-775-3692
www.ncta.com

National Exchange Carrier Association
(NECA)
80 South Jefferson Rd.
Whippany, NJ 07981-8597
Tel: 800-228-8597
Fax: 973-884-8469
www.neca.org

National Fire Protection Association (NFPA)
1 Batterymarch Park
Quincey, MA 02169
Tel: 617-770-3000
Fax: 617-770-0700
www.nfpa.org

National Institute of Standards and
Technology (NIST)
100 Bureau DR Stop 3460,
Gaithersburg, MD 20899
Tel: 301-975-6478
www.nist.gov

National Technical Information Service
(NTIS)
U.S. Department of Commerce
Springfield, VA 22161
Tel: 703-605-6000
Fax: 703-321-8547
www.ntis.gov

National Telecommunications and
Information Administration (NTIA)
U.S. Department of Commerce 1401
Constitution Ave. NW
Washington, DC 20230
Tel: 202-482-7002
www.ntia.doc.gov

National Telephone Co-op Association
(NTCA)
4121 Wilson Blvd., Tenth Floor
Arlington, VA 22203
Tel: 703-351-2000
Fax: 703-351-2001
www.ntca.org

Network Professional Association (NPA)
17 South High Street, Suite 200
Columbus, OH 43215714-573-4780
Tel: 888 NPA-NPA0
Fax: 614-221-1989
www.npanet.org

Office of the Federal Register (OFR)
National Archives & Record Administration
8601 Adelphi Road
College Park, MD 20740-6001
Tel: 1-866-272-6272
www.nara.gov

Open Mobile Alliance (OMA) & (Wap Forum)
4275 Executive Square
Suite 240s
La Jolla, Ca 92037
Tel: 858-623-0742
www.openmobilealliance.org

Optical Storage Technology Association
(OSTA)
19925 Stevens Blvd.
Cupertino, CA 95014
Tel: 408-253-3695
Fax: 408-253-9938
www.osta.org

Organization for Promotion and
Advancement of Small Telecom Companies
(OPASTCO)
21 Dupont Circle, NW, Suite 700
Washington, DC 20036
Tel: 202-659-5990
Fax: 202-659-4619
www.opastco.org

Pacific Telecommunications Council (PTC)
2454 S. Beretania ST., Suite 302
Honolulu, HI 96826
Tel: 808-941-3789
Fax: 808-944-4874
www.ptc.org

PCI Industrial Computer Manufacturers
Group (PICMG)
401 Edgewater Place, Suite 600
Wakefield, MA 01880
Tel: 781-246-9318
Fax: 781-224-1239
www.picmg.org

Personal Communications Industy
Association (PCIA)
500 Montgomery St., Suite 700
Alexandria, VA 22314-1561
Tel: 703-739-0300
Fax: 703-836-1608
www.pcia.com

Portable Computer and Communications
Association (PCCA)
980A University Ave
Los Gatos, CA 95032
Tel: 541-490-5140
Fax: 419-831-4799
www.pcca.org

Rural Cellular Association (RCA)
2579 Western Trails Blvd.
Suite 100
Austin, Tx 78745
Tel: 800-722-1872
Fax: 512-472-1071
www.rca-usa.org

Satellite Broadcasting & Communications
Association (SBCA)
225 Reinekers Lane, Suite 600
Alexandria, VA 22314
Tel: 703-549-6990
Fax: 703-549-7640
www.sbca.com

Satellite Industry Association (SIA)
225 Reinekers Lane, Suite 600
Alexandria, VA 22314
Tel: 703-739-8358
Fax: 703-549-9188
www.sia.org

Silicon Valley-China Wireless Technology
Association
P.O.Box 360184,
Milpitas, CA 95036-0184
www.svcwireless.org/cms

Society of Cable Telecommunications
Engineers Inc. (SCTE)
140 Phillips Road
Exton, PA 19341-1318
Tel: 800-542-5040
Fax: 610-363-5898
www.scte.org

Society of Motion Pictures & Television
Engineers (SMPTE)
595 W. Hartsdale Avenue
White Plains, NY 10607-1824
Tel: 914-761-1100
Fax: 914-761-3115
www.smpte.org

Telecommunications Industry Association
(TIA)
2500 Wilson Blvd, Suite 300
Arlington, VA 22201
Tel: 703-907-7700
Fax: 703-907-7727
www.tiaonline.org

TeleManagement Forum (TMF)
89 Headquarters Plaza North
Suite 350
Morristown, NJ 07960-6628
Tel: 973-292-1901
Fax: 973-993-3131
www.tmforum.org

The Association for Telecommunications
Professionals in Higher Education (ACUTA)
152 W. Zandale Drive, Suite 200
Lexington, KY 40503
Tel: 606-278-3338
Fax: 606-278-3268
www.acuta.org

The Computing Technology Industry
Association (CompTIA)
1815 S. Myers Rd.
Oakbrook Terrace, IL 60181
Tel: 630-678-8300
Fax: 630-268-1384
www.comptia.org

The Consumer Electronics Association (CEA)
2500 Wilson Blvd.
Arlington, VA 22201
Tel: 703-907-7600
Fax: 703-907-7675
www.ce.org

The Electronic Frontier Foundation (EFF)
454 Sohotwell St.
San Francisco, CA 94110-4832
Tel: 415-436-9333
Fax: 415-436-9993
www.eff.org

The Internet Engineering Task Force
www.ietf.org

The Open Group
44 Montgomery St. Ste. 960
San Francisco, CA 94104
Tel: 415-374-8280
Fax: 415-374-8293
www.opengroup.org

United States Internet Service Provider
Association (USIPSA)
1330 Connecticut Avenue, NW
Washington, DC 20036
Tel: 202-862-3816
Fax: 202-261-0604
www.cix.org

United States Telecom Association (USTA)
1401 H St. NW, Suite 600
Washington, DC 20005-2164
Tel: 202-326-7300
Fax: 202-326-7333
www.usta.org

United States Telecommunications Training
Institute (USTTI)
1150 Connecticut Avenue, NW Suite 702
Washington, DC 20036
USA
Tel: 202.785.7373
Fax: 202.785.1930
www.ustti.org

United Telecom Council (UTC)
1901 Pennsylvania Avenue, NW 5th Floor
Washington, DC 20006
Tel: 202-872-0030
Fax: 202-872-1331
www.utc.org

US Internet Industry Association (USIIA)
5810 Kingstowne Center Drive
Suite 120, PMB 212
Alexandria, VA 22315-5711
Tel: 703-924-0006
Fax: 703-924-4203
www.usiia.org

Wall Street Telecommunications Association
(WSTA)
241 Maple Ave.
Red Bank, NJ 07701
Tel: 732-530-8808
Fax: 731-530-0020
www.wsta.org

WiMax Forum
15220 NW Greenbrier Pkwy
Suite 310
Beaverton, OR 97006
Tel: 503-924-2922
Fax: 503-924-3063
www.wimaxforum.com

Wireless Communications Association
International (WCA)
1333 H Street, NW
Suite 700 West
Washington, DC 20005-4754
Tel: 202-452-7823
Fax: 202-452-0041
www.wcai.com

Wireless Industry Association (WIA)
9746 Tappenbeck Drive
Houston, TX 77055
Tel: 800-624-6918
Fax: 800-820-2284
www.wirelessindustry.com

Wireless LAN Alliance (WLANA)
P.O. Box 9097
San Jose, CA 95157
Tel: 650-352-4709
Fax: 650-649-2305
www.wlana.org

Billing Related Magazines

Billing/OSS World
Virgo Publishing
P.O. Box 40079
Phoenix AZ 85067-0079
www.BillingWorld.com

Broadband Week Magazine
PO box 466008
Highlands Ranch, CO80216
303 470-4800
www.broadbandweek.com

Business Leader
3801 Wake Forest Road, Suite 102
Raleigh, NC27609
919 872-7077
www.businessleader.com

Call Center Magazine
11 West 19th Street, 3rd Floor
New York, New York 10011
212-600-3000
www.commweb.com

EE Times.com
600 Community Drive
Manhasset, NY11030
650 513-4306
www.eetimes.com

InformationWeek
600 Community Drive
Manhasset, NY11030
516 562-7911
www.informationweek.com

Intele-Card News
523 N. Sam Houston Parkway,East, Suite 300
Houston, TX77060
281 272-2744
www.intelecard.com

Internet Telephony
One Technology Plaza
Norwalk, CT.06854
United States
801 320-7654
www.tmcnet.com/it/

Lightwave
98 Spit Brook Rd.
Nashua, NH03062-5737
United States
603 891-0123
lw.pennnet.com

Network Computing
600 Community Drive
Manhasset, NY11030
516 562-5000
www.networkcomputing.com

New Architect Magazine
600 Harrison Street
San Francisco, CA94107
415 947-6000
www.webtechniques.com

New Telephony
927 18th Street, Suite A,
Santa Monica, CA90403
310 453-1231
www.newtelephony.com

Phone+
P.O. BOX 40079
Phoenix, AZ85067
480 990 1101
www.phoneplusmag.com

Billing Related Magazines

Telephony
One IBM Plaza, Suite 2300
Chicago, IL 60611
(312) 595-1080
www.telephonyonline.com

Vanilla Plus
Prestige Media Ltd,
Suite 117, 70 Churchill Square,
Kings Hill, West Malling,
Kent ME19 4YU, UK.
Tel: +44 (0)1634 243 869
www.vanillaplus.com

Wireless Week
6041 S. Syracuse Way
Suite 310
Greenwood Village, CO 80111
www.wirelessweek.com

X-Change
P.O. Box 40079
Phoenix AZ 85067-0079
480 990-1100
www.xchangemag.com

Currency Index

Country	Currency	Symbol
Afghanistan	afghani	Af
Albania	lek	L
Algeria	dinar	DA
American Samoa	dollar	$
Andorra	euro	€
Angola	Kwanza	Kz
Anguilla	Dollar	EC$
Antigua and Barbuda	Dollar	EC$
Argentina	Peso	$
Armenia	Dram	
Aruba	Guilder	Afl.
Australia	Australian dollar	A$
Austria	Euro	€
Azerbaijan	Manta	
Bahamas	Dollar	B$
Bahrain	Dinar	BD
Bangladesh	Taka	Tk
Barbados	Dollar	Bds$
Belarus	Ruble	BR
Belgium	Euro	€
Belize	Dollar	BZ$
Belorussia	Belarus	
Benin	Franc	CFAF
Bermuda	Dollar	Bd$
Bhutan	ngultrum	Nu
Bolivia	boliviano	Bs
Bosnia-Herzegovina	convertible mark	KM
Botswana	pula	P
Brazil	real	R$
British Indian Ocean Territory	GBP	
British Virgin Islands	dollar	$
Brunei	ringgit	B$
Bulgaria	new lev	Lv
Burkina Faso	franc	CFAF
Burma	now Myanmar.	
Burundi	franc	FBu
Cambodia	new riel	CR
Cameroon	franc	CFAF
Canada	dollar	Can$
Canton and Enderbury Islands	see Kiribati	
Cape Verde Island	escudo	C.V.Esc.
Cayman Islands	dollar	CI$
Central African Republic	franc	CFAF
Chad	franc	CFAF
Chile	peso	Ch$
China	yuan renminbi	Y
Christmas Island	Australian dollar	A$
Cocos (Keeling) Islands	see Australia	
Colombia	peso	Col$
Comoros	franc	CF
Congo	franc	CFAF
Congo, Dem. Rep. (former Zaire)	franc	
Congo, Dem. Rep. (2001-)	franc	
Cook Islands	see New Zealand	
Costa Rica	colon	slashed C
Côte d'Ivoire	franc	CFAF
Croatia	kuna	HRK

Country	Currency	Symbol
Cuba (external)	convertible peso	Cuc$
Cuba (internal)	peso	Cu$
Cyprus	pound	£C
Cyprus (Northern)	see Turkey	
Czech Republic	koruna	Kc (with hacek on c)
Denmark	krone (pl. kroner)	Dkr
Djibouti	franc	DF
Dominica	dollar	EC$
Dominican Rep.	peso	RD$
Dronning Maud Land	see Norway	
East Timor	see Timor-Leste	
Ecuador	sucre	S/
Ecuador	US Dollar	$
Egypt	pound	£E
El Salvador	colon	¢
Equatorial Guinea	franc	CFAF
Eritrea	nakfa	Nfa
Estonia	kroon (pl. krooni)	KR
Ethiopia	birr	Br
European Union	euro	€
Faeroe Islands (Føroyar)	see Denmark	
Falkland Islands	pound	£F
Fiji	dollar	F$
Finland (1999-)	euro	&euro
France (1999-)	euro	&euro
French Guiana	see France	
French Polynesia	franc	CFPF
Gabon	franc	CFAF
Gambia	dalasi	D
Gaza	see Israel and Jordan	
Georgia	lari	
Germany (1999-)	euro	&euro
Ghana (-July 2007)	old cedi	¢
Ghana (Aug 2007-)	new cedi	¢
Gibraltar	pound	£G
Great Britain	see United Kingdom	
Greece (2001-)	euro	&euro
Greenland	see Denmark	
Grenada	dollar	EC$
Guadeloupe	see France	
Guam	see United States	
Guatemala	quetzal	Q
Guernsey	see United Kingdom	
Guinea-Bissau (-Apr1997)	peso	PG
Guinea-Bissau (May1997-)	franc	CFAF
Guinea	syli	FG
Guinea	franc	
Guyana	dollar	G$
Haiti	gourde	G
Heard and McDonald Islands	see Australia	
Honduras	lempira	L
Hong Kong	dollar	HK$
Hungary	forint	Ft
Iceland	króna	IKr
India	rupee	Rs
Indonesia	rupiah	Rp
International Monetary Fund	Special Drawing Right	SDR
Iran	rial	Rls

Country	Currency	Symbol
Iraq (-2003)	dinar	ID
Iraq (2004-)	new dinar	ID
Ireland (1999-)	euro	&euro
Isle of Man	see United Kingdom	
Israel	new shekel	NIS
Italy (-1998)	lira (pl. lire)	Lit
Italy (1999-)	euro	&euro
Ivory Coast	see Côte d'Ivoire	
Jamaica	dollar	J$
Japan	yen	¥
Jersey	see United Kingdom	
Johnston Island	United States	
Jordan	dinar	JD
Kampuchea	see Cambodia	
Kazakhstan	tenge	
Kenya	shilling	K Sh
Kiribati	see Australia.	
Korea, North	won	Wn
Korea, South	won	W
Kuwait	dinar	KD
Kyrgyzstan	som	
Laos	new kip	KN
Latvia	lat	Ls
Lebanon	pound (livre)	L.L.
Lesotho	loti, pl., maloti	L, pl., M
Liberia	dollar	$
Libya	dinar	LD
Liechtenstein	see Switzerland	
Lithuania	litas, pl., litai	
Luxembourg (-1998)	franc	LuxF
Luxembourg (1999-)	euro	€
Macao (Macau)	pataca	P
Macedonia (Former Yug. Rep.)	denar	MKD
Madagascar	franc	FMG
Madagascar	ariayry = 5 francs	MGA
Malawi	kwacha	MK
Malaysia	ringgit	RM
Maldives	rufiyaa	Rf
Mali	franc	CFAF
Malta	lira, pl., liri	Lm
Martinique	see France	
Mauritania	ouguiya	UM
Mauritius	rupee	Mau Rs
Micronesia	see United States	
Midway Islands	see United States	
Mexico	peso	Mex$
Moldova	leu, pl., lei	
Monaco	see France	
Mongolia	tugrik (tughrik?)	Tug
Montserrat	dollar	EC$
Morocco	dirham	DH
Mozambique	metical	Mt
Myanmar	kyat	K
Nauru	see Australia	
Namibia	dollar	N$
Nepal	rupee	NRs
Netherlands Antilles	guilder (a.k.a. florin or gulden)	Ant.f. or NAf.
Netherlands (-1998)	guilder (a.k.a. florin or gulden)	f.

Country	Currency	Symbol
Netherlands (1999-)	euro	€
New Caledonia	franc	CFPF
New Zealand	dollar	NZ$
Nicaragua	gold cordoba	C$
Niger	franc	CFAF
Nigeria	naira	double-dashed N
Niue	see New Zealand	
Norfolk Island	see Australia	
Norway	krone (pl. kroner)	NKr
Oman	rial	RO
Pakistan	rupee	Rs
Palau	see United States	
Panama	balboa	B
Panama Canal Zone	see United States	
Papua New Guinea	kina	K
Paraguay	guarani	slashed G
Peru	inti	
Peru	new sol	S/.
Philippines	peso	dashed P
Pitcairn Island	see New Zealand	
Poland	zloty	z dashed l
Portugal (-1998)	escudo	Esc
Portugal (1999-)	euro	€
Puerto Rico	see United States	
Qatar	riyal	QR
Reunion	see France	
Romania (-June 2005)	leu (pl. lei)	L
Romania (July 2005-)	[new] leu (pl. lei)	L
Russia (-1997)	ruble	R
Russia (1998-)	ruble	R
Rwanda	franc	RF
Samoa (Western)	see Western Samoa	
Samoa (America)	US dollar	$
San Marino	Italy	
Sao Tome & Principe	dobra	Db
Saudi Arabia	riyal	SRls
Senegal	franc	CFAF
Serbia	dinar	Din
Seychelles	rupee	SR
Sierra Leone	leone	Le
Singapore	dollar	S$
Slovakia	koruna	Sk
Slovenia	tolar	SIT
Solomon Island	dollar	SI$
Somalia	shilling	So. Sh.
South Africa	rand	R
Spain (-1998)	peseta	Ptas
Spain (1999-)	euro	€
Sri Lanka	rupee	SLRs
St. Helena	pound	£S
St. Kitts and Nevis	dollar	EC$
St. Lucia	dollar	EC$
St. Vincent and the Grenadines	dollar	EC$
Sudan (-1992)	pound	
Sudan (1992-2007)	dinar	
Sudan (2007-)	pound	
Suriname	Dollar	$Sur
Svalbard and Jan Mayen Islands	see Norway	

Country	Currency	Symbol
Swaziland	lilangeni, pl., emalangeni	L, pl., E
Sweden	krona (pl. kronor)	kr or Sk
Switzerland	franc	SwF
Syria	pound	£S
Tahiti	French Polynesia	
Taiwan	new dollar	NT$
Tajikistan	ruble	
Tajikistan	somoni	100 dirams
Tanzania	shilling	TSh
Thailand	baht	Bht or Bt
Timor-Leste (May2002-)	uses the US Dollar	
Togo	franc	CFAF
Tokelau	see New Zealand	
Tonga	pa'anga	PT or T$
Trinidad and Tobago	dollar	TT$
Tunisia	dinar	TD
Turkey (2005-)	new lira	YTL
Turkmenistan	manat	
Tuvalu	see Australia.	
Uganda	shilling	USh
Uganda	shilling	USh
Ukraine	Hryvnia	
United Arab Emirates	dirham	Dh
United Kingdom	pound	£
United States of America	dollar	$
Upper Volta	now Burkina Faso	
Uruguay	peso uruguayo	$U
Uzbekistan	som	
Vanuatu	vatu	VT
Vatican	see Italy	
Venezuela	bolivar	Bs
Vietnam	(new) dong	dashed d, or D
Virgin Islands	US dollar	$
Wake Island	US dollar	$
Wallis and Futuna Islands	franc	
Western Sahara	see Spain	
Western Samoa	tala	WS$
Yemen	rial	YRls
Zambia	kwacha	ZK
Zimbabwe	dollar	Z$

ALTHOS

Althos Publishing Book List

Product ID	Title	# Pages	ISBN	Price	Copyright
Billing					
BK7781338	Billing Dictionary	644	1932813381	$39.99	2006
BK7781339	Creating RFPs for Billing Systems	94	193281339X	$19.99	2007
BK7781373	Introduction to IPTV Billing	60	193281373X	$14.99	2006
BK7781384	Introduction to Telecom Billing, 2nd Edition	68	1932813845	$19.99	2007
BK7781343	Introduction to Utility Billing	92	1932813438	$19.99	2007
BK7769438	Introduction to Wireless Billing	44	097469438X	$14.99	2004
IP Telephony					
BK7781361	Tehrani's IP Telephony Dictionary, 2nd Edition	628	1932813616	$39.99	2005
BK7781311	Creating RFPs for IP Telephony Communication Systems	86	193281311X	$19.99	2004
BK7780530	Internet Telephone Basics	224	0972805303	$29.99	2003
BK7727877	Introduction to IP Telephony, 2nd Edition	112	0974278777	$19.99	2006
BK7780538	Introduction to SIP IP Telephony Systems	144	0972805389	$14.99	2003
BK7769430	Introduction to SS7 and IP	56	0974694304	$12.99	2004
BK7781309	IP Telephony Basics	324	1932813098	$34.99	2004
BK7780532	Voice over Data Networks for Managers	348	097280532X	$49.99	2003
IP Television					
BK7781334	IPTV Dictionary	652	1932813349	$39.99	2006
BK7781362	Creating RFPs for IP Television Systems	86	1932813624	$19.99	2007
BK7781355	Introduction to Data Multicasting	68	1932813551	$19.99	2006
BK7781340	Introduction to Digital Rights Management (DRM)	84	1932813403	$19.99	2006
BK7781351	Introduction to IP Audio	64	1932813519	$19.99	2006
BK7781335	Introduction to IP Television	104	1932813357	$19.99	2006
BK7781341	Introduction to IP Video	88	1932813411	$19.99	2006
BK7781352	Introduction to Mobile Video	68	1932813527	$19.99	2006
BK7781353	Introduction to MPEG	72	1932813535	$19.99	2006
BK7781342	Introduction to Premises Distribution Networks (PDN)	68	193281342X	$19.99	2006
BK7781357	IP Television Directory	154	1932813578	$89.99	2007
BK7781356	IPTV Basics	308	193281356X	$39.99	2007
BK7781389	IPTV Business Opportunities	232	1932813896	$24.99	2007
Legal and Regulatory					
BK7781378	Not so Patently Obvious	224	1932813780	$39.99	2006
BK7780533	Patent or Perish	220	0972805338	$39.95	2003
BK7769433	Practical Patent Strategies Used by Successful Companies	48	0974694339	$14.99	2003
BK7781332	Strategic Patent Planning for Software Companies	58	1932813322	$14.99	2004
Telecom					
BK7781316	Telecom Dictionary	744	1932813160	$39.99	2006
BK7781313	ATM Basics	156	1932813136	$29.99	2004
BK7781345	Introduction to Digital Subscriber Line (DSL)	72	1932813454	$14.99	2005
BK7727872	Introduction to Private Telephone Systems 2nd Edition	86	0974278726	$14.99	2005
BK7727876	Introduction to Public Switched Telephone 2nd Edition	54	0974278769	$14.99	2005
BK7781302	Introduction to SS7	138	1932813020	$19.99	2004
BK7781315	Introduction to Switching Systems	92	1932813152	$19.99	2007
BK7781314	Introduction to Telecom Signaling	88	1932813144	$19.99	2007
BK7727870	Introduction to Transmission Systems	52	097427870X	$14.99	2004
BK7780537	SS7 Basics, 3rd Edition	276	0972805370	$34.99	2003
BK7780535	Telecom Basics, 3rd Edition	354	0972805354	$29.99	2003
BK7780539	Telecom Systems	384	0972805397	$39.99	2006

For a complete list please visit
www.AlthosBooks.com

ALTHOS

Althos Publishing Book List

Product ID	Title	# Pages	ISBN	Price	Copyright
	Wireless				
BK7769431	Wireless Dictionary	670	0974694312	$39.99	2005
BK7769434	Introduction to 802.11 Wireless LAN (WLAN)	62	0974694347	$14.99	2004
BK7781374	Introduction to 802.16 WiMax	116	1932813748	$19.99	2006
BK7781307	Introduction to Analog Cellular	84	1932813071	$19.99	2006
BK7769435	Introduction to Bluetooth	60	0974694355	$14.99	2004
BK7781305	Introduction to Code Division Multiple Access (CDMA)	100	1932813055	$14.99	2004
BK7781308	Introduction to EVDO	84	193281308X	$14.99	2004
BK7781306	Introduction to GPRS and EDGE	98	1932813063	$14.99	2004
BK7781370	Introduction to Global Positioning System (GPS)	92	1932813705	$19.99	2007
BK7781304	Introduction to GSM	110	1932813047	$14.99	2004
BK7781391	Introduction to HSPDA	88	1932813918	$19.99	2007
BK7781390	Introduction to IP Multimedia Subsystem (IMS)	116	193281390X	$19.99	2006
BK7769439	Introduction to Mobile Data	62	0974694398	$14.99	2005
BK7769432	Introduction to Mobile Telephone Systems	48	0974694320	$10.99	2003
BK7769437	Introduction to Paging Systems	42	0974694371	$14.99	2004
BK7769436	Introduction to Private Land Mobile Radio	52	0974694363	$14.99	2004
BK7727878	Introduction to Satellite Systems	72	0974278785	$14.99	2005
BK7781312	Introduction to WCDMA	112	1932813128	$14.99	2004
BK7727879	Introduction to Wireless Systems, 2nd Edition	76	0974278793	$19.99	2006
BK7781337	Mobile Systems	468	1932813373	$39.99	2007
BK7780534	Wireless Systems	536	0972805346	$34.99	2004
BK7781303	Wireless Technology Basics	50	1932813039	$12.99	2004
	Optical				
BK7781365	Optical Dictionary	712	1932813659	$39.99	2007
BK7781386	Fiber Optic Basics	316	1932813861	$34.99	2006
BK7781329	Introduction to Optical Communication	132	1932813292	$14.99	2006
	Marketing				
BK7781323	Web Marketing Dictionary	688	1932813233	$39.99	2007
BK7781318	Introduction to eMail Marketing	88	1932813187	$19.99	2007
BK7781322	Introduction to Internet AdWord Marketing	92	1932813225	$19.99	2007
BK7781320	Introduction to Internet Affiliate Marketing	88	1932813209	$19.99	2007
BK7781317	Introduction to Internet Marketing	104	1932813292	$19.99	2006
BK7781317	Introduction to Search Engine Optimization (SEO)	84	1932813179	$19.99	2007
	Programming				
BK7781300	Introduction to xHTML:	58	1932813004	$14.99	2004
BK7727875	Wireless Markup Language (WML)	287	0974278750	$34.99	2003
	Datacom				
BK7781331	Datacom Basics	324	1932813314	$39.99	2007
BK7781355	Introduction to Data Multicasting	104	1932813551	$19.99	
BK7727873	Introduction to Data Networks, 2nd Edition	64	0974278734	$19.99	2006
	Cable Television				
BK7781371	Cable Television Dictionary	628	1932813713	$39.99	2007
BK7780536	Introduction to Cable Television, 2nd Edition	96	0972805362	$19.99	2006
BK7781380	Introduction to DOCSIS	104	1932813802	$19.99	2007
	Business				
BK7781368	Career Coach	92	1932813683	$14.99	2006
BK7781359	How to Get Private Business Loans	56	1932813594	$14.99	2005
BK7781369	Sales Representative Agreements	96	1932813691	$19.99	2007
BK7781364	Efficient Selling	156	1932813640	$24.99	2007

For a complete list please visit
www.AlthosBooks.com

Printed in the United States
218159BV00004B/9/P